FUNCTIONAL APPROACH TO PRECALCULUS

SECOND EDITION

FUNCTIONAL APPROACH TO PRECALCULUS

MUSTAFA A. MUNEM

JAMES P. YIZZE

MACOMB COUNTY COMMUNITY COLLEGE

WORTH PUBLISHERS, INC.

PREFACE TO THE SECOND EDITION

WHY A SECOND EDITION? We have received many helpful suggestions concerning the first edition of *Precalculus* from users throughout the country. We have each taught from the book continually since its publication and have tried to be alert to any sources of difficulty. Although it is gratifying that the first edition seems to be satisfying the needs of so many students and teachers, the time has come, we feel, to improve it. We are confident that this second edition is a much better textbook.

PURPOSE: The second edition, like the first, provides the preparation necessary for students who intend to take calculus or other specialized courses in college mathematics. It also gives students in general education an opportunity to investigate and understand that level of mathematics usually referred to as precalculus mathematics.

PREREQUISITES: It is assumed that the students who use this book have had the equivalent of at least one year of plane geometry as taught in high schools and that they have the level of competence usually acquired in one and a half years of high school algebra.

NEW FEATURES: Color has been used to highlight definitions, theorems, important statements, and parts of graphs. Examples have been reworked and reorganized in order to relate them more closely to the text. The problems have been coordinated more closely with the examples, and the problem sets have been arranged so that the odd-numbered problems strive for the level of understanding desired by most users. The answers to odd-numbered problems are given in the book. The even-numbered problems probe for a deeper understanding of the concepts being covered. This arrangement of problems should simplify the task of making assignments.

OBJECTIVES: The objectives of the first edition have been maintained: first, that the student be able to learn from the book itself; second, that he or she gain the ability to apply the principles learned to specific situations. In order to accomplish these objectives the text has been interspersed with many examples and illustrations. Geometric interpretations supplement explanations whenever possible. Definitions and theorems have been stated carefully, and there is a reasonable balance between theory, on the one hand, and technique, drill, and application, on the other. Review problem sets at the end of each chapter will help students to gain confidence in their understanding of the material covered or will indicate areas requiring additional study.

Our primary objective, that the presentation be clear and accessible to students, has led us to write some passages in ways that may be questioned by more theoretically oriented mathematicians. Aware of these dilemmas we have nonetheless chosen to be guided by what our experience in the classroom has taught us is most appropriate for students.

CONTENTS: The function concept continues to serve as the central theme.

Chapter 1 sets forth some important preliminary material dealing with real number sets. Set interval notation is introduced here and used throughout the text.

Chapter 2 presents some general notations dealing with the concept of a function. Included here are discussions of domains, ranges, graphs, symmetry, properties of functions, composite functions, and inverse functions. This chapter has been rearranged so that the algebra of functions and the composition of functions are covered together in a new section. Also the symmetry of the graphs of functions has been related more clearly to the concept of even and odd functions.

Chapter 3 has also been rearranged so that the rational functions are covered in the chapter instead of in an appendix, as was the case in the first edition. The chapter, which deals with polynomial functions, also includes a brief discussion of the solution of linear systems. The section on quadratic functions has been rewritten to incorporate suggestions from many users.

Chapter 4 deals with exponential and logarithmic functions. In this edition, the proofs of the properties of logarithms have been made easier for the students to understand. The chapter includes mathematical induction, which is used to prove the binomial theorem. Finite sums and geometric series are also covered here.

Chapters 5 and 6, which deal with circular functions and trigonometric functions, include the study of periodic functions, of analytic trigonometry, and the traditional topics of triangle trigonometry.

In chapter 7, vectors in the plane and elementary properties of rotations are presented as applications of geometry and trigonometry.

Chapter 8 covers complex numbers, DeMoivre's theorem, and the fundamental theorem of algebra.

Chapter 9 deals with the standard topics of analytic geometry related to the circle, the parabola, the ellipse, and the hyperbola.

PACE: The pace of the course, as well as the choice of topics, depends on the particular teacher, on the curriculum, and on the academic calendar. The following suggested pace, which was used for a four-credit one-semester course, is meant only as a general guide:

Chapter 1: 4 lectures

Chapter 2: 10 lectures

Chapter 3: 10 lectures (Sections 3, 4, and 6 are optional)

Chapter 4: 8 lectures (Sections 2, 6, 7, and 8 are optional)

Chapters 5 and 6: 20 lectures

Chapter 8: 4 lectures (Section 6 is optional)

Other options are possible. For example, if the students are well enough prepared, Chapter 1 can be reviewed briefly in one lecture so that the topics in Chapters 7 and 9 can be covered.

This text can also be used in a trigonometry course that meets for 40 hours. The following schedule could be employed:

Chapter 1: 4 lectures

Chapter 2: 8 lectures

Chapters 5 and 6: 20 lectures

Chapter 7: 4 lectures

Chapter 8: 4 lectures

ADDITIONAL AIDS: There is an accompanying *Study Guide* available for students who need more drill or more assistance. The *Study Guide* is written in a semi-programmed format, it conforms with the arrangement of topics in the book, and it contains a great many carefully graded fill-in statements and problems broken down into simpler units, chapter reviews, and tests for each chapter. All answers are provided in the *Study Guide* to encourage self-testing at each student's own pace.

ACKNOWLEDGEMENTS: By the time a book reaches a second edition, so many people have contributed to its development that a list of acknowledgements would be quite long, but we would like to thank those who were particularly helpful in the writing of this edition: Professors Garret J. Etgen of the University of Houston, Adam J. Hulin of Louisiana State University at New Orleans, and Mary Nichols of the University of Detroit.

We would like to express our thanks to our colleagues at Macomb College for their contributions. Finally, we thank the staff of Worth Publishers, especially Robert C. Andrews who was active in bringing us the reviews and comments.

Mustafa A. Munem
James P. Yizze

Warren, Michigan
January 1974

CONTENTS

CHAPTER 1

Sets and Numbers

1 Sets

The primary objective in this section is to present enough about set theory so that the language of sets can be used later to describe mathematical concepts. *Sets* are collections of objects. For example, we can speak of the set of students in a particular course or the set of automobiles in the parking lot or the set of all books in the school library or the set of all letters in the word "Florida." In geometry we speak of a set of lines passing through a fixed point P in the plane (Figure 1), or we may refer to the set of all points that are equidistant from a fixed point C (Figure 2) or to the set of points of intersection of two circles in a plane (Figure 3). We also speak of sets of

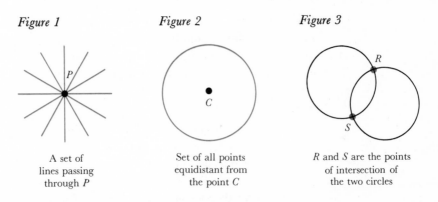

Figure 1   *Figure 2*   *Figure 3*

A set of lines passing through P

Set of all points equidistant from the point C

R and S are the points of intersection of the two circles

numbers, such as the set of all counting numbers $(1, 2, 3, \text{etc.})$ or the set of prime numbers greater than 2 and less than 75.

The set that has no members is called the *null set* or *empty set* and is denoted by $\varnothing$ or by { }. For example, the set of all women presidents in the United States is the empty set, since no woman has been elected to the presidency of the United States. It is important to keep in mind that 0 (zero) and $\varnothing$ are not the same.

A set is said to be *finite* if it is possible to list or enumerate *all* the members of the set; a set that is neither finite nor empty is an *infinite set*. For example, if A is the set of all 20 students in a particular class, A is a finite set since *all* its elements can be enumerated. If we use $N(A)$ to denote "number of elements in A," then $N(A) = 20$. On the other hand, if C is the set of all counting numbers, C is an infinite set since it is not possible to enumerate all the elements in this set.

Set descriptions are usually included between braces. Finite sets can be described by enumeration. For example, $A = \{a, b, c, d\}$ denotes that A is a finite set containing elements a, b, c, and d and no others. We use the notation "$a \in A$" to indicate that "a is an element of set A."

It is important to realize that $\varnothing$ is different from $\{0\}$, since $\{0\}$ is a set with one element, 0, whereas $\varnothing$ is a set that contains no elements.

Besides enumeration, another set description, *set builder notation*, takes the form $A = \{x \mid x$ has property $P\}$, which is read "A is the set of all elements x such that x has property P." For example, $E = \{x \mid x$ is an even counting number$\}$ is read "E is the set of all x such that x is an even counting number." Notice that in this case, $2 \in E$, $4 \in E$, $6 \in E$, etc. Since E is an infinite set, it is impossible to enumerate all the elements of E; however, we can use the fact that the members of E form a generally known pattern to write E as $E = \{2, 4, 6, ...\}$, where the three dots mean the same as "etc." We use the symbol $\notin$ to mean "is not a member of"; hence $1 \notin E$, $3 \notin E$, and $5 \notin E$.

EXAMPLE

Use set notation to describe each of the following sets.
a) A, the set of all counting numbers less than 7
b) B, the set of all counting numbers greater than 3
c) C, the set of all students less than 2 inches tall

SOLUTION

a) $A = \{x \mid x$ is a counting number less than 7$\}$ or, equivalently, $A = \{1, 2, 3, 4, 5, 6\}$. Note that $7 \notin A$.
b) $B = \{x \mid x$ is a counting number greater than 3$\}$. B cannot be described by enumeration because B is an infinite set; however, the known pattern of the elements of B suggests that B can be written as $B = \{4, 5, 6, 7, ...\}$.
c) Since C has no members, $C = \varnothing$.

1.1   Set Relations

Suppose that F is the set of all Ford automobiles and that M is the set of all motor vehicles. Clearly, all the members of F are also found in M. We say that F is a subset of M or, symbolically, $F \subseteq M$, which is read "F is contained in M." The set of all girls in a biology class is a subset of the set of all students in the class. In general, set A is a *subset* of set B, written $A \subseteq B$, if every element of A is an element of B. *The empty set is considered to be a subset of every set.*

EXAMPLES

1   Let $A = \{1, 2, 3, 4, 5, 6\}$ and $B = \{2, 5, 3\}$; then $B \subseteq A$.
2   If $A = \{1, 2, 3\}$ and $B = \{x \mid x$ is a counting number less than 4$\}$, then $A \subseteq B$ and $B \subseteq A$.

Example 2 illustrates the definition of *equality of sets*; for, if $A \subseteq B$ and $B \subseteq A$, we consider A and B to be different names for the same sets and we write $A = B$.

In Example 1 we have $B \subseteq A$, but $B \neq A$. B is an example of a proper subset of A. In general, B is said to be a *proper subset* of a set A, written $B \subset A$ (notice that the horizontal bar is left off), if all members of B are in A and A has at least one member not in B; that is, $B \subseteq A$ but $B \neq A$.

EXAMPLES

1   If $A = \{1, 2, 3\}$, $B = \{2, 3, 1, 7, 9\}$, and $C = \{2, 3, 1\}$, then $A \subseteq B$, $C \subseteq B$, $A \subseteq C$, and $C \subseteq A$. More informatively, $A \subset B$, $C \subset B$, and $A = C$.
2   List all subsets of $\{a, b, c\}$.

SOLUTION. $\{a\}$, $\{b\}$, $\{c\}$, $\{a, b\}$, $\{a, c\}$, $\{b, c\}$, $\{a, b, c\}$, and $\emptyset$ are the subsets of $\{a, b, c\}$. Note that all the subsets, with the exception of $\{a, b, c\}$ itself, are proper subsets of $\{a, b, c\}$.

3   If $A = \{2, 3, 4\}$, $B = \{1, 2, 3, 4, 7, 8\}$, and $C = \{7, 8\}$, then $A \subset B$ and $C \subset B$.

In Example 3, we see that A and C have no members in common. The set of all girls taking biology has no member in common with the set of all boys taking the same class. Such sets are called disjoint sets. In general, two sets that have no members in common are *disjoint sets*. For example, the sets $\{1, 2, 3\}$ and $\{4, 8, 10\}$ are disjoint sets.

When the selection of elements of subsets is limited to a fixed set, the limiting set is called a *universal set* or a *universe*. A universal set represents the complete set or the largest set from which all other sets in that same discussion are formed. The choice of the universal set is dependent upon the situation being considered. For example, in one case it may be the set of all people in the United States, and in another, it may be the set of all people in Michigan.

EXAMPLE

Describe set A, where $A = \{x \mid x$ is a number greater than 2 and x is a member of the universal set $U\}$ for each of the following.
a) $U = \{1, 2, 3, \frac{4}{3}, \frac{1}{8}\}$
b) U is the set of all counting numbers.
c) $U = \{0, 1, 2\}$

SOLUTION

a) $A = \{3\}$

b) $A = \{x \mid x$ is a counting number greater than 2$\}$ or, equivalently,
$A = \{3, 4, 5, ...\}$

c) $A = \varnothing$

Subsets can be represented pictorially by drawings called *Venn diagrams*. These diagrams often help in understanding set concepts. If we let U be the universal set, an arbitrary set $A \subseteq U$ can be represented as another closed region within the closed region representing U (Figure 4). Each of the three set relations discussed above can be represented by one of three Venn diagrams (Figures 5a, b, and c).

Figure 4

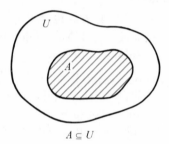

$A \subseteq U$

Figure 5

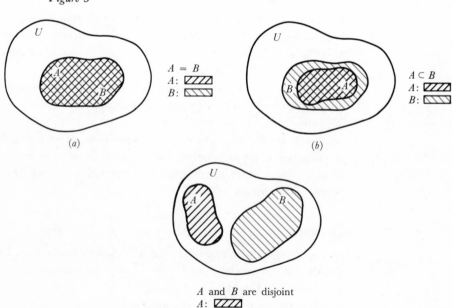

(a)   (b)

A and B are disjoint

(c)

EXAMPLE

Let N be the set of counting numbers and assume that

$$A = \{x \mid x = 3n,\ n \in N\} \qquad \text{and} \qquad B = \{y \mid y = 6m,\ m \in N\}.$$

Use a Venn diagram to illustrate the set relationship between A and B.

SOLUTION. $A = \{3, 6, 9, 12, ...\}$ and $B = \{6, 12, 18, 24, ...\}$ are infinite sets. Since each member of set B is also found in set A, but $A \neq B$, it follows that $B \subset A$ (Figure 6).

Figure 6

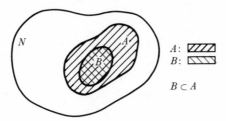

1.2 Set Operations

Consider a universal set $U = \{1, 2, 3, 4, 5, 6, 7, 8\}$. From U we can form $A = \{1, 2, 3, 4\}$ and $B = \{1, 3, 7\}$. How can sets A and B be used to form other sets? One way is simply to combine all the elements of A and B to form $\{1, 2, 3, 4, 7\}$. The operation suggested by this example is that of *set union*.

A union B, written $A \cup B$, and represented by the entire shaded region in Figure 7, is defined as

$$A \cup B = \{x \mid x \in A \text{ or } x \in B \text{ (or both)}\}$$

Figure 7

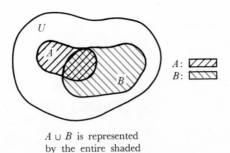

$A \cup B$ is represented by the entire shaded region.

Hence, in the example,

$$\{1,2,3,4\} \cup \{1,3,7\} = \{1,2,3,4,7\}$$

Another way to use $A = \{1,2,3,4\}$ and $B = \{1,3,7\}$ to form another set is to form set $\{1,3\}$, the set of all elements common to A and B. This is an example of *set intersection*.

A intersect B, written $A \cap B$, and represented by the shaded region in Figure 8, is defined as

$$A \cap B = \{x \mid x \in A \text{ and (simultaneously) } x \in B\}$$

Figure 8

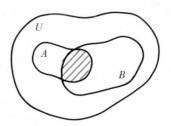

$A \cap B$ is represented by the shaded region.

Thus

$$\{1,2,3,4\} \cap \{1,3,7\} = \{1,3\}$$

If $A = \{1,2,3,4\}$ and $B = \{5,6,7\}$, then $A \cap B = \emptyset$ and A and B are disjoint sets. In general, A and B are disjoint sets whenever $A \cap B = \emptyset$.

The union of two sets, then, is simply the set that results when all the elements of the two sets are combined; the intersection is merely the set of all elements common to the two sets. Note that when the union of two sets containing common elements is described, the common elements are *not* listed twice; hence $\{2,3,4\} \cup \{1,4,8\}$ is *not* written as $\{2,3,4,1,4,8\}$ but rather as $\{2,3,4,1,8\}$, since the listing of 4 twice is superfluous.

EXAMPLES

1 Determine $A \cup B$ and $A \cap B$ if $A = \{1,2,3,4,5\}$ and $B = \{2,5,6,7\}$.

SOLUTION

$$A \cup B = \{1,2,3,4,5,6,7\} \quad \text{and} \quad A \cap B = \{2,5\}$$

2 Let $A = \{x \mid x \text{ is a counting number}\}$ and let $B = \{x \mid x \text{ is an even counting number}\}$; that is, $B = \{2,4,6,8,...\}$. Form $A \cup B$ and $A \cap B$.

SOLUTION.   Note that $B \subset A$ and U is the set of all counting numbers (Figure 9).

Figure 9

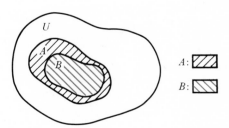

$A \cup B = \{x \,|\, x$ is a counting number or x is an even counting number$\}$

Therefore,

$A \cup B = \{x \,|\, x$ is a counting number$\} = A$

$A \cap B = \{x \,|\, x$ is a counting number and (simultaneously) x is an even counting number$\}$

Therefore,

$A \cap B = \{x \,|\, x$ is an even counting number$\} = B$

3   Use Venn diagrams to illustrate that $A \cap (B \cup C)$ and $(A \cap B) \cup (A \cap C)$ are equal sets.

SOLUTION.   The shaded area of Figure 10*a* represents $B \cup C$, and the shaded area of Figure 10*b* represents $A \cap (B \cup C)$. The shaded areas of Figures 10*c* and *d* represent $A \cap B$ and $A \cap C$, respectively, and the shaded area of Figure 10*e* represents $(A \cap B) \cup (A \cap C)$. Clearly, Figure 10*b* and *e* have the same shaded areas, so that the Venn diagrams illustrate the fact that

$$A \cap (B \cup C) = (A \cap B) \cup (A \cap C)$$

Figure 10

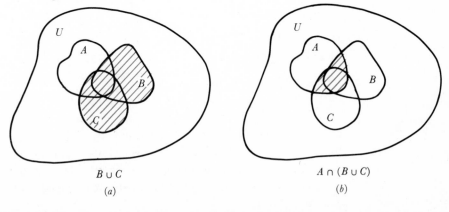

$B \cup C$

(a)

$A \cap (B \cup C)$

(b)

Figure 10 (continued)

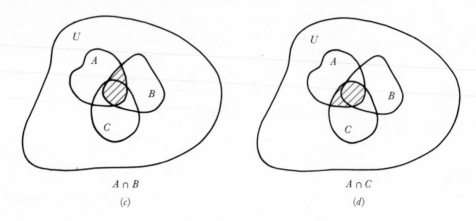

$A \cap B$

(c)

$A \cap C$

(d)

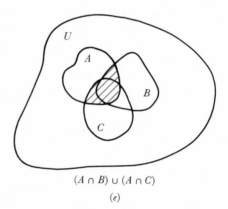

$(A \cap B) \cup (A \cap C)$

(e)

PROBLEM SET 1

1   Indicate which of the following statements are true and which are false.
a) $2 \in \{2\}$ b) $\{\varnothing\}$ is empty.
c) $\{2, 3\} \in \{2, 3, 7, 8\}$ d) $5 = \{5\}$
e) $\{2, 3, 4\}$ and $\{3, 2, 4\}$ are unequal sets.
f) $3 \in \{3, 4\}$ g) $\{3\} \in \{3, 4\}$
h) $\{3\} \subseteq \{3, 4\}$ i) $\{3\} \subset \{3, 4\}$
j) $\{1, 3\}$ and $\{1, 5, 15\}$ are disjoint.
k) $\{\{x, y\}\} = \{y, x\}$ l) $\{x, y\} = \{y, x\}$

2   Use set builder notation, $\{x \mid x$ has property $P\}$, to describe each of the
following sets. Also describe the set by enumeration, if possible. Indicate
which of the sets are finite and which are infinite.
a) A is the set of all even counting numbers.
b) A is the set of all counting numbers greater than 7 and less than 29.
c) A is the set of all counting numbers divisible by 3, that is, all counting
numbers that have a zero remainder when divided by 3.

3 A is the set of numbers greater than 2 but less than 8. Describe A by enumeration if the universal set from which A is formed is
   a) $\{10, 11, 12\}$ b) $\{1, 4, 7, 10, 13, 16\}$
   c) The set of all counting numbers
   d) $\{1, 2, 3, 4, 5, 6, 7, 8\}$ e) $\{7, 3, 10, 2, 5\}$

4 List all the subsets of each of the following sets. Indicate which of the subsets are proper subsets.
   a) $\{3\}$ b) $\{3, 8\}$
   c) $\{1, 2, 3\}$ d) $\{a, b, c, d\}$

5 Use $A = \{1, 4, 7\}$, $B = \{1, 2, 5, 7\}$, and $C = \{5, 6, 7, 8\}$ to form each of the following sets.
   a) $A \cup C$ b) $B \cap C$
   c) $A \cap C$ d) $A \cup (B \cap C)$
   e) $A \cap (B \cup C)$

6 Tabulate the number of distinct subsets of a set having
   a) 0 elements   b) 1 element (see Problem 4a)
   c) 2 elements (see Problem 4b)   d) 3 elements (see Problem 4c)
   e) 4 elements (see Problem 4d)
   Can you generalize your result? (That is, if a set has n elements, how many subsets can be formed?)

7 Indicate which of the set relations (proper subset, equal, disjoint), if any, holds between each pair of the following sets.
   a) $\{12, 13\}$; $\{2, 3, 4\}$
   b) The set of all distinct letters in the word "mathematics"; the set of all letters in the alphabet
   c) The set of all counting numbers greater than 3; The set of all counting numbers less than 8
   d) The set of all counting numbers greater than 3; The set of all counting numbers greater than 8
   e) The set of all even counting numbers less than 11; $\{2, 4, 6, 8, 10\}$

8 a) Why is it true that any set is a subset of itself?
   b) Use examples to illustrate that if A is a proper subset of B, then A is a subset of B, but if A is a subset of B, then A is not necessarily a proper subset of B.
   c) *Theorem (transitive law).* If $A \subseteq B$ and $B \subseteq C$, then $A \subseteq C$. Give two examples of this theorem.

9 a) Use Venn diagrams to illustrate the validity of each of the following properties of the set operations.
   i *Commutativity of set union:*

$$A \cup B = B \cup A$$

ii   *Commutativity of set intersection:*

$$A \cap B = B \cap A$$

iii *Associativity of set union:*

$$(A \cup B) \cup C = A \cup (B \cup C)$$

iv   *Associativity of set intersection:*

$$(A \cap B) \cap C = A \cap (B \cap C)$$

v   *Intersection distributes over union:*

$$A \cap (B \cup C) = (A \cap B) \cup (A \cap C)$$

vi   *Union distributes over intersection:*

$$A \cup (B \cap C) = (A \cup B) \cap (A \cup C)$$

b) Give two examples to illustrate each of the six properties of part a.

10 a) A is a set and x is a number. Fill in the table. Compare the role of $\emptyset$ in the algebra of sets with the role of 0 in the algebra of numbers.

$A \cup \emptyset =$	$x + 0 =$
$A \cap \emptyset =$	$x \cdot 0 =$

b) A is a set formed from a universal set U and x is a number. Fill in the table. Does $x+1$ simplify? Compare the role of U in the algebra of sets to the role of 1 in the algebra of numbers.

$A \cup U =$	$x + 1 =$
$A \cap U =$	$x \cdot 1 =$

2 Real Numbers

The language of sets will be used to describe some of the number sets of algebra. By repeatedly adding 1 to itself, we can generate the *set of positive integers* $I_p = \{1, 2, 3, 4, ...\}$. The *set of negative integers*, $I_n = \{-1, -2, -3, ...\}$, consists of the negatives of the positive integers. The set $I_p \cup I_n \cup \{0\} = \{..., -3, -2, -1, 0, 1, 2, 3, 4, ...\}$ is called the *set of integers* and will be denoted by I.

A *rational number* is any number that *can be* expressed in the form x/y, where x is an integer and y is a positive integer. For example, 3, $2\frac{1}{2}$, $-\frac{5}{7}$, and 53 percent are considered to be rational numbers, since they can be written as $\frac{3}{1}$, $\frac{5}{2}$, $\frac{-5}{7}$, and $\frac{53}{100}$, respectively. Q is generally used to denote the set of all rational numbers; hence

$$Q = \left\{ q \mid q = \frac{x}{y}, x \in I, y \in I_p \right\}$$

Since x is considered to be the same as $x/1$, we identify an integer as a rational number, so that $I \subseteq Q$. More precisely, since $I \neq Q$ ($\frac{1}{2} \in Q$, but $\frac{1}{2} \notin I$), I is a proper subset of Q. That is, all integers are rational numbers; however, not all rational numbers are integers.

The rational numbers can also be described by investigating their decimal representations. For example, let us consider the specific rational number $\frac{3}{7}$. The number $\frac{3}{7}$ can be interpreted as $3 \div 7$, as shown in Figure 1.

Figure 1

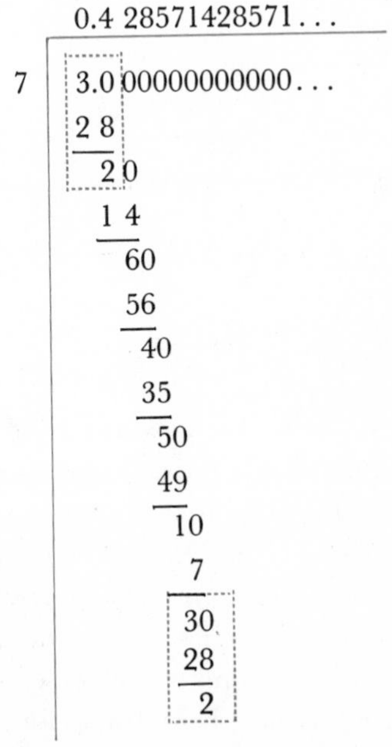

```
       0.4 28571428571...
      ┌─────────────────────
  7   │ 3.0 00000000000...
      │ 2 8
      │ ───
      │   2 0
      │   1 4
      │   ───
      │     60
      │     56
      │     ──
      │      40
      │      35
      │      ──
      │       50
      │       49
      │       ──
      │        10
      │         7
      │        ──
      │        30
      │        28
      │        ──
      │         2
```

Notice the pattern. In each step within the division, the remainder must be either 0, 1, 2, 3, 4, 5, or 6. (Why?) Therefore, if enough zeros are added after the decimal of the dividend (this does not change the value of the dividend) and the division by 7 is performed more than seven times, one of the remainders must reoccur; but as soon as a remainder appears again (in this example it is 3), the digits in the quotient repeat. In this example,

$$\tfrac{3}{7} = 0.428571428571\overline{428571}$$

where the bar identifies the block of digits that repeats infinitely often.

This concept can be generalized; for if x/y is a rational number, where $x \in I$ and $y \in I_p$, the division

$$y\overline{)x.0000\cdots}$$

can be performed until a remainder repeats. [There are at most y remainders possible. (Why?)] When the remainder repeats, the digits in the quotient repeat. Hence it follows that *every rational number can be represented by an eventually repeating decimal.* The converse of this statement also holds; that is, *every eventually repeating decimal represents a rational number.*

This means that any decimal number which "eventually" has a repeating block of digits in its decimal part can be represented by a ratio of two integers.

EXAMPLE

Find the repeating block for each of the following rational numbers.

a) 2   b) $\tfrac{2}{3}$ c) $\tfrac{1310}{99}$

SOLUTION

a) $2 = \tfrac{2}{1} = 2.\overline{0}$
b) $\tfrac{2}{3} = 0.666\overline{6}$
c) $\tfrac{1310}{99} = 13.2\overline{323}$

Rational numbers, such as 2 and $\tfrac{14}{2}$, in which the repeating block is the digit 0, are sometimes called *terminating decimals.*

In summary, a rational number is a number that can be considered from two viewpoints: as a ratio of two integers or as a repeating decimal. A more formal treatment of repeating decimals is given in Chapter 4, Section 8, when geometric series are considered.

If a rational number is represented in either of the two forms, it can be converted to the other form, as illustrated in the first two examples below.

EXAMPLES

1 Show that each of the following numbers is a rational number by examining
   the decimal representation of the number.
   a) $\frac{8}{2}$ b) 17 percent   c) $\frac{10}{3}$
   d) $\frac{1310}{3}$ e) 7.142845$\overline{845}$

SOLUTION

   a) $\frac{8}{2} = 4.0$ is a terminating decimal (the repeating block is the digit 0).
   b) 17 percent $= 0.17$ is a terminating decimal.
   c) $\frac{10}{3} = 3.33\overline{3}$ has a repeating block (the digit 3).
   d) $\frac{1310}{3} = 436.66\overline{6}$ has a repeating block (the digit 6).
   e) The number 7.142845$\overline{845}$ is a rational number since 845 is a repeating
       block.

2 Express each of the following rational numbers as the ratio of two integers.
   a) 2.03   b) 0.99$\overline{9}$
   c) 0.23232$\overline{23}$ d) 7.25134$\overline{134}$

SOLUTION

   a) $2.03 = 2 + \frac{3}{100} = \frac{203}{100}$
   b) Let $x = 0.99\overline{9}$. Multiplying both sides of this equation by 10 moves
       one of the repeating blocks of decimals to the left of the decimal point
       so that

       $$10x = 9.99\overline{9} \qquad \text{and} \qquad x = 0.99\overline{9}$$

       Subtracting the corresponding sides of these latter two equations
       results in

       $$9x = 9 \qquad \text{or} \qquad x = 1$$

       which is equivalent to $x = 0.99\overline{9}$.
   c) Let $x = 0.23232\overline{323}$. Multiplying both sides of the equation by 100
       moves one of the repeating blocks of the decimal to the left of the
       decimal point, so that

       $$100x = 23.2323\overline{23}$$
       $$x = 0.23232\overline{323}$$

       Subtracting the corresponding sides of the latter two equations we have

       $$99x = 23 \qquad \text{or} \qquad x = \frac{23}{99}$$

       which is equivalent to $x = 0.2323\overline{23}$.

d) Let $x = 7.25134\overline{134}$. Here we multiply both sides of the equation by 100,000 in order to move one of the repeating blocks of decimals to the left of the decimal point to get

(1) $100,000x = 725,134.\overline{134134}$

Next we multiply both sides of $x = 7.25134\overline{134}$ by 100 in order to put *only* repeating blocks to the right of the decimal point to get

(2) $100x = 725.\overline{134134}$

(In Examples 2b and c the latter multiplication was not necessary, since x already had only repeating blocks to the right of the decimal.) Subtracting equation (2) from equation (1) yields

$$99,900x = 724,409$$

so that

$$x = \frac{724,409}{99,900}$$

which is equivalent to $x = 7.25134\overline{134}$.

3   Show that there exists no rational number whose square is equal to 2.

PROOF. We will show that the assumption "a rational r exists such that $r^2 = 2$" leads to a contradiction.

Let $r = x/y$ (x and y are positive integers) be in its lowest terms. If we assume that $r^2 = 2$, then

$$2 = \frac{x^2}{y^2} \quad \text{or} \quad 2y^2 = x^2$$

The number $2y^2$ is even; therefore, x^2 is even and x must be even.

Since x is even, x can be written in the form $x = 2k$, where k is a positive integer. Hence

$$2y^2 = (2k)^2 \quad \text{or} \quad y^2 = 2k^2$$

Since y^2 is even, y must be even. But both x and y cannot be even numbers, since x/y was assumed to be in lowest terms. Therefore, the assumption that r is rational is false, and r is not a rational number.

We denote the number r by $\sqrt{2}$ (Figure 2). By the Pythagorean theorem $r^2 = 1^2 + 1^2 = 2$, so the length of the hypotenuse of the isosceles right triangle with two legs of length 1 is $\sqrt{2}$.

Figure 2

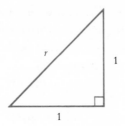

We have seen that the set of rational numbers is the set of numbers represented by repeating decimals, and repeating decimals represent rational numbers. But there are decimals that do not repeat, for example the decimal 1.01001000100001..., where there is one more "0" after each "1" than there is before the "1." Another example of a nonrepeating decimal is $\pi = 3.14159265358....$ $\sqrt{2}$ is not rational, and it can be shown that $\sqrt{2}$ has a nonrepeating decimal representation, $\sqrt{2} = 1.4142135....$ The numbers $1 + \sqrt{2}$, $3 - \sqrt{2}$, $5\sqrt{2}$, $3 + \sqrt{2}$, $\sqrt{3}$, $\sqrt[3]{2}$, and $\sqrt{5}$ also have nonrepeating decimal forms. Such numbers are called *irrational numbers.*

The set of *real numbers,* in a sense, is the set of all numbers that can be written as decimal numbers. Consequently, the set of real numbers consists of two sets of numbers: the rational numbers or repeating decimals, and the irrational numbers or nonrepeating decimals. If we use R to denote the set of real numbers and L to denote the set of irrational numbers, then we can express R as

$$R = Q \cup L \qquad \text{(note that } Q \cap L = \varnothing\text{)}$$

[It is recommended that students who require a review of the algebra of polynomials and rational expressions, including factoring, complete such a review before proceeding to the next section. One possible source of such material is the review chapter in the *Study Guide to Accompany Functional Approach to Precalculus* by Mustafa A. Munem, William Tschirhart, and James P. Yizze (Worth Publishers).]

2.1 One-to-One Correspondence

The set of real numbers can be represented geometrically as the set of all points on a straight line. This geometric representation is possible because the set of real numbers can be put in "one-to-one correspondence" with points on a line, a correspondence in which each real number is associated with *exactly* one point on the line and, conversely, a correspondence that associates each point on the line with exactly one real number.

Since the concept of a one-to-one correspondence between two sets will be used again later, it is worthwhile to give special attention to the topic here before we apply it to real numbers.

Suppose that a dinner party is being arranged. For each guest there is provided a place setting; conversely, for each place setting there is a guest. We say that the place settings are in one-to-one correspondence with the guests. Similarly, at a party attended only by married couples, we could establish a one-to-one correspondence between the females and the males in attendance. For each female there is a male (her husband, for example), and for each male there is a female. In general, two sets A and B can be put in *one-to-one correspondence* if it is possible to associate each member of set A with exactly one member of set B, and, conversely, if it is possible to associate each member of set B with exactly one member of set A.

The concept of a one-to-one correspondence is not difficult to understand when the sets are finite sets. Quite simply, two finite sets can be placed in a one-to-one correspondence whenever both sets have the same number of members. For example, it is *not* possible to establish a one-to-one correspondence between set $A = \{1, 3, 4\}$ and set $B = \{1, 2\}$; for, although it is true that for each member of set B we can associate exactly one member of set A, it is not possible to associate each member of set A with exactly one member of set B.

$$
\begin{array}{cc}
B & A \\
1 \longleftrightarrow & 1 \\
2 \longleftrightarrow & 3 \\
\leftarrow ? \rightarrow & 4
\end{array}
$$

EXAMPLE

Show that $A = \{1, 2, 3\}$ and $B = \{a, b, c\}$ can be put in one-to-one correspondence in more than one way.

SOLUTION

$$
\begin{array}{lll}
1 \longleftrightarrow a & 1 \longleftrightarrow b & 1 \longleftrightarrow c \\
2 \longleftrightarrow b \quad \text{or} & 2 \longleftrightarrow c \quad \text{or} & 2 \longleftrightarrow a \\
3 \longleftrightarrow c & 3 \longleftrightarrow a & 3 \longleftrightarrow b
\end{array}
$$

The concept of one-to-one correspondence is not so easy to understand, however, if the sets have infinitely many members. Let us consider two examples of this type.

EXAMPLES

1 The set of positive integers $\{1, 2, 3, 4, \ldots\}$ can be put in one-to-one correspondence with the even numbers $\{2, 4, 6, 8, \ldots\}$ using the following scheme:

$$
\begin{array}{c}
1 \longleftrightarrow 2 \\
2 \longleftrightarrow 4 \\
3 \longleftrightarrow 6 \\
\cdots \cdots \cdots \\
n \longleftrightarrow 2n
\end{array}
$$

2 Consider triangle ABC with line segment $\overline{DE}$ (Figure 3). Show how the points of segment $\overline{DE}$ can be put in one-to-one correspondence with the points of segment $\overline{AC}$.

Figure 3

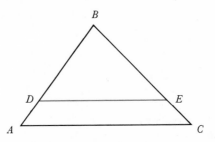

SOLUTION. Consider all line segments from point B to line segment $\overline{AC}$ (Figure 4). Each such line segment associates a point on DE with exactly one point on $\overline{AC}$; conversely, each point of $\overline{AC}$ is associated with exactly one point on $\overline{DE}$. Hence segments $\overline{DE}$ and $\overline{AC}$ are, as point sets, in one-to-one correspondence.

Figure 4

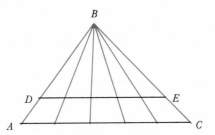

2.2 Real Line

Let us now examine a one-to-one correspondence between the real numbers and points on a line. The resulting "numbered line" is called a *real line* or *real axis* or *number line*.

An arbitrary point on the line is selected to represent 0 and another arbitrary point to the right of 0 is selected to represent 1. The point 0 is called the *origin* and the *line segment* determined by the point 0 and the point 1 is called the *scale unit* (Figure 5). By repeating the scale unit, moving from left to right, starting at 0, we can associate the set of positive integers $I_p = \{1, 2, 3, 4, \ldots\}$, with (equispaced) points on the line. Moving from right to left, starting at 0, we can associate the set of negative integers

$I_n = \{-1, -2, -3, -4, \ldots\}$, with (equispaced) points on the line (Figure 5). The remaining real numbers can be "located" or "plotted" on the real line by using decimal representations or by using the geometry of the number line.

Figure 5

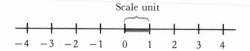

Scale unit

EXAMPLES

1   Locate 2.3 and 2.38 on the real line.

SOLUTION.   We can locate 2.3 by subdividing the portion of the number line between 2 and 3 into 10 equal parts; then, starting at 2, we move 3 parts to the right to 2.3 (Figure 6). Next, 2.38 can be located by subdividing the segment between 2.3 and 2.4 into 10 equal parts; then, starting at 2.3, we move 8 parts to the right (Figure 7).

Figure 6

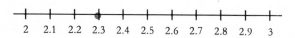

2   2.1   2.2   2.3   2.4   2.5   2.6   2.7   2.8   2.9   3

Figure 7

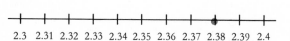

2.3 2.31 2.32 2.33 2.34 2.35 2.36 2.37 2.38 2.39 2.4

2   Locate $\frac{17}{8}$ on the real line by using the decimal representation of $\frac{17}{8}$.

SOLUTION

$$\frac{17}{8} = 2.125 \qquad \text{(Figure 8)}$$

Figure 8

3   Locate $\sqrt{2}$ on the number line.

SOLUTION. We have seen that $\sqrt{2}$ is an irrational number, so $\sqrt{2}$ has a decimal representation that is a nonterminating decimal ($\sqrt{2} = 1.41421\ldots$). If we were to attempt to locate $\sqrt{2}$ by using the decimal representation, we would become involved in an unending process in which we would "approach" but never actually locate the point. (Why?) We can, however, locate $\sqrt{2}$ by using the geometry of the number line, as illustrated in Figure 9.

Figure 9

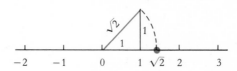

4 What number corresponds to the point midway between 2.7 and 2.8?

SOLUTION. The point 2.75 is midway between the points 2.7 and 2.8 on the real line (Figure 10).

Figure 10

2.7   2.71   2.72   2.73   2.74   2.75   2.76   2.77   2.78   2.79   2.8

Notice that when a real number is plotted on the number line, the number is not the point nor is the point the number; the point *represents* the number. It is customary, however, to use the words *real number* and *point* interchangeably. Thus we speak of the point $\frac{3}{4}$ rather than the point corresponding to the real number $\frac{3}{4}$.

PROBLEM SET 2

1 Show that each of the following numbers are rational numbers, first by examining the decimal representation, and second by representing the numbers as ratios of integers.

a) -5 b) 0
c) 0.2   d) $5\frac{7}{15}$
e) $3.46464\overline{6}$ f) $\frac{2}{9}$
g) $\frac{4}{13}$ h) 33 percent
i) $0.499\overline{9}$ j) $7.3621\overline{621}$
k) $0.9797\overline{97}$ l) $4.888\overline{8}$

2 Show that there exists no rational number whose square is 3. (*Hint:* See the proof of Example 3 in Section 2 on page 14.)

3 Use decimal representations to construct a rational number between each of the following pairs of real numbers.
a) 8.3 and 8.4   b) 3 and π
c) $\sqrt{2}$ and $\sqrt{3}$

4 Using I_p, I_n, I, Q, L, and R as defined in Section 2, describe each of the following number sets.
a) $I_p \cap I_n$ b) $I_p \cup I_n$
c) $L \cap Q$ d) $I \cap L$
e) $I \cup Q$

5 Establish, if possible a one-to-one correspondence between the two sets in each of the following parts.
a) $\{1,2,3,4,6\}$; $\{a,c,d,e,f\}$
b) Positive integers; Negative integers
c) $\{\ \}$; $\{0\}$ d) $\{7,8,9\}$; $\{8,9,91\}$
e) $\{1,2,4\}$; $\{a,b,c,4\}$ f) Positive integers; Integers

6 Is it possible for a set to be put in one-to-one correspondence with a proper subset of itself? (Consider Problem 5f.)

7 Locate each of the following real numbers on the real line.
a) $-2\frac{1}{7}$ b) π
c) $\sqrt{5}$ d) 1.93 percent
e) $\frac{15}{7}$ f) $-\frac{1}{2}$
g) $-\sqrt{2}$ h) $\frac{7}{3}$
i) 6.99   j) 6.999.

3   Order

The purpose here is to use an intuitive understanding of the real line to motivate and develop some of the properties of the inequality relation.

3.1   Positive and Negative Numbers

All real numbers represented by points on the real line that lie to the "right" of the zero point are *positive numbers*, whereas all real numbers "left" of the zero point are *negative numbers*. Zero, then, is neither positive nor negative. Figure 1 illustrates the following principle:

1 TRICHOTOMY PRINCIPLE

If a is a real number, then one and only one of the following conditions must hold (see Figure 1):

a is positive   (a is to the right of 0)

a is negative   (a is to the left of 0)

a is zero

Figure 1

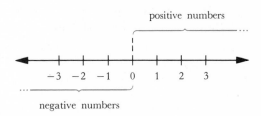

The next axiom establishes the basic assumptions that are made to characterize the set of positive real numbers R_p.

2 POSITIVE NUMBER AXIOM

The set of positive numbers is *closed under addition*; that is, the sum of any two positive numbers is always a positive number. The set of positive numbers is *closed under multiplication*; that is, the product of two positive numbers is always a positive number.

Using the notion of positive numbers, it is possible to give a precise "algebraic" characterization of the order of real numbers that is suggested by the "geometry" of the real line.

Geometrically, a real number a is less than a real number b if the point associated with a is to the "left" of the point associated with b on the real line (Figure 2).

Figure 2

a

b

a is less than b

The next definition formalizes this ordering of the real numbers.

DEFINITION   ORDER

Assume that a and b are real numbers. We say that *a is less than b*, written $a < b$, or, equivalently, *b is greater than a*, written $b > a$, if $b - a$ is a positive number.

Thus $2 < 3$ or $3 > 2$ because $3 - 2 = 1$ is a positive number; $-3 < -2$ because $-2 - (-3) = 1$; $-2 < 3$ since $3 - (-2) = 5$ (Figure 3).

Figure 3

If $b > 0$, then $b - 0 = b$ is positive; and, if $a < 0$, then $0 - a = -a$ is positive or a is negative (Figure 4).

Figure 4

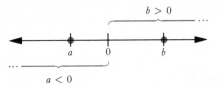

Notice that *−x is not necessarily a negative number*. In fact, $-x < 0$ only when $0 - (-x) = x$ is positive. Is $-(-2)$ negative? What about $-(-x)$?

3.2   Properties of Order Relations

The positive number axiom and the definition of order can be used to prove some important properties of the order relation.

THEOREM 1   TRANSITIVE PROPERTY

If a, b, and c are real numbers such that $a < b$ and $b < c$, then $a < c$.

GEOMETRIC INTERPRETATION.   If a is left of b and b is left of c, then a is left of c (Figure 5). For example, if $x < y$ and $y < 7$, then $x < 7$.

Figure 5

PROOF OF THEOREM.   By the definition of order, $a < b$ implies that $b - a = p$ is positive, and $b < c$ implies that $c - b = q$ is positive, so that

$$(b - a) + (c - b) = p + q$$

which is positive because the positive numbers are closed under addition. After simplifying, we get

$$c - a = p + q$$

so $c - a$ is positive, from which we can conclude, by the definition of order, that $a < c$.

THEOREM 2   ADDITION PROPERTY

If a and b are real numbers with $a < b$, then $a + c < b + c$ for any real number c.

GEOMETRIC INTERPRETATION.   If a is to the left of b and if any number c is added to both a and b, then the result $a + c$ is to the left of $b + c$. In other words, $a + c$ and $b + c$ can be obtained by "shifting" or "translating" a and b to the right c units if c is positive (Figure 6), or to the left if c is negative.

Figure 6

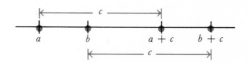

For example, let us examine the situation in which 5 is added to both sides of the inequality $-3 < 2$. Obviously $-3 < 2$, so that -3 is left of 2 (Figure 7).

Figure 7

Geometrically, after adding 5 to each side, -3 and 2 are "translated" 5 units to the right, so the two results 2 and 7 have the same relative positions as -3 and 2 (Figure 8).

Figure 8

PROOF OF THEOREM. $a < b$ implies that $b - a = p$ is positive. (Why?) But

$$b - a = b - a + 0$$

$$= b - a + c - c$$

$$= (b + c) - (a + c)$$

so that

$$(b + c) - (a + c) = p$$

is positive. Hence, by the definition of order,

$$a + c < b + c$$

THEOREM 3

If a, b, and c are real numbers such that $c > 0$ and $a < b$, then $ac < bc$.

GEOMETRIC INTERPRETATION.   If a is to the left of b and each number is multiplied by a *positive* number c, then ac is to the left of bc. For example, if $2 < 4$ and $x > 0$, then $2x < 4x$ (Figure 9).

Figure 9

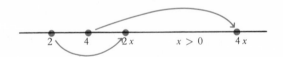

$$2 \quad 4 \quad 2x \qquad x > 0 \qquad 4x$$

PROOF OF THEOREM.   Since $a < b$, $b - a = p$, a positive number. But $c > 0$ and $p > 0$ implies that $pc > 0$, since the positive numbers are closed under multiplication. Hence $(b - a)c = pc$ is positive, so $bc - ac$ is positive; that is,

$$ac < bc$$

THEOREM 4

If a, b, and c are real numbers such that $c < 0$ (c is negative) and $a < b$, then $ac > bc$.

GEOMETRIC INTERPRETATION.   If a is to the left of b and each number is multiplied by a *negative* number c, the order is reversed; that is, ac is right of bc. For example, let us see what happens to the inequality $-3 < 2$ when we multiply each side by -3. -3 is to the left of 2; however, $(-3)(-3) = 9$ is to the right of $(2)(-3) = -6$ (Figure 10).

Figure 10

PROOF OF THEOREM.   By the definition of order $a < b$ implies that $b-a = p$, a positive number. But c is negative, so that by the trichotomy axiom $-c$ is positive. Hence

$p(-c)$ is a positive number

since the positive numbers are closed under multiplication. Consequently,

$$p(-c) = (b-a)(-c) = ac - bc$$

is positive, so, by the definition of order,

$$bc < ac$$

In other words, Theorems 2, 3, and 4 tell us that an inequality maintains the same sense if the same number is added or subtracted (see Problem 8) on both sides or if the same *positive* number is multiplied or divided (see Problem 8) on both sides, whereas if the same *negative* number is multiplied or divided (see Problem 8) on both sides, the sense of the inequality is reversed.

EXAMPLES

1   $-4 < 2$ because $2 - (-4) = 6$ is positive. Hence $-4 + 3 < 2 + 3$; $(-4)(5) < (2)(5)$; $[(-4)/7] < \frac{2}{7}$; $(-4)(-1) > (2)(-1)$.

2   Prove that if a and b are real numbers such that $a < b$, then $-a > -b$.

   PROOF.   Let $c = -1$. Then, by Theorem 4, $(-1)a > (-1)b$ or, equivalently, $-a > -b$.

3   Find an example to disprove the following assertion. (Such an example is called a *counterexample*.)

   If $a < b$ then $a^2 < b^2$

   SOLUTION.   Let $a = -3$ and $b = 2$. Then $b - a = 2 - (-3) = 5 > 0$ so that $a < b$. However, $b^2 - a^2 = 4 - 9 = -5$ implies that

   $b^2 < a^2$ or $a^2 \not< b^2$ ($\not<$ is read "is not less than")

4 Prove that if $a < b$ and $a > 0$, then $a^2 < b^2$. (Compare this statement to the statement of Example 3.)

PROOF. $a < b$ implies that $b - a > 0$. But $0 < a$ and $a < b$ implies, by the transitive property, that $b > 0$. Since the positive numbers are closed under addition, $b + a > 0$, so that

$$b^2 - a^2 = (b-a)(b+a) > 0$$

because the positive numbers are closed under multiplication. Hence

$$a^2 < b^2$$

5 If $a \neq 0$, then $a^2 > 0$.

PROOF. By the trichotomy axiom, $a > 0$ or $a < 0$. If $a > 0$, then, by Theorem 3, $a^2 > 0 \cdot a$; that is, $a^2 > 0$. If $a < 0$, then, by Theorem 4, $a^2 > 0 \cdot a$; that is, $a^2 > 0$.

3.3 Notation for Order Relations

There is some standard shorthand notation that is used in connection with the order relation to describe subsets of the number line.

1 *Betweenness.* $a < b < c$ means that $a < b$ *and simultaneously* $b < c$ (Figure 11). For example, $-1 < 4 < 5$ is correct, since $-1 < 4$ and, simultaneously, $4 < 5$. But $2 < a < 1$ is an incorrect use of this notation, since $2 < a < 1$ suggests that $2 < a$ and, *simultaneously*, $a < 1$, from which we conclude, by the transitive property, that $2 < 1$, which is false. In other words, $a < b < c$ means that b is *between* a and c on the real line and $a < c$.

Figure 11

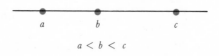

$$a < b < c$$

2 $a \not< b$ means that a is not less than b. Because of trichotomy, this implies that $a = b$ or $a > b$. For example, if $3 \not< x$, then $3 = x$ or $3 > x$ (Figure 12).

Figure 12

3   $a \leqq b$ means that either $a = b$ or $a < b$. For example, $x \leqq 2$ suggests that $x < 2$ or $x = 2$ (Figure 13).

Figure 13

In view of statements 1, 2, and 3, the meanings of $a \not> b$, $a \geqq b$, $a \leqq b$, and $a \not\geqq b$ should be obvious. Notice that different notations may be used to represent the same relation. For example, $a \leqq b$ means the same as $a \not> b$ and $a \geqq b$ means the same as $a \not< b$.

Now we can combine concepts from three areas—set theory, order relations, and the geometric interpretation of order relations—to introduce some convenient notation.

4   *Bounded intervals.*   We shall assume here that a and b are real numbers such that $a < b$. The *open interval from a to b*, denoted by (a, b), is defined as

$$(a, b) = \{x \mid a < x < b\} \qquad \text{(Figure 14)}$$

Figure 14

a excluded   b excluded

Notice that $a \notin (a, b)$ and $b \notin (a, b)$.

The *closed interval from a to b*, denoted by $[a, b]$ is defined as

$$[a, b] = \{x \mid a \leqq x \leqq b\} \qquad \text{(Figure 15)}$$

Figure 15

a included   b included

The closed interval includes the end points, whereas the open interval does not include the end points. For example, $[0, 1]$ is the set of all real numbers between 0 and 1, including 0 and 1, whereas $(0, 1)$ is the set of all real numbers between 0 and 1, excluding 0 and 1 (Figures 16*a* and *b*).

Figure 16

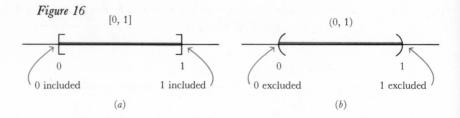

(*a*)   (*b*)

An interval from *a* to *b* including one end point but excluding the other end point is said to be *half-open* (or *half-closed*). This can happen in one of two ways:

$$[a, b) = \{x \mid a \leq x < b\} \qquad \text{(Figure 17a)}$$

Figure 17a

or

$$(a, b] = \{x \mid a < x \leq b\} \qquad \text{(Figure 17b)}$$

Figure 17b

5   *Unbounded intervals.*   We use the symbols ∞ and $-\infty$ (∞ and $-\infty$ are not real numbers) to describe unbounded intervals as follows. If *a* is a real number, then

$$[a, \infty) = \{x \mid x \geq a\} \qquad \text{(Figure 18a)}$$

Figure 18a

$$(-\infty, a] = \{x \mid x \leq a\} \qquad \text{(Figure 18b)}$$

Figure 18b

$$(a, \infty) = \{x \mid x > a\} \qquad \text{(Figure 18c)}$$

Figure 18c

$$(-\infty, a) = \{x \mid x < a\} \qquad \text{(Figure 18d)}$$

Figure 18d

For example,

$$(-\infty, 3) \cap (0, \infty) = (0, 3) \qquad \text{(Figure 19)}$$

Figure 19

Finally, we use interval notation $(-\infty, \infty)$ to denote the set of all real numbers R.

EXAMPLES

1   If $2 \leq x \leq 4$, the set of values of x determines the interval in Figure 20, and this set can be denoted by $[2, 4]$.

Figure 20

2   $[3, 3] = \{3\}$. (Why?)

3   $(2, 4) \subset [2, 4]$, since $2 \in [2, 4]$ but $2 \notin (2, 4)$ (Figure 21).

Figure 21

4    $(-\infty, 2) \cap (-3, \infty) = (-3, 2)$ (Figure 22)

Figure 22

5    $(-3, 0] \cup [2, 5)$ (Figure 23)

Figure 23

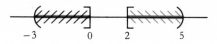

6    $(-\infty, 3) \cap [3, 4) = \varnothing$ (Figure 24)

Figure 24

7    $[2, 3] \cap (2, 3) = (2, 3)$ (Figure 25)

Figure 25

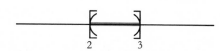

8    $\{x \mid x < -2 \text{ or } x > 1\} = \{x \mid x < -2\} \cup \{x \mid x > 1\}$
$= (-\infty, -2) \cup (1, \infty)$ (Figure 26)

Figure 26

The *complement* A^c of a set A of real numbers is the set of all real numbers that are not contained in A. Thus, if A is a set of real numbers, $A^c = \{x \mid x \notin A\}$. For example, if A is the interval set $[2, \infty)$, then the complement of the set A is $(-\infty, 2)$ (Figure 27). If Q is the set of rational numbers, then the complement of Q, Q^c, relative to the set of numbers R, is the set of irrational numbers.

Figure 27

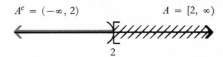

$$A^c = (-\infty, 2) \qquad\qquad A = [2, \infty)$$

EXAMPLE

Find the complement of each of the following sets relative to the set of real numbers R.

a) R b) $(3, \infty)$ c) $(8, 9)$

SOLUTION

a) $R^c = \varnothing$

b) $(3, \infty)^c = (-\infty, 3]$ (Figure 28)

Figure 28

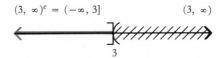

$$(3, \infty)^c = (-\infty, 3] \qquad\qquad (3, \infty)$$

c) $(8, 9)^c = (-\infty, 8] \cup [9, \infty)$ (Figure 29)

Figure 29

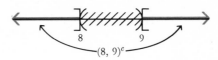

$$(8, 9)^c$$

3.4 Linear Inequalities in One Unknown

Linear inequalities in one unknown can be solved in much the same manner as linear equations. We replace an inequality by an equivalent inequality which has a solution that is obvious. As with equations, two *inequalities* are *equivalent* if their solution sets are the same. The properties of inequalities enable us to convert a given inequality to an equivalent one. A few examples will help to clarify the method for solving linear inequalities in one unknown.

EXAMPLES

Solve each of the following inequalities and represent the solution in set notation. Finally, show the solution set on the number line.

1     $3x - 2 < 7$

SOLUTION

$$
\begin{aligned}
\{x \mid 3x - 2 < 7\} &= \{x \mid 3x - 2 + 2 < 7 + 2\} && \text{(Theorem 2)} \\
&= \{x \mid 3x < 9\} \\
&= \{x \mid (\tfrac{1}{3})(3x) < (\tfrac{1}{3})(9)\} && \text{(Theorem 3)} \\
&= \{x \mid x < 3\} \\
&= (-\infty, 3) && \text{(Figure 30)}
\end{aligned}
$$

Figure 30

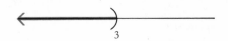

2     $x + 2 < 7x - 1$

SOLUTION

$$
\begin{aligned}
\{x \mid x + 2 < 7x - 1\} &= \{x \mid x - 7x + 2 < 7x - 1 - 7x\} && \text{(Theorem 2)} \\
&= \{x \mid -6x + 2 < -1\} \\
&= \{x \mid -6x + 2 - 2 < -1 - 2\} && \text{(Theorem 2)} \\
&= \{x \mid -6x < -3\} \\
&= \{x \mid (-\tfrac{1}{6})(-6x) > (-\tfrac{1}{6})(-3)\} && \text{(Theorem 4)} \\
&= \{x \mid x > \tfrac{1}{2}\} \\
&= (\tfrac{1}{2}, \infty) && \text{(Figure 31)}
\end{aligned}
$$

Figure 31

3     $3x + 5 \leq 12$

SOLUTION

$$
\begin{aligned}
\{x \mid 3x + 5 \leq 12\} &= \{x \mid 3x + 5 - 5 \leq 12 - 5\} && \text{(Theorem 2)} \\
&= \{x \mid 3x \leq 7\} \\
&= \{x \mid (\tfrac{1}{3})(3x) \leq (\tfrac{1}{3})(7)\} && \text{(Theorem 3)} \\
&= \{x \mid x \leq \tfrac{7}{3}\} \\
&= (-\infty, \tfrac{7}{3}] && \text{(Figure 32)}
\end{aligned}
$$

Figure 32

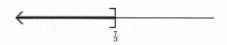

4    $3-x > -4$

SOLUTION

$$\{x \mid 3-x > -4\} = \{x \mid 3-3-x > -4-3\} \qquad \text{(Theorem 2)}$$
$$= \{x \mid -x > -7\}$$
$$= \{x \mid (-1)(-x) < (-1)(-7)\} \qquad \text{(Theorem 4)}$$
$$= \{x \mid x < 7\}$$
$$= (-\infty, 7) \qquad\qquad \text{(Figure 33)}$$

Figure 33

7

PROBLEM SET 3

1   If we were to replace the positive number set with the negative number set, would the positive number axiom hold? Give numerical examples to support your assertion.

2   a)   Under what condition is $-x < 0$?
    b)   Under what condition is $-x > 0$?
    c)   Give two numerical examples to support your assertion in parts a and b and represent the examples on the number line.

3   *Theorem 1   Transitive Property.*   If a, b, and c are real numbers such that a is less than b and b is less than c, then a is less than c. For example, if $-2 < 3$ and $3 < 5$, then $-2 < 5$. Restate Theorems 2, 3, and 4 *in words* and give two numerical examples of each theorem. Illustrate the examples on the real line.

4   *Principle of Duality.*   Consider Theorems 1 and 2 in Section 3.2. If we replace $<$ with $>$, then the theorems still hold. Rewrite the theorems with these substitutions, and then give two examples of each theorem and illustrate them geometrically.

5   In each of the following parts, state the appropriate axiom, definition, or theorem that proves the statement if it is valid. If the statement is not valid, give a counterexample to disprove it. R_p is used to denote the set of positive numbers.
    a) $-a$ is a negative number.
    b)   If $a \in R_p$, then $3a \in R_p$.
    c)   If $a \in R$, then $a+b \in R_p$.
    d) $0 \in R_p$
    e)   If $a < 4$ and $b > 4$, then $a < b$.
    f) $3+a < 8+a$
    g)   If $a < b$, then $-7a > -7b$.
    h) $-\frac{3}{4} < -\frac{1}{2}$
    i)   If $a < 0$, then $a^3 < 0$.
    j)   If $a > 1$ and $b > a$, then $b > 0$.
    k)   If $a < b$, then $(a/-10) > (b/-10)$.
    l)   If $a^2 < 4$, then $a < 2$.

6 In each of the following parts, restate the assertion *in words,* and then use Theorem 3 or 4 to prove the statement. Give two examples of each statement.
   a) If $a < 0$ and $b > 0$, then $ab < 0$.
   b) If $a < 0$ and $b < 0$, then $ab > 0$.

7 a) Prove that if $a < b$, then $a < [(b+a)/2] < b$.
    [*Hint:* $a+a < b+a < b+b$. (Why?)]
   b) Prove that if $0 < a < 1$, then $a^2 < a$, $a^3 < a^2$, etc. (*Hint:* Use Theorem 3 repeatedly.)
   c) Prove that if $a < b$ and $c < d$, then $a+c < b+d$.

8 a) Theorem 2, together with the fact that $a-c = a+(-c)$, implies that if the same number is subtracted from both sides of an equality, the order of the inequality remains the same; that is, if $a < b$, then $a-c < b-c$ for any real number c. Prove this implication and give two examples.
   b) Using the fact that $a \div c = a(1/c)$ for $c \neq 0$, restate Theorems 3 and 4 for division. Prove each statement and give two examples of each.

9 Sketch each of the following sets on a real line.
   a) $[-1, 2)$ b) $(-4, 5] \cap [-8, 3)$
   c) $[-1, 0] \cup [-\frac{1}{2}, 0]$ d) $(-2, 1) \cup (3, 8)$
   e) $(0, \infty) \cap (2, \infty)$ f) $[3, 10] \cup (7, 9)$
   g) $(0, 5) \cup (2, 7)$ h) $(-\infty, -1) \cap (-\infty, -4)$

10 Use interval notation and set operations to represent each of the following sets.

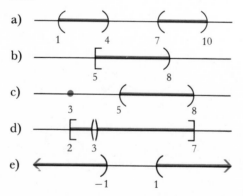

11 a) A car was driven 10 miles, and the speed during the entire trip was between 70 and 75 miles per hour. Use set interval notation to find all possible times required for such a trip.
   b) If one of the dimensions of a rectangular room is 13 feet and its area is less than 432 square feet, what can be concluded about the other dimension of the room?

12 Find the complement of each of the following sets relative to the set of real numbers R.

a) $A = (-\infty, 3)$

b) $B = [3, 4)$

c) $C = (-1, 5)^c$

d) $D = [0, \infty)$

13 Find the solution set of each of the following inequalities and represent the solution in set form and in interval form. Also show the solution set on the number line.

a) $7x > \frac{3}{2}x + 1$

b) $4x + 3 \geq 12$

c) $x + 6 \leq 4 - 3x$

d) $5 \leq 3x < 17$

e) $3 \leq 5x \leq 2x + 11$

f) $3x - 2 > 7$

g) $x + 1 \leq 3x + 2 < 5 + x$

h) $5 - x < -x + 3$

i) $2x < 0$

j) $3 + x > -1 + x$

14 a)   If a and b are real numbers such that $a > 0$ and $b > 0$, show that
$[(a+b)/2] \geq \sqrt{ab}$.

4   Absolute Value

If a and b are real numbers such that $a \leq b$, then the *distance* between a and b is considered to be the nonnegative number $b - a$ (Figure 1). For example,

Figure 1

the distance between -1 and 4 is given by $4 - (-1) = 5$ (Figure 2). The distance between -5 and -2 is equal to $(-2) - (-5) = 3$ (Figure 3).

Figure 2

Figure 3

Suppose that we are interested in finding the distance between 0 and any real number x. For convenience, we will use the notation $|x|$, which is read the *absolute value of x*, to represent the distance between x and 0. Then $|3| = 3$, $|0| = 0$, and $|-5| = 5$.

In fact, if $x < 0$, we have

$$|x| = 0 - x = -x \qquad \text{(Figure 4}a\text{)}$$

Figure 4a

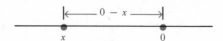

If $x > 0$,

$$|x| = x - 0 = x \qquad \text{(Figure 4}b\text{)}$$

Figure 4b

Finally, if $x = 0$,

$$|x| = 0 - 0 = 0$$

DEFINITION   ABSOLUTE VALUE

If x is a real number, the *absolute value of x*, denoted by $|x|$, is defined as follows:

$$|x| = \begin{cases} x & \text{if } x \geq 0 \\ -x & \text{if } x < 0 \end{cases}$$

It follows from the definition that $|3| = 3$ because $3 > 0$; $|0| = 0$; $|-5| = -(-5) = 5$ because $-5 < 0$.

EXAMPLE

If $x = 4$ and $y = -7$, compute each of the following expressions.

a) $|x+2y|$ b) $|x| + |2y|$

c) $|xy|$ d) $|x| \cdot |y|$

e) $|x-y|$ f) $|x| - |y|$

g) $\left| \dfrac{x}{y} \right|$ h) $\dfrac{|x|}{|y|}$

i) $|x|^2$ j) $|y|^2$

SOLUTION

a) $|x+2y| = |4+2(-7)| = |4-14| = |-10| = 10$
b) $|x|+|2y| = |4|+|2(-7)| = |4|+|-14| = 4+14 = 18$
c) $|xy| = |4(-7)| = |-28| = 28$
d) $|x|\cdot|y| = |4|\cdot|-7| = (4)(7) = 28$
e) $|x-y| = |4-(-7)| = |11| = 11$
f) $|x|-|y| = |4|-|-7| = 4-7 = -3$
g) $\left|\dfrac{x}{y}\right| = \left|\dfrac{4}{-7}\right| = \dfrac{4}{7}$
h) $\dfrac{|x|}{|y|} = \dfrac{|4|}{|-7|} = \dfrac{4}{7}$
i) $|x|^2 = |4|^2 = 4^2 = 16$
j) $|y|^2 = |-7|^2 = 7^2 = 49$

4.1 Properties of the Absolute Value Equality

We can use the definition to prove some properties of the absolute value.

THEOREM 1

Given any two real numbers x and y, $|x-y|$ represents the distance between x and y.

PROOF. The trichotomy axiom indicates that there are three possible situations:

$$x < y \qquad x > y \qquad \text{or} \qquad x = y$$

CASE 1. If $x < y$ (Figure 5), then

$$y - x > 0 \qquad \text{or} \qquad x - y < 0$$

so that, by the definition of absolute value,

$$|(x-y)| = -(x-y) = y - x$$

the distance between x and y.

Figure 5

CASE 2. If $x > y$ (Figure 6), then $x-y > 0$, so that by the definition of absolute value

$$|(x-y)| = x - y$$

the distance between x and y.

Figure 6

CASE 3.   Finally, if $x = y$, $|x-y| = |0| = 0$, which is clearly the distance between x and y.

THEOREM 2

$|x| \geq 0$ for any real number x.

PROOF.   Using the definition of absolute value, if $x \geq 0$,

$$|x| = x \geq 0$$

and if $x < 0$,

$$|x| = -x > 0 \qquad \text{(why?)}$$

THEOREM 3

$|-x| = |x|$ for any real number x.

GEOMETRIC INTERPRETATION. $|-x| = |x|$ means that the distance between 0 and x is the same as the distance between 0 and $-x$. (See Figure 7 for the case in which $x > 0$.) For example, $|-7| = |7| = 7$.

Figure 7

PROOF OF THEOREM.   If $x \geq 0$, then $-x \leq 0$, so that

$$|x| = x \qquad \text{and} \qquad |-x| = -(-x) = x$$

On the other hand, if $x < 0$, then $-x > 0$, so that

$$|x| = -x \qquad \text{and} \qquad |-x| = -x$$

In any case,

$$|x| = |-x|$$

THEOREM 4

$|x|^2 = x^2$ for any real number x.

PROOF. For $x \geq 0$, $|x| = x$, so that $|x|^2 = (x)^2 = x^2$. For $x < 0$, $|x| = -x$, so that $|x|^2 = (-x)^2 = x^2$. Hence for all possible values of x, $|x|^2 = x^2$.

EXAMPLES

1 Prove that $|x-y| = |y-x|$ for any real numbers x and y.

PROOF. By Theorem 3,

$$|x-y| = |-(x-y)| = |y-x|$$

2 Solve $|x-3| = 4$.

SOLUTION. Since $|4| = |-4| = 4$, $x-3 = 4$ or $x-3 = -4$ (Figure 8), so that $x = 7$ or $x = -1$; that is, the solution set is $\{-1, 7\}$.

Figure 8

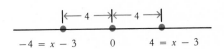

$$-4 = x - 3 \qquad 0 \qquad 4 = x - 3$$

3 Solve $|3x-4| = 5$.

SOLUTION. Since $|-5| = |5| = 5$, $3x-4 = 5$ or $3x-4 = -5$ (Figure 9), so that $x = 3$ or $x = -\frac{1}{3}$; that is, the solution set is $\{-\frac{1}{3}, 3\}$.

Figure 9

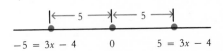

$$-5 = 3x - 4 \qquad 0 \qquad 5 = 3x - 4$$

4 Simplify $x/|x|$ if x is a nonzero real number.

SOLUTION. If $x > 0$, $|x| = x$, so that $x/|x| = x/x = 1$. If $x < 0$, $|x| = -x$, so that $x/|x| = x/-x = -1$. Consequently,

$$\frac{x}{|x|} = \begin{cases} 1 & \text{if } x > 0 \\ -1 & \text{if } x < 0 \end{cases}$$

5 Solve $|x+2| = |x-7|$.

SOLUTION 1. By Theorem 4, $|x+2|^2 = |x-7|^2$ yields
$$x^2 + 4x + 4 = x^2 - 14x + 49$$
so that $18x = 45$; that is, $x = \frac{5}{2}$.

SOLUTION 2.   We can also solve the equation by considering the various cases suggested by the definition of absolute value.

i $|x+2| = x+2$ if $x+2 \geq 0$, that is, if $x \geq -2$; and $|x+2| = -x-2$ if $x+2 < 0$, that is, if $x < -2$ (Figure 10a).

Figure 10a

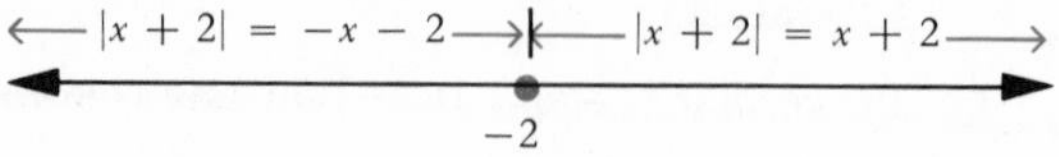

$$\longleftarrow |x+2| = -x-2 \longrightarrow\!\!\vdash\!\!\longleftarrow |x+2| = x+2 \longrightarrow$$

$$-2$$

ii $|x-7| = x-7$ if $x-7 \geq 0$, that is, if $x \geq 7$; and $|x-7| = -x+7$ if $x-7 < 0$, that is, if $x < 7$ (Figure 10b).

Figure 10b

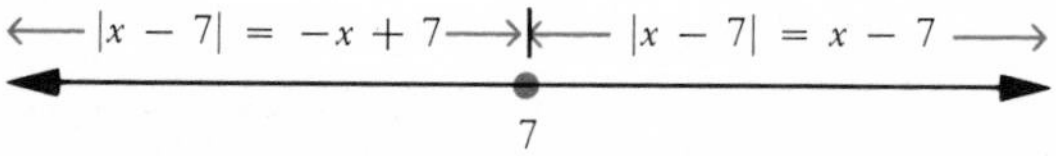

$$\longleftarrow |x-7| = -x+7 \longrightarrow\!\!\vdash\!\!\longleftarrow |x-7| = x-7 \longrightarrow$$

$$7$$

Combining the above cases, we have the following possibilities shown in Figure 11.

Figure 11

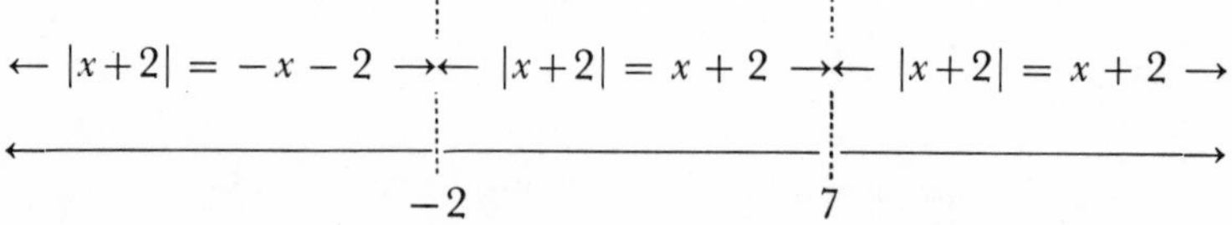

$$\longleftarrow |x+2| = -x-2 \longrightarrow\!\!\longleftarrow |x+2| = x+2 \longrightarrow\!\!\longleftarrow |x+2| = x+2 \longrightarrow$$

$$-2 \qquad\qquad 7$$

and	and	and
$|x-7| = -x+7$	$|x-7| = -x+7$	$|x-7| = x-7$
so that	so that	so that
$-x-2 = -x+7$	$x+2 = -x+7$	$x+2 = x-7$
which is impossible.	$2x = 5$	which is impossible.
	$x = \frac{5}{2}$	

Hence the solution set is $\{\frac{5}{2}\}$.

4.2 Properties of Absolute Value Inequalities

We know that $|x|$ represents the distance between 0 and x as shown in Figure 12, where x is illustrated as a positive number. Now we can use this geometric interpretation, together with the results above, to get a clear understanding of absolute value inequalities of the forms $|x| < a$ or $|x| > a$, where a is a positive number.

Figure 12

THEOREM 1

If $|x| < a$, then $-a < x < a$, where $a > 0$; in fact,

$$\{x \,|\, |x| < a\} = \{x \,|\, -a < x < a\} = (-a, a)$$

GEOMETRIC INTERPRETATION. Quite simply, $|x| < a$ means that the distance between 0 and x is less than a units; or, equivalently, x is within a units of 0 (Figure 13). Using interval notation, $x \in (-a, a)$.

Figure 13

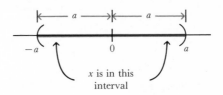

Using inequalities, this means that $-a < x < a$.

PROOF OF THEOREM. If $|x| < a$, then $-a < -|x|$. (Why?) By the definition of absolute value,

$$|x| = x \qquad \text{or} \qquad |x| = -x$$

Hence

$$-a < -|x| \leq x \leq |x| < a \qquad \text{(see Problem 6a)}$$

so that, by the transitive property of inequalities,

$$-a < x < a$$

and

$$\{x \,|\, |x| < a\} = \{x \,|\, -a < x < a\} = (-a, a)$$

EXAMPLES

1   Solve $|x| < 3$.

SOLUTION.   By Theorem 1,

$$\{x \,|\, |x| < 3\} = \{x \,|\, -3 < x < 3\} = (-3, 3) \qquad \text{(Figure 14)}$$

Figure 14

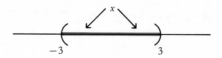

2   Solve $|3x - 2| < 8$.

SOLUTION.   By Theorem 1,

$$\begin{aligned}
\{x \,|\, |3x-2| < 8\} &= \{x \,|\, -8 < 3x - 2 < 8\} \\
&= \{x \,|\, -6 < 3x < 10\} \\
&= \{x \,|\, -2 < x < \tfrac{10}{3}\} \\
&= (-2, \tfrac{10}{3}) \qquad \text{(Figure 15)}
\end{aligned}$$

Figure 15

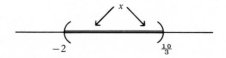

THEOREM 2

If $|x| > a$, then $x < -a$ or $x > a$, where $a > 0$; in fact,

$$\begin{aligned}
\{x \,|\, |x| > a\} &= \{x \,|\, x < -a\} \cup \{x \,|\, x > a\} \\
&= (-\infty, -a) \cup (a, \infty)
\end{aligned}$$

GEOMETRIC INTERPRETATION. $|x| > a$ means that the distance between 0 and x is more than a units; or, equivalently, x is more than a units from 0 (Figure 16).

Figure 16

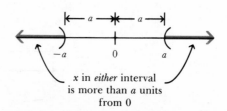

x in *either* interval
is more than a units
from 0

PROOF OF THEOREM.   By the definition of absolute value either

$$|x| = x \quad \text{or} \quad |x| = -x.$$

Hence

$$|x| = x > a \quad \text{or} \quad |x| = -x > a$$

That is,

$$x > a \quad \text{or} \quad -x > a$$

But $-x > a$ implies that $x < -a$ (why?), so that

$$x > a \quad \text{or} \quad x < -a$$

and

$$\{x \mid |x| > a\} = \{x \mid x < -a\} \cup \{x \mid x > a\}$$

$$= (-\infty, -a) \cup (a, \infty)$$

EXAMPLES

1   Solve $|x| > 7$.

SOLUTION.   By Theorem 2,

$$\{x \mid |x| > 7\} = \{x \mid x < -7\} \cup \{x \mid x > 7\} = (-\infty, -7) \cup (7, \infty)$$

$$\text{(Figure 17)}$$

Figure 17

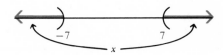

2   Solve $|2x - 3| \geqq 5$.

SOLUTION. $|2x - 3| \geqq 5$ suggests that

$$2x - 3 \leqq -5 \quad \text{or} \quad 2x - 3 \geqq 5 \qquad \text{(Figure 18)}$$

Figure 18

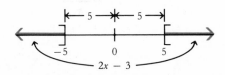

so that

$$2x \leqq -2 \qquad \text{or} \qquad 2x \geqq 8$$

That is,

$$x \leqq -1 \qquad \text{or} \qquad x \geqq 4 \qquad \text{(Figure 19)}$$

Figure 19

Hence

$$\{x \,|\, |2x-3| \geqq 5\} = \{x \,|\, x \leqq -1\} \cup \{x \,|\, x \geqq 4\} = (-\infty, -1] \cup [4, \infty)$$

3   Solve $|2x+7| \geqq 11$.

SOLUTION. $|2x+7| \geqq 11$ suggests that

$$2x+7 \leqq -11 \qquad \text{or} \qquad 2x+7 \geqq 11 \qquad \text{(Figure 20)}$$

Figure 20

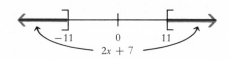

so that

$$2x \leqq -18 \qquad \text{or} \qquad 2x \geqq 4$$

That is,

$$x \leqq -9 \qquad \text{or} \qquad x \geqq 2$$

Hence

$$\{x \,|\, |2x+7| \geqq 11\} = \{x \,|\, x \leqq -9\} \cup \{x \,|\, x \geqq 2\} = (-\infty, -9] \cup [2, \infty)$$
$$\text{(Figure 21)}$$

Figure 21

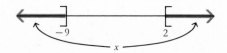

THEOREM 3 TRIANGLE INEQUALITY

If a and b are real numbers, then $|a+b| \leq |a| + |b|$.

PROOF. We know that $-|a| \leq a \leq |a|$ and $-|b| \leq b \leq |b|$ (see Problem 6a). Adding the two inequalities (see Problem 7c of Problem Set 3), we get

$$-(|a| + |b|) \leq a + b \leq (|a| + |b|)$$

Now, if $a+b \geq 0$,

$$|a+b| = a + b \leq |a| + |b|$$

whereas if $a+b < 0$,

$$|a+b| = -(a+b)$$

We have, after multiplying each side of $-(|a|+|b|) \leq a+b$ by -1,

$$|a+b| = -(a+b) \leq |a| + |b|$$

In either case,

$$|a+b| \leq |a| + |b|$$

EXAMPLES

1 Show that if $|x| < 3$ and $|y| < 1$, then $|x+y| < 4$.

SOLUTION. By the triangle inequality,

$$|x+y| \leq |x| + |y| < 3 + 1$$

so that, by transitivity,

$$|x+y| < 4$$

2 If x, y, and z are real numbers, then $|x-y| \leq |x-z| + |y-z|$.

PROOF. Let $a = x-z$ and $b = z-y$ in Theorem 3 to get

$$|(x-z) + (z-y)| \leq |x-z| + |z-y|$$

That is,

$$|x-y| \leq |x-z| + |z-y|$$

Since $|z-y| = |y-z|$ (see Example 1, Section 4.1)

$$|x-y| \leqq |x-z| + |y-z|$$

3   Use the triangle inequality to find a real number c such that

$$|x^3 + 3x^2 - 2x + 5| \leqq c$$

for all values of x such that $|x| \leqq 2$.

SOLUTION.   By repeated application of the triangle inequality, we have

$$\begin{aligned}
|x^3 + 3x^2 - 2x + 5| &\leqq |x^3 + 3x^2 - 2x| + |5| \\
&\leqq |x^3 + 3x^2| + |-2x| + |5| \\
&\leqq |x^3| + |3x^2| + |-2x| + |5| \\
&= |x^3| + 3x^2 + 2|x| + 5 \\
&\leqq 8 + 3(4) + 2(2) + 5 \\
&= 29
\end{aligned}$$

so that $c = 29$.

4   Express $\{x \,|-2 < x < 4\}$ as an absolute value inequality.

SOLUTION.   Interval $(-2, 4)$ is of length 6 and has midpoint 1 (Figure 22). Since $|x-1|$ represents the distance between x and 1, $|x-1| < 3$. That is,

$$\{x \,|-2 < x < 4\} = \{x \,\big|\,|x-1| < 3\}$$

Figure 22

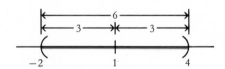

PROBLEM SET 4

1   If $x = 3$ and $y = -4$, compute each of the following numbers.
a) $|x| + |y|$ b) $|x+y|$
c) $|x-y|$ d) $|x| - |y|$
e) $|xy|$ f) $|x| \cdot |y|$
g) $|y|^2$ h) $|x/y|$
i) $|3x| + |-4y|$ j) $3|x| - 4|y|$

2   a) Give two examples of each of the four theorems of Section 4.1.
b) Give two examples of each of the three theorems of Section 4.2.

3   Solve each of the following equations.
a) $|x| = 4$ b) $|3x+2| = 5$
c) $|x-7| = |3x+1|$ d) $|2x-9| = |5x-3|$

$\quad$ e) $\;|x| = -3$ $\qquad\qquad\qquad$ f) $\;|x| = |-3|$

$\quad$ g) $\;|x-5| = |-3x+7|$ $\qquad$ h) $\;3|x| = 15$

$\quad$ i) $\;|3x| = 15$ $\qquad\qquad\qquad$ j) $\;|-3x| = 15$

4 $\quad$ a) $\;$ Compute $|x/|x||$ if x is a nonzero real number.

$\qquad$ b) $\;$ For what values of x, if any, is $|x^3| = x^3$?

$\qquad$ c) $\;$ Under what conditions does $|x+y| = |x|+|y|$?

$\qquad$ d) $\;$ Under what conditions does $|x| = |y|$?

$\qquad$ e) $\;$ Prove that $|xy| = |x| \cdot |y|$.

$\qquad$ f) $\;$ Prove that $|x/y| = |x|/|y|$.

5 $\quad$ Solve each of the following inequalities. Represent your answer in interval notation. Illustrate the solution on the number line.

$\qquad$ a) $\;|x| > 1$ $\qquad\qquad\qquad\qquad$ b) $\;|x| < 2$

$\qquad$ c) $\;|3-2x| < 5$ $\qquad\qquad\qquad$ d) $\;|x-1| \leq 3$

$\qquad$ e) $\;|x+2| \geq 5$ $\qquad\qquad\qquad$ f) $\;|3+2x| < 1$

$\qquad$ g) $\;|3x-1| > 0.1$ $\qquad\qquad\;$ h) $\;|4-7x| \leq 0.0001$

$\qquad$ i) $\;|x+5| < |x+1|$ $\qquad\qquad$ j) $\;|4x-2| \leq 1$

$\qquad$ k) $\;|5-1/x| < 1$

6 $\quad$ Prove each of the following assertions and give two examples of each.

$\qquad$ a) $\;-|x| \leq x \leq |x|$ $\qquad\qquad$ b) $\;|x| + |y| = ||x| + |y||$

$\qquad$ c) $\;|x-y| \leq |x| + |y|$ $\qquad\quad$ d) $\;||x| - |y|| \leq |x-y|$

$\qquad$ e) $\;$ If $|x-a| < \frac{1}{10}$ and $|a-y| < \frac{1}{10}$, show that $|x-y| < \frac{1}{5}$.

$\qquad$ (*Hint:* Use the triangle inequality for parts c, d, and e.)

7 $\quad$ Write each of the following inequalities as an absolute value inequality.

$\qquad$ a) $\;-3 < x < 3$ $\qquad\qquad\qquad$ b) $\;1.99 < x < 2.0$

$\qquad$ c) $\;-3.1 < x < -3$

8 $\quad$ a) $\;$ Use Theorem 1, Section 4.2, to prove that if $|x-c| < a$, then $-a+c < x < a+c$, where $a > 0$.

$\qquad$ b) $\;$ Use Theorem 2, Section 4.2, to prove that if $|x-c| > a$, then $x < c-a$ or $x > a+c$, where $a > 0$.

5 $\;$ Cartesian Coordinate System and Distance Formula

We have seen that the number line provides us with a geometric representation of the real numbers as points on a line. This geometric representation was used in investigating the order of the real numbers and the notion of distance between points on a line. In this section we shall investigate a method of representing "ordered pairs" of real numbers as points in a plane. Then this geometric representation will be used to determine a method of finding the distance between two points in a plane.

5.1   Ordered Pairs

The elements in the set $\{a, b\}$ do not have to be listed in any particular order. This set could be written either as $\{a, b\}$ or as $\{b, a\}$; in other words, $\{a, b\} = \{b, a\}$.

By contrast, (a, b) is an *ordered pair* consisting of *first the element a* and *second the element b*. [This notation is the same as that which is used to denote open intervals. The text will be sufficiently clear to indicate what (a, b) represents.]

Two ordered pairs are considered to be *equal* only when the two ordered pairs have equal first members *and* equal second members. For example, $(1, 2) \neq (2, 1)$, even though each pair contains the same entries. Likewise, $(4, 3) \neq (4, 4)$ and $(7, 8) \neq (-7, 8)$, whereas $(9, x) = (y, 8)$ only if $x = 8$ and $y = 9$.

If $A = \{1, 2\}$ and $B = \{a, b\}$, the set of all possible ordered pairs formed by selecting the first member of the ordered pair from A and the second member of the ordered pair from B is denoted by $A \times B$ and is given by

$$A \times B = \{(1, a), (1, b), (2, a), (2, b)\}$$

A simple method of grouping the elements in order to form the ordered pairs of two sets is to arrange the elements in a rectangular pattern as follows:

	Set B	
$\times$	a	b
Set A 1	$(1, a)$	$(1, b)$
2	$(2, a)$	$(2, b)$

$$A \times B$$

By contrast,

$$B \times A = \{(a, 1), (b, 1), (a, 2), (b, 2)\}$$

In general, if A and B are any two sets, $A \times B$, which is read "A cross B," denotes the set of all ordered pairs of the form (x, y), with x an element of set A and y an element of set B. Symbolically,

$$A \times B = \{(x, y) \mid x \in A \text{ and } y \in B\}$$

The set $A \times B$ is called the *Cartesian product* of sets A and B.

EXAMPLE

Let $A = \{2, 3\}$ and $B = \{2, 7, 5\}$.

a) Determine $A \times A$ and $B \times B$.

b) Form $(A \times B) \cup (B \times A)$ and $(A \times B) \cap (B \times A)$.

c) Discuss the set relation between $A \times B$ and $B \times A$.

SOLUTION

a) $A \times A = \{(2,3),(3,2),(2,2),(3,3)\}$

and

$B \times B = \{(2,2),(2,7),(2,5),(7,2),(7,7),(7,5),(5,2),(5,7),(5,5)\}$

b) $A \times B = \{(2,2),(2,7),(2,5),(3,2),(3,7),(3,5)\}$

whereas

$B \times A = \{(2,2),(7,2),(5,2),(2,3),(7,3),(5,3)\}$

Hence

$(A \times B) \cup (B \times A)$
$= \{(2,2),(2,7),(7,2),(2,5),(5,2),(3,2),(2,3),(3,7),(7,3),(5,3),(3,5)\}$

and

$(A \times B) \cap (B \times A) = \{(2,2)\}$

c) We have from part b, that $(2,2) \in [(A \times B) \cap (B \times A)]$. Also, $(2,7) \in A \times B$, but $(2,7) \notin B \times A$; and $(5,2) \in B \times A$, but $(5,2) \notin A \times B$, so that $A \times B$ and $B \times A$ are not disjoint; neither is a subset of the other; and, $A \times B \neq B \times A$.

5.2 Cartesian Coordinate System

$R \times R$, the set of all ordered pairs of real numbers, can be represented as the set of points in a plane by using a two-dimensional indexing system called the *Cartesian coordinate system*, which is constructed as follows.

First, two perpendicular lines L_1 and L_2 are constructed (Figure 1). The

Figure 1

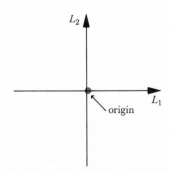

point of intersection of the two lines is called the *origin*. Next L_1 and L_2 are scaled as number lines by using the origin as the O point (zero point) for each of the two lines. The portion of L_2 above the origin is the positive direction and the portion below the origin is the negative direction of the number line; the portion of L_1 to the right of the origin is the positive direction and the portion to the left is the negative direction (Figure 2). The resulting two number lines are called the *coordinate axes*. The coordinate axes, L_1 and L_2, are often referred to as the *horizontal axis* or *x* axis and the *vertical axis* or *y* axis, respectively.

Figure 2

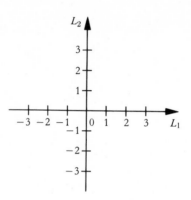

Given an ordered pair of real numbers (x,y) (the first member of the pair, x, is called the *abscissa*; the second member of the pair, y, is called the *ordinate*; x and y are called the *coordinates*), we can use the coordinate system to represent or *locate* (x, y) as a point in the plane as follows.

The abscissa x is located on the horizontal axis. Then a line is drawn perpendicular to this axis at point x; the ordinate y is located on the vertical axis and a second line is drawn perpendicular to this axis at point y. The intersection of the two lines that have just been constructed is the point in the plane used to represent the ordered pair (x, y) (Figure 3).

Figure 3

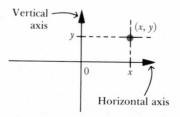

Thus for each ordered pair of real numbers in $R \times R$ we can associate a point in the plane. Conversely, for each point in the plane, we can associate an ordered pair in $R \times R$. Hence there is a one-to-one correspondence between $R \times R$ and the points in the plane.

For example, the ordered pair $(1, 1)$ is located by moving 1 unit to the right of O, on the x axis, then 1 unit up from the x axis. Similarly, $(7, 5)$ is located by moving 7 units to the right of O and 5 units up; $(-\frac{1}{2}, 0)$ is located by moving $\frac{1}{2}$ unit to the left of O and no units up from the x axis (Figure 4).

Figure 4

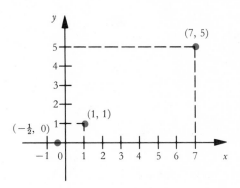

EXAMPLES

1 Locate points $(1, -2)$, $(3, 4)$, $(0, 3)$, $(-2, -3)$, and $(\pi, 0)$.

SOLUTION. These points are located in Figure 5.

2 If $A = \{1, 2, 3\}$ and $B = \{-1, -2\}$, locate the members of $A \times B$ in the plane.

Figure 5

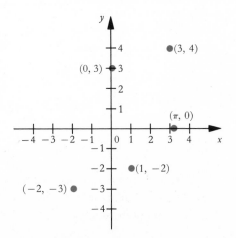

SOLUTION

$$A \times B = \{(1, -1), (1, -2), (2, -1), (2, -2), (3, -1), (3, -2)\} \quad \text{(Figure 6)}$$

Figure 6

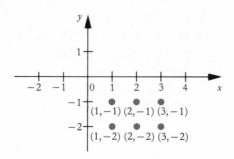

3   Locate all points with abscissas greater than 1 and ordinates less than or equal to -2; that is, locate the region in the plane containing all points corresponding to $\{(x,y) \mid x > 1 \text{ and } y \leq -2\}$.

SOLUTION.   The region is the shaded region in Figure 7. Note that the solid line is part of the region, whereas the dashed line is not part of the region.

Figure 7

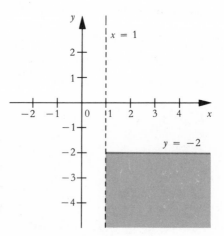

The coordinate axes divide the plane into four disjoint regions called *quadrants*, which are described in Figure 8. Notice that the coordinate axes have no point in common with the four quadrants.

Figure 8

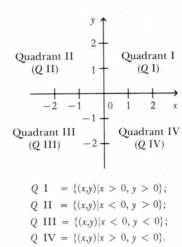

$Q\ \mathrm{I} \ = \{(x,y)|x > 0, y > 0\};$
$Q\ \mathrm{II} = \{(x,y)|x < 0, y > 0\};$
$Q\ \mathrm{III} = \{(x,y)|x < 0, y < 0\};$
$Q\ \mathrm{IV} = \{(x,y)|x > 0, y < 0\}.$

EXAMPLE

Indicate which quadrant, if any, contains each of the following points.

a) $(\frac{1}{2}, -\frac{1}{2})$ b) $(\pi, 1/\pi)$
c) $(-2, -6)$ d) $(-3, 4)$
e) $(-\sqrt{2}, 0)$ f) $(0, -5)$

SOLUTION.   After locating the points (Figure 9), we see that

a) $(\frac{1}{2}, -\frac{1}{2})$ lies in quadrant IV.
b) $(\pi, 1/\pi)$ lies in quadrant I.
c) $(-2, -6)$ lies in quadrant III.
d) $(-3, 4)$ lies in quadrant II.
e) $(-\sqrt{2}, 0)$ is not in any quadrant.
f) $(0, -5)$ is not in any quadrant.

Figure 9

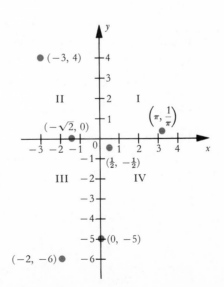

5.3   Distance Between Points

Suppose that a Cartesian coordinate system is established using the same scale for the x axis and the y axis. We could then find the distance between any two points, say P_1 and P_2, by connecting the two points with a line segment and then "measuring" or marking off the number of scale units along this segment. The total number of such units would be the length of the segment or the distance between the two points (Figure 10). This process is purely geometric and the accuracy of the result is dependent upon physical measurement. The coordinates of P_1 and P_2 have played no part in this measurement.

Figure 10

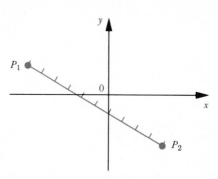

The question arises: Is it possible to use the coordinates of the two given points to get an *exact* value for the distance between them? This question will be answered in the affirmative by deriving the *distance formula*.

THEOREM 1   DISTANCE FORMULA

Given any two points P_1 and P_2 with coordinates (x_1, y_1) and (x_2, y_2), respectively, the distance d between P_1 and P_2 is given by the formula $d = \sqrt{(x_1 - x_2)^2 + (y_1 - y_2)^2}$.

PROOF.   The distance d between P_1 and P_2 in terms of coordinates x_1, y_1, x_2, and y_2 can be derived by considering three cases:

i   If the two points lie on the same vertical line, that is, $x_1 = x_2$, then $|y_1 - y_2| = d$ (Figure 11).

Figure 11

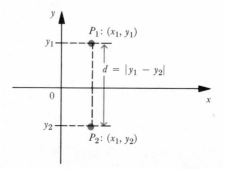

ii   If the two points lie on the same horizontal line, that is, $y_1 = y_2$, then $|x_1 - x_2| = d$ (Figure 12).

Figure 12

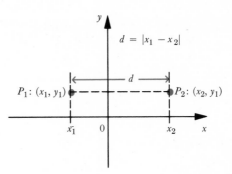

iii   If the two points lie on a line that is neither horizontal or vertical, a right triangle, that is, a triangle with a 90° angle, is determined (Figure 13).

Figure 13

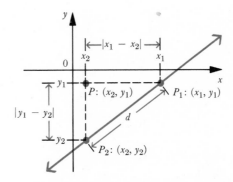

By using the Pythagorean theorem we have

$$(\text{length of segment } \overline{PP_1})^2 + (\text{length of segment } \overline{PP_2})^2 = d^2$$

That is,

$$|x_1 - x_2|^2 + |y_1 - y_2|^2 = d^2$$

so that

$$(x_1 - x_2)^2 + (y_1 - y_2)^2 = d^2 \qquad (\text{why?})$$

Hence the *distance formula*

$$d = \sqrt{(x_1 - x_2)^2 + (y_1 - y_2)^2}$$

Notice that this formula is also applicable in the special cases where P_1 and P_2 are on the same vertical line or same horizontal line. (Why?) Since $(a-b)^2 = (b-a)^2$, the "order" of subtracting the abscissas or the ordinates is irrelevant.

EXAMPLES

1   Find the distance between $(-1, -2)$ and $(3, -4)$.

SOLUTION.   The distance d is given by

$$d = \sqrt{[3 - (-1)]^2 + [-4 - (-2)]^2} = \sqrt{4^2 + 2^2} = \sqrt{20} = 2\sqrt{5}$$
(Figure 14)

Figure 14

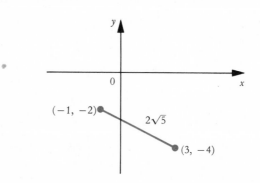

2   Derive a formula for the distance between the origin and any point (x,y) in the plane.

SOLUTION.   The distance d between $(0,0)$ and (x,y) is given by

$$d = \sqrt{(x-0)^2 + (y-0)^2} \qquad \text{(Figure 15)}$$

Figure 15

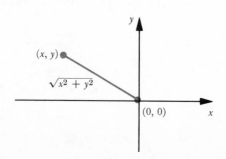

so that the formula is

$$d = \sqrt{x^2 + y^2}$$

3 Derive a formula for the distance between the x axis and any point (a, b).

SOLUTION. The distance is given by $\sqrt{0^2 + b^2} = |b|$ (Figure 16).

Figure 16

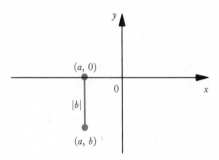

4 The *unit circle* is the circle with center at the origin and radius 1. Use the distance formula to derive the equation of the unit circle.

SOLUTION. Suppose that (x, y) is any point on the unit circle. For illustrative purposes (x, y) will be sketched as a point in the first quadrant (Figure 17). Then, by the distance formula,

$$1 = \sqrt{(x-0)^2 + (y-0)^2}$$

so that $1 = x^2 + y^2$. In other words $\{(x, y) \mid x^2 + y^2 = 1\}$ is the set of all points on the unit circle and $x^2 + y^2 = 1$ is the equation of the unit circle.

Figure 17

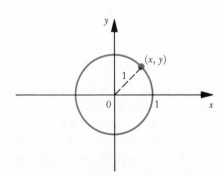

PROBLEM SET 5

1 $S = \{2, 4, 6\}$ and $T = \{a, b\}$. Form sets $S \times T$, $S \times S$, $T \times T$, and $T \times S$.
   a) How are each of these latter four sets related to the others?
   b) For each of the four sets of ordered pairs, form the union with each of the others.
   c) Do the same for the intersection.

2 a) $A \times B = B \times A$. Does $A = B$? Give an example to illustrate this situation.
   b) $A = B$. Does $A \times B = B \times A$? Explain.

3 Let $S = \{1, 2, 3, 4\}$ and $T = \{1, 2, 3\}$.
   a) Form $S \times T$.
   b) Locate the points of $S \times T$.
   c) Form the set $A = \{(x, y) \mid (x, y) \in S \times T, x = y\}$.
   d) Form $B = \{(x, y) \mid (x, y) \in S \times T, x < y\}$.
   e) Form $C = \{(x, y) \mid (x, y) \in S \times T, x > y\}$.
   f) Form $A \cup B \cup C$.

4 Given $A = \{-2, 0, 1\}$ and $B = \{-1, 3, 4\}$, form $A \times B$. Locate the members of $A \times B$.

5 Locate each of the following points and indicate what quadrant, if any, contains the point.
   a) $(3, 3)$ b) $(-2, 4)$ c) $(\pi, \sqrt{2})$
   d) $(0, 7)$ e) $(-1, -5)$ f) $(-3, 0)$

6 a) Give the coordinates of any five points on the x axis.
   b) What is common to the coordinates of all points on the x axis?
   c) Repeat parts a and b for the y axis.

7 Sketch the region in the plane corresponding to the set $\{(x, y) \mid x \leq -1$ and $y > 2\}$.

8 a) Draw the line segment between points $(1, 3)$ and $(-2, 5)$. Do all the points lie in either quadrant I or II?
   b) Find the intersection of the line segment and the vertical axis.

9 Find the distance between each of the following pair of points.
   a) $(-\frac{1}{2}, 1)$; $(2, 3)$ b) $(-3, -4)$; $(-5, -7)$
   c) $(5, 0)$; $(-7, 3)$ d) $(t, 8)$; $(t, 7)$

10 Give the coordinates of four points that are exactly 1 unit from the origin (see Example 4 on page 57).

11 Given points $P_1 : (-3, -2)$; $P_2 : (1, 2)$; and $P_3 : (3, 4)$, do the following:
   a) Locate the points.
   b) Find the lengths of segments $\overline{P_1 P_2}$, $\overline{P_2 P_3}$, and $\overline{P_1 P_3}$.
   c) Three points are said to be *collinear* if they all lie on the same straight line. Are P_1, P_2, and P_3 collinear? Explain.

12 Given points $P_1:(a, b)$; $P_2:(c, d)$; and $P_3:[(a+c)/2, (b+d)/2]$, do the following:

a) Find the lengths of segments $\overline{P_1 P_3}$ and $\overline{P_2 P_3}$ in terms of a, b, c, and d.

b) How do these two lengths compare?

c) What can you conclude about the geometric position of P_3 with respect to P_1 and P_2?

d) Give a specific example to illustrate this situation.

13 Use the fact that $((a+c)/2, (b+d)/2)$ represents the midpoint of the line segment with end points (a, b) and (c, d) to answer each of the following parts.

a) What is the midpoint of the line segment with end points $(1, -3)$ and $(5, 8)$?

b) A line segment has $(-2, -4)$ as one end point and $(1, -2)$ as the midpoint. What are the coordinates of the other end point?

REVIEW PROBLEM SET

1 Let A be a set that contains five elements and let B be a set that contains three elements. Which of the following are always true?

a) $A \cap B$ contains exactly five elements.

b) $A \cup B$ contains at least five elements.

c) $A \cup B$ contains exactly four elements.

d) $A \cap B$ is a subset of A.

e) B is a subset of A.

f) $A \cup B$ can contain no more than eight elements.

g) If $A \cap B = \emptyset$, then $A \cup B = \emptyset$.

h) If $x \in A$ and $x \in B$, then $A \cap B$ is not an empty set.

i) If $A \cap B$ contains three elements, then B is a subset of A.

j) If $A \cup B$ contains six elements, then $A \cap B$ contains two elements.

2 What set could you select as a universal set U from which each given pair of sets has been formed?

a) {All single women}; {All married women}

b) {All even integers}; {All odd integers}

c) {All Southern states}; {All Western states}

3 Use $A = \{x, y, z\}$ and $B = \{c, y, x\}$ to decide whether the following statements are true or false.

a) $x \in \{x\} \cup B$ b) $c \in A \cup B$

c) $A = B$ d) $y \in A \cap B$

e) $A \cap B = \{y, z\}$ f) $\{x\} = A \cap B$

4 Let $U = \{x, y, u, v\}$. If $A = \{x, y\}$, $A \cap B = \{x\}$, and $A \cup B = U$, find the set B.

5 A positive integer m is said to be a multiple of a positive integer n if there exists a positive integer k such that $m = kn$. For example, 18 is a multiple of 6 since $18 = 3 \cdot 6$. Describe $A \cap B$ in each of the following cases, where x is assumed to be a positive integer.
   a) $A = \{x \mid x$ is a multiple of 3$\}$ and $B = \{x \mid x$ is a multiple of 5$\}$
   b) $A = \{x \mid x$ is a multiple of 3$\}$ and $B = \{x \mid x$ is a multiple of 6$\}$
   c) $A = \{x \mid x$ is a multiple of 9$\}$ and $B = \{x \mid x$ is a multiple of 12$\}$

6 Use the following sets for each of the parts below.
   I_p, the set of positive integers
   I, the set of integers
   F, the set of quotient numbers (rational numbers that are not integers)
   Q, the set of rational numbers
   L, the set of irrational numbers
   R, the set of real numbers
   $\emptyset$, the empty set
   Each description below corresponds to at least one of the above sets. Identify these sets.
   a) A set that contains 49 but not -49.
   b) A set that contains both $\frac{2}{3}$ and π.
   c) A set that contains $-\frac{3}{4}$ but not π.
   d) $I_p \cup I$
   e) $F \cap Q$
   f) $L \cup R$
   g) A subset of the quotient numbers that contains $\frac{3}{2}$ but not 3.
   h) A subset of the real numbers that does not contain 1 but is not the empty set.
   i) A subset of the irrational numbers that does not contain π.
   j) A set that is disjoint from the positive integers and is not a subset of the integers.

7 Notice that $\frac{1}{9} = 0.11\overline{1}$ repeats in one digit blocks, and $\frac{1}{3} = 0.333\overline{3}$ repeats in one digit blocks. Compare the decimal representations of $\frac{1}{9} \cdot \frac{1}{3}$ and $\frac{1}{9} + \frac{1}{3}$ to the decimal representations of $\frac{1}{9}$ and $\frac{1}{3}$.

8 Indicate which of the following are true and which are false. If the assertion is false, give a counterexample to support your answer. Assume that a and b are real numbers such that $a < b$.
   a) $a < 3b$ b) $2a > -(-2)b$
   c) $a + c < b + c, \, c \in R$ d) $a + 3 < b + 4$
   e) $a - 4 < b - 4$ f) $a - 2 < b - 3$
   g) $1/a < 1/b$ h) $|a| < b$

9 Solve the following inequalities, show the solution on the number line, and write the solution using interval notation.

a) $5x-9 > 2x+3$

b) $2x/3 + \frac{1}{5} > \frac{7}{15} + 4x/5$

c) $x/2 + \frac{5}{6} \le 2x/3 + \frac{1}{12}$

d) $5x-2 > 6x+5$

e) $5x-11 \le 8x-5$

10 Let x, y, and z be real numbers. Which of the following is true? Give a counterexample to disprove any false statement.

a) If $-\frac{1}{2} < x < \frac{1}{2}$, then $-1 < x < 1$.

b) If $0 < x < \frac{1}{2}$, then $-\frac{1}{3} < x < \frac{1}{3}$.

c) If $x > y$, then $x-z > y-z$.

d) If $x > y$ and $z > 0$, then $x/z > y/z$.

e) If $x > y$, $x > 0$ and $y > 0$, then $x^2 > y^2$ and $1/x < 1/y$.

f) If $x^3 > y^3$, $x > 0$ and $y > 0$, then $x < y$.

11 Solve each of the following; indicate your solution set by using interval notation and represent the solution on the number line.

a) $|x-5| \ge \frac{3}{2}$

b) $|3x+1| > 8$

c) $|x-3| = 0$

d) $|x-2| < 3$

e) $|x-2| = |2-x|$

f) $|2x-5| < 9$

g) $|x-2| \ge -5$

h) $|2/x-3| < 1$

12 Use the triangle inequality to find a value of m that satisfies the following inequality.

$$|x^4 + 3x^3 - 5x^2 + x - 2| < m \qquad \text{if } |x| < 3$$

13 Let $A = \{a, b, c\}$ and $B = \{x, y\}$.

a) List all the ordered pairs of $A \times B$.

b) List all the ordered pairs of $B \times A$.

14 Let $A = \{1, 2, 3, 4, 5, 6, 7, 8\}$ and $B = \{1, 2, 3, 4, 5, 6, 7, 8, 9\}$.

a) How many elements of $A \times B$ are of the form (x, x)?

b) How many elements (x, y) of $A \times B$ satisfy the condition that $x < y$?

15 Locate the points and then find the distance between each of the following pair of points.

a) $(2, 1)$ and $(4, -5)$

b) $(-3, 2)$ and $(6, -1)$

c) $(-6, -3)$ and $(2, 1)$

d) $(0, -7)$ and $(4, 3)$

e) $(-2, 6)$ and $(-2, -1)$

16 Show that the points $(-5, 7)$, $(2, 6)$, and $(1, -1)$ lie on the same circle which has center $(-2, 3)$.

CHAPTER 2

Relations and Functions

1 Relations

The statements "Joe is married to Rose," "Barbara is a sister of Jim," "New York is larger than Michigan," and "5 is less than 8" involve what is commonly understood to be a relationship. Expressions of the type "is married to," "is a sister of," "is larger than," and "is less than" are classified as relations. Hence a *relation* suggests a correspondence or an association between the elements of two sets. For example, Table 1 suggests a relation between the numbers in the x column with the numbers in the y column. Here the correspondence between the numbers in the x column and the numbers in the y column is given by the formula $y = 2x$, where x is a positive integer.

Table 1

$$x \mapsto y$$

$$
\begin{array}{ccc}
1 & \mapsto & 2 \\
2 & \mapsto & 4 \\
3 & \mapsto & 6 \\
4 & \mapsto & 8 \\
5 & \mapsto & 10 \\
& \cdots & \\
x & \mapsto & 2x
\end{array}
$$

Similarly, Table 2 suggests a correspondence between the numbers in two sets. Here the correspondence between the numbers is given by the formula $y = 2x + 1$, where x is a positive integer.

Table 2

$$x \mapsto y$$

$$
\begin{array}{ccc}
1 & \mapsto & 3 \\
2 & \mapsto & 5 \\
3 & \mapsto & 7 \\
4 & \mapsto & 9 \\
& \cdots & \\
x & \mapsto & 2x + 1
\end{array}
$$

In each of the above examples there are three main ingredients: a first set, a second set, and a correspondence between the members of the two sets. These two examples suggest a relation between two sets of numbers. In order to describe a relation so that the corresponding members of the two sets are clearly identified, the ordered pair notation is used.

DEFINITION   RELATION

A *relation* is a set of ordered pairs (see Chapter 1, Section 5.1). The *domain* of a relation is the set of all first members of the ordered pairs and the *range* of a relation is the set of all second members of the ordered pairs.

Hence, recalling that I_p represents the set of positive integers, $\{(x,y) \mid y = 2x, \, x \in I_p\}$ is the relation suggested by Table 1, and $\{(x,y) \mid y = 2x+1, \, x \in I_p\}$ is the relation suggested by Table 2.

Since we will consider only those relations which are formed from real numbers, we can use the Cartesian coordinate system to represent relations as points in a plane. This representation of a relation is called the *graph* of the relation; the graph provides us with a "geometric picture" of the relation. It should be noted here that there is a difference between the relation (a set of ordered pairs) and the graph (a set of points in the plane) which we use to represent the relation geometrically.

EXAMPLES

1   Assume that $U = \{1,2,3\}$ for each of the following relations. Enumerate the members of the relation, indicate the domain and range, and graph the relation.
a) $R_1 = \{(x,y) \mid y = x, \text{ where } x \in U, y \in U\}$
b) $R_2 = \{(x,y) \mid y > x, \text{ where } x \in U, y \in U\}$

SOLUTION

a) $R_1 = \{(1,1),(2,2),(3,3)\}$, so the domain and range are the same set $\{1,2,3\}$. The graph is composed of three points (Figure 1).

Figure 1

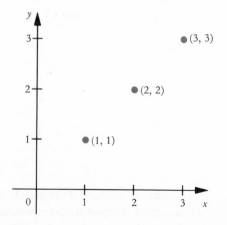

b) $R_2 = \{(x, y) \mid y > x\} = \{(1,2),(1,3),(2,3)\}$, so that the domain of R_2 is $\{1,2\}$ and the range of R_2 is $\{2,3\}$. The graph has three points (Figure 2).

Figure 2

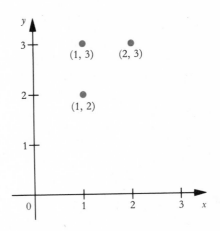

2 Graph the relation $\{(x,y) \mid y = 5x, x \in R\}$.

SOLUTION.   It is impossible here to list all members of this relation. The best that we can do is to plot some of the points to discover the "pattern" of the graph (Figure 3). Tables are often used to list members of a relation; $(0,0)$, $(1,5)$, $(\frac{1}{5}, 1)$, $(-\frac{1}{5}, -1)$, and $(-1, -5)$ are recorded in the table next to Figure 3. If we were to continue to plot members of this relation, the points would form a *linear* pattern. In fact, the graph would actually turn out to be a line (Figure 3).

Figure 3

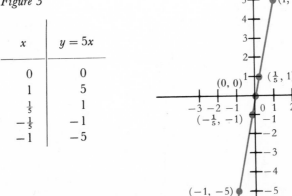

x	$y = 5x$
0	0
1	5
$\frac{1}{5}$	1
$-\frac{1}{5}$	-1
-1	-5

In general, the *graph* of any relation that is defined by a linear equation such as $y = 5x$, $y = 1 - 8x$, and $y = 3x + 7$ is a *straight line*. (This topic will be covered more extensively in Chapter 3.)

3   Use a table of values to graph $\{(x,y) \mid x^2 + y^2 = 4,\ x \in \{-2, -1, 0, 1, 2\}\}$. Indicate the domain and range.

SOLUTION.   The graph is given in Figure 4.

Figure 4

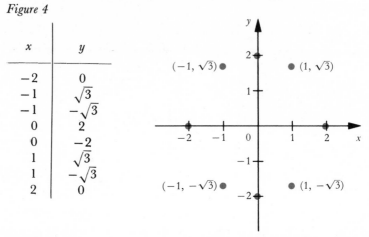

x	y
-2	0
-1	$\sqrt{3}$
-1	$-\sqrt{3}$
0	2
0	-2
1	$\sqrt{3}$
1	$-\sqrt{3}$
2	0

The domain is $\{-2, -1, 0, 1, 2\}$ and the range is $\{0, \sqrt{3}, -\sqrt{3}, 2, -2\}$.

Quite often set notation is not used to describe a relation, instead the relation is defined by giving only its *rule of correspondence*. For example, $x < y$ defines the relation $\{(x, y) \mid x < y,\ x \in R\}$; $y = 2x$ defines the relation $\{(x, y) \mid y = 2x\}$ and $y = 1/x$ defines the relation $\{(x, y) \mid y = 1/x\}$. Assuming that x and y represent real numbers, the domain of $y = 1/x$ does *not* include 0 (Figure 5), since division by zero is not defined. Here the rule of correspondence of the relation $y = 1/x$ *implies* that 0 must be excluded from the domain.

Figure 5

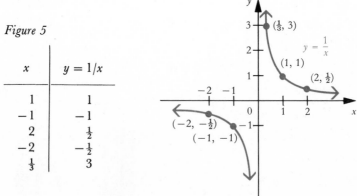

x	$y = 1/x$
1	1
-1	-1
2	$\frac{1}{2}$
-2	$-\frac{1}{2}$
$\frac{1}{3}$	3

In general, the domain and/or range of a relation is not always given. If this is the case, we determine the domain by inspection as we did with

$y = 1/x$. Also, if the universal set is not given, we will assume the universal set for the relation to be $R \times R$, where R is the set of real numbers.

Note that if (x,y) is a member of a relation, we can consider the real numbers x and y from two viewpoints. On the one hand, x is a member of the domain and y is the corresponding member of the range of the relation. On the other hand, x represents the abscissa and y the corresponding ordinate of a point in a plane. On the graph, then, the members of the domain are the abscissas, and the members of the range are the ordinates. Any restriction on the domain is a restriction on "the horizontal position of the graph," and any restriction on the range is a restriction on "the vertical position of the graph."

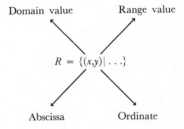

EXAMPLES

1 Graph $y = x^2$ and indicate the domain and range.

SOLUTION. $y = x^2$ implies the relation $\{(x,y) \mid y = x^2\}$ with universal set $R \times R$. Since any real number can be squared and since the square of a real number is always nonnegative, the domain is R and the range is set $[0, \infty)$. This means that the graph "lies" in the region above and on the x axis (Figure 6).

Figure 6

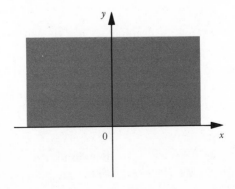

Finally, we can use a table of values to determine the graph (Figure 7).

Figure 7

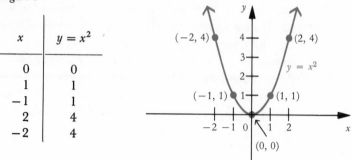

x	$y = x^2$
0	0
1	1
-1	1
2	4
-2	4

2   Assume that the graph of a relation is given in Figure 8. Use the graph to identify the domain and the range of the relation.

Figure 8

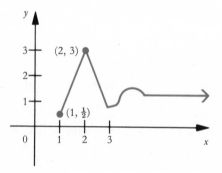

SOLUTION.   It can be seen from the graph in Figure 8 that each of the abscissas x satisfies $x \geq 1$ and each of the ordinates y satisfies $\frac{1}{2} \leq y \leq 3$; consequently, the domain is set $[1, \infty)$ and the range is set $[\frac{1}{2}, 3]$.

3   Sketch the graph of each of the following relations on the same coordinate system.
   a) $R_1 = \{(x,y) \mid y = 4x\}$
   b) $R_2 = \{(x,y) \mid y < 4x\}$
   c) $R_3 = \{(x,y) \mid y > 4x\}$

SOLUTION
   a)   We will list a few points and plot them to determine the graph of $y = 4x$ (Figure 9).

b) The graph of $y < 4x$ is the shaded area R_2 (Figure 9).
c) The graph of $y > 4x$ is the shaded area R_3 (Figure 9).

Figure 9

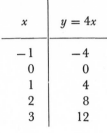

x	$y = 4x$
-1	-4
0	0
1	4
2	8
3	12

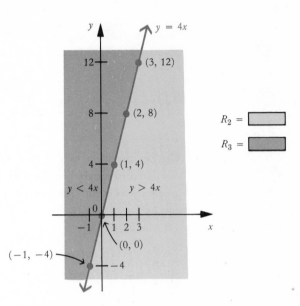

PROBLEM SET 1

1 Which of the following sets are relations?
  a) $\{(1,2),(2,3),(3,4)\}$ b) $\{1,5,9\}$
  c) $\{\{1,2\},\{2,3\}\}$ d) $\{(1,2),(3,4),(2,2),(2,1)\}$

2 Let $U \times U$ be the universal set, where $U = \{0,1,2,3\}$ for each of the following relations. Tabulate each of the relations. State the domain and the range for each and graph the relation.
  a) $R_1 = \{(x,y)\,|\,y = x\}$ b) $R_2 = \{(x,y)\,|\,y > x\}$
  c) $R_3 = \{(x,y)\,|\,y < x\}$ d) $R_4 = \{(x,y)\,|\,x+y = 1\}$
  e) $R_5 = \{(x,y)\,|\,x^2 + y^2 = 1\}$

3 Shade the smallest region in the plane that contains the graph of a relation with domain D and range R_0 for each of the following.
  a) $D = [-1,1]$ and $R_0 = [3,5]$
  b) $D = [0,\infty)$ and $R_0 = \{5\}$
  c) $D = (-\infty,0]$ and $R_0 = [0,\infty)$
  d) $D = \{2,3\}$ and $R_0 = [-2,3]$
  e) $D = [-1,0)$ and $R_0 = (1,4)$

4 Suppose that a relation has the graph in Figure 10. What is the domain? What is the range?

Figure 10

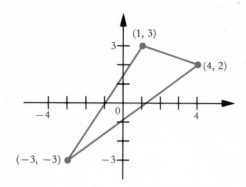

5 Use set notation to describe each of the following relations. State the
domain and range. List five members of the relation in table form, and then
graph the relation. Assume x represents the domain values and y the range
values.

a) $y = 2x + 1$ b) $y = -2$

c) $y^2 = x$ d) $2x + 2y = 3$

e) $y < 2x + 1$ f) $x^2 = y^2$

g) $x = 7$ h) $x^2 + y^2 = 1$

i) $y = |x + 1|$ j) $|x| + |y| = 1$

6 Give two examples to show that it is possible for two different ordered pairs
of a relation to have the same first member.

7 Let $U = \{1, 2\}$. Indicate the different relations that are subsets of $U \times U$.

2 Functions

We have already encountered the usage of functions in everyday living.
For example, the amount of sales tax charged for a purchase of $5.00 is a
function of the sales tax rate; the number of books to be ordered for a course
is a function of the number of students in the course; the number of Con-
gressional representatives for a particular state is a function of the
population of the state.

Suppose that there is a 5 percent sales tax on all purchases. A purchase
of $1.00 would yield 5 cents tax; $2.50 would yield 13 cents tax; in general,
a purchase of x dollars would have a sales tax of $0.05x$ rounded off to the
nearest cent. The sales tax is a function of cost.

Intuitively, a function suggests some kind of a correspondence. In each
of the examples above, there is an established correspondence between
numbers—the amount of sales tax corresponds with the cost, the number
of books with the number of students, the number of representatives with
the number of people. In general, we have the following definition.

DEFINITION 1 FUNCTION AS A CORRESPONDENCE

A *function* is a correspondence that assigns to each member in a certain set, called the *domain* of the function, *exactly one member* in a second set, called the *range* of the function.

For example, let us consider the situation in which as each student registers for classes at the beginning of the term, the tuition charge is recorded with the student account number, as illustrated by the following partial table.

Student account number	Tuition charge
895	315.00
475	323.50
182	260.90
743	315.00
234	370.00

Here the set of account numbers is the domain; the set of tuition charges is the range; the correspondence between the domain members and range members is suggested by the following table.

Domain	"Corresponds to"	Range
895	"Corresponds to"	315.00
475	"Corresponds to"	323.50
182	"Corresponds to"	260.90
743	"Corresponds to"	315.00
234	"Corresponds to"	370.00

In the table, the tuition charge is a function of the student account number.

Ordered pair notation is also used to represent functions; for if x is a member of the domain of a function and y is the member of the range corresponding to x, we can represent the correspondence between x and y as the ordered pair (x,y). In fact, we can say that the function is the set of all such ordered pairs. In the example above, then, we can represent the function as the set of ordered pairs $\{\dots, (895, 315.00), (475, 323.50), (182, 260.90), (743, 315.00), (234, 370.00), \dots\}$. In this sense, we define a function a second way, recalling that a relation is a set of ordered pairs.

DEFINITION 2 FUNCTION AS A RELATION

A *function* is a relation in which no two different pairs have the same first member. The set of all first members (of the ordered pairs) is called the *domain* of the function. The set of all second members (of the ordered pairs) is called the *range* of the function. For each ordered pair (x, y) of the function we say that x in the domain has y as the corresponding member of the range.

We have seen above that a 5 percent sales tax on any purchase can be considered as a function. We can describe this function using ordered pair notation as $\{(x, 0.05x) \mid x$ is the amount of the purchase and $0.05x$ is the tax rounded to the nearest cent$\}$; a purchase of x dollars corresponds to a $0.05x$ sales tax. We could also display this correspondence by a table, as follows.

Purchase in dollars (domain)	Sales tax in dollars (range)
0.25	0.01
0.50	0.03
1.00	0.05
. . . .	. . . .
x	$0.05x$

From the definition of a function (Definition 2) we conclude that all functions are relations; however, *not all relations are functions*. For example, $\{(1, 1), (2, 2), (3, 7), (3, 5)\}$ is a relation, with domain $\{1, 2, 3\}$ and range $\{1, 2, 7, 5\}$, but it is not a function because $(3, 7)$ and $(3, 5)$ have the same first members (Definition 2). By contrast, $\{(1, 2), (3, 4), (4, 4)\}$ is a relation that is a function with domain $\{1, 3, 4\}$ and range $\{2, 4\}$. Note that in the latter example, two pairs have the same second member; this does *not* violate the definition of a function.

Figure 1

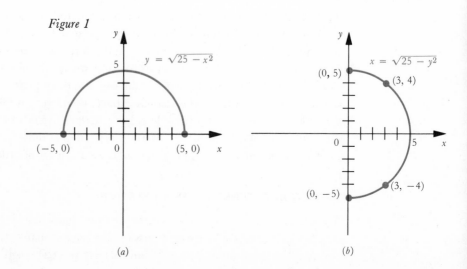

(a)   (b)

Hence, if at least two different ordered pairs of a relation have the same first member, the relation is not a function. In other words, if a domain member appears with more than one range member, the relation is not a function. Geometrically, this means that if the graph of a relation has more than one point with the same abscissa, the relation is not a function.

Consider the graphs of the relations in Figure 1a and b. The relation $\{(x,y)\,|\,y = \sqrt{25-x^2}\}$ in Figure 1a represents a function since the graph does not have two different points with the same abscissas, whereas the relation $\{(x,y)\,|\,x = \sqrt{25-y^2}\}$ in Figure 1b is not a function, since there are two different points with the same abscissas; for example, $(3,4)$ and $(3,-4)$ are two such points.

EXAMPLES

1 In each of the following parts, indicate whether the relation is a function. What is the domain and range? Graph the relation.
   a) $\{(1,2),(2,3),(3,4),(4,4),(5,6)\}$ b) $\{(x,y)\,|\,y > x\}$
   c) $y = 3$ d) $x = 1$
   e) $y = 3x$ f) $y = 3x,\ x \in \{-1,0,2,3\}$

SOLUTION

   a) $\{(1,2),(2,3),(3,4),(4,4),(5,6)\}$ is a function with domain $\{1,2,3,4,5\}$ and range $\{2,3,4,6\}$ (Figure 2). Note that we do not draw lines connecting these points, since they are the only points that belong to the relation.

Figure 2

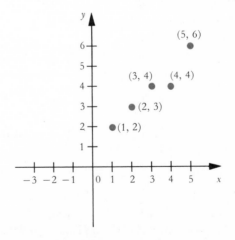

   b) $\{(x,y)\,|\,y > x\}$ is not a function because, for example, $(1,2)$ and $(1,\frac{5}{2})$ are two different ordered pairs with the same first members. The domain is the set of all real numbers and the range is also the set of all real numbers. The shaded region in Figure 3 is the graph of $\{(x,y) \mid y > x\}$. Notice that the graph does *not* include the points on the dashed line.

Figure 3

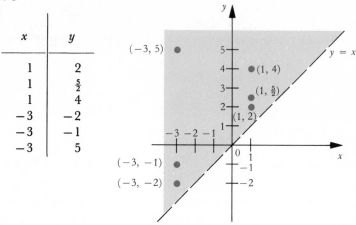

x	y
1	2
1	$\frac{5}{2}$
1	4
-3	-2
-3	-1
-3	5

c) $y = 3$ is an abbreviated way of writing the relation $\{(x,y) \mid y = 3\}$. The domain is the set of all real numbers because there is no restriction on x, whereas the range is $\{3\}$. The relation is a function because no two different ordered pairs have the same first members. The graph is the line parallel to the x axis in Figure 4.

Figure 4

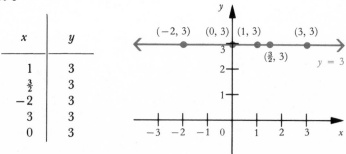

x	y
1	3
$\frac{3}{2}$	3
-2	3
3	3
0	3

d) $x = 1$ is the relation $\{(x,y) \mid x = 1\}$. This relation is not a function because $(1, 3)$ and $(1, 4)$ are members of the set. The domain of the relation is $\{1\}$ and the range is the set of all real numbers. The graph is the line parallel to the y axis in Figure 5. We can see geometrically that $x = 1$ is not a function because there are at least two (actually an infinite number of) points with the same abscissa.

Figure 5

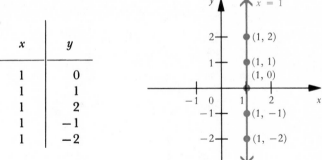

x	y
1	0
1	1
1	2
1	−1
1	−2

e) $y = 3x$ is the relation $\{(x,y) \mid y = 3x\}$. It is a function; the domain and range is the set of all real numbers. The graph is given in Figure 6. The graph is a line since $y = 3x$ is a linear equation. It can be seen geometrically that this relation is a function since no two points have the same abscissa.

Figure 6

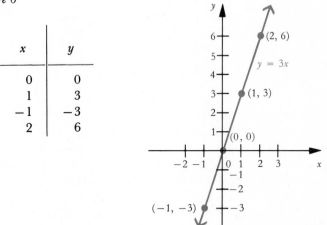

x	y
0	0
1	3
−1	−3
2	6

f) $y = 3x$ with $x \in \{-1, 0, 2, 3\}$ is the relation $\{(-1, -3), (0, 0), (2, 6),$ $(3, 9)\}$. It is a function with domain $\{-1, 0, 2, 3\}$ and range $\{-3, 0, 6, 9\}$ and the graph is composed of the four points in Figure 7. Notice that the equation used in this function is the same as the equation used in part e; however, the functions are different. (They have different domains.)

Figure 7

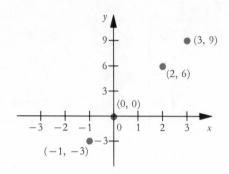

2   Examine the graphs of each of the relations given in Figures 8*a* and *b* to decide whether or not the relation is a function.

Figure 8

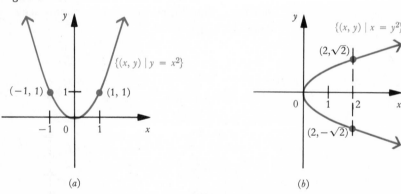

(a)   (b)

SOLUTION.   Since the graph in Figure 8*a* does not have two different points with equal abscissas, no two different ordered pairs of the relation have equal first members, and the relation $\{(x,y) \mid y = x^2\}$ is a function.

It can be seen from the graph in Figure 8*b* that there are two different points with abscissa 2, so the relation $\{(x,y) \mid x = y^2\}$ is not a function.

2.1   Function Notation

Let $y = 2x - 1$ be a function with domain $\{0, 1, 2\}$. The range is $\{-1, 1, 3\}$. This function could also be described as $\{(0, -1), (1, 1), (2, 3)\}$. The latter representation gives a more complete picture of the function than the first representation in the sense that the ordered pairs indicate precisely which member of the domain is associated with each member of the range.

A function f can also be described as a "mapping" of members of the domain to corresponding members of the range because f "maps" x to y. In the example $y = 2x - 1$, 0 is "mapped" to -1, 1 is "mapped" to 1, and 2 is "mapped" to 3.

In general, if (a, b) is a member of a function f, then f associates a in the domain with b in the range and we say that f *maps a to b* or *a is mapped to b by f* or, equivalently, *b is the image of a under f*. This mapping is symbolized in various ways:

$$f: a \mapsto b \quad \text{or} \quad a \overset{f}{\mapsto} b \quad \text{or} \quad a \mapsto f(a)$$

The notation $f: A \mapsto B$ or $A \overset{f}{\mapsto} B$ is used to denote the fact that the function f "maps" the domain set A into set B, where B is a set containing the range of the function f.

EXAMPLES

Interpret each of the following functions as mappings.

1   $f = \{(1, 2), (3, 4), (4, 4)\}$

SOLUTION. $1 \overset{f}{\mapsto} 2$; $3 \overset{f}{\mapsto} 4$; $4 \overset{f}{\mapsto} 4$.

2   $y = 3x$, where x represents members of the domain.

SOLUTION.   Here 6 is the image of 2, 0 is the image of 0, and $-\frac{3}{2}$ is the image of $-\frac{1}{2}$. In mapping notation we write $2 \overset{f}{\mapsto} 6$, $0 \overset{f}{\mapsto} 0$, and $-\frac{1}{2} \overset{f}{\mapsto} -\frac{3}{2}$. In general, if x is any real number, then $x \overset{f}{\mapsto} 3x$.

Now, if a function is to be described by use of an equation, we must have additional information to define the function. For example, $3r + 5t = 3$ might have either r or t representing the members of the domain. If t represents members of the domain, then, after solving for r in terms of t, $t \mapsto (3 - 5t)/3$ would be the function, and we say that r is a function of t; whereas if r represents members of the domain, after solving for t in terms of r, we get $r \mapsto (3 - 3r)/5$, so that t is a function of r.

If x represents members of the domain of $y = x^2$, then y is considered to be a function of x and we indicate this by writing $f(x) = x^2$, which reads "f of x equals x squared" and means "the value of the function f at the number x is the number x^2." Hence $f(x) = x^2$ is another way of defining the function $\{(x, y) \mid y = x^2\}$.

In other words, $y = f(x)$ means that (x, y) is a member of the function. For example, if $4x - 2y = 1$ and $y = f(x)$, then after solving for y in terms of x, we could write the function either as $\{(x, y) \mid y = 2x - \frac{1}{2}\}$ or as $f(x) = 2x - \frac{1}{2}$.

It is important to realize that letters other than x, y, or f can be used to denote functions. For example, $h(r) = r^2 - 1$ defines the function given by $\{(r, h(r)) \mid h(r) = r^2 - 1\}$; $3r + 5t = 3$ with $t = g(r)$ defines the function given by $\{(r, t) \mid t = (3 - 3r)/5\}$; and $c(d) = \pi d$ defines the function given by $\{(d, c(d)) \mid c(d) = \pi d\}$.

EXAMPLES

1   Let $f(x) = x + 1$. Then $f(2) = 3$ means that $(2, 3)$ is a member of the function, or, equivalently, $2 \mapsto 3$.

2    $f(x) = 1/x$ represents the function given by $\{(x, y) \mid y = 1/x\}$ with the implied domain and range the same, namely, the set of nonzero numbers.

When functional notation is used, sometimes it is helpful to think of the variable that represents the members of the domain as a "blank." For example, $g(t) = t^2$ can be thought of as $g(\ ) = (\ )^2$; hence if any expression (representing a real number) is used to represent a member of the domain, it is easy to see where this same expression is to be substituted in the equation describing the function.

Using $g(t) = t^2$ again, $g(x + h)$ can be determined by first writing the function as

$$g(\ ) = (\ )^2$$

so that, after substituting the expression $x + h$ into the blank, we get

$$g(x + h) = (x + h)^2 = x^2 + 2xh + h^2$$

Similarly,

$$g(3 - 5x) = (3 - 5x)^2 = 9 - 30x + 25x^2$$

EXAMPLES

1   Assume that $f(x) = 7x + 2$. Determine the values $f(1)$, $f(2)$, $f(3)$, and $f(4)$ and then describe these function values using the function notation, the mapping interpretation, and the ordered pair interpretation.

SOLUTION

Since $f(x) = 7x + 2$, we have

$$f(1) = 7(1) + 2 = 7 + 2 = 9$$

$$f(2) = 7(2) + 2 = 14 + 2 = 16$$

$$f(3) = 7(3) + 2 = 21 + 2 = 23$$

$$f(4) = 7(4) + 2 = 28 + 2 = 30$$

Functional notation $y = f(x)$	Mapping $f: x \mapsto 7x + 2$	Ordered pairs (x, y)
$f(1) = 9$	$1 \to 9$	$(1, 9)$
$f(2) = 16$	$2 \to 16$	$(2, 16)$
$f(3) = 23$	$3 \to 23$	$(3, 23)$
$f(4) = 30$	$4 \to 30$	$(4, 30)$

2 Given that $f(x) = x^2 - 1$, what is the domain and the range of f? Find each of the following expressions.

a) $f(3) + f(5)$ b) $f(x+1)$

c) $f(2a+4)$ d) $2f(b-2)$

SOLUTION. The domain of the function f is the set of real numbers and the range of f is the set $[-1, \infty)$ since $y = x^2 - 1$ implies that $y \geq -1$.

a) $f(3) = 3^2 - 1 = 8$ and $f(5) = 5^2 - 1 = 24$, so that

    $f(3) + f(5) = 8 + 24 = 32$

b) $f(x+1) = (x+1)^2 - 1 = x^2 + 2x$

c) $f(2a+4) = (2a+4)^2 - 1 = 4a^2 + 16a + 15$

d) $2f(b-2) = 2[(b-2)^2 - 1] = 2(b^2 - 4b + 3)$

          $= 2b^2 - 8b + 6$

3 Determine which of the following points lie on the graph of the function $f(x) = \sqrt{1+x}$: $(3, 2)$, $(8, -3)$, $(15, 4)$, and $(9, 10)$.

SOLUTION. If the ordered pairs $(3, 2)$, $(8, -3)$, $(15, 4)$, and $(9, 10)$ lie on the graph of f, then each should satisfy the equation $f(x) = \sqrt{1+x}$. For $(3, 2)$, we have $f(3) = \sqrt{1+3} = \sqrt{4} = 2$, so the point $(3, 2)$ lies on the graph of f. For $(8, -3)$, we have $f(8) = \sqrt{1+8} = \sqrt{9} = 3$, so the point $(8, -3)$ does *not* lie on the graph of f. For $(15, 4)$, we have $f(15) = \sqrt{1+15} = \sqrt{16} = 4$, so the point $(15, 4)$ lies on the graph of f. For $(9, 10)$, we have $f(9) = \sqrt{1+9} = \sqrt{10}$, so the point $(9, 10)$ does *not* lie on the graph of f.

4 The *difference quotient* of a function $y = f(x)$ is defined as

$$\frac{f(x+h) - f(x)}{h} \qquad h \neq 0$$

Compute the difference quotient for each of the following functions.

a) $f(x) = 2x + 1$ b) $f(x) = x^2$

SOLUTION

a) $f(x+h) = 2(x+h) + 1$, so that

$$\frac{f(x+h) - f(x)}{h} = \frac{[2(x+h) + 1] - (2x+1)}{h}$$

$$= \frac{2x + 2h + 1 - 2x - 1}{h} = \frac{2h}{h} = 2 \qquad h \neq 0$$

b) $f(x+h) = (x+h)^2 = x^2 + 2xh + h^2$ implies that

$$\frac{f(x+h) - f(x)}{h} = \frac{x^2 + 2xh + h^2 - x^2}{h}$$

$$= \frac{h(2x+h)}{h}$$

$$= 2x + h \qquad h \neq 0$$

5 Use functional notation to express the area of a square as a function of the length of its diagonal.

SOLUTION.   By the Pythagorean theorem, $x^2 + x^2 = z^2$, where z is the length of the diagonal and x is the length of a side (Figure 9). Hence $2x^2 = z^2$. The area of the square is x^2 or $z^2/2$, from which it follows that the area can be expressed as a function of z by $A(z) = z^2/2$.

Figure 9

PROBLEM SET 2

1   a)   Let $R_1 = \{(1,2),(3,2)\}$ be a relation. Is R_1 a function? Explain.
    b)   Give an example of a relation that is not a function. Explain.
    c)   Given an example of two *different* functions that have the same domain and the same range.

2   Which of the graphs in Figure 10 represent functions?

Figure 10

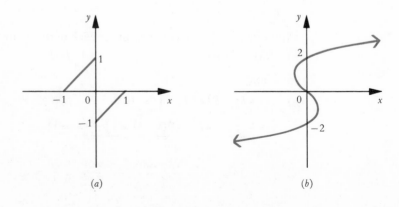

(a)   (b)

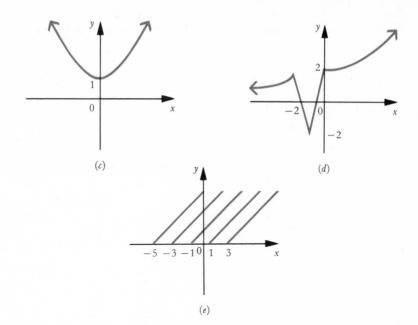

(c)

(d)

(e)

3   In each of the following parts, identify which of the relations is a function and which is not a function. If it is not a function, state the reason. In any case, identify the domain and range and graph each relation.
   a) $\{(-8,0),(-7,1),(-6,2),(-5,3),(-4,0),(-3,1),(-2,2)\}$
   b) $\{(1,1),(2,1),(3,1)\}$
   c) $\{(x,y)\mid y=5x+1\}$
   d) $|y|=|x|$, where $x \in \{1,2,3,4\}$ and x represents domain values
   e) $|y|=|x|$, where x represents domain values
   f) $y=\sqrt{x}$ where x represents domain values (Note that $\sqrt{}$ represents the *principal root*; that is, it is only positive or zero.)
   g) $\{(x,y)\mid -3x^2 = y\}$

4   Let $f(x) = 3|x|-x$. What is the domain and range of f?
   a)   Determine each of the values $f(-2), f(-1), f(0), f(1)$, and $f(2)$.
   b)   Describe these function values using the function notation, the mapping interpretation, and the ordered pair interpretation.

5   Let $f(x) = -5x + 7$. What is the domain and range of f? Determine each of the following values.
   a) $f(2)$ b) $f(-2)$ c) $3f(2)$
   d) $f(a)$ e) $f(3a)$ f) $f(b)$
   g) $f(\sqrt{a})$ h) $f(a^2)$ i) $\dfrac{f(5b)}{5}$

   j) $f(a+b)$ k) $f(a+b)-f(a)$ l) $\sqrt{f(a)+1}$

6   Determine which of the following points lie on the graph of $f(x) = x + 1/x$:
   $(1,2), (2,\frac{7}{2}), (-1,0), (-2, -\frac{5}{2})$, and $(3, \frac{10}{3})$.

7   Assume that $g(t) = t^2 + 1$. Determine each of the following expressions.

  a) $g(2)$ b) $g(x+1)$
  c) $g(x)+1$ d) $g(x^2-y^2)$
  e) $g(x^2)-g(y^2)$ f) $g(3x+y)$

8   a) If $f(x) = (x-3)/(5x+7)$, find $f(a/5)$ and $f(4/x)$.
  b) If $f(x) = (3x-1)/(1+2x)$, find $f(a^2)$, $[f(a)]^2$, $f(1/a)$, and $1/f(a)$.

9   Form the difference quotient $[f(x+h)-f(x)]/h$, $h \neq 0$, and then simplify the resulting expression for each of the following functions.

  a) $f(x) = 2$ b) $f(x) = 3x+5$
  c) $f(x) = x^2/2$ d) $f(x) = (x+2)^2$
  e) $f(x) = 1/x$

10 Use $f(x) = x+2$, $g(x) = 2x-1$, and $h(x) = x-1$ to answer each of the following parts.

  a) Is $f(1/x) = 1/f(x)$?
  b) Does $g(x) \cdot [1/g(x)] = 1$ for all x?
  c) Is $h(x-1) = h(x)-h(1)$?
  d) Is $f(x+2) = f(x)+f(2)$?
  e) Is $g(2x) = 2 \cdot g(x)$?

11 Let $f(x) = x(x+1)(x+2)(x+3)$. Show that

$$\frac{f(a+1)}{a+1} = \frac{f(a+2)}{a+5}$$

12 a) Write an equation that expresses the radius r of a circle as a function of its circumference c.

  b) Find an expression for the length l of the diagonal of a rectangular prism as a function of the width s if the length is twice the width and the height is three times the width.

  c) Equal squares are cut from the four corners of a rectangular piece of cardboard 10 inches by 14 inches. An open box is then formed by folding up the flaps. Express the volume V of the box as a function of x, where x is the length of the side of the squares removed.

3   Some Properties of Functions

Besides determining the domain, range, and graph of functions, there will be occasions when we will investigate other properties of functions.

3.1   Symmetry: Even and Odd Functions

The recognition of *symmetry* often simplifies the graphing of functions. Suppose that we have graphed a function and assume that the part of the graph that lies to the right of the y axis is drawn with wet ink. Now "fold" the plane along the y axis. If the "inked" part of the graph coincides with the part of the graph that lies to the left of the y axis, we say that the graph of f is *symmetric with respect to the y axis*. For example, the graph in Figure 1 is symmetric with respect to the y axis.

Figure 1

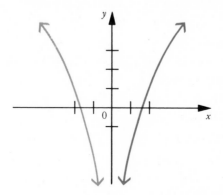

What does symmetry with respect to the y axis mean in terms of the points that are on the graph? Very simply, a graph is symmetric with respect to the y axis if, whenever (x,y) is on the graph, $(-x,y)$ is also on the graph (Figure 2).

Figure 2

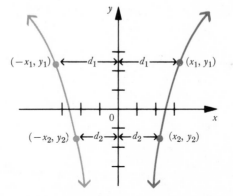

If an equation represents a function f, we can test for symmetry of the graph of f with respect to the y axis by substituting $-x$ for x. If the new equation, when simplified, is the same as the given equation, this means that either x or $-x$ results in the same y; that is, $f(-x) = f(x)$ and the graph of the function f is symmetric with respect to the y axis. A function f possessing the property that $f(-x) = f(x)$ for all x in the domain of f is called an *even function*.

EXAMPLES

1 Suppose that the graph of a function is symmetric with respect to the y axis, and $(1,2)$, $(3,4)$, $(7,0)$, $(2,-3)$, and $(4,-2)$ are members of the function. What other points are on the graph of the function? Is the function even?

SOLUTION. The graph of the function also contains $(-1,2)$, $(-3,4)$, $(-7,0)$, $(-2,-3)$, and $(-4,-2)$ (Figure 3). The function is even.

Figure 3

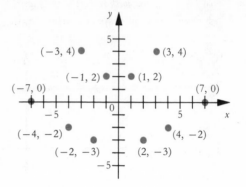

2   Determine if the function defined by $f(x) = x^2$ is an even function. Is the graph of f symmetric with respect to the y axis? Use the results to sketch the graph.

SOLUTION.   Replace x with $-x$ to get $(-x)^2 = x^2$ so that $f(-x) = f(x)$ and f is an even function. The graph of f is symmetric with respect to the y axis, since (x,y) and $(-x,y)$ both lie on the graph of f simultaneously. Therefore, it is only necessary to graph the function for nonnegative abscissas. The remainder of the graph is determined by "reflection" across the y axis (Figures 4a and b).

Figure 4

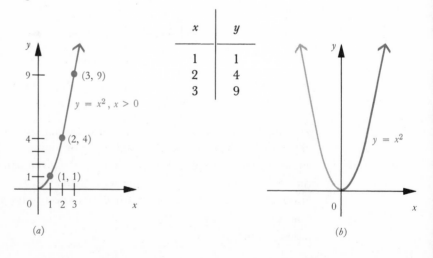

x	y
1	1
2	4
3	9

(a)   (b)

The graph of a function f is *symmetric with respect to the origin* if, whenever the point (x,y) is on the graph of f, then the point $(-x, -y)$ is also on the graph of f (Figure 5).

Figure 5

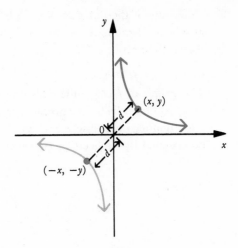

If an equation represents a function f, then we can test for symmetry of the graph of f with respect to the origin merely by substituting $-x$ for x and $-y$ for y simultaneously. If the new equation, when simplified, is the same as the original, this means that $(-x, -y)$ is on the graph of f whenever (x, y) is on the graph of f; that is, $f(-x) = -f(x)$ and the graph will be symmetric with respect to the origin. A function f possessing the property that $f(-x) = -f(x)$ for all x in the domain of f is called an *odd function*.

EXAMPLES

1 If the graph of a function is symmetric with respect to the origin and contains points $(2, 3)$, $(-2, -5)$, $(5, 0)$, and $(4, -4)$, what other points are on the graph of the function? Is the function odd?

SOLUTION. The graph of the function also contains the points $(-2, -3)$, $(2, 5)$, $(-5, 0)$, and $(-4, 4)$ (Figure 6). Notice that if each pair of corresponding points is connected by a line, the line passes through the origin. The function is odd.

Figure 6

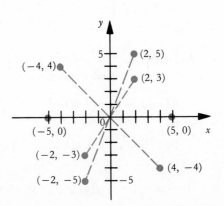

2   Determine if the function defined by $f(x) = x^3$ is an odd function. Is the graph of f symmetric with respect to the origin? Use the symmetry to sketch the graph.

SOLUTION. $f(-x) = (-x)^3 = -(x^3) = -f(x)$, so that f is an odd function. The graph of f is symmetric with respect to the origin since (x, y) and $(-x, -y)$ both lie on the graph of f. Therefore, we can graph the function f for nonnegative abscissas (Figure 7a). The remainder of the graph is determined by "reflection" across the origin (Figure 7b).

Figure 7

x	y
0	0
1	1
2	8

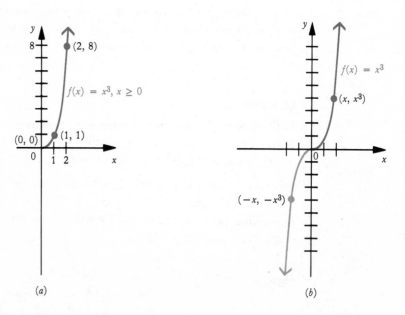

(a)   (b)

Most functions are neither even nor odd. For example, $f(x) = 2x + 1$ is neither even nor odd because

$$f(-x) = 2(-x) + 1 = -2x + 1$$

so that

$$2x + 1 = f(x) \neq f(-x) = -2x + 1$$

and

$$-2x - 1 = -f(x) \neq f(-x) = -2x + 1$$

For example, $f(2) = 2(2)+1 = 5$ and $f(-2) = 2(-2)+1 = -3$, therefore, $f(2) \neq f(-2)$; and, $-f(2) \neq f(-2)$ since $-5 \neq -3$.

This fact can also be seen from the graph of $f(x) = 2x+1$ (Figure 8) in that there is no symmetry with respect to either the y axis or the origin.

Figure 8

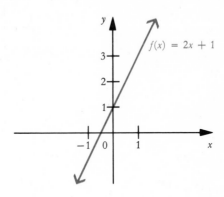

EXAMPLES

Determine whether the following functions are even, odd, or neither. Discuss the symmetry of each graph and sketch the graph.

1 $f(x) = 4x^2$

SOLUTION. $f(-x) = 4(-x)^2 = 4x^2 = f(x)$, so that f is an even function. Consequently, the graph of f is symmetric with respect to the y axis (Figure 9).

Figure 9

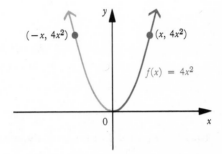

2 $f(x) = -5x^3$

SOLUTION. $f(-x) = -5(-x)^3 = -(-5x^3) = -f(x)$, so that f is an odd function and the graph of f is symmetric with respect to the origin (Figure 10).

Figure 10

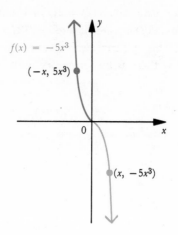

3   $f(x) = x^2 + 4x + 4$

SOLUTION. $f(-x) = (-x)^2 + 4(-x) + 4 = x^2 - 4x + 4 \neq x^2 + 4x + 4 = f(x)$, and f is not an even function. Also $f(-x) \neq -f(x)$ for all x; for example, $f(-1) = 1$ and $-f(1) = -9$. Hence f is not an odd function. Therefore, f is neither even nor odd (Figure 11).

Figure 11

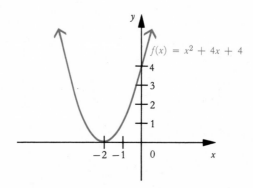

3.2   Increasing and Decreasing Functions

The concept of increasing and decreasing functions can be motivated by the graphs of $f(x) = x^3$, $f(x) = -3x + 1$, and $f(x) = 7$ (Figure 12). If x increases, how do the corresponding values of $f(x)$ change? In the first case, where $f(x) = x^3$, we see from the graph that as x increases (varies from left to right) the corresponding y values given by $y = x^3$ increase (rise); for $f(x) = -3x + 1$, as x increases (varies from left to right) the corresponding y values given by $y = -3x + 1$ decrease (fall); finally, as x increases,

$f(x) = 7$ neither increases nor decreases. Thus "$f(x) = x^3$ is an increasing function in R," "$f(x) = -3x + 1$ is a decreasing function in R," and "$f(x) = 7$ is neither increasing nor decreasing."

Figure 12

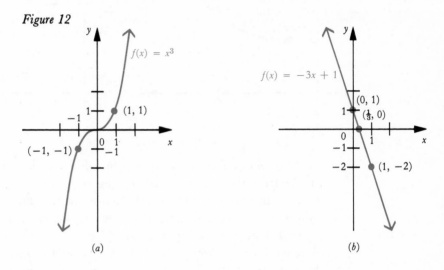

(a) (b)

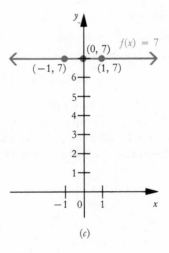

(c)

More formally, a function f is said to be a (*strictly*) *increasing function* in an interval I if, whenever a and b are two numbers in I such that $a < b$, we have $f(a) < f(b)$ (Figure 13a). A function f is said to be a (*strictly*) *decreasing function* in an interval I, if, whenever a and b are two numbers in I such that $a < b$, we have $f(a) > f(b)$ **(Figure 13b).**

Figure 13

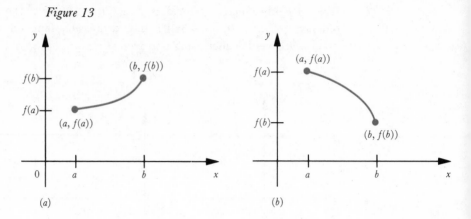

(a)   (b)

If f is an increasing or decreasing function in R, the entire real line, we shall refer to the function f merely as an increasing or decreasing function without mentioning the interval. Hence, for the three examples above, $f(x) = x^3$ is an increasing function, since for any two real numbers x_1 and x_2 such that $x_1 < x_2$, we have $x_1^3 < x_2^3$; $f(x) = -3x + 1$ is a decreasing function, since for any two real numbers x_1 and x_2 such that $x_1 < x_2$, we have $-3x_1 + 1 > -3x_2 + 1$; and $f(x) = 7$ is neither increasing nor decreasing, since $x_1 < x_2$ yields $f(x_1) = f(x_2)$.

EXAMPLES

Indicate the intervals where f is increasing or decreasing for each of the following functions.

1    $f(x) = x^2$ with domain the set $[-2, 2]$.

SOLUTION.   The graph of f (Figure 14) indicates that f is increasing in $[0, 2]$; for example, $\frac{1}{2} < 1$ implies that $f(\frac{1}{2}) < f(1)$ since $f(\frac{1}{2}) = \frac{1}{4}$ and $f(1) = 1$, whereas f is decreasing in $[-2, 0]$; for example, $f(-2) > f(-1)$ since $f(-2) = 4$ and $f(-1) = 1$.

Figure 14

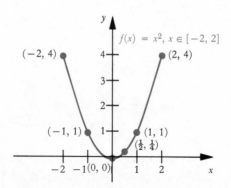

2   $f(x) = 3x + 2.$

SOLUTION. The graph of f (Figure 15) indicates that f is an increasing function in R. For example, for -2 and 7, we have $-2 < 7$ and $f(-2) < f(7)$, since $3(-2) + 2 < 3(7) + 2$ or $-4 < 23$.

Figure 15

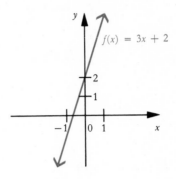

3   $f(x) = 2.$

SOLUTION. The graph of f (Figure 16) indicates that f is neither an increasing nor a decreasing function on any subset of R. For example, for 1 and 5, $1 < 5$, but $f(1) = f(5)$. (Why?)

Figure 16

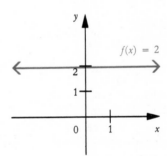

3.3   Special Functions

The properties of functions can now be used to investigate the following special functions.

1   Identity Function

The *identity function* is $f(x) = x$, and $f = \{(x,y) \mid y = x\}$ is the ordered pair representation of the identity function. The domain is R, the set of all real numbers, and the range is also R. Since $f(-x) = -x = -f(x)$, the graph of f is symmetric with respect to the origin; in other words, f is an *odd* function. Also, $a < b$ implies that $f(a) < f(b)$ (why?), so that f is an increasing function (Figure 17).

Figure 17

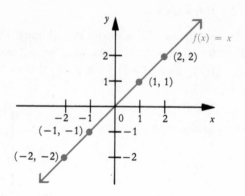

2 Absolute Value Function

The *absolute value function* is $f(x) = |x|$, or, equivalently, in ordered pair notation, $f = \{(x,y) \mid y = |x|\}$. The domain is R and the range is the set of all nonnegative real numbers. Since $|x| = |-x|$, it follows that $f(x) = f(-x)$; that is, f is an even function and its graph is symmetric with respect to the y axis. Also, f is increasing in the interval $[0, \infty)$ and it is decreasing in the interval $(-\infty, 0]$ (Figure 18).

Figure 18

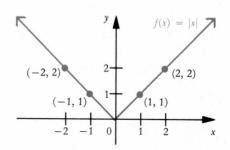

x	y
-2	2
-1	1
1	1
2	2

3 Greatest Integer Function

The *greatest integer of a real number x*, written $[\![x]\!]$, is the integer n that satisfies $n \le x < n+1$. In other words, $[\![x]\!]$ is the "nearest" integer less than or equal to x. Thus $[\![5\tfrac{1}{4}]\!] = 5$, $[\![-2\tfrac{1}{2}]\!] = -3$, $[\![-\tfrac{1}{2}]\!] = -1$, and $[\![\sqrt{2}]\!] = 1$. The function $f(x) = [\![x]\!]$ is the *greatest integer function*. In ordered pair notation the greatest integer function is $\{(x,y) \mid y = [\![x]\!]\}$. The domain is R and the range is I, the set of integers.

Since $f(-3\frac{1}{2}) = -4$ and $f(3\frac{1}{2}) = 3$, the greatest integer function is neither even nor odd (Figure 19).

Figure 19

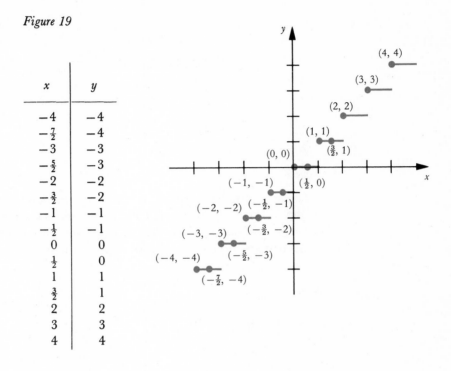

x	y
-4	-4
$-\frac{7}{2}$	-4
-3	-3
$-\frac{5}{2}$	-3
-2	-2
$-\frac{3}{2}$	-2
-1	-1
$-\frac{1}{2}$	-1
0	0
$\frac{1}{2}$	0
1	1
$\frac{3}{2}$	1
2	2
3	3
4	4

Notice that the graph of $f(x) = [\![x]\!]$ has "jumps" or "gaps" at each of the integers; it is an example of a *discontinuous function*. By contrast, $f(x) = x$ and $f(x) = |x|$ are *continuous functions*; their graphs are unbroken and can be drawn without lifting the pencil off the paper.

The term "continuous function" is used to describe such functions because intuitively one can move "continuously," that is, without breaks or jumps, from one point on the graph to another. However, this statement is very imprecise and, in fact, is not correct in all cases. For the functions considered in this text, however, it will be sufficient. (It is beyond our scope to examine continuous functions using the formal definition accepted today.)

EXAMPLE

$f(x) = 1/x$ is not continuous at $x = 0$, and the graph of $f(x) = 1/x$ displays a "jump" at $x = 0$ (Figure 20).

Figure 20

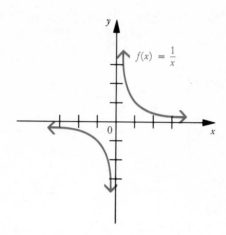

4   Sequences

A *sequence* is a function whose domain is the set of positive integers. For example, $f(n) = 2/n$, where n is a positive integer, is a sequence. The graph of f consists of discrete points (Figure 21). Notice that in graphing f, the points that are displayed in Figure 21 are not to be connected.

Figure 21

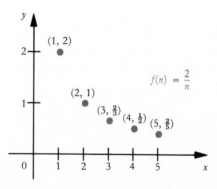

n	$f(n)$
1	2
2	1
3	$\frac{2}{3}$
4	$\frac{1}{2}$
5	$\frac{2}{5}$

Quite often, subscript notation is used to describe sequences. For example, $f(n) = 2/n$ can be written in the form $S_n = 2/n$, so that

$$S_1 = f(1) = 2$$
$$S_2 = f(2) = 1$$
$$S_3 = f(3) = \tfrac{2}{3}$$
$$S_4 = f(4) = \tfrac{1}{2}$$
$$S_5 = f(5) = \tfrac{2}{5}$$

EXAMPLES

1   If $f(x) = [\![3x]\!]$, find $f(-\tfrac{3}{2})$, $f(\tfrac{7}{2})$, and $f(0)$.

SOLUTION.   If $f(x) = [\![3x]\!]$, then $f(-\tfrac{3}{2}) = [\![-\tfrac{9}{2}]\!] = -5$, $f(\tfrac{7}{2}) = [\![\tfrac{21}{2}]\!] = 10$, and $f(0) = [\![0]\!] = 0$.

2   If $f(x) = [\![2x]\!] - 2x$, find $f(\tfrac{1}{3})$, $f(-\tfrac{1}{5})$, and $f(\tfrac{9}{2})$.

SOLUTION.   If $f(x) = [\![2x]\!] - 2x$, $f(\tfrac{1}{3}) = [\![2(\tfrac{1}{3})]\!] - 2(\tfrac{1}{3}) = -\tfrac{2}{3}$,
$f(-\tfrac{1}{5}) = [\![2(-\tfrac{1}{5})]\!] - 2(-\tfrac{1}{5}) = -\tfrac{3}{5}$, and $f(\tfrac{9}{2}) = [\![2(\tfrac{9}{2})]\!] - 2(\tfrac{9}{2}) = 9 - 9 = 0$.

3   For each of the following sequences, write the first five terms, describe the range, and plot five points of the graph.
    a) $S_n = (-1)^n$ b) $S_n = 3 - 1/n$

SOLUTION

a) We have $S_n = (-1)^n$, $S_1 = -1$, $S_2 = 1$, $S_3 = -1$, $S_4 = 1$, and $S_5 = -1$. The range is the set $\{-1, 1\}$. Figure 22 contains five points of the graph.

Figure 22

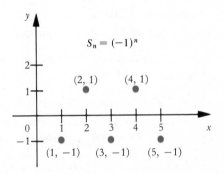

b) We have $S_n = 3 - 1/n$, $S_1 = 2$, $S_2 = \tfrac{5}{2}$, $S_3 = \tfrac{8}{3}$, $S_4 = \tfrac{11}{4}$, and $S_5 = \tfrac{14}{5}$. The range is a subset of the rational numbers. Figure 23 contains five points of the graph.

Figure 23

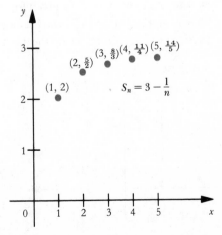

PROBLEM SET 3

1   Indicate whether each of the given functions is even or odd or neither.
    a) $f(x) = x^4 + 3$ b) $f(x) = \sqrt{x} + 4$

c) $f(x) = x^3 + x$

d) $f(x) = (x^2 + 1)^3$

e) $f(x) = x^3 + 1/x$

f) $f(x) = x^2 - 2x + 3$

2 Let $f(x) = x^3 - 2x + 1/x$.
  a) Show that $f(a) + f(-a) = 0$.
  b) Is f an even or odd function?

3 Determine if the graph of f is symmetric with respect to the y axis or the origin. Use the symmetry to sketch the graph.
  a) $f(x) = -2x$
  b) $f(x) = x^4$
  c) $f(x) = -4x^3$
  d) $f(x) = -\sqrt{4-x^2}$
  e) $f(x) = -3x^4$
  f) $f(x) = -3x^2 + 6$

4 Let $f(x) = x^4 - 3x^2 + 10$.
  a) Find the value of $f(2a) - 2f(a)$.
  b) Show that f is an even function.

5 Let $f(x) = \sqrt{x}$. Sketch the graph for $x \in [1, 9]$. Is f increasing or decreasing in the interval $[1, 9]$?

6 a) If f is an increasing function, is $2f$ necessarily increasing?
  b) If f is an increasing function, is $-f$ necessarily increasing?
  c) If f is a decreasing function, is $-2f$ necessarily decreasing?

7 Sketch the graph of f and indicate the intervals where the given function is increasing or decreasing.
  a) $f(x) = -5x + 3$
  b) $f(x) = |x| + x$
  c) $f(x) = \sqrt{x-1}$
  d) $f(x) = \sqrt{x+2}$
  e) $f(x) = x^3 + 1$
  f) $f(x) = -\sqrt[3]{x}$

8 Decide whether each of the given functions whose graph is given in Figure 24 is even, odd, or neither. Does the graph of the function have symmetry? Indicate those intervals where the function is either increasing or decreasing.

Figure 24

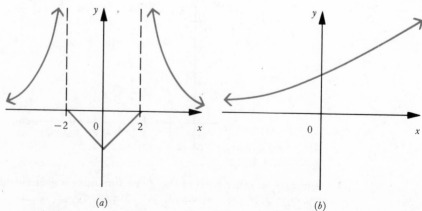

(a)   (b)

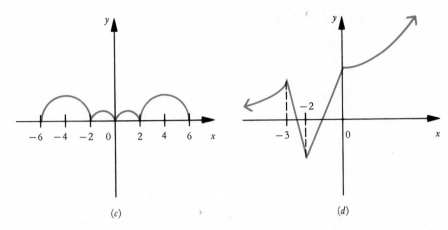

(c)   (d)

9   Answer these questions for each of the following functions. What is the domain? What is the range? Is f even or odd? Does the graph of f have symmetry? Where is f increasing? Decreasing? Graph f.

a) $f(x) = 5x + 2$ b) $f(x) = x^{1/3}$

c) $f(x) = |x - 1|$ d) $f(x) = 1/(x^2 - 1)$

e) $f(x) = [\![5x]\!]$ f) $f(x) = -2x + 2$

g) $f = \{(x,y) \mid x^2 = y^3, y \le 0\}$ h) $f = \{(x,y) \mid y = x^2 + 3\}$

10   In each of the following spaces answer true if the equation holds for all real numbers. If the equation is false, give a counterexample.

$f(x)$	$[f(x)]^3 = f(x^3)$	$f(2x) = 2f(x)$	$f(x+y) = f(x) + f(y)$	$f(xy) = f(x) \cdot f(y)$
x				
$\|x\|$				
$[\![x]\!]$				

11   For each of the following sequences, write the first five terms, describe the range, and plot five points of the graph.

a) $S_n = 1/(n+1)$ b) $S_n = 2/(n^2 + 1)$

c) $S_n = 1 + (-1)^n$ d) $S_n = 1/[n(n+1)]$

4   Algebra of Functions and Composition of Functions

In this section we will see how new functions can be formed from other functions by using certain algebraic rules. We will examine ways of combining two functions under the operations of *addition, subtraction, multiplication,* and *division* to form new functions. In addition, we shall introduce the concept of the *composition* of two functions.

4.1 Algebra of Functions

Let us first consider the addition and subtraction of the functions $f(x) = 2x+1$ and $g(x) = x^2-1$ to form two new functions. The domain of both f and g is the set of real numbers R. If we sketch the graphs of both f and g on the same coordinate axes (Figure 1), then the graphs of the new functions, denoted by $f+g$ and $f-g$, are obtained by adding and subtracting the ordinates of the points on the graphs of f and g (Figure 2).

x	$f(x)$	$g(x)$	$f(x)+g(x)$	$f(x)-g(x)$
-2	-3	3	0	-6
-1	-1	0	-1	-1
0	1	-1	0	2
1	3	0	3	3
2	5	3	8	2

Figure 1

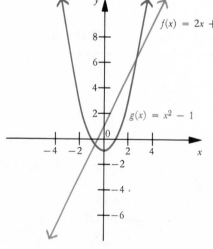

Figure 2

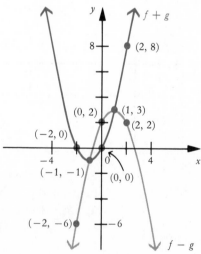

Notice that the sum and difference functions of f and g are formed as follows:

$$(f+g)(x) = f(x) + g(x) = (2x+1) + (x^2-1) = 2x + x^2$$

$$(f-g)(x) = f(x) - g(x) = (2x+1) - (x^2-1) = 2x - x^2 + 2$$

The graphs of the product and quotient of two functions are not easily constructed from the separate graphs of f and g. Since these types of functions will be considered later in calculus, we will survey the product and quotient functions, denoted by $f \cdot g$ and f/g, respectively, to determine the domains of these two functions from each of the domains of f and g. For example, if $f(x) = 3x - 2$ and $g(x) = -5x$, then the domain of the product function $(f \cdot g)(x) = (3x - 2)(-5x)$ is the set of real numbers R. The domain of the quotient function f/g is $\{x \mid x \in R,\ x \neq 0\}$, since $(3x - 2)/(-5x)$ is not defined whenever the denominator is zero. Notice in this example that the domain of f and g along with the domain of $f \cdot g$ is the set of real numbers. However, the domain of f/g is $\{x \mid x \in R,\ x \neq 0\}$. The formation of the sum, difference, product, and quotient function is generalized in the following definition.

DEFINITION SUM, DIFFERENCE, PRODUCT, AND QUOTIENT FUNCTIONS

Let f and g be two functions and suppose that D_f and D_g denote the domains of f and g, respectively; then we define the functions $f+g$, $f-g$, $f \cdot g$, and f/g, called the *sum*, the *difference*, the *product*, and the *quotient*, respectively, as follows:

$$f + g = \{(x,y) \mid y = f(x) + g(x) \text{ and } x \in D_f \cap D_g\}$$
$$f - g = \{(x,y) \mid y = f(x) - g(x) \text{ and } x \in D_f \cap D_g\}$$
$$f \cdot g = \{(x,y) \mid y = f(x) \cdot g(x) \text{ and } x \in D_f \cap D_g\}$$
$$\frac{f}{g} = \left\{(x,y) \mid y = \frac{f(x)}{g(x)},\ g(x) \neq 0 \text{ and } x \in D_f \cap D_g\right\}$$

where $g(x) \neq 0$ means that f/g has meaning only if we exclude any $x \in D_g$ which yields $g(x) = 0$.

For example, if $f(x) = \sqrt{x}$ and $g(x) = x + 3$, then

$$(f+g)(4) = f(4) + g(4) = \sqrt{4} + 4 + 3 = 9$$

$$(f-g)(4) = f(4) - g(4) = \sqrt{4} - 4 - 3 = -5$$

$$(f \cdot g)(4) = f(4) \cdot g(4) = \sqrt{4}(4+3) = 14$$

$$\left(\frac{f}{g}\right)(4) = \frac{f(4)}{g(4)} = \frac{\sqrt{4}}{4+3} = \frac{2}{7}$$

Notice that $(f+g)(-4)$, $(f-g)(-4)$, $(f \cdot g)(-4)$, and $(f/g)(-4)$ are not defined, since $-4 \notin D_f \cap D_g$. (Why?) In general, since the domain of f is $[0, \infty)$ and the domain of g is R, then the domain of each of $f+g$, $f-g$, and $f \cdot g$ is $D_f \cap D_g = [0, \infty)$. The domain of f/g is also $[0, \infty)$ since -3, the number that yields a zero value for $g(x)$, is not in the set $[0, \infty)$.

EXAMPLES

1 Use the graphs of $g(x) = 3x^2$ and $h(x) = 1$ to graph $f(x) = g(x) + h(x) = 3x^2 + 1$.

SOLUTION. If we place the graphs of g and h on the same coordinate system, then we can obtain the graph of $f(x) = g(x) + h(x) = 3x^2 + 1$ by adding the ordinates of g and h at various points (Figure 3).

Figure 3

x	$g(x)$ $3x^2$	$h(x)$ 1	$f(x)$ $3x^2 + 1$
0	0	1	1
1	3	1	4
−1	3	1	4
2	12	1	13
−2	12	1	13

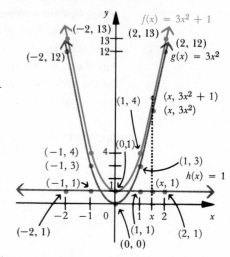

2 Use the graphs of $g(x) = x^2$ and $h(x) = 2x$ to graph $f(x) = g(x) - h(x) = x^2 - 2x$.

SOLUTION. If we place the graphs of g and h on the same coordinate system, then we can obtain the graph of $f(x) = g(x) - h(x)$ by subtracting the ordinates of g and h at various points (Figure 4).

Figure 4

x	$g(x)$ x^2	$h(x)$ $2x$	$f(x)$ $x^2 - 2x$
0	0	0	0
−1	1	−2	3
2	4	4	0
−2	4	−4	8

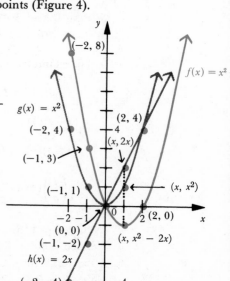

3   Use $f(x) = x^2$ and $g(x) = x - 1$ to form each of the following functions.

   a) $f + g$ b) $f - g$
   c) $f \cdot g$ d) f/g

SOLUTION.   The domain of both f and g is the set of all real numbers, so that the domain of $f + g, f - g$, and $f \cdot g$ is also the set of all real numbers. The domain of f/g, however, is the set of real numbers except 1, since $g(1) = 0$.

   a) $(f + g)(x) = f(x) + g(x) = x^2 + x - 1$
   b) $(f - g)(x) = f(x) - g(x) = x^2 - x + 1$
   c) $(f \cdot g)(x) = f(x) g(x) = x^2 (x - 1) = x^3 - x^2$
   d) $(f/g)(x) = [f(x)/g(x)] = x^2/(x - 1)$

4   Use $f = \{(1, 2), (2, 2), (3, 5), (4, 6)\}$ and $g = \{(2, 2), (-1, 3), (3, 7)\}$ to form each of the following functions.

   a) $f + g$ b) $f - g$
   c) $f \cdot g$ d) f/g

SOLUTION.   Since $D_f \cap D_g = \{1, 2, 3, 4\} \cap \{2, -1, 3\} = \{2, 3\}$ and since $g(2) \neq 0$ and $g(3) \neq 0$, $\{2, 3\}$ is the domain of each of the four functions.

   a) Since $(f + g)(2) = f(2) + g(2) = 2 + 2 = 4$ and $(f + g)(3) = f(3) + g(3) = 5 + 7 = 12, f + g = \{(2, 4), (3, 12)\}$.
   b) Since $(f - g)(2) = f(2) - g(2) = 2 - 2 = 0$ and $(f - g)(3) = f(3) - g(3) = 5 - 7 = -2, f - g = \{(2, 0), (3, -2)\}$.
   c) Since $(f \cdot g)(2) = f(2) \cdot g(2) = 2 \cdot 2 = 4$ and $(f \cdot g)(3) = f(3) \cdot g(3) = 5 \cdot 7 = 35, f \cdot g = \{(2, 4), (3, 35)\}$.
   d) Since $(f/g)(2) = [f(2)/g(2)] = \frac{2}{2} = 1$ and $(f/g)(3) = [f(3)/g(3)] = \frac{5}{7}$, $f/g = \{(2, 1), (3, \frac{5}{7})\}$.

5   Suppose that f is an even function and g is an even function. Which of the following functions are even functions?

   a) $f + g$ b) $f \cdot g$

SOLUTION. f and g are even functions, so that $f(-x) = f(x)$ and $g(-x) = g(x)$.

   a) $(f + g)(-x) = f(-x) + g(-x)$ (why?)
   $$= f(x) + g(x) = (f + g)(x)$$

   Therefore, $f + g$ is an even function.

   b) $(f \cdot g)(-x) = f(-x) \cdot g(-x)$ (why?)
   $$= f(x) \cdot g(x) = (f \cdot g)(x)$$

   Therefore, $f \cdot g$ is an even function.

4.2   Composition Of Functions

We know from solid geometry that the volume V of a sphere of radius r is given by the formula $V = \frac{4}{3}\pi r^3$. V can be interpreted as a function of r, and we can use function notation to write

$$V = V(r) = \tfrac{4}{3}\pi r^3$$

Now, suppose that air is being pumped into a spherical balloon so that at the end of t seconds, the radius r satisfies $r = t^2 + 1$. Assuming that r is a function of t, we can write

$$r = r(t) = t^2 + 1$$

We see here that given a specific value of t, we can determine the volume V of the balloon by first finding $r(t)$, followed by determining $V(r)$. That is,

$$t \mapsto r(t) \mapsto V(r) = V[r(t)]$$

For example,

$$1 \overset{r}{\mapsto} 2 \overset{V}{\mapsto} \frac{32\pi}{3}$$

$$2 \overset{r}{\mapsto} 5 \overset{V}{\mapsto} \frac{500\pi}{3}$$

$$3 \overset{r}{\mapsto} 10 \overset{V}{\mapsto} \frac{4000\pi}{3}$$

In general,

$$t \overset{r}{\mapsto} t^2 + 1 \overset{V}{\mapsto} \tfrac{4}{3}\pi(t^2+1)^3$$

The volume, then, is determined by applying two functions in succession, "r followed by V," and we say the function that maps $t \mapsto \frac{4}{3}\pi(t^2+1)^3$ is "composed of the function r followed by the function V," written $V \circ r$. Hence $(V \circ r)(t) = \frac{4}{3}\pi(t^2+1)^3$.

Let us consider another example. Assume that $f(x) = 2x + 7$ and $g(x) = x^2 + 2$. We can find the image of 3 by applying the functions in succession—first f, then g—that is, by finding $g \circ f$ of 3 as follows:

$$3 \overset{f}{\mapsto} f(3) \overset{g}{\mapsto} g[f(3)]$$

Thus

$$g[f(3)] = g(13) = 13^2 + 2 = 171$$

In general, the image of x under the "successive function," f followed by g, written $g \circ f$, is formed as follows:

$$x \xmapsto{f} f(x) \xmapsto{g} g[f(x)]$$

and is given by

$$(g \circ f)(x) = g[f(x)] = g(2x+7) = (2x+7)^2 + 2$$
$$= 4x^2 + 28x + 51$$

The function $g \circ f$ which has been constructed here is called the composition of f by g.

DEFINITION   COMPOSITE FUNCTION

The *composite function of f by g*, denoted by $g \circ f$, is the function defined as $(g \circ f)(x) = g[f(x)]$. The *domain of $g \circ f$* is the subset of the domain of f containing those values for which $g \circ f$ is defined.

Schematically, $g \circ f$ is shown in Figure 5.

Figure 5

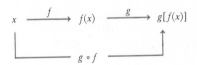

EXAMPLES

1   Let $f(x) = x^2$ and $g(x) = 2x - 3$.
   a)   Compare $(g \circ f)(2)$ with $(f \circ g)(2)$.
   b)   Compare $(g \circ f)(2)$ with $(g \cdot f)(2)$, where $g \cdot f$ denotes the product.

SOLUTION
   a) $(g \circ f)(2) = g[f(2)]$
$$= g(4)$$
$$= 5$$

whereas

$$(f \circ g)(2) = f[g(2)]$$
$$= f(1)$$
$$= 1$$

so that
$$(g \circ f)(2) \neq (f \circ g)(2)$$

b) We have seen that $(g \circ f)(2) = 5$. On the other hand,

$$(g \cdot f)(2) = g(2) \cdot f(2)$$
$$= 1 \cdot 4$$
$$= 4$$

Hence

$$(g \circ f)(2) \neq (g \cdot f)(2)$$

Example 1a illustrates the fact that the composition of functions is *not* commutative; that is, $(g \circ f)(x)$ and $(f \circ g)(x)$ are not always equal. Example 1b emphasizes the fact that the composed function $g \circ f$ is different from the product $g \cdot f$.

2   Use $f(x) = 2x^2$ and $g(x) = 4x + 1$ to determine an expression to represent each of the following composed functions. Also, indicate the domain of each composed function.

a) $(g \circ g)(x)$ b) $(f \circ g)(x)$ c) $(g \circ f)(x)$

SOLUTION

a) $x \xmapsto{g} g(x) \xmapsto{g} g[g(x)]$
$\qquad \underset{g \circ g}{\underline{\qquad\qquad}}$

$$(g \circ g)(x) = g[g(x)]$$
$$= g(4x + 1)$$
$$= 4(4x + 1) + 1$$
$$= 16x + 5$$

Since the domain of g (considered as the "first" function) is R and since g (considered as the "second" function) is defined for all real numbers, the domain of $g \circ g$ is also R.

b) $x \xmapsto{g} g(x) \xmapsto{f} f[g(x)]$
$\qquad \underset{f \circ g}{\underline{\qquad\qquad}}$

$$(f \circ g)(x) = f[g(x)]$$
$$= f(4x + 1)$$
$$= 2(4x + 1)^2$$
$$= 32x^2 + 16x + 2$$

The domain of g is R, and since $(f \circ g)(x) = 32x^2 + 16x + 2$ is defined for all real numbers, the domain of $f \circ g$ is also R.

c) $x \xmapsto{f} f(x) \xmapsto{g} g[f(x)]$
$\qquad \underset{g \circ f}{\underline{\qquad\qquad}}$

$$(g \circ f)(x) = g[f(x)]$$
$$= g(2x^2)$$
$$= 4(2x^2) + 1$$
$$= 8x^2 + 1$$

The domain of f is R. Since $(g \circ f)(x) = 8x^2 + 1$ is defined for all reals, the domain of $g \circ f$ is also R.

3   Let $f(x) = x - 1$ and $g(x) = \sqrt{x}$. Determine the expression and domain for the following composed functions.

a) $(f \circ g)(x)$ b) $(g \circ f)(x)$

SOLUTION

a) $x \xmapsto{g} g(x) \xmapsto{f} f[g(x)]$
$\qquad \underset{f \circ g}{\underline{\qquad\qquad}}$

$$(f \circ g)(x) = f[g(x)]$$
$$= f(\sqrt{x})$$
$$= \sqrt{x} - 1$$

The domain of g is the set $[0, \infty)$, and since $(f \circ g)(x) = \sqrt{x} - 1$ is defined for all numbers in $[0, \infty)$, the domain of $f \circ g$ is also $[0, \infty)$.

b) $x \xmapsto{f} f(x) \xmapsto{g} g[f(x)]$

$\underset{g \circ f}{\underline{\qquad\qquad\uparrow}}$

$(g \circ f)(x) = g[f(x)]$
$\qquad\quad = g(x-1)$
$\qquad\quad = \sqrt{x-1}$

The domain of f is R; however, $(g \circ f)(x) = \sqrt{x-1}$ is defined only when $x - 1$ represents a nonnegative number, so that the domain of $g \circ f$ is $\{x \mid x - 1 \geq 0\} = [1, \infty)$, which is a (proper) subset of the domain of f (reexamine the definition of a composite function).

4 Suppose that f is an even function and g is an odd function. Indicate which of the following functions are even and which are odd.

a) $f \circ f$ b) $f \circ g$

SOLUTION. f is an even function, so that $f(-x) = f(x)$ and g is an odd function; therefore, $g(-x) = -g(x)$.

a) $(f \circ f)(-x) = f[f(-x)] = f[f(x)] = (f \circ f)(x)$
Therefore, $f \circ f$ is an even function.

b) $(f \circ g)(-x) = f[g(-x)] = f[-g(x)] = f[g(x)] = (f \circ g)(x)$
Therefore, $f \circ g$ is an even function.

5 The function $I(f) = 144f$ enables us to convert from f square feet to I square inches; and the function $f(y) = 9y$ enables us to convert from y square yards to f square feet. Use the composition of functions to determine a function that will convert y square yards to I square inches.

SOLUTION

$y \xmapsto{f} 9y \xmapsto{I} 144(9y)$

$\underset{I \circ f}{\underline{\qquad\qquad\uparrow}}$

so that $(I \circ f)(y) = I[f(y)] = I(9y) = (144)(9y) = 1296y$ is the composed function that converts square yards into square inches.

PROBLEM SET 4

1 Use the graphs of the functions f and g to sketch the graph of $f + g$ for each of the following pairs of functions.

a) $f(x) = x$ and $g(x) = -5$
b) $f(x) = x^2$ and $g(x) = 1$
c) $f(x) = x^3$ and $g(x) = 2x^2$

2 Let $f(x) = x^2$ for $-2 \leq x \leq 1$ and $g(x) = 3x + 4$ for $0 \leq x \leq 3$.

a) Find $(f+g)(\frac{1}{2})$, $(f-g)(\frac{1}{2})$, $(f \cdot g)(1)$, and $(f/g)(\frac{1}{3})$.
b) Form each of the following functions and find their domains: $f+g$, $f-g$, $f \cdot g$, and f/g.

3 Let $f(x) = 2x + 5$ and $g(x) = 7 - 3x$.
   a) Find $(f+g)(3)$, $(f-g)(2)$, $(f \cdot g)(-2)$, and $(f/g)(-5)$.
   b) Form each of the following functions: $f+g, f-g, f \cdot g,$ and f/g.
   c) Are there any values in the domain of f and g excluded from the domains of $f+g, f-g, f \cdot g,$ or f/g?

4 Let $f = \{(1,1),(2,4),(3,9),(-1,1),(-2,4),(-3,6)\}$, and $g = \{(1,4),(2,1),(4,5),(-2,4)\}$. Form each of the following functions and indicate the domain and range of each.
   a) $f+g$ b) $f-g$
   c) $f \cdot g$ d) f/g

5 Let $f = \{(x,y) \mid y = x - 2\}$ and $g = \{(x,y) \mid y = x^2 + 7\}$. Form each of the following functions.
   a) $f+g$ b) $f-g$
   c) $f \cdot g$ d) f/g

6 Suppose that f and g are odd functions. Which of the following functions are odd functions?
   a) $f+g$ b) $f-g$ c) $f \cdot g$ d) f/g

7 Let $A = \{2,3,4,5,6\}$ and let the functions $f: A \to A$ and $g: A \to A$ be defined by $f(2) = 3$, $f(3) = 4$, $f(5) = 2$, $f(4) = 6$, $f(6) = 5$, $g(2) = 5$, $g(3) = 6$, $g(5) = 3$, $g(4) = 5$, and $g(6) = 2$. Find each of the following values.
   a) $(f \circ g)(2)$ b) $(f \circ g)(3)$
   c) $(f \circ g)(5)$ d) $(f \circ g)(4)$
   e) $(f \circ g)(6)$ f) $(g \circ f)(2)$
   g) $(g \circ f)(3)$ h) $(g \circ f)(5)$
   i) $(g \circ f)(4)$ j) $(g \circ f)(6)$

8 If $f(x) = 2x^2 + 6$ and $g(x) = 7x + 2$, find each of the following.
   a) $(g \circ f)(4)$ b) $(f \circ f)(3)$
   c) $(g \circ g)(2)$ d) $(f \circ g)(5)$
   e) $(g \circ f)(5)$ f) $(f \circ f)(x)$ and the domain of $f \circ f$
   g) $(g \circ f)(x)$ and the domain of $g \circ f$
   h) $(f \circ g)(x)$ and the domain of $f \circ g$

9 a) Let $f(x) = 3x - 7$ and $g(x) = 2x + k$. Determine k so that $(f \circ g)(x) = (g \circ f)(x)$.
   b) Suppose that $f(x) = (2x^3 + 7)^3$. Find a function g such that $(f \circ g)(x) = (g \circ f)(x)$.

10 a) Let $f(x) = x^2 + 2x$ and $g(x) = 3x + 4$. Find the domains of $f \circ g$ and $g \circ f$.
   b) If $f(x) = x^2$ and $g(x) = 1/x$, find the domains and ranges of $g \circ f$ and $f \circ g$.

11   Let $f(x) = 3x + 1$ and $g(x) = -5x + 2$.
   a) Find $(f \circ g)(x)$ and $(g \circ f)(x)$.
   b) Find $(f \cdot g)(x)$ and $(g \cdot f)(x)$ and compare these products with the results of part a.
   c) Solve for x if $(f \circ g)(x) = 2$.

12   Verify that $(f \circ f \circ f)(x) = x$ if $f(x) = 1 - 1/x$.

13   Suppose that a right cylindrical vessel has a circular base of radius 4 inches.
   a) Express the volume V of the vessel as a function of the height h.
   b) Express the height h as a function of time if, after t seconds, the height is $2t + 4$.
   c) Use parts a and b to construct a composed function that expresses the volume of the vessel as a function of time.

14   Assume that f is an odd function and g is an even function. Indicate whether each of the following functions is even or odd.
   a) $f \circ f$ b) $f \circ g$

15   Suppose that the functions f, g, and h map real numbers into real numbers.
   a) Give an example to display that $(f + g) \circ h = f \circ h + g \circ h$ is true.
   b) Give a counterexample to prove that $f \circ (g + h) = f \circ g + f \circ h$ is false.

5   Inverse Functions

If $f(x) = x + 5$ and $g(x) = x - 5$, then

$$\begin{aligned}(g \circ f)(x) &= g[f(x)] \\ &= g(x + 5) \\ &= (x + 5) - 5 \\ &= x\end{aligned}$$

and

$$\begin{aligned}(f \circ g)(x) &= f[g(x)] \\ &= f(x - 5) \\ &= (x - 5) + 5 \\ &= x\end{aligned}$$

Schematically,

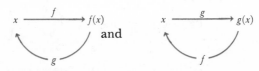

and

Both of the compositions of f and g result in the identity function; that is,
$$(f \circ g)(x) = (g \circ f)(x) = x.$$

Let us consider another example. Suppose that $f = \{(1, 2), (3, 4), (-1, 0)\}$ and $g = \{(2, 1), (4, 3), (0, -1)\}$. Then

$$(g \circ f)(1) = g[f(1)] = g(2) = 1$$
$$(g \circ f)(3) = g[f(3)] = g(4) = 3$$

and

$$(g \circ f)(-1) = g[f(-1)] = g(0) = -1$$

Similarly,

$$(f \circ g)(2) = 2$$
$$(f \circ g)(4) = 4$$

and

$$(f \circ g)(0) = 0$$

Here we have $(g \circ f)(x) = x$ for $x \in \{1, 3, -1\}$, the domain of f, and $(f \circ g)(x) = x$ for $x \in \{2, 4, 0\}$, the domain of g.

The functions f and g, which have been defined in each of the two examples above, are called "invertible functions." We can formalize this concept as follows.

DEFINITION INVERSE FUNCTION

Let f and g be two functions such that $(g \circ f)(x) = x$ for every element x in the domain of f and $(f \circ g)(x) = x$ for every element x in the domain of g; then f and g are said to be *invertible*, and each is said to be the *inverse* of the other. We use the notation

$$g = f^{-1} \qquad \text{or} \qquad f = g^{-1}$$

Hence in the first example above we can write $f^{-1}(x) = x - 5$ for $f(x) = x + 5$, or we could write $g^{-1}(x) = x + 5$ for $g(x) = x - 5$; in the second example, $f^{-1} = g = \{(2, 1), (4, 3), (0, -1)\}$ or $g^{-1} = f = \{(1, 2), (3, 4), (-1, 0)\}$.

EXAMPLES

1 Suppose that $f(x) = 5x$ and $g(x) = x/5$. Show that $f = g^{-1}$.

SOLUTION

$$(f \circ g)(x) = f[g(x)] = f\left(\frac{x}{5}\right) = 5\left(\frac{x}{5}\right) = x$$

and

$$(g \circ f)(x) = g[f(x)] = g(5x) = \tfrac{1}{5}(5x) = x$$

Since $(f \circ g)(x) = (g \circ f)(x) = x$, it follows that $f = g^{-1}$.

2 Suppose that $f(x) = 3x + 7$ and $g(x) = (x-7)/3$. Show that f and g are invertible.

SOLUTION

$$(f \circ g)(x) = f[g(x)] = f\left(\frac{x-7}{3}\right) = 3\left(\frac{x-7}{3}\right) + 7 = x$$

and

$$(g \circ f)(x) = g[f(x)] = g(3x+7) = \frac{(3x+7) - 7}{3} = x$$

Hence f and g are invertible.

3 Suppose that $f(x) = \sqrt{x}$ and $g(x) = x^2$ for x in the set of nonnegative real numbers. Show that f and g are invertible.

SOLUTION

$$(f \circ g)(x) = f[g(x)] = f(x^2) = \sqrt{x^2} = x$$

since x is nonnegative, and

$$(g \circ f)(x) = g[f(x)] = g(\sqrt{x}) = (\sqrt{x})^2 = x$$

Hence f and g are invertible.

5.1 One-to-One Functions

Let us examine the functions $f_1 = \{(1,2), (3,4), (-1,0)\}$ and $f_2 = \{(1,2), (4,2)\}$. Since f_1 and f_2 are functions, it follows from the definition of a function that for each member of the domain there is one and only one corresponding member of the range.

$$f_1 : 1 \to 2 \qquad f_2 : 1 \to 2$$
$$3 \to 4 \qquad\qquad 4 \to 2$$
$$-1 \to 0$$

Now, it is also true that for each member of the range of f_1 there is one and only one corresponding member of the domain of f_1, whereas for f_2, 2 in the range corresponds to more than one member of the domain. We say that f_1 is a "one-to-one function," whereas f_2 is not one-to-one.

In general, a *function f is one-to-one* if each member of the range of the function is the image of one and only one member of the domain.

Although it is possible to investigate one-to-one functions in a more formal way, we will use the graph of a function to determine whether or not a function is one-to-one. If $y = f(x)$, we know from the definition of a function that for each x there is one and only one y; hence each of all possible vertical lines (representing all possible values of x) intersects the graph no more than once. Now, f is one-to-one if it is also true that for each y there is one and only one x; this property holds for a function if each of all possible horizontal lines (representing all possible values of y) intersects the graph of $y = f(x)$ no more than once.

EXAMPLES

Use the graph of the given function to decide whether or not the function is one-to-one.

1        $f = \{(x,y) \mid y = x^2\}$

SOLUTION.   Clearly, any horizontal line above the x axis intersects the graph twice. Hence the function is not one-to-one. For example, 1 in the range is the image of both -1 and 1 in the domain (Figure 1).

Figure 1

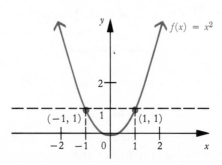

2        $f(x) = 3x - 5$

SOLUTION.   From Figure 2 it can be seen that no horizontal line intersects the graph more than once; hence for each range number there is one and only one corresponding domain number, thus f is one-to-one.

Figure 2

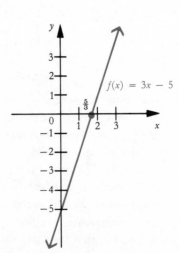

3      $$f = \{(x,y) \mid y = \sqrt{4-x^2}\}$$

SOLUTION.   Figure 3 shows that f is not one-to-one. For example, $\sqrt{15}/4$ is the image of both $-\tfrac{7}{4}$ and $\tfrac{7}{4}$.

Figure 3

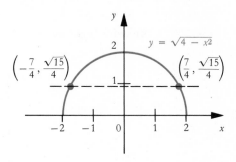

The next theorem, given without proof, tells us that invertible functions and functions that are one-to-one are the same.

THEOREM 1

If a function is one-to-one, then the function has an inverse; conversely, if a function has an inverse, then the function is one-to-one.

This means that we can use the graph to determine whether or not a function is one-to-one (Examples 1, 2, and 3 above), and in doing so we also discover whether or not the function has an inverse.

Also, if a one-to-one function is defined by $y = f(x)$, then the equation that defines the inverse can be constructed by interchanging the roles of x and y by expressing x in terms of y to obtain $x = f^{-1}(y)$. For example, if $f(x) = 3x+2$, then we can find the value of f^{-1} at a given number y by solving the equation $y = f(x)$; namely, $y = 3x+2$ for x in terms of y. Thus we have $x = \tfrac{1}{3}y - \tfrac{2}{3}$; in other words, $f^{-1}(y) = \tfrac{1}{3}y - \tfrac{2}{3}$. The letter that we use to denote a number in the domain of the inverse function is immaterial, so the latter equation can be written as

$$f^{-1}(u) = \tfrac{1}{3}u - \tfrac{2}{3} \quad \text{or} \quad f^{-1}(t) = \tfrac{1}{3}t - \tfrac{2}{3} \quad \text{or} \quad f^{-1}(x) = \tfrac{1}{3}x - \tfrac{2}{3}$$

and it will still define the same function f^{-1}. This result can be verified as follows:

$$(f \circ f^{-1})(x) = f[f^{-1}(x)] = f(\tfrac{1}{3}x - \tfrac{2}{3}) = 3(\tfrac{1}{3}x - \tfrac{2}{3}) + 2 = x - 2 + 2 = x$$

and

$$(f^{-1} \circ f)(x) = f^{-1}[f(x)] = f^{-1}(3x+2) = \tfrac{1}{3}(3x+2) - \tfrac{2}{3} = x + \tfrac{2}{3} - \tfrac{2}{3} = x$$

Geometrically, we should note that the graph of f^{-1} can be obtained from the graph of f by reflecting (flipping) the graph of f across the line $y = x$.

For example, the reflection of the graph of $f(x) = 3x+2$ across the line $y = x$ is the graph of $f^{-1}(x) = \frac{1}{3}x - \frac{2}{3}$.

Figure 4 illustrates the reflections of the points $(1, 5)$, $(0, 2)$, $(2, 8)$ and $(-2, -4)$ on the graph of f, onto the points $(5, 1)$, $(2, 0)$, $(8, 2)$ and $(-4, -2)$, respectively, of the graph of f^{-1}.

Figure 4

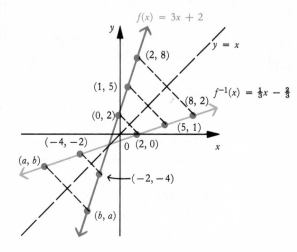

EXAMPLES

Examine the graph of each of the given functions in 1, 2, 3, 4, and 5 to determine whether or not the function has an inverse f^{-1}. If f^{-1} exists, find it and graph f^{-1} on the same coordinate system as one containing the graph of the given function f.

1   $f(x) = 2x - 3$

SOLUTION.   Figure 5 shows $f(x) = 2x-3$ to be a one-to-one function; hence, by Theorem 1, f^{-1} exists. If we let $y = 2x-3$, then $x = (y+3)/2$ results from solving for x in terms of y; therefore, $f^{-1}(y) = (y+3)/2$, or, after changing notation, $f^{-1}(x) = (x+3)/2$. This can be verified as follows.

$$(f \circ f^{-1})(x) = f[f^{-1}(x)] = f\left(\frac{x+3}{2}\right) = 2\left(\frac{x+3}{2}\right) - 3 = x$$

and

$$(f^{-1} \circ f)(x) = f^{-1}[f(x)] = f^{-1}(2x-3) = \frac{(2x-3)+3}{2} = x$$

After graphing $f(x) = 2x-3$ and $f^{-1}(x) = (x+3)/2$ on the same coordinate system, we can observe that the graph of the inverse function f^{-1} can be obtained by reflecting the graph of f across the line $y = x$, since $(x, y) \in f$ implies that $(y, x) \in f^{-1}$ (Figure 5).

Figure 5

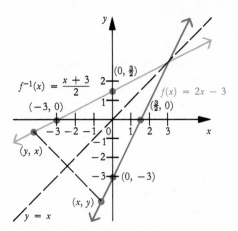

2   $f(x) = x^3$

SOLUTION.   The graph of $f(x) = x^3$ (Figure 6) shows that the function is one-to-one, so we know that f^{-1} exists. Next, we can find f^{-1} by first letting $y = x^3$ and solving for x in terms of y to get $x = \sqrt[3]{y}$. Hence $f^{-1}(y) = \sqrt[3]{y}$, or, after changing notation, $f^{-1}(x) = \sqrt[3]{x}$, which can be verified as follows.

$$(f \circ f^{-1})(x) = f[f^{-1}(x)] = f(\sqrt[3]{x}) = (\sqrt[3]{x})^3 = x$$

and

$$(f^{-1} \circ f)(x) = f^{-1}[f(x)] = f^{-1}(x^3) = \sqrt[3]{x^3} = x$$

When both $f(x) = x^3$ and $f^{-1}(x) = \sqrt[3]{x}$ are graphed on the same coordinate system, we discover again that the graph of f^{-1} can be obtained by reflecting the graph of f across the line $y = x$ (Figure 6).

Figure 6

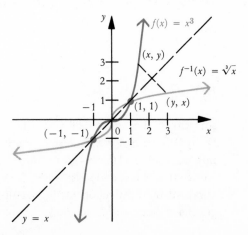

3   $f(x) = x^2$

SOLUTION.   The graph of $f(x) = x^2$ (see Figure 7, page 68) shows that f is not one-to-one [for example, $(2,4)$ and $(-2,4)$ are members of the function]; hence, by Theorem 1, f^{-1} does not exist.

4   $f(x) = x^2, \quad x \in [0, \infty)$

SOLUTION.   The graph of $f(x) = x^2$, $x \in [0, \infty)$ (Figure 7) shows that f is one-to-one; hence, by Theorem 1, f^{-1} exists.
    If we let $y = x^2$, we can solve for x in terms of y to obtain $x = \sqrt{y}$ so that $f^{-1}(y) =\bullet \sqrt{y}$, or, after changing notation, $f^{-1}(x) = \sqrt{x}$. The graph of f^{-1} is a reflection of the graph of f across $y = x$ (Figure 7).

Figure 7

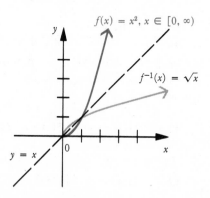

Notice that in Examples 3 and 4 the function expression is the same; however, by restricting the domain to $[0, \infty)$, we have an invertible function, whereas for domain R, the function $f(x) = x^2$ is not invertible.

5   $f(x) = \dfrac{1}{x}$

SOLUTION.   The graph of $f(x) = 1/x$ (Figure 8) shows that f^{-1} exists. (Why?) Solving for x in terms of y in $y = 1/x$, we obtain $x = 1/y$. This implies that $f^{-1}(y) = 1/y$ or, after changing notation, $f^{-1}(x) = 1/x$. Notice that this function is its own inverse. Again the graph of f^{-1} can be obtained by a reflection of the graph of f across $y = x$ (Figure 8). (The graphs of f and f^{-1} coincide.)

Figure 8

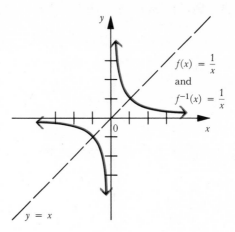

$$f(x) = \frac{1}{x}$$
and
$$f^{-1}(x) = \frac{1}{x}$$

$y = x$

6 Prove that every increasing function has an inverse.

PROOF.   Assume that f is an increasing function. If x_1 and x_2 are different members of the domain of f and $x_1 < x_2$, then, since f is increasing, $f(x_1) < f(x_2)$ (Figure 9). This means that no two different ordered pairs of the function f have the same second members. In other words, each member of the range is the image of one and only one member of the domain; that is, f is one-to-one, from which we can conclude by Theorem 1 that f^{-1} exists.

Figure 9

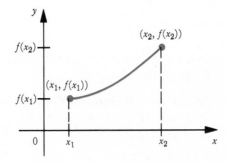

$(x_2, f(x_2))$

$f(x_2)$

$(x_1, f(x_1))$

$f(x_1)$

$0 \quad x_1 \quad\quad\quad x_2 \quad\quad x$

PROBLEM SET 5

1 Verify that $g = f^{-1}$ for each of the following pairs of functions.
   a) $f(x) = 3x + 2$ and $g(x) = x/3 - \frac{2}{3}$.
   b) $f(x) = 1 - 5x$ and $g(x) = \frac{1}{5} - \frac{1}{5}x$.
   c) $f(x) = x^4$, where x is a positive real number and $g(x) = \sqrt[4]{x}$.

2 Graph each pair of functions in Problem 1 on the same coordinate system to see that the graph of f^{-1} is a reflection of the graph of f across the line $y = x$.

3 Decide whether each function whose graph is given in Figure 10 has an inverse.

Figure 10

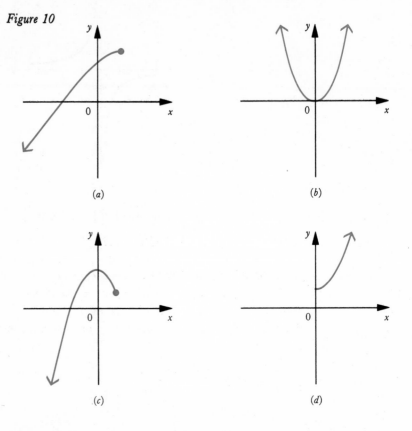

(a)

(b)

(c)

(d)

4 Does every decreasing function have an inverse? Support your assertion by giving an example.

5 Each of the following functions has an inverse; find the inverse f^{-1} and verify that $(f \circ f^{-1})(x) = x$ and $(f^{-1} \circ f)(x) = x$.

a) $f(x) = 3x - 7$
b) $f(x) = 5 - 11x$
c) $f(x) = \frac{3}{4}x + 5$
d) $f(x) = 8x^3$
e) $f(x) = x^2$, $x \in (-\infty, 0]$
f) $f(x) = -x^2$, $x \in [0, \infty)$

6 Show that $f(x) = |x + 1|$ is not one-to-one. Does f^{-1} exist? Explain.

7 a) Show that $f(x) = 2x + 1$ is one-to-one. Find f^{-1}.
b) Show that $f(x) = -5x + 2$ is decreasing. Find f^{-1}.

8 a) Use the graph of $f(x) = -3x^2 + 1$ to show that f^{-1} does not exist.
b) Show that $f(x) = -3x^2 + 1$, $x \in (-\infty, 0]$, has an inverse function f^{-1}. Find f^{-1}.
c) Explain the difference between the two functions in parts a and b.

9 Examine the graphs of each of the following functions to determine whether or not f^{-1} exists. If f^{-1} exists, find it and graph f^{-1} on the same coordinate system as f.

a) $f(x) = 3$

b) $f(x) = 7x + 5$

c) $f(x) = 3/x$

d) $f = \{(x,y) \mid y = -2\,|x|\}$

e) $f(x) = 1 - 3x$

f) $f(x) = x^3 + 5$

g) $f = \{(x,y) \mid y = x^2 + 2,\ x \geq 0\}$

h) $f(x) = [\![2x]\!]$

REVIEW PROBLEM SET

1 Indicate which of the following relations is a function. Indicate the domain and the range of each.

a) $f(x) = 7x - 2$

b) $f(x) = 25 - x^2$

c) $f = \{(1,2),(2,3),(1,5),(6,7)\}$

d) $\{(x,y) \mid |y| > |x|\}$

2 Figure 11 contains graphs of relations. Which graphs represent functions? Which of the functions have inverses?

Figure 11

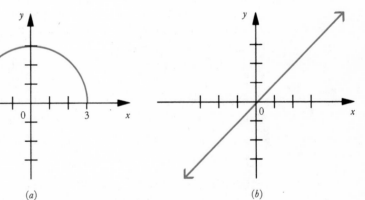

(a)

(b)

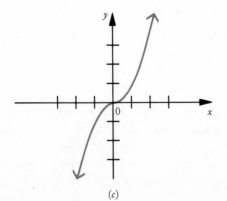

(c)

(d)

3 Let $f(x) = 3x^2 + 2$. Find each of the following expressions.

a) $f(-1)$ 　　　　　　　　　　　　b) $f(0)$

c) $f(1)$ 　　　　　　　　　　　　　d) $f(\frac{2}{3})$

e) $f(a)$ 　　　　　　　　　　　　　f) $f(b)$

g) $f(a+b)$ 　　　　　　　　　　　h) $[f(a+b)-f(a)]/b$

4 A rectangle of sides x and y units is inscribed in a circle of radius 3 units. Express y as a function of x. Express the area A of the rectangle as a function of x.

5 Determine which of the following points lie on the graph of $f(x) = x^{3/2} - x$: $(1, 0)$, $(-1, 0)$, $(4, 4)$, and $(9, 17)$.

6 Decide which of the following statements are true and which are false, if $f(x) = 2x + 1$.

a) $f(x+3) = f(x) + 3$ 　　　　　　b) $f(x+a) = f(x) + a$

c) $f(x) + a = f(x) + f(a)$ 　　　　d) $f(x+a) = f(x) + f(a)$

e) $f(ax) = af(x)$

7 Indicate whether each of the following functions has a graph that is symmetric with respect to the y axis or origin. Graph the function. Indicate the domain and range. Are the functions increasing or decreasing? Are the functions even or odd?

a) $f(x) = -2x + 3$ 　　　　　　　b) $f(x) = |2x + 5|$

c) $f(x) = |x|^2$ 　　　　　　　　　d) $f(x) = 3|x| - 3x$

e) $f(x) = -2x^3$ 　　　　　　　　　f) $f(x) = 2\sqrt[3]{x}$

g) $f(x) = -x^2/2$ 　　　　　　　　h) $f(x) = x^4 + 1$

8 Use the graphs of the functions $f(x) = x^3$ and $g(x) = -x$ to graph $f+g$ and $f-g$. What is the domain of $f+g$ and $f-g$?

9 Sketch the graphs of $f \cdot g$ (product) and $f \circ g$ (composition) for each of the following pairs of functions on the same coordinate system.

a) $f(x) = x+2$ and $g(x) = x-1$

b) $f(x) = 1-x$ and $g(x) = x$

c) $f(x) = \sqrt{4-x^2}$, $-2 \leq x \leq 2$, and $g(x) = 4$

d) $f(x) = 3x-2$ and $g(x) = 2x+3$

10 Let f be an even function and g be an odd function. Which of the following functions are even or odd?

a) $f+g$ 　　　　　　　　　　　　　b) $f-g$

c) $f \cdot g$ 　　　　　　　　　　　　d) f/g

11 Let $f(x) = 7-x^2$ and $g(x) = 1+5x$. Find an expression for each of the following functions. Also, indicate the domain.

a) $(f \circ f)(x)$ 　　　　　　　　　b) $(g \circ f)(x)$

c) $(g \circ g)(x)$ 　　　　　　　　　d) $(f \circ g)(x)$

e) $f[1/g(x)]$ 　　　　　　　　　　f) $g[1/f(x)]$

12 Let $f(x) = 7 - 13$ and $g(x) = 3x + 7$.
  a) Find f^{-1} and g^{-1}.
  b) Show that $(f \circ g)^{-1} = g^{-1} \circ f^{-1}$.
  c) Show that $f[f^{-1}(x)] = x$.
  d) Show that $g^{-1}[g(x)] = x$.

13 Examine the graphs of the given functions to determine whether or not the function has an inverse f^{-1}. If f^{-1} exists, find it and graph f^{-1} on the same coordinate system as one containing the graph of the given function f.
  a) $f(x) = \tfrac{1}{2}x + 2$
  b) $f(x) = \sqrt{x}$
  c) $f(x) = 3|x|$
  d) $f = \{(x,y) \mid y = -3x^4\}$
  e) $f = \{(x,y) \mid y = 1/(x-1)\}$

CHAPTER 3

Polynomial Functions and Rational Functions

1 Introduction

In Chapter 2 some of the general properties of functions were considered. In this chapter we shall investigate particular types of functions, called polynomial functions and rational functions, by investigating the various properties of functions which were presented in Chapter 2. These properties are listed here for convenience:

1 The domain and the range
2 The graph
3 Increasing or decreasing or neither
4 Symmetry
5 Even or odd or neither
6 Inverse if it exists

DEFINITION POLYNOMIAL FUNCTION

Any function expressible in the form

$$f(x) = a_n x^n + a_{n-1} x^{n-1} + a_{n-2} x^{n-2} + \cdots + a_1 x + a_0$$

where n is a positive integer and $a_n, a_{n-1}, \ldots, a_1$ and a_0 are real numbers with $a_n \neq 0$, is called a *polynomial function of degree n in x*. The numbers $a_n, a_{n-1}, \ldots, a_1$, and a_0 are called the *coefficients* of the polynomial function. If $n = 0$, then $f(x) = a$, where $a \neq 0$, is called a *zero-degree* polynomial function. $f(x) = 0$ is called the *zero polynomial* and no degree is assigned to it.

For example, $f(x) = 2x^3 + 5x^2 - 9x + 3$ is a polynomial function of degree 3 since it can be written in the form $f(x) = 2x^3 + 5x^2 + (-9)x + 3$. The coefficients of this polynomial function are 2, 5, -9, and 3, where $a_3 = 2$, $a_2 = 5$, $a_1 = -9$, and $a_0 = 3$.

$f(x) = 4$ is an example of a zero-degree polynomial with one coefficient, $a_0 = 4$, whereas $f(x) = 2x + 1/x^2 = 2x + x^{-2}$ is not a polynomial function in x because of the negative 2 exponent.

EXAMPLES

Express each of the following functions in polynomial form and then identify the degree and the coefficients of the polynomial functions.

1 $f(x) = -5x^3 + 7x^2 + 3x - 4$

SOLUTION. $f(x)$ can be written in the polynomial form as $f(x) = (-5)x^2 + 7x^2 + 3x + (-4)$, so that the degree is 3 and the coefficients are $a_3 = -5$, $a_2 = 7$, $a_1 = 3$, and $a_0 = 4$.

2 $f(x) = x^4 + \sqrt{2}x^2 + 5$

SOLUTION. $f(x)$ can be written in polynomial form as

$$f(x) = 1 \cdot x^4 + 0 \cdot x^3 + \sqrt{2}x^2 + 0 \cdot x + 5$$

so that the degree is 4 and the coefficients are given by $a_4 = 1$, $a_3 = 0$, $a_2 = \sqrt{2}$, $a_1 = 0$, and $a_0 = 5$.

2 Linear Functions

Suppose that we have two thermometers which are used simultaneously to measure temperature—one graduated according to the Fahrenheit scale, the other graduated according to the Celsius scale. If x represents the Celsius reading and y represents the Fahrenheit reading, a functional relationship between x and y can be found as follows.

The freezing point (of water) is 0° Celsius or 32° Fahrenheit, whereas the boiling point (of water) is 100° Celsius or 212° Fahrenheit. The number of Fahrenheit degrees is the same as $\frac{9}{5}$ times the number of Celsius degrees. Now, if x is the Celsius reading, x represents the "directed" number (x may be negative) of Celsius degrees, so $\frac{9}{5}x$ represents the corresponding directed number of Fahrenheit degrees. Since the "starting point" of the Fahrenheit

Figure 1

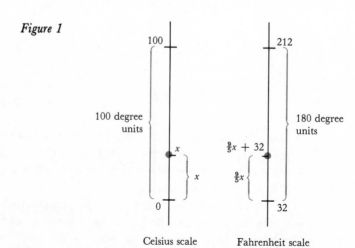

Celsius scale   Fahrenheit scale

scale is 32, as compared with 0 on the Celsius scale, it follows that a Fahrenheit reading of y corresponding to a Celsius reading of x is given by $y = \frac{9}{5}x + 32$ (Figure 1).

If x is assumed to represent members of the domain, then $y = \frac{9}{5}x + 32$ determines the function $f(x) = \frac{9}{5}x + 32$ (Figure 2).

Figure 2

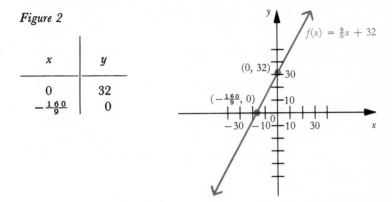

x	y
0	32
$-\frac{160}{9}$	0

This is an example of a linear function. More formally, a linear function is defined as follows.

DEFINITION   LINEAR FUNCTION

A polynomial function of the form $f(x) = mx + b$ is called a *linear function*, where m and b are constant real numbers.

Such functions are called *linear* functions because their graphs are straight lines. Before proving this assertion let us consider the following example.

EXAMPLE

Show that the three points of the linear function $f(x) = 2x + 1$ with abscissas $x = 0$, 1, and 2 lie on the same straight line; that is, show that these three points are *collinear*.

SOLUTION.   Since $f(0) = 1$, $f(1) = 3$, and $f(2) = 5$, the three points that belong to the graph of f are $P_1 : (0, 1)$, $P_2 : (1, 3)$, and $P_3 : (2, 5)$ (Figure 3).

Figure 3

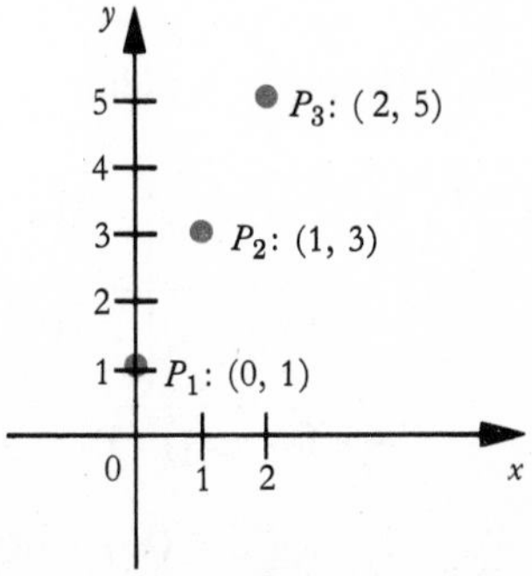

The three points P_1, P_2, and P_3 are collinear if $\overline{P_1P_2}+\overline{P_2P_3} = \overline{P_1P_3}$, since the shortest distance between two points is a straight line.

By the distance formula,

$$\overline{P_1P_2} = \sqrt{(1-0)^2 + (3-1)^2} = \sqrt{1+4} = \sqrt{5}$$

$$\overline{P_2P_3} = \sqrt{(2-1)^2 + (5-3)^2} = \sqrt{1+4} = \sqrt{5}$$

$$\overline{P_1P_3} = \sqrt{(2-0)^2 + (5-1)^2} = \sqrt{4+16} = \sqrt{20} = 2\sqrt{5}$$

Since $\sqrt{5}+\sqrt{5} = 2\sqrt{5}$, $\overline{P_1P_2}+\overline{P_2P_3} = \overline{P_1P_3}$, and the three points are collinear.

The argument used in the above example is exactly what we need to prove that *any three points* on the graph of a linear function are collinear, which establishes the fact that the graph of a linear function is a straight line.

THEOREM 1

The graph of a linear function $f(x) = mx+b$ is a straight line.

PROOF.   Consider any three points on the graph of f and denote them by $P_1:(x_1,y_1)$, $P_2:(x_2,y_2)$, and $P_3:(x_3,y_3)$ such that $x_1 < x_2 < x_3$ (Figure 4). These three points are collinear if $\overline{P_1P_2}+\overline{P_2P_3} = \overline{P_1P_3}$.

Figure 4

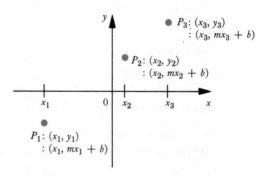

By the distance formula,

$$\overline{P_1P_3} = \sqrt{(x_3-x_1)^2 + (y_3-y_1)^2}$$

$$= \sqrt{(x_3-x_1)^2 + [(mx_3+b) - (mx_1+b)]^2}$$

$$= \sqrt{(x_3-x_1)^2 + m^2(x_3-x_1)^2} \qquad \text{(why?)}$$

$$= \sqrt{(x_3-x_1)^2(1+m^2)}$$

Since $x_3 > x_1, x_3 - x_1$ is positive and

$$\overline{P_1 P_3} = (x_3 - x_1)\sqrt{1 + m^2}$$

Similarly, we can show that

$$\overline{P_1 P_2} = (x_2 - x_1)\sqrt{1 + m^2}$$

and

$$\overline{P_2 P_3} = (x_3 - x_2)\sqrt{1 + m^2}$$

Therefore,

$$
\begin{aligned}
\overline{P_1 P_2} + \overline{P_2 P_3} &= (x_2 - x_1)\sqrt{1 + m^2} + (x_3 - x_2)\sqrt{1 + m^2} \\
&= [(x_2 - x_1) + (x_3 - x_2)]\sqrt{1 + m^2} \\
&= (x_3 - x_1)\sqrt{1 + m^2} \\
&= \overline{P_1 P_3}
\end{aligned}
$$

Hence any three points on the graph of f are collinear. That is, the graph of $f(x) = mx + b$ is always a straight line.

Because of this theorem, it will be enough to locate two points in order to determine the graph of a linear function, since a straight line is uniquely determined by any two of its points.

If $m = 0$, then the linear function $f(x) = mx + b$ becomes $f(x) = b$. In this case f is called a *constant function*. Its graph is the set of all points with an ordinate of b; it is a line parallel to or coincidental to the x axis through the point $(0, b)$. If, for example, $b = 3$, then $f(x) = 3$ has the set representation $f = \{(x, y) \mid y = 3\}$ and its graph is shown in Figure 5.

Figure 5

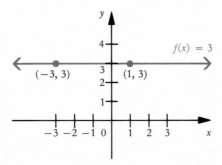

EXAMPLES

1 Find the domain and the range of the linear function $f(x) = 2x + 5$ and sketch the graph. Explain why the function f has an inverse f^{-1}. Determine f^{-1} and graph it.

SOLUTION. Since for any real number x there is a corresponding real number $f(x)$, the domain of f is the set of real numbers R and the range is also R. The graph can be determined by locating two points (Figure 6). f^{-1} exists since f is increasing (Figure 6). f^{-1} can be constructed by using the method of Chapter 2, Section 5.

First, let $y = f(x)$, so that $y = 2x + 5$. After solving for x in terms of y, we get $x = (y-5)/2$, so $f^{-1}(y) = (y-5)/2$, or, after changing notation, it follows that $f^{-1}(x) = (x-5)/2$ (Figure 6).

Figure 6

x	$f(x)$
$-\frac{5}{2}$	0
0	5

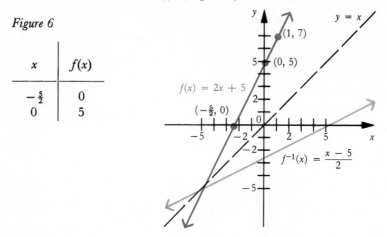

2 Suppose that $y = f(x)$ is a linear function whose graph contains $(3, -4)$ and $(-2, 5)$. Find f in equation form.

SOLUTION. Since $(3, -4)$ and $(-2, 5)$ both lie on the same line, the ordered pairs of numbers must both satisfy the functional relationship $y = mx + b$ simultaneously. In other words, m and b must satisfy both equations

$$-4 = 3m + b \quad \text{and} \quad 5 = -2m + b$$

Subtracting the corresponding sides of the equations yields $-9 = 5m$ or $m = -\frac{9}{5}$. Substituting $m = -\frac{9}{5}$ into $-4 = 3m + b$ results in $b = \frac{7}{5}$; hence $f(x) = -\frac{9}{5}x + \frac{7}{5}$ (Figure 7).

Figure 7

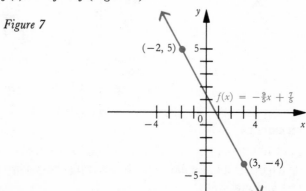

3   Show that a linear function of the form $f(x) = mx$ is an odd function.

SOLUTION.   Since $f(x) = mx$, $f(-x) = m(-x) = -mx$, so that $f(-x) = -f(x)$; hence the function is odd and the graph is symmetric with respect to the origin.

2.1   Slope of a Line

Figure 8 displays the graphs of the functions $y = x$, $y = 2x$, and $y = 5x$.

 Figure 8

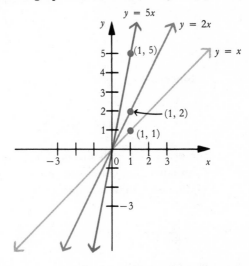

Note that each of these functions is "increasing at a different rate." For example, as x increases from 1 to 2,

$y = x$ increases from 1 to 2 (1-unit increase)

$y = 2x$ increases from 2 to 4 (2-unit increase)

$y = 5x$ increases from 5 to 10 (5-unit increase)

What we would like to do next is derive some way of measuring the "inclination" or "rate of change" of lines and to relate such a measure to the equation of the line. The easiest way to do this is by what is called the *slope of a line.*

DEFINITION   SLOPE OF A LINE

Suppose that (x_1, y_1) and (x_2, y_2) are any two points of a line such that $x_1 \neq x_2$ (Figure 9). The number s that is defined by the equation

$$s = \frac{y_2 - y_1}{x_2 - x_1} \qquad x_1 \neq x_2$$

is called the *slope* of the line.

Figure 9

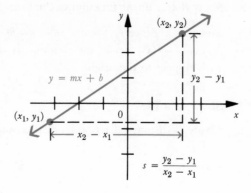

Since

$$\frac{y_1-y_2}{x_1-x_2} = \frac{-(y_1-y_2)}{-(x_1-x_2)} = \frac{y_2-y_1}{x_2-x_1}$$

the order in which the two points are taken when the coordinates are subtracted does not change the value of the slope. For example, the slope of the line containing the points $(-2, 5)$ and $(3, -4)$ (Figure 10) is

$$\frac{-4-5}{3-(-2)} = -\frac{9}{5}$$

or, equivalently,

$$\frac{5-(-4)}{-2-3} = -\frac{9}{5}$$

Figure 10

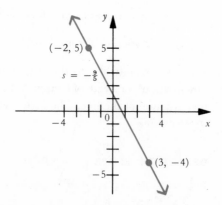

EXAMPLE

Determine the slope of the line containing the two points whose coordinates are given by

a) $(6, 2)$ and $(3, 7)$ b) $(3, -2)$ and $(5, -6)$

SOLUTION

a) The slope of the line is given by $s = (y_2 - y_1)/(x_2 - x_1)$, so that if $P_1 : (6, 2)$ and $P_2 : (3, 7)$, then

$$s = \frac{7-2}{3-6} = \frac{5}{-3} = -\frac{5}{3}$$

b) If $P_1 : (3, -2)$ and $P_2 : (5, -6)$, then

$$s = \frac{-6-(-2)}{5-3} = \frac{-6+2}{5-3} = -\frac{4}{2} = -2$$

Notice that the values of the slopes of the lines in the examples above appear to depend on the two points selected. Will the value of the slope of a line be the same no matter what two points on the line are used to compute the slope? This question is answered by the following theorem.

THEOREM 1

The slope of the line determined by the function $f(x) = mx + b$ is m. In other words, no matter which points are selected to compute the slope, the results will be the same value, m, the coefficient of x.

PROOF. Consider any x_1 and x_2 such that $x_1 \neq x_2$. Then $(x_1, f(x_1))$ and $(x_2, f(x_2))$ are two points of the line, so that the slope s can be computed as follows.

$$s = \frac{f(x_2) - f(x_1)}{x_2 - x_1} = \frac{(mx_2 + b) - (mx_1 + b)}{x_2 - x_1}$$

$$= \frac{mx_2 - mx_1}{x_2 - x_1}$$

$$= \frac{m(x_2 - x_1)}{x_2 - x_1} = m$$

For example, the slope of the line defined by $y = 2x + 1$ is 2; the slope of the line defined by $y = -\frac{1}{2}x$ is $-\frac{1}{2}$.

Because of Theorem 1, we usually use the letter m to represent the slope of a straight line.

If $x_1 = x_2$ for all points on a line, then the slope of the line

$$m = \frac{y_2 - y_1}{x_2 - x_1} = \frac{y_2 - y_1}{0}$$

is not defined because of the division by 0. This situation occurs whenever the line contains points with the same abscissas; such a line is parallel to the y axis or coincides with the y axis and its graph is not a function (see Problem 4). For example, $x = 4$ is a line parallel to the y axis (Figure 11); the slope of this line is not defined.

Figure 11

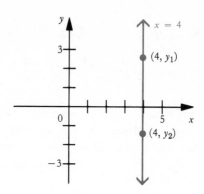

If the slope of a line is zero, then the linear function $f(x) = mx + b$ that defines the line assumes the form $f(x) = b$. That is, the function is constant and its graph is a horizontal line. $f(x) = 3$, which was graphed in Figure 5 on page 125, is an example of such a function.

The next theorem provides a way of determining whether a linear function is increasing or decreasing by using the value of the slope of the line.

THEOREM 2

The linear function $f(x) = mx + b$ is an increasing function if m is positive and a decreasing function if m is negative.

PROOF.   Let (x_1, y_1) and (x_2, y_2) be any two points on the graph of f with $x_1 < x_2$. Then

$$m = \frac{y_2 - y_1}{x_2 - x_1}$$

so that

(1) $m(x_2 - x_1) = y_2 - y_1$ with $x_2 - x_1 > 0$

If $m > 0$, then $y_2 - y_1 > 0$ in equation (1), so $y_1 < y_2$. Hence the function is increasing and its graph rises from left to right (Figure 12).

Figure 12

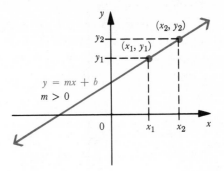

If $m < 0$, then $y_2 - y_1 < 0$ in equation (1) (why?), so $y_2 < y_1$; hence the function is decreasing and the graph falls from left to right (Figure 13).

Figure 13

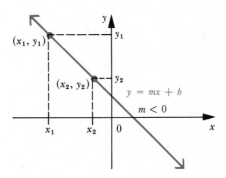

Consequently, if $m \neq 0, f(x) = mx + b$ is either increasing or decreasing, and f has an inverse.

EXAMPLES

Determine the slope of each of the following lines. Use the slope to decide whether the function $y = f(x)$ is increasing, decreasing, or constant. Find f^{-1} if it exists and graph both f and f^{-1} on the same coordinate axes.

1 $y + 3x = 7$

SOLUTION. $y + 3x = 7$ is equivalent to $y = -3x + 7$, so the slope is -3. Since the slope is negative, the function is decreasing and it has an inverse. The inverse function f^{-1} can be found by first solving for x in terms of y to get $x = (y - 7)/(-3)$ so that $f^{-1}(x) = -\frac{1}{3}x + \frac{7}{3}$. Figure 14 contains the graphs of f and f^{-1}.

Figure 14

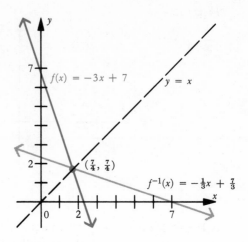

2   $5x - y + 13 = 0$

SOLUTION. $5x - y + 13 = 0$ is equivalent to $y = 5x + 13$; hence the slope is 5. The function is increasing and the inverse is $f^{-1}(x) = \frac{1}{5}x - \frac{13}{5}$ (why?) (Figure 15).

Figure 15

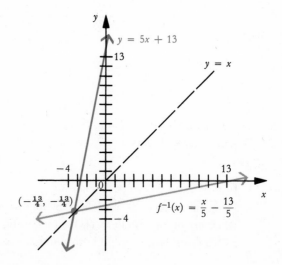

3   $y = -2$

SOLUTION. $y = -2$ is a constant function. Its graph is a horizontal line (Figure 16). The function does not have an inverse because it is not one-to-one.

Figure 16

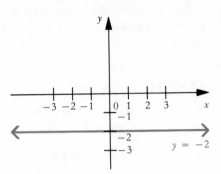

2.2 Forms of Equations of Lines

Since $(0, b)$ is the point on the graph of $f(x) = mx + b$ which is on the y axis, we say that $(0, b)$ is the y *intercept* of the line **(Figure 17)**. The form $f(x) = mx + b$ is called the *slope-intercept form* of the line defined by the equation $y = mx + b$.

Figure 17

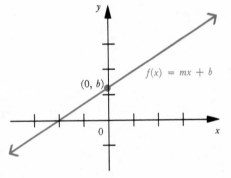

Suppose that the slope of a line is m and that (x_1, y_1) is a point on the line. If $(x, y) \neq (x_1, y_1)$ is used to represent any point on the line, then $m = (y - y_1)/(x - x_1)$, and $y - y_1 = m(x - x_1)$.

The form of the equation of the line, $y - y_1 = m(x - x_1)$, where m is the slope and (x_1, y_1) is any point on the line, is called the *point-slope form* (Figure 18). Notice that (x_1, y_1) also satisfies the equation $y - y_1 = m(x - x_1)$.

Figure 18

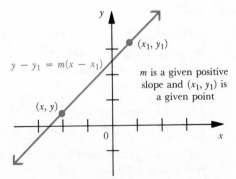

For example, the point-slope form of the equation of the line that contains the point $(-1, 2)$ with slope 3 is $y - 2 = 3[x-(-1)]$, or $y - 2 = 3(x+1)$.

EXAMPLES

1 Find both forms of the equation of a line with slope 5 and containing the point $(2, 3)$ and graph the line.

SOLUTION.   Using the point-slope form, that is,

$$y - y_1 = m(x - x_1) \qquad \text{with } m = 5 \quad \text{and} \quad (x_1, y_1) = (2, 3)$$

we have $y - 3 = 5(x - 2)$, or, equivalently, $y = 5x - 7$, which is the slope-intercept form of the line (Figure 19).

Figure 19

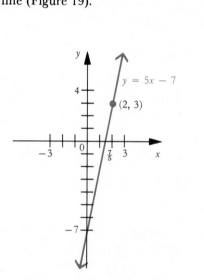

2 Find both forms of the equation of the line that contains the points $(-2, 5)$ and $(3, -4)$.

SOLUTION.   The slope of the line is given by $[5-(-4)]/(-2-3) = 9/-5$ so that the point-slope equation of the line is $y - 5 = -\frac{9}{5}(x+2)$ or, equivalently, the slope-intercept form, obtained by solving for y in terms of x, is $y = -\frac{9}{5}x + \frac{7}{5}$.

3 Find the slope-intercept form of the equation of the line defined by the linear function f if the slope is -2 and $f(3) = 1$ and graph f.

SOLUTION. $f(x) = mx + b$, so that $f(3) = -2(3) + b = 1$; hence $b = 7$ and $f(x) = -2x + 7$ (Figure 20).

Figure 20

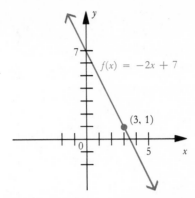

$f(x) = -2x + 7$

$(3, 1)$

4   Let Q be the amount of heat required to change 1 gram of solid ice at $0°$ Celsius to water at t degrees Celsius. Assume that Q is a linear function of t, the temperature of water, and also assume that $0 \leq t \leq 100$. If $Q = 70$ at $t = 15°$ and $Q = 140$ at $t = 85°$, what is the amount of heat required to transform the ice into $5°$ Celsius water?

SOLUTION.   Since the function is linear, it has the form

$$Q = f(t) = mt + b$$

Since $Q = 70$ when $t = 15$ and $Q = 140$ when $t = 85$,

$$70 = 15m + b \quad \text{and} \quad 140 = 85m + b$$

Solving these latter two equations simultaneously results in $m = 1$ and $b = 55$; therefore, $Q = t + 55$, so that when $t = 5$, $Q = 60$.

2.3   Geometry of Two Lines

If two different lines are graphed on the same coordinate axis, either they intersect at one point or they are parallel. We can determine whether two lines intersect or are parallel by examining the relationship between their slopes.

Let $y = m_1 x + b_1$ and $y = m_2 x + b_2$ be the equations of two distinct lines. If the graphs of the two lines intersect, there must be a value of x that satisfies both equations simultaneously; that is,

$$m_1 x + b_1 = m_2 x + b_2$$
$$m_1 x - m_2 x = b_2 - b_1$$

or

$$(m_1 - m_2)x = b_2 - b_1$$

If $m_1 \neq m_2$, then $x = (b_2 - b_1)/(m_1 - m_2)$ satisfies both of the original equations and the two lines intersect. For example, the two lines $x - 2y = 4$ and $3x + 2y = 4$ intersect, since the slope of the first is $\frac{1}{2}$, and the slope of the second is $-\frac{3}{2}$ (Figure 21). (In Section 3 we will study methods of finding the point of intersection of the two lines.)

Figure 21

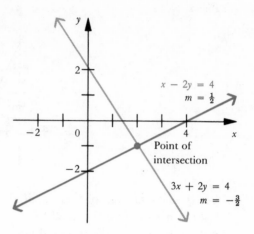

On the other hand, if $m_1 = m_2$ and $b_1 \neq b_2$, $x = (b_2 - b_1)/(m_1 - m_2)$ is undefined because the denominator would be zero and the two lines are parallel. In general, we have the following property.

PROPERTY 1

Two different lines with slopes m_1 and m_2 intersect if and only if $m_1 \neq m_2$ and are parallel if and only if $m_1 = m_2$.

For example, the line containing the points $P_1 : (3, 3)$ and $P_2 : (5, 6)$ is parallel to the line containing the points $P_3 : (-1, 1)$ and $P_4 : (1, 4)$ since their slopes are respectively,

$$m_1 = \frac{6-3}{5-3} = \frac{3}{2} \quad \text{and} \quad m_2 = \frac{4-1}{1+1} = \frac{3}{2} \quad \text{(Figure 22)}$$

Figure 22

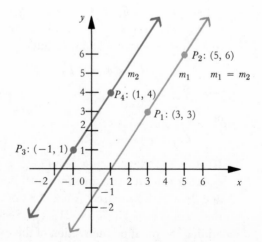

Finally, we state without proof (see Problem 12b) the following property.

PROPERTY 2

Two lines with slopes m_1 and m_2 are perpendicular if and only if $m_1 m_2 = -1$.

For example, consider the line containing the points $P_1:(3,3)$ and $P_2:(5,6)$. It has slope

$$m_1 = \frac{6-3}{5-3} = \frac{3}{2}$$

The line containing the points $Q_1:(1,4)$ and $Q_2:(-2,6)$ has slope

$$m_2 = \frac{6-4}{-2-1} = -\frac{2}{3}$$

Since $m_1 m_2 = (\frac{3}{2})(-\frac{2}{3}) = -1$, the two lines are perpendicular (Figure 23).

Figure 23

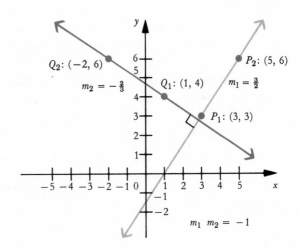

EXAMPLES

1 Find the equation of a line containing the point $(3, -2)$ and
   a) parallel to the x axis   b) parallel to the y axis

SOLUTION
   a) The slope of the line containing the point $(3, -2)$ and parallel to the
       x axis is 0. Hence the equation of that line is $y = -2$ (Figure 24).
   b) The slope of the line containing the point $(3, -2)$ and parallel to the y
       axis is undefined and the equation of that line is $x = 3$ (Figure 24).

Figure 24

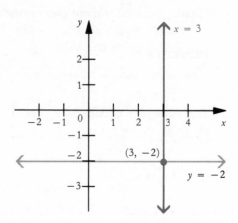

2   Find a value for a so that the line defined by $3x + ay = 5$ is parallel to the line defined by $y = -2x + 1$.

SOLUTION. $y = -3x/a + 5/a$, so that the slope is given by $-3/a$. Since the two lines are parallel, the slopes are equal, that is, $-3/a = -2$, and $a = \frac{3}{2}$.

3   a)   Find the equation of the line containing $(3, 2)$ and perpendicular to the line containing the points $(-4, -2)$ and $(-2, 2)$.
     b)   Also find the equation of the line containing $(3, 2)$ and parallel to the line containing the points $(-4, -2)$ and $(-2, 2)$.

SOLUTION

a)   The slope of the line containing points $(-4, -2)$ and $(-2, 2)$ is $[2 - (-2)]/[-2 - (-4)] = \frac{4}{2} = 2$, so that the slope of the perpendicular line is $-\frac{1}{2}$. Hence the equation of the line perpendicular to the given line and containing the point $(3, 2)$ is $y - 2 = -\frac{1}{2}(x - 3)$, or, equivalently, $y = -\frac{1}{2}x + \frac{7}{2}$ (Figure 25).

Figure 25

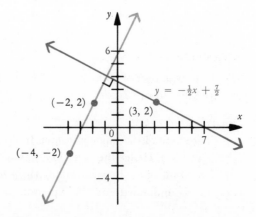

b) Since the line is parallel to the given line, its slope must be equal to 2. Hence the equation of the line is $y-2 = 2(x-3)$ or, equivalently, $y = 2x-4$ (Figure 26).

Figure 26

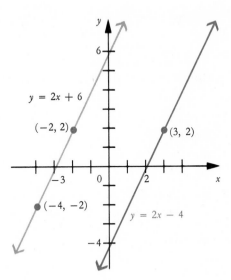

$y = 2x + 6$

$(-2, 2)$

$(3, 2)$

$(-4, -2)$

$y = 2x - 4$

PROBLEM SET 1

1 Which of the following functions are polynomial functions? Indicate the degree of each polynomial function, put it in polynomial form, and identify the coefficients.

a) $\{(x,y) \mid y = 7x - 5x^2 + \pi\}$

b) $f(x) = x^{-2} + x$

c) $f(x) = 5^{-1}$

d) $f(x) = 1/x^3 + 2x^2 + x - 2$

e) $f(x) = 2x - 3^{1/2}x^2 + 5x^3 - 7$

f) $f(x) = x^5 - x^3 + 3x - 7x^6$

2 In each of the following parts, determine a linear function f that satisfies the given conditions.

a) $f(1) = -1$ and $f(3) = 4$

b) $f(5x+2) = 5f(x)+2$

c) $f(7x) = 7f(x)$

d) $f(x_1+x_2) = f(x_1)+f(x_2)$

3 a) Use the distance formula to determine whether or not $(2,5)$, $(4,3)$, and $(3,4)$ are collinear.

b) Use slopes to decide whether or not $(1, 1)$, $(2, 4)$, and $(3, 2)$ are collinear.

4 Let k be a constant. Discuss the relation $\{(x,y) \mid x = k\}$. What is the domain? The range? Is the relation a function? Graph it.

5 Discuss the following linear functions. Indicate the domain and range. Use the slope to decide if the function is increasing or decreasing. Graph the function. Is it even or odd? Find the inverse and graph it if it exists. [Assume that $y = f(x)$.]

a) $\{(x, 3) \mid x \in R\}$

b) $f(x) = 5 - 3x$

c) $y + 1 = 2(x - 1)$

d) $y/2 - 3x/5 = 1$

e) $3x - y + 5 = 0$

6 a) Graph on the same coordinate system all the linear functions

$$y = mx \quad \text{where } m \in \{-10, -5, -2, -1, 0, 1, 2, 5, 10\}$$

What happens to the position of a line as the slope increases from -10 to 10?

b) Prove that the difference quotient for $f(x) = mx + b$ is m, the slope of the line.

7 Find the slope and the y intercept of each of the following lines. Sketch the graphs.

a) $2y = 3x - 12$

b) $y = 3 - x$

c) $3x + 5y - 6 = 0$

d) $2x - y + 5 = 0$

8 a) Show that the equation of a line with y intercept $b \neq 0$ and with x intercept $a \neq 0$ can be written in the form $x/a + y/b = 1$. (This equation is called the *intercept form* of the equation of a line.)

b) Use the equation of part a to write the equation of the line that contains the points $(4, 0)$ and $(0, -6)$.

9 Find the linear function that satisfies each of the following conditions, and then graph the line. Indicate the form of the equation of the line used.

a) The slope is 3 and the line contains the point $(1, 1)$.

b) The line contains the points $(-2, 5)$ and $(2, -3)$.

c) The slope is 0 and the line contains the point $(-3, -2)$.

d) $f(1) = 2$ and the slope is -3.

e) $f(3) = 3$ and $f(1) = -2$.

f) The line is parallel to the line $3x + 2y + 2 = 0$ and it contains the point $(-3, -1)$.

g) The line is perpendicular to $3x + 2y + 2 = 0$ and it contains the point $(-3, -5)$.

10 a) Show that if a, b, and c are constant real numbers such that a and b are not both zero, then the graph of the equation $ax + by + c = 0$ is a line. (This equation is called the *general form* of the equation of a line.)

b) Find the slope and the y intercept of the line whose equation is $2x - 3y + 5 = 0$.

c) Show that $ax + by + c = 0$, where $a \neq 0$ and $b \neq 0$, determines a line that intersects the x and y axes at the points $(-c/a, 0)$ and $(0, -c/b)$, respectively.

11 Given $f(x) = -7x + 13$, find a linear function (i) whose graph is parallel to the graph of f, (ii) whose graph is perpendicular to the graph of f and which contains the point

a) $(2, 3)$

b) $(-2, 5)$

12 a) Given the lines $L_1 = \{(x,y) \mid y = m_1 x + b_1\}$ and $L_2 = \{(x,y) \mid y = m_2 x + b_2\}$, discuss (geometrically) all the possibilities for $L_1 \cap L_2$. How do the slopes "predict" each of the possible geometric situations?

b) Prove that if the product of the slopes of two lines is -1, then the two lines are perpendicular; conversely, prove that if the two lines are perpendicular, then the product of the slopes is -1. (*Hint:* Use the distance formula.)

c) Use slopes to prove that the triangles with the following vertices are right triangles.

   i) $(-4, -2)$, $(2, -8)$, and $(4, 6)$ ii) $(2, 3)$, $(6, 0)$, and $(5, 7)$

13 Let the temperature k feet above the surface of the earth be $t°$ Celsius and assume that the function that relates t and k is linear. If the temperature on the surface of the earth is $80°$ Celsius and the temperature at 2500 feet is $56°$ Celsius, what is the temperature at 5000 feet?

3 Systems of Linear Equations

The graphs of linear functions can be used to motivate the study of solving systems of linear equations. A set of linear equations is called a *system of linear equations*. The *solution of a linear system* containing two variables is the set of all ordered pairs of numbers that satisfy all the equations in the system simultaneously.

Geometrically, the solution of a system containing two linear equations with two variables is the set of ordered pairs of numbers corresponding to the points of intersection of the graphs of the two linear equations (provided that such points exist). For example, let us consider the solution of the linear system

$$\begin{cases} 3x + 4y = 12 \\ 3x - 8y = 0 \end{cases}$$

The graphs of the two corresponding linear functions $y = -\frac{3}{4}x + 3$ and $y = \frac{3}{8}x$ are shown in Figure 1. The solution of the system is the ordered pair $(\frac{8}{3}, 1)$, where $x = \frac{8}{3}$ and $y = 1$, and the point of intersection of the two linear equations is the same ordered pair $(\frac{8}{3}, 1)$.

Figure 1

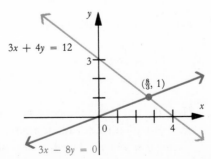

$3x + 4y = 12$

$(\frac{8}{3}, 1)$

$3x - 8y = 0$

It may happen that the graphs of two linear equations are parallel lines. In such a case the intersection is the null set, and the system is called *inconsistent.* **For example, consider the linear system**

$$\begin{cases} x + 2y = 4 \\ 3x + 6y = -3 \end{cases}$$

Note that if the two lines are expressed in the slope-intercept form, the system becomes

$$\begin{cases} y = -\frac{1}{2}x + 2 \\ y = -\frac{1}{2}x - \frac{1}{2} \end{cases}$$

The slopes are equal but the intercepts are different, so we have two parallel lines (Figure 2), the solution set is empty, and the system is inconsistent.

Figure 2

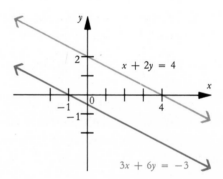

If the graphs of two linear equations coincide, the intersection is every point of the line. In this case the system is said to be *dependent.* **For example,** the system

$$\begin{cases} 3x - y = -1 \\ 6x - 2y = -2 \end{cases}$$

can be written in the slope-intercept form as

$$\begin{cases} y = 3x + 1 \\ y = 3x + 1 \end{cases}$$

Here the lines have the same slope with the same intercepts; consequently, each equation represents the same line (Figure 3), and the solution set is $\{(x,y) \mid y = 3x + 1\}$. This is a dependent system.

Figure 3

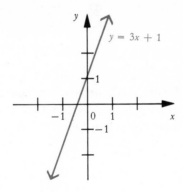

In general, a linear system

$$\begin{cases} a_1 x_1 + b_1 x_2 = c_1 \\ a_2 x_1 + b_2 x_2 = c_2 \end{cases}$$

can fall in one of three categories:

1 If the two lines $a_1 x_1 + b_1 x_2 = c_1$ and $a_2 x_1 + b_2 x_2 = c_2$ intersect in a unique (one) point, the solution is an ordered pair of numbers.

2 If the two lines $a_1 x_1 + b_1 x_2 = c_1$ and $a_2 x_1 + b_2 x_2 = c_2$ are parallel, the solution set is the null set. In this case the system is said to be *inconsistent*.

3 If the two lines $a_1 x_1 + b_1 x_2 = c_1$ and $a_2 x_1 + b_2 x_2 = c_2$ are coincident, the solution set is the set of all points on the line. The system in this case is said to be *dependent*.

3.1 Elimination Method

Now we shall investigate an *algebraic* procedure for solving a system of two linear equations. We will see that this method can be extended to systems containing more than two linear equations and more than two variables.

Let us consider a specific example of the method by solving the following system:

$$\begin{cases} x - y = 1 \\ 3x + 2y = 8 \end{cases}$$

First, multiply the first equation by -3 and add it to the second equation, so that the system becomes

$$\begin{cases} x - y = 1 \\ 0 + 5y = 5 \end{cases}$$

Next, multiply the second equation by $\frac{1}{3}$ to get

$$\begin{cases} x - y = 1 \\ \phantom{x - {}} y = 1 \end{cases}$$

Substituting $y = 1$ into the first equation yields $x = 2$. Thus the solution set is $\{(2, 1)\}$, since $x = 2$ and $y = 1$ satisfy each equation in the system simultaneously (Figure 4).

Figure 4

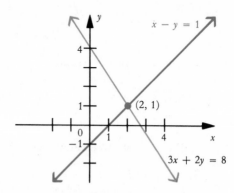

This procedure for solving a system of linear equations is called the *elimination method* or, equivalently, the *process of elimination*.

The process of elimination may also be applied to systems containing more than two unknowns. This application is illustrated in the following example.

EXAMPLE

Use the elimination method to solve the system (A) of linear equations:

$$(A) \quad \begin{cases} 2x_1 + x_2 - 2x_3 = 10 \\ 3x_1 + 2x_2 + 2x_3 = 1 \\ 5x_1 + 4x_2 + 3x_3 = 4 \end{cases}$$

SOLUTION.   First, in system (A) replace the second equation by the sum of -3 times the first equation and 2 times the second equations to obtain system (B):

$$(B) \quad \begin{cases} 2x_1 + x_2 - 2x_3 = 10 \\ \phantom{2x_1 + {}} x_2 + 10x_3 = -28 \\ 5x_1 + 4x_2 + 3x_3 = 4 \end{cases}$$

System (B) is *equivalent* to system (A), since any solution of system (B) is also a solution of system (A). Next, replace the third equation of (B) by the sum of -5 times the first equation and 2 times the third to obtain the equivalent system (C):

$$(C) \quad \begin{cases} 2x_1 + x_2 - 2x_3 = 10 \\ x_2 + 10x_3 = -28 \\ 3x_2 + 16x_3 = -42 \end{cases}$$

Now replace the third equation of (C) by the sum of -3 times the second equation and the third equation to get the equivalent system (D):

$$(D) \quad \begin{cases} 2x_1 + x_2 - 2x_3 = 10 \\ x_2 + 10x_3 = -28 \\ -14x_3 = 42 \end{cases}$$

Next multiply the third equation by $-\frac{1}{14}$ and add -1 times the second equation to the first equation to obtain the equivalent system (E):

$$(E) \quad \begin{cases} 2x_1 - 12x_3 = 38 \\ x_2 + 10x_3 = -28 \\ x_3 = -3 \end{cases}$$

Finally, multiply the first equation by $\frac{1}{2}$ to obtain the equivalent system (F):

$$(F) \quad \begin{cases} x_1 - 6x_3 = 19 \\ x_2 + 10x_3 = -28 \\ x_3 = -3 \end{cases}$$

From the third equation, $x_3 = -3$, so after substituting into the second equation we get $x_2 = 2$; then, after substituting into the first equation, we obtain $x_1 = 1$. Thus the solution is $\{(1, 2, -3)\}$, since $x_1 = 1$, $x_2 = 2$, and $x_3 = -3$ satisfy each equation in the system simultaneously.

In order to see that the six systems given above are equivalent, observe that each system is obtained from the preceding one by performing one or more of the following equation operations:

1 Interchange the positions of two equations in the system.

2 Replace an equation in the system by a nonzero multiple of itself.

3 Replace an equation in the system by the sum of a nonzero multiple of that equation and a nonzero multiple of another equation of the system.

3.2   Matrices and Row Reduction

Notice that in the elimination process there is little reason to continue writing the variables. All we need to do is to maintain a record of the coefficients that belong to each variable in each equation. The standard device used for doing this is a *matrix*.

A *matrix* is a rectangular array of numbers. The *size* of a matrix is described by specifying the number of rows and columns. The numbers occurring in a matrix are called the *entries* of the matrix. For example, the matrix

$$A = \begin{bmatrix} 1 & 2 \\ 3 & -1 \end{bmatrix}$$

has two rows and two columns and we say that A is a 2×2 matrix and 1, 2, 3, and -1 are the entries of A. The matrix

$$B = \begin{bmatrix} 1 & 3 & 5 \\ -2 & 5 & 6 \end{bmatrix}$$

has two rows and three columns, so B is a 2×3 matrix. B has entries 1, 3, 5, -2, 5, and 6. The matrix

$$C = \begin{bmatrix} 3 \\ 1 \\ 2 \end{bmatrix}$$

is a 3×1 matrix with entries 3, 1, and 2. Notice that when the size of a matrix is specified, the number of rows is given first and then the number of columns.

The entry of a matrix is identified by using subscripts to indicate its row and its column position. For example, the entry of the second row and the second column of a matrix D is denoted by d_{22}; the entry of the first row and the fourth column of a matrix D is d_{14}. In general, the entry of the ith row and the jth column of matrix D is denoted by d_{ij}.

For example, if

$$A = \begin{bmatrix} 3 & 2 & -1 \\ 4 & 1 & 5 \end{bmatrix}$$

$a_{11} = 3$, $a_{12} = 2$, $a_{13} = -1$, $a_{21} = 4$, $a_{22} = 1$, and $a_{23} = 5$.

Now, let us see how matrix notation, together with the elimination method, can be used to solve linear systems.

Suppose that we are to solve the linear system

$$\text{(A)} \quad \begin{cases} 3x - y = 1 \\ x + 2y = 0 \end{cases}$$

The matrix form of (A) is written as

$$A = \begin{bmatrix} 3 & -1 & \vdots & 1 \\ 1 & 2 & \vdots & 0 \end{bmatrix}$$

A is called an *augmented matrix*. The augmented matrix can be used to "maintain a record" of the coefficients of the linear equations when the elimination method is applied to solve (A). ($\downarrow$ will be used to indicate that the systems are equivalent.)

System	*Matrix form of system*
(A) $\begin{cases} 3x - y = 1 \\ x + 2y = 0 \end{cases}$	$A = \begin{bmatrix} 3 & -1 & \vdots & 1 \\ 1 & 2 & \vdots & 0 \end{bmatrix}$
$\downarrow$ First, multiply the second equation by -3	$\downarrow$ First, multiply the second row by -3
(B) $\begin{cases} 3x - y = 1 \\ -3x - 6y = 0 \end{cases}$	$B = \begin{bmatrix} 3 & -1 & \vdots & 1 \\ -3 & -6 & \vdots & 0 \end{bmatrix}$
$\downarrow$ Next, replace equation two with the sum of equations one and two	$\downarrow$ Next, replace row two with the sum of rows one and two.
(C) $\begin{cases} 3x - y = 1 \\ - 7y = 1 \end{cases}$	$C = \begin{bmatrix} 3 & -1 & \vdots & 1 \\ 0 & -7 & \vdots & 1 \end{bmatrix}$
$\downarrow$ Multiply equation two by $-\frac{1}{7}$.	$\downarrow$ Multiply row two by $-\frac{1}{7}$.
(D) $\begin{cases} 3x - y = 1 \\ y = -\frac{1}{7} \end{cases}$	$D = \begin{bmatrix} 3 & -1 & \vdots & 1 \\ 0 & 1 & \vdots & -\frac{1}{7} \end{bmatrix}$
$\downarrow$ Replace equation one with the sum of equations one and two.	$\downarrow$ Replace row one with the sum of rows one and two.
(E) $\begin{cases} 3x = \frac{6}{7} \\ y = -\frac{1}{7} \end{cases}$	$E = \begin{bmatrix} 3 & 0 & \vdots & \frac{6}{7} \\ 0 & 1 & \vdots & -\frac{1}{7} \end{bmatrix}$
$\downarrow$ Multiply equation one by $\frac{1}{3}$.	$\downarrow$ Multiply row one by $\frac{1}{3}$.
(F) $\begin{cases} x = \frac{2}{7} \\ y = -\frac{1}{7} \end{cases}$	$F = \begin{bmatrix} 1 & 0 & \vdots & \frac{2}{7} \\ 0 & 1 & \vdots & -\frac{1}{7} \end{bmatrix}$

Hence the solution of (A) is $\{(\frac{2}{7}, -\frac{1}{7})\}$.

Realizing that the numbers to the left of the dashed line represent the coefficients of x and y, we have $\{(\frac{2}{7}, -\frac{1}{7})\}$ as the solution set.

EXAMPLE

Solve the system of linear equations

$$(A) \quad \begin{cases} x_1 - 2x_2 + 3x_3 = -1 \\ 2x_1 - x_2 + 2x_3 = 2 \\ 3x_1 + x_2 + 2x_3 = 3 \end{cases}$$

by using the elimination method and displaying the corresponding matrix.

SOLUTION.   The matrix form of system (A) is written as

$$A = \begin{bmatrix} 1 & -2 & 3 & \vdots & -1 \\ 2 & -1 & 2 & \vdots & 2 \\ 3 & 1 & 2 & \vdots & 3 \end{bmatrix}$$

The elimination method proceeds as follows.

System	*Matrix form of system*
$(A) \quad \begin{cases} x_1 - 2x_2 + 3x_3 = -1 \\ 2x_1 - x_2 + 2x_3 = 2 \\ 3x_1 + x_2 + 2x_3 = 3 \end{cases}$	$A = \begin{bmatrix} 1 & -2 & 3 & \vdots & -1 \\ 2 & -1 & 2 & \vdots & 2 \\ 3 & 1 & 2 & \vdots & 3 \end{bmatrix}$

First, replace the second equation by the sum of -2 times the first equation and the second equation.

First, replace the second row by the sum of -2 times the first row and the second row.

| $(B) \quad \begin{cases} x_1 - 2x_2 + 3x_3 = -1 \\ 3x_2 - 4x_3 = 4 \\ 3x_1 + x_2 + 2x_3 = 3 \end{cases}$ | $B = \begin{bmatrix} 1 & -2 & 3 & \vdots & -1 \\ 0 & 3 & -4 & \vdots & 4 \\ 3 & 1 & 2 & \vdots & 3 \end{bmatrix}$ |

Next, replace equation three with the sum of -3 times equation one and the third equation.

Next, replace row three with the sum of -3 times row one and row three.

(C) $\begin{cases} x_1 - 2x_2 + 3x_3 = -1 \\ \quad\quad 3x_2 - 4x_3 = 4 \\ \quad\quad 7x_2 - 7x_3 = 6 \end{cases}$

$$C = \begin{bmatrix} 1 & -2 & 3 & \vdots & -1 \\ 0 & 3 & -4 & \vdots & 4 \\ 0 & 7 & -7 & \vdots & 6 \end{bmatrix}$$

↓ Multiply equation two by $\frac{1}{3}$. ↓ Multiply row two by $\frac{1}{3}$.

(D) $\begin{cases} x_1 - 2x_2 + 3x_3 = -1 \\ \quad\quad x_2 - \frac{4}{3}x_3 = \frac{4}{3} \\ \quad\quad 7x_2 - 7x_3 = 6 \end{cases}$

$$D = \begin{bmatrix} 1 & -2 & 3 & \vdots & -1 \\ 0 & 1 & -\frac{4}{3} & \vdots & \frac{4}{3} \\ 0 & 7 & -7 & \vdots & 6 \end{bmatrix}$$

Now, replace the third equation with -7 times the second equation plus the third equation. Now, replace the third row with -7 times the second row plus the third row.

(E) $\begin{cases} x_1 - 2x_2 + 3x_3 = -1 \\ \quad\quad x_2 - \frac{4}{3}x_3 = \frac{4}{3} \\ \quad\quad\quad \frac{7}{3}x_3 = -\frac{10}{3} \end{cases}$

$$E = \begin{bmatrix} 1 & -2 & 3 & \vdots & -1 \\ 0 & 1 & -\frac{4}{3} & \vdots & \frac{4}{3} \\ 0 & 0 & \frac{7}{3} & \vdots & -\frac{10}{3} \end{bmatrix}$$

Next replace equation one with 2 times equation two plus equation one. Next replace row one with 2 times row two plus row one.

(F) $\begin{cases} x_1 \quad\quad + \frac{1}{3}x_3 = \frac{5}{3} \\ \quad\quad x_2 - \frac{4}{3}x_3 = \frac{4}{3} \\ \quad\quad\quad \frac{7}{3}x_3 = -\frac{10}{3} \end{cases}$

$$F = \begin{bmatrix} 1 & 0 & \frac{1}{3} & \vdots & \frac{5}{3} \\ 0 & 1 & -\frac{4}{3} & \vdots & \frac{4}{3} \\ 0 & 0 & \frac{7}{3} & \vdots & -\frac{10}{3} \end{bmatrix}$$

↓ Multiply equation three by $\frac{3}{7}$. ↓ Multiply row three by $\frac{3}{7}$.

(G) $\begin{cases} x_1 \quad\quad + \frac{1}{3}x_3 = \frac{5}{3} \\ \quad\quad x_2 - \frac{4}{3}x_3 = \frac{4}{3} \\ \quad\quad\quad x_3 = -\frac{10}{7} \end{cases}$

$$G = \begin{bmatrix} 1 & 0 & \frac{1}{3} & \vdots & \frac{5}{3} \\ 0 & 1 & -\frac{4}{3} & \vdots & \frac{4}{3} \\ 0 & 0 & 1 & \vdots & -\frac{10}{7} \end{bmatrix}$$

Now replace equation one with the sum of $-\frac{1}{3}$ times equation three and equation one. Now replace row one with the sum of $-\frac{1}{3}$ times row three and row one.

(H) $\begin{cases} x_1 \quad\quad = \frac{15}{7} \\ \quad\quad x_2 - \frac{4}{3}x_3 = \frac{4}{3} \\ \quad\quad\quad x_3 = -\frac{10}{7} \end{cases}$

$$H = \begin{bmatrix} 1 & 0 & 0 & \vdots & \frac{15}{7} \\ 0 & 1 & -\frac{4}{3} & \vdots & \frac{4}{3} \\ 0 & 0 & 1 & \vdots & -\frac{10}{7} \end{bmatrix}$$

Finally, replace equation two with the sum of $\frac{4}{3}$ times equation three and equation two. Finally, replace row two with the sum of $\frac{4}{3}$ times row three and row two.

$$\text{(I)} \quad \begin{cases} x_1 & = \frac{15}{7} \\ & x_2 & = -\frac{4}{7} \\ & & x_3 = -\frac{10}{7} \end{cases} \qquad I = \begin{bmatrix} 1 & 0 & 0 & \vdots & \frac{15}{7} \\ 0 & 1 & 0 & \vdots & -\frac{4}{7} \\ 0 & 0 & 1 & \vdots & -\frac{10}{7} \end{bmatrix}$$

Hence the solution is
$\{(\frac{15}{7}, -\frac{4}{7}, -\frac{10}{7})\}$.

Hence the solution set can be read from the augmented column as
$\{(\frac{15}{7}, -\frac{4}{7}, -\frac{10}{7})\}$.

Notice that in solving the system of linear equations (A), the system was replaced by an augmented matrix A whose elements are the coefficients and constants occurring in the equations. Notice also that we can work with the augmented matrix instead of the actual equations by performing the following operations on the augmented matrix [compare with equation operations (1) through (3) on page 145]:

1′ Interchange two rows of the matrix $(R_i \leftrightarrow R_j)$.

2′ Multiply the elements in a row of the matrix by a nonzero number $(R_i \rightarrow kR_i, k \neq 0)$.

3′ Replace a row with the sum of a nonzero multiple of itself and a nonzero multiple of another row $(R_i \rightarrow kR_j + cR_i, c \neq 0$ and $k \neq 0)$.

The operations (1′), (2′), and (3′) are called the *elementary row operations*. The resulting matrix I in the above example is called a **row-reduced echelon matrix**. In general, a matrix is a *row-reduced echelon* matrix if all of the following conditions hold:

i   The first nonzero entry in each row is 1; all other entries in that column are zeros.

ii   Each row that consists entirely of zeros is below each row that contains a nonzero entry.

iii   The first nonzero entry in each row is to the right of the first nonzero entry in the preceding row.

EXAMPLES

Solve the linear system (A) by performing the elementary row operations if the system has a unique solution for each of the following systems.

1   (A) $\begin{cases} 4x + y = 2 \\ x + 4y = 1 \end{cases}$

SOLUTION.   The matrix form of the system is

$$\begin{bmatrix} 4 & 1 & \vdots & 2 \\ 1 & 4 & \vdots & 1 \end{bmatrix}$$

First, multiply row one by $\frac{1}{4}$ $(R_1 \rightarrow \frac{1}{4}R_1)$ to get

$$\left[\begin{array}{cc:c} 1 & \frac{1}{4} & \frac{1}{2} \\ 1 & 4 & 1 \end{array}\right]$$

Next, replace row two with the sum of row two and -1 times row one $[R_2 \rightarrow R_2 + (-1)R_1]$ to get

$$\left[\begin{array}{cc:c} 1 & \frac{1}{4} & \frac{1}{2} \\ 0 & \frac{15}{4} & \frac{1}{2} \end{array}\right]$$

Multiply row two by $\frac{4}{15}$ $(R_2 \rightarrow \frac{4}{15}R_2)$ to get

$$\left[\begin{array}{cc:c} 1 & \frac{1}{4} & \frac{1}{2} \\ 0 & 1 & \frac{2}{15} \end{array}\right]$$

Finally, replace row one with the sum of row one and $-\frac{1}{4}$ times row two $[R_1 \rightarrow R_1 + (-\frac{1}{4})R_2]$ to get

$$\left[\begin{array}{cc:c} 1 & 0 & \frac{7}{15} \\ 0 & 1 & \frac{2}{15} \end{array}\right]$$

Hence the solution set is $\{(\frac{7}{15}, \frac{2}{15})\}$.

2 (A) $\quad \begin{cases} x_1 + 2x_2 - 3x_3 = 6 \\ 2x_1 - x_2 + 4x_3 = 2 \\ 4x_1 + 3x_2 - 2x_3 = 14 \end{cases}$

SOLUTION. The matrix form of the system is

$$\left[\begin{array}{ccc:c} 1 & 2 & -3 & 6 \\ 2 & -1 & 4 & 2 \\ 4 & 3 & -2 & 14 \end{array}\right]$$

First, replace the second row by the sum of -2 times the first row and the second row $[R_2 \rightarrow (-2)R_1 + R_2]$ to obtain the equivalent matrix

$$\left[\begin{array}{ccc:c} 1 & 2 & -3 & 6 \\ 0 & -5 & 10 & -10 \\ 4 & 3 & -2 & 14 \end{array}\right]$$

Next, replace the third row by the sum of -4 times the first row and the third row $[R_3 \to (-4)R_1 + R_3]$ to obtain

$$\begin{bmatrix} 1 & 2 & 3 & \vdots & 6 \\ 0 & -5 & 10 & \vdots & -10 \\ 0 & -5 & 10 & \vdots & -10 \end{bmatrix}$$

Since the second row R_2 and the third row R_3 are identical, the corresponding linear system will have two identical equations,

$$-5x_2 + 10x_3 = -10$$

In this case, the system is a dependent system, and the system does not have a unique solution.

3   (A) $\begin{cases} 3x_1 - x_2 + x_3 = 1 \\ 7x_1 + x_2 - x_3 = 6 \\ 2x_1 + x_2 - x_3 = 2 \end{cases}$

SOLUTION.   The augmented matrix that corresponds to the linear system is given by

$$\begin{bmatrix} 3 & -1 & 1 & \vdots & 1 \\ 7 & 1 & -1 & \vdots & 6 \\ 2 & 1 & -1 & \vdots & 2 \end{bmatrix}$$

which can be reduced by the following row operations:

$$\begin{bmatrix} 3 & -1 & 1 & \vdots & 1 \\ 7 & 1 & -1 & \vdots & 6 \\ 2 & 1 & -1 & \vdots & 2 \end{bmatrix} \xrightarrow[R_3 \to R_1 + R_3]{} \begin{bmatrix} 3 & -1 & 1 & \vdots & 1 \\ 7 & 1 & -1 & \vdots & 6 \\ 5 & 0 & 0 & \vdots & 3 \end{bmatrix}$$

$$\xrightarrow[R_2 \to R_1 + R_2]{} \begin{bmatrix} 3 & -1 & 1 & \vdots & 1 \\ 10 & 0 & 0 & \vdots & 7 \\ 5 & 0 & 0 & \vdots & 3 \end{bmatrix}$$

$$\xrightarrow[R_2 \to R_2 + (-2)R_3]{} \begin{bmatrix} 3 & -1 & 1 & \vdots & 1 \\ 0 & 0 & 0 & \vdots & 1 \\ 5 & 0 & 0 & \vdots & 3 \end{bmatrix}$$

But R_2 implies that $0 \cdot x_1 + 0 \cdot x_2 + 0 \cdot x_3 = 1$, that is, $0 = 1$, which, of course, is not possible. Hence the system has no solution, and the system is inconsistent.

In summary, the procedure for solving a linear system is to form the augmented matrix and proceed to reduce it to echelon form:

1 If a resulting augmented echelon matrix has no row with its first nonzero entry in the last column, then the system has one (unique) or more solutions (dependent).

2 If a resulting augmented echelon matrix has a row with its first nonzero entry appearing in the last column, then the system has no solution, and the system is said to be inconsistent.

PROBLEM SET 2

1 Solve each of the following systems of linear equations by use of the elimination method. In each step of the process show the corresponding matrix form of the system. Also, sketch the graph of each system in parts a) through f) on the same coordinate system.

a) $\begin{aligned} 3x + y &= 14 \\ 2x - y &= 1 \end{aligned}$

b) $\begin{aligned} 4x + 3y &= 15 \\ 3x + 5y &= 14 \end{aligned}$

c) $\begin{aligned} -2x + 3y &= 8 \\ 2x - y &= 5 \end{aligned}$

d) $\begin{aligned} x - 2y &= 5 \\ 3x - 6y &= 4 \end{aligned}$

e) $\begin{aligned} \tfrac{1}{2}x + \tfrac{1}{6}y &= \tfrac{2}{3} \\ 3x + y &= 4 \end{aligned}$

f) $\begin{aligned} x - 2y - 4 &= 0 \\ x + y + 3 &= 0 \end{aligned}$

g) $\begin{aligned} x + y + z &= 6 \\ 3x - y + 2z &= 7 \\ 2x + 3y - z &= 5 \end{aligned}$

h) $\begin{aligned} 2x + 3y + z &= 6 \\ x - 2y + 3z &= -3 \\ 3x + y - z &= 8 \end{aligned}$

i) $\begin{aligned} x + y + 2z &= 4 \\ x + y - 2z &= 0 \\ x - y &= 0 \end{aligned}$

j) $\begin{aligned} x + y + z &= 4 \\ x - y + 2z &= 8 \\ 2x + y - z &= 3 \end{aligned}$

k) $\begin{aligned} 2x + y - z &= 7 \\ y - x &= 1 \\ z - y &= 1 \end{aligned}$

l) $\begin{aligned} 3x + 2y + 2z &= 6 \\ x - 5y + 6z &= 2 \\ 6x - 8z &= 12 \end{aligned}$

2 Suppose that

$$A = \begin{bmatrix} -4 & 0 & 1 \\ 2 & 3 & -1 \\ 5 & 2 & 8 \end{bmatrix}$$

a) What is the size of matrix A?

b) Use subscript notation to identify each of the entries of matrix A.

3 Indicate which of the following augmented matrices are in row-reduced echelon form. Assuming that the matrix represents a linear system, what can be concluded about the solution of the system?

a) $\begin{bmatrix} 1 & 0 & 1 & \vdots & 3 \\ 0 & 1 & 0 & \vdots & 4 \\ 0 & 0 & 0 & \vdots & 0 \end{bmatrix}$

b) $\begin{bmatrix} 1 & 1 & 0 & \vdots & 0 \\ 0 & 1 & 0 & \vdots & 0 \\ 0 & 0 & 1 & \vdots & 2 \end{bmatrix}$

c) $\begin{bmatrix} 1 & 0 & 0 & \vdots & 3 \\ 0 & 0 & 1 & \vdots & 0 \\ 0 & 0 & 0 & \vdots & 5 \end{bmatrix}$

d) $\begin{bmatrix} 1 & 1 & \vdots & 2 \\ 0 & 0 & \vdots & 2 \end{bmatrix}$

e) $\begin{bmatrix} 1 & 0 & \vdots & 1 \\ 0 & 1 & \vdots & 1 \end{bmatrix}$

f) $\begin{bmatrix} 1 & 0 & 0 & 0 & \vdots & 3 \\ 0 & 1 & 0 & 0 & \vdots & -2 \\ 1 & 0 & 0 & 1 & \vdots & 1 \\ 0 & 0 & 1 & 0 & \vdots & 5 \end{bmatrix}$

g) $\begin{bmatrix} 0 & 1 & \vdots & -3 \\ 1 & 0 & \vdots & 4 \end{bmatrix}$

4 For each of the following systems find the augmented matrix, reduce it to a row-reduced echelon matrix, and determine the solution if it is unique. If the solution is not unique, indicate whether the system is dependent or inconsistent.

a) $\begin{aligned} x - 2y + z &= -1 \\ 3x + y - 2z &= 4 \\ y - z &= 1 \end{aligned}$

b) $\begin{aligned} x + y - 2z &= 3 \\ 3x - y + z &= 5 \\ 3x + 3y - 6z &= 9 \end{aligned}$

c) $\begin{aligned} 2x + y + z &= 1 \\ 4x + 2y + 3z &= 1 \\ -2x - y + z &= 2 \end{aligned}$

d) $\begin{aligned} x + y + z &= 0 \\ 2x - y - 4z &= 15 \\ x - 2y - z &= 7 \end{aligned}$

e) $\begin{aligned} 2x - 3y + z &= 4 \\ x - 4y - z &= 3 \\ x - 9y - 4z &= 5 \end{aligned}$

f) $\begin{aligned} 2x + 3y - z &= -2 \\ x - y + 2z &= 4 \end{aligned}$

4 Determinants

The study of linear functions has led us into an investigation of solving systems of linear equations. Before returning to the main topic of this chapter—the study of polynomial functions—we will study a function called the determinant, which can be used to solve linear systems that have unique solutions. The determinant is a function that associates with each square matrix a real number.

The determinant is quite difficult for the beginning student to understand if it is presented in the precise, formal way that is usually reserved for an advanced presentation of linear algebra. Since it is so difficult to digest the general definition, we will restrict our attention to the manipulative properties of the determinant and its application to solving linear systems.

The *determinant* is a function that has the *set of square matrices as its domain* and the *set of real numbers as its range.* If A is a square matrix, the determinant of A is denoted by $\det A$ or $|A|$.

The determinant of the 2×2 matrix

$$\begin{bmatrix} a_{11} & a_{12} \\ a_{21} & a_{22} \end{bmatrix}$$

is defined to be the number $a_{11}a_{22} - a_{21}a_{12}$, and we write

$$\det \begin{bmatrix} a_{11} & a_{12} \\ a_{21} & a_{22} \end{bmatrix} = a_{11}a_{22} - a_{21}a_{12}$$

Thus

$$\det \begin{bmatrix} 5 & 3 \\ -3 & -6 \end{bmatrix} = 5(-6) - (-3)(3) = -30 + 9 = -21$$

The determinant of the 3×3 matrix

$$\begin{bmatrix} a_{11} & a_{12} & a_{13} \\ a_{21} & a_{22} & a_{23} \\ a_{31} & a_{32} & a_{33} \end{bmatrix}$$

is defined as follows:

$$\det \begin{bmatrix} a_{11} & a_{12} & a_{13} \\ a_{21} & a_{22} & a_{23} \\ a_{31} & a_{32} & a_{33} \end{bmatrix}$$

$$= a_{11} \det \begin{bmatrix} a_{22} & a_{23} \\ a_{32} & a_{33} \end{bmatrix} - a_{12} \det \begin{bmatrix} a_{21} & a_{23} \\ a_{31} & a_{33} \end{bmatrix} + a_{13} \det \begin{bmatrix} a_{21} & a_{22} \\ a_{31} & a_{32} \end{bmatrix}$$

$$= a_{11}(a_{22}a_{33} - a_{32}a_{23}) - a_{12}(a_{21}a_{33} - a_{31}a_{23})$$
$$+ a_{13}(a_{21}a_{32} - a_{31}a_{22})$$

For example,

$$\det \begin{bmatrix} 3 & 2 & 7 \\ -1 & 5 & 3 \\ 2 & -3 & -6 \end{bmatrix}$$

$$= 3 \det \begin{bmatrix} 5 & 3 \\ -3 & -6 \end{bmatrix} - 2 \det \begin{bmatrix} -1 & 3 \\ 2 & -6 \end{bmatrix} + 7 \det \begin{bmatrix} -1 & 5 \\ 2 & -3 \end{bmatrix}$$

$$= 3(-30 + 9) - 2(6 - 6) + 7(3 - 10)$$
$$= 3(-21) - 2(0) + 7(-7)$$
$$= -63 - 49 = -112$$

(Although this function can be generalized to any square matrix, we will restrict ourselves here to 2×2 and 3×3 matrices only.)

4.1 Properties of Determinants

The determinant possesses properties that can be used to simplify the task of its evaluation. Although we will be restricting the investigation of these properties to 2×2 and 3×3 matrices, it is important to realize that the properties hold for evaluating the determinant of *any* square matrix.

THEOREM 1

A common factor that appears in all entries in a row of a matrix can be factored out of the row when evaluating the determinant of the matrix.

For example,

$$\begin{vmatrix} 6 & 9 \\ 1 & 4 \end{vmatrix} = 3 \begin{vmatrix} 2 & 3 \\ 1 & 4 \end{vmatrix}$$

and

$$\begin{vmatrix} 15 & 45 & 60 \\ 1 & 2 & -1 \\ 2 & 4 & 8 \end{vmatrix} = 15 \begin{vmatrix} 1 & 3 & 4 \\ 1 & 2 & -1 \\ 2 & 4 & 8 \end{vmatrix}$$

$$= (15)(2) \begin{vmatrix} 1 & 3 & 4 \\ 1 & 2 & -1 \\ 1 & 2 & 4 \end{vmatrix}$$

PROOF.   (For $n = 2$.)

$$\begin{vmatrix} a_{11} & a_{12} \\ a_{21} & a_{22} \end{vmatrix} = a_{11} a_{22} - a_{21} a_{12}$$

so that

$$\begin{vmatrix} ka_{11} & ka_{12} \\ a_{21} & a_{22} \end{vmatrix} = ka_{11} a_{22} - ka_{21} a_{12}$$

$$= k(a_{11} a_{22} - a_{21} a_{12})$$

$$= k \begin{vmatrix} a_{11} & a_{12} \\ a_{21} & a_{22} \end{vmatrix}$$

THEOREM 2

If two (not necessarily adjacent) rows of a square matrix are interchanged, the values of the determinants of the two matrices differ only in the algebraic sign.

For example,

$$\begin{vmatrix} 2 & 3 \\ 4 & 1 \end{vmatrix} = -10$$

whereas

$$\begin{vmatrix} 4 & 1 \\ 2 & 3 \end{vmatrix} = 10$$

and

$$\begin{vmatrix} 1 & -1 & 0 \\ 3 & 0 & 4 \\ 2 & 1 & 5 \end{vmatrix} = 3 \quad \text{(why?)}$$

whereas

$$\begin{vmatrix} 2 & 1 & 5 \\ 3 & 0 & 4 \\ 1 & -1 & 0 \end{vmatrix} = -3 \quad \text{(why?)}$$

PROOF. (For $n = 2$.)

$$\begin{vmatrix} a_{11} & a_{12} \\ a_{21} & a_{22} \end{vmatrix} = a_{11}a_{22} - a_{21}a_{12}$$

$$= -(a_{21}a_{12} - a_{11}a_{22})$$

$$= - \begin{vmatrix} a_{21} & a_{22} \\ a_{11} & a_{12} \end{vmatrix}$$

THEOREM 3

If any nonzero multiple of one row is added to any other row of a square matrix, the value of the determinant is unaltered.

For example, consider

$$\begin{vmatrix} 1 & 0 & 2 \\ 4 & 6 & 1 \\ -1 & 0 & -1 \end{vmatrix}$$

If we multiply the third row by 2 and add the result to the first row, we get

$$\begin{vmatrix} -1 & 0 & 0 \\ 4 & 6 & 1 \\ -1 & 0 & -1 \end{vmatrix}$$

and we are assured by the theorem that the latter determinant has the same value as the original. Note that this operation effects only the first row, whereas the other two rows remain the same. But we need not stop here; indeed, we can add the third row to the second row in the latter determinant to obtain

$$\begin{vmatrix} -1 & 0 & 0 \\ 3 & 6 & 0 \\ -1 & 0 & -1 \end{vmatrix}$$

Again, this does not change the value of the determinant. Determinants such as the last one, which contain many zero entries, are relatively easy to evaluate; hence the above theorem simplifies the task of calculating determinants.

For the proof of this theorem see Problem 4a.

EXAMPLES

Use the three theorems above to evaluate each of the following determinants.

$$1 \quad \begin{vmatrix} 1 & 0 & 2 \\ 4 & 6 & 1 \\ -1 & 0 & -1 \end{vmatrix}$$

SOLUTION

$$\begin{vmatrix} 1 & 0 & 2 \\ 4 & 6 & 1 \\ -1 & 0 & -1 \end{vmatrix} = \begin{vmatrix} -1 & 0 & 0 \\ 4 & 6 & 1 \\ -1 & 0 & -1 \end{vmatrix} \qquad \text{(Theorem 3)} \\ (R_1 \rightarrow R_1 + 2R_3)$$

$$= \begin{vmatrix} -1 & 0 & 0 \\ 3 & 6 & 0 \\ -1 & 0 & -1 \end{vmatrix} \qquad \text{(Theorem 3)} \\ (R_2 \rightarrow R_2 + R_3)$$

$$= \begin{vmatrix} -1 & 0 & 0 \\ 3 & 6 & 0 \\ 0 & 0 & -1 \end{vmatrix} \qquad \text{(Theorem 3)} \\ [R_3 \rightarrow R_3 + (-1)R_1]$$

$$= 3 \begin{vmatrix} -1 & 0 & 0 \\ 1 & 2 & 0 \\ 0 & 0 & -1 \end{vmatrix} \qquad \text{(Theorem 1)}$$

$$= (3)(-1) \begin{vmatrix} 1 & 0 & 0 \\ 1 & 2 & 0 \\ 0 & 0 & -1 \end{vmatrix} \qquad \text{(Theorem 1)}$$

$$= (3)(-1)(-1) \begin{vmatrix} 1 & 0 & 0 \\ 1 & 2 & 0 \\ 0 & 0 & 1 \end{vmatrix} \qquad \text{(Theorem 1)}$$

$$= 3 \begin{vmatrix} 1 & 0 & 0 \\ 0 & 2 & 0 \\ 0 & 0 & 1 \end{vmatrix} \qquad \text{(Theorem 3)} \\ [R_2 \to R_2 + (-1)R_1]$$

$$= (3)(2) \begin{vmatrix} 1 & 0 & 0 \\ 0 & 1 & 0 \\ 0 & 0 & 1 \end{vmatrix} \qquad \text{(Theorem 1)}$$

$$= 6 \cdot 1 \qquad \text{(why?)}$$

2   $\begin{vmatrix} 3 & 1 & -1 \\ 0 & 2 & 4 \\ -1 & 4 & 2 \end{vmatrix}$

SOLUTION

$$\begin{vmatrix} 3 & 1 & -1 \\ 0 & 2 & 4 \\ -1 & 4 & 2 \end{vmatrix} = 2 \begin{vmatrix} 3 & 1 & -1 \\ 0 & 1 & 2 \\ -1 & 4 & 2 \end{vmatrix} \qquad \text{(Theorem 1)}$$

$$= 2 \begin{vmatrix} 0 & 13 & 5 \\ 0 & 1 & 2 \\ -1 & 4 & 2 \end{vmatrix} \qquad \begin{array}{l} \text{(Theorem 3)} \\ (R_1 \to R_1 + 3R_3) \end{array}$$

$$= 2 \begin{vmatrix} 0 & 13 & 5 \\ 0 & 1 & 2 \\ -1 & 3 & 0 \end{vmatrix} \qquad \begin{array}{l} \text{(Theorem 3)} \\ [R_3 \to R_3 + (-1)R_2] \end{array}$$

$$= (2)(-1) \begin{vmatrix} 0 & 13 & 5 \\ 0 & 1 & 2 \\ 1 & -3 & 0 \end{vmatrix} \qquad \text{(Theorem 1)}$$

$$= (2)(-1)(-1) \begin{vmatrix} 1 & -3 & 0 \\ 0 & 1 & 2 \\ 0 & 13 & 5 \end{vmatrix} \qquad \text{(Theorem 2)}$$

$$= 2 \begin{vmatrix} 1 & -3 & 0 \\ 0 & 1 & 2 \\ 0 & 0 & -21 \end{vmatrix} \qquad \begin{array}{l} \text{(Theorem 3)} \\ [R_3 \to R_3 + (-13)R_2] \end{array}$$

$$= (2)(-21) \begin{vmatrix} 1 & -3 & 0 \\ 0 & 1 & 2 \\ 0 & 0 & 1 \end{vmatrix} \qquad \text{(Theorem 1)}$$

$$= (-42)\left(1 \begin{vmatrix} 1 & 2 \\ 0 & 1 \end{vmatrix} + 3 \begin{vmatrix} 0 & 2 \\ 0 & 1 \end{vmatrix} + 0 \begin{vmatrix} 0 & 1 \\ 0 & 0 \end{vmatrix} \right)$$

$$= -42$$

4.2   Cramer's Rule

We have seen what the determinant function is, and we have investigated methods for computing determinants. *Cramer's rule* provides us with a technique for using determinants to solve systems of linear equations. Although Cramer's rule is not the most practical way to solve linear systems, we shall investigate the rule as an example of an application of determinants.

Before stating Cramer's rule, let us establish some useful notation. Suppose that we are given a linear system (S) containing the same number of equations as unknowns.

$$(S) \quad \begin{cases} a_{11}x_1 + a_{12}x_2 + \cdots + a_{1n}x_n = c_1 \\ a_{21}x_1 + a_{22}x_2 + \cdots + a_{2n}x_n = c_2 \\ \cdots\cdots\cdots\cdots\cdots\cdots\cdots\cdots\cdots \\ a_{n1}x_1 + a_{n2}x_2 + \cdots + a_{nn}x_n = c_n \end{cases}$$

The determinant of the matrix of coefficients occurring in the system is called the *determinant of the coefficient matrix* and will be denoted by D. Hence

$$D = \begin{vmatrix} a_{11} & a_{12} & \cdots & a_{1n} \\ a_{21} & a_{22} & \cdots & a_{2n} \\ \cdots\cdots\cdots\cdots\cdots \\ a_{n1} & a_{n2} & \cdots & a_{nn} \end{vmatrix}$$

D_j will be used to denote the determinant of the matrix obtained by replacing the jth column in D by the column of constant terms in the system, so that

$$\overset{\displaystyle\downarrow j\text{th column}}{D_j = \begin{vmatrix} a_{11} & a_{12} & \cdots & c_1 & \cdots & a_{1n} \\ a_{21} & a_{22} & \cdots & c_2 & \cdots & a_{2n} \\ \cdots\cdots\cdots\cdots\cdots\cdots\cdots \\ a_{n1} & a_{n2} & \cdots & c_n & \cdots & a_{nn} \end{vmatrix}}$$

For example, for

$$\begin{cases} 2x + 3y = 1 \\ x - y = 2 \end{cases}$$

$$D = \begin{vmatrix} 2 & 3 \\ 1 & -1 \end{vmatrix} = -5$$

$$D_1 = \begin{vmatrix} 1 & 3 \\ 2 & -1 \end{vmatrix} = -7$$

and

$$D_2 = \begin{vmatrix} 2 & 1 \\ 1 & 2 \end{vmatrix} = 3$$

and for

$$\begin{cases} 3x - y + 3z = 1 \\ x + y = 4 \\ -5x + 7y - 2z = -2 \end{cases}$$

$$D = \begin{vmatrix} 3 & -1 & 3 \\ 1 & 1 & 0 \\ -5 & 7 & -2 \end{vmatrix} = 28$$

$$D_1 = \begin{vmatrix} 1 & -1 & 3 \\ 4 & 1 & 0 \\ -2 & 7 & -2 \end{vmatrix} = 80$$

$$D_2 = \begin{vmatrix} 3 & 1 & 3 \\ 1 & 4 & 0 \\ -5 & -2 & -2 \end{vmatrix} = 32$$

and

$$D_3 = \begin{vmatrix} 3 & -1 & 1 \\ 1 & 1 & 4 \\ -5 & 7 & -2 \end{vmatrix} = -60$$

THEOREM 1 CRAMER'S RULE

Let (S) be the system of n linear equations in n unknowns described on page 160. Let D be the determinant of the coefficient matrix of (S). If $D \neq 0$, then the system (S) has exactly one solution given by

$$x_j = \frac{D_j}{D}, \qquad j = 1, 2, \ldots, n$$

where D_j is the determinant defined on page 160.

The proof of Cramer's rule is beyond the scope of this text. However we shall consider its application to cases in which $n = 2$ or 3.

For $n = 2$, Cramer's rule indicates that the solution of

$$\begin{cases} a_{11}x_1 + a_{12}x_2 = c_1 \\ a_{21}x_1 + a_{22}x_2 = c_2 \end{cases}$$

is given by

$$x_1 = \frac{\begin{vmatrix} c_1 & a_{12} \\ c_2 & a_{22} \end{vmatrix}}{\begin{vmatrix} a_{11} & a_{12} \\ a_{21} & a_{22} \end{vmatrix}} \quad \text{and} \quad x_2 = \frac{\begin{vmatrix} a_{11} & c_1 \\ a_{21} & c_2 \end{vmatrix}}{\begin{vmatrix} a_{11} & a_{12} \\ a_{21} & a_{22} \end{vmatrix}}$$

if

$$\begin{vmatrix} a_{11} & a_{12} \\ a_{21} & a_{22} \end{vmatrix} \neq 0$$

For $n = 3$, Cramer's rule can be applied to

$$\begin{cases} a_{11}x_1 + a_{12}x_2 + a_{13}x_3 = c_1 \\ a_{21}x_1 + a_{22}x_2 + a_{23}x_3 = c_2 \\ a_{31}x_1 + a_{32}x_2 + a_{33}x_3 = c_3 \end{cases}$$

as follows.

First, we determine

$$D = \begin{vmatrix} a_{11} & a_{12} & a_{13} \\ a_{21} & a_{22} & a_{23} \\ a_{31} & a_{32} & a_{33} \end{vmatrix} \qquad D_1 = \begin{vmatrix} c_1 & a_{12} & a_{13} \\ c_2 & a_{22} & a_{23} \\ c_3 & a_{32} & a_{33} \end{vmatrix}$$

$$D_2 = \begin{vmatrix} a_{11} & c_1 & a_{13} \\ a_{21} & c_2 & a_{23} \\ a_{31} & c_3 & a_{33} \end{vmatrix} \qquad \text{and} \qquad D_3 = \begin{vmatrix} a_{11} & a_{12} & c_1 \\ a_{21} & a_{22} & c_2 \\ a_{31} & a_{32} & c_3 \end{vmatrix}$$

If $D \neq 0$, then $x_1 = D_1/D$, $x_2 = D_2/D$, and $x_3 = D_3/D$.

Notice that Cramer's rule has nothing to say about the existence of solutions in the case in which $D = 0$. Actually, if $D = 0$, either the system has no solution (inconsistent) or the system has an infinite number of different solutions (dependent).

EXAMPLES

Use Cramer's rule to solve each of the following linear systems, if possible.

1 $\begin{cases} 3x + 4y = 12 \\ 3x - 8y = 0 \end{cases}$

SOLUTION. Here

$$D = \begin{vmatrix} 3 & 4 \\ 3 & -8 \end{vmatrix} = -24 - 12 = -36$$

$$D_1 = \begin{vmatrix} 12 & 4 \\ 0 & -8 \end{vmatrix} = -96 \qquad \text{and} \qquad D_2 = \begin{vmatrix} 3 & 12 \\ 3 & 0 \end{vmatrix} = -36$$

Since $D \neq 0$, this system has one solution $\{(\frac{8}{3}, 1)\}$, since $x = D_1/D = -96/-36 = \frac{8}{3}$ and $y = D_2/D = -36/-36 = 1$ satisfy both equations simultaneously.

2
$$\begin{cases} x + 2y = 4 \\ 3x + 6y = -3 \end{cases}$$

SOLUTION. Here

$$D = \begin{vmatrix} 1 & 2 \\ 3 & 6 \end{vmatrix} = 0$$

Since $D = 0$, Cramer's rule is not applicable. This situation is consistent with the geometry of the two lines (they are parallel).

3
$$\begin{cases} x_1 + x_2 + x_3 = 2 \\ 2x_1 - x_2 + x_3 = 0 \\ x_1 + 2x_2 - x_3 = 4 \end{cases}$$

SOLUTION. Here

$$D = \begin{vmatrix} 1 & 1 & 1 \\ 2 & -1 & 1 \\ 1 & 2 & -1 \end{vmatrix} = 7 \qquad D_1 = \begin{vmatrix} 2 & 1 & 1 \\ 0 & -1 & 1 \\ 4 & 2 & -1 \end{vmatrix} = 6$$

$$D_2 = \begin{vmatrix} 1 & 2 & 1 \\ 2 & 0 & 1 \\ 1 & 4 & -1 \end{vmatrix} = 10 \quad \text{and} \quad D_3 = \begin{vmatrix} 1 & 1 & 2 \\ 2 & -1 & 0 \\ 1 & 2 & 4 \end{vmatrix} = -2$$

Since $D \neq 0$, this system has exactly one solution $\{(\frac{6}{7}, \frac{10}{7}, -\frac{2}{7})\}$. That is, $x_1 = D_1/D = \frac{6}{7}$, $x_2 = D_2/D = \frac{10}{7}$, and $x_3 = D_3/D = -\frac{2}{7}$ satisfy the system.

PROBLEM SET 3

1 Evaluate each of the following determinants.

a) $\begin{vmatrix} -1 & 3 \\ -7 & 4 \end{vmatrix}$

b) $\begin{vmatrix} 2 & 3 \\ 9 & 4 \end{vmatrix}$

c) $\begin{vmatrix} 2 & -1 & 3 \\ 9 & -7 & 4 \\ 11 & -6 & 2 \end{vmatrix}$

d) $\begin{vmatrix} 3 & -1 & 2 \\ 0 & 1 & -5 \\ 6 & 7 & 4 \end{vmatrix}$

e) $\begin{vmatrix} 2 & 2 & 2 \\ 3 & 3 & 3 \\ 4 & 4 & 4 \end{vmatrix}$

f) $\begin{vmatrix} \frac{1}{2} & 4 & 7 \\ 1 & -1 & 2 \\ 3 & 2 & 5 \end{vmatrix}$

2 Solve the following for x.

a) $\begin{vmatrix} x & -x \\ 5 & 3 \end{vmatrix} = 2$

b) $\begin{vmatrix} x & 4 & 5 \\ 0 & 1 & x \\ 5 & 2 & 1 \end{vmatrix} = 7$

c) $\begin{vmatrix} x & 0 & 0 \\ 3 & 1 & 2 \\ 0 & 4 & 1 \end{vmatrix} = 5$

d) $\begin{vmatrix} x & 0 & 1 \\ 2x & 1 & 2 \\ 3x & 2 & 3 \end{vmatrix} = 0$

3 Show why each of the following is true, not by evaluating each side but by citing the appropriate theorems of Section 4.1 that have been used.

a) $\begin{vmatrix} 4 & 5 \\ 3 & -2 \end{vmatrix} = - \begin{vmatrix} 3 & -2 \\ 4 & 5 \end{vmatrix}$

b) $\begin{vmatrix} 3 & 0 & 1 \\ 1 & 1 & 2 \\ 3 & 0 & 1 \end{vmatrix} = \begin{vmatrix} 0 & 0 & 0 \\ 1 & 1 & 2 \\ 3 & 0 & 1 \end{vmatrix}$

c) $\begin{vmatrix} 3 & -6 & 2 \\ 5 & -3 & 0 \\ 0 & 9 & 18 \end{vmatrix} = 9 \begin{vmatrix} 3 & -6 & 2 \\ 5 & -3 & 0 \\ 0 & 1 & 2 \end{vmatrix}$

d) $\begin{vmatrix} 2 & 4 & 12 \\ -1 & 0 & 3 \\ 1 & 0 & 6 \end{vmatrix} = 18 \begin{vmatrix} 1 & 2 & 6 \\ -1 & 0 & 3 \\ 0 & 0 & 1 \end{vmatrix}$

e) $\begin{vmatrix} 1 & 1 & 1 \\ 3 & 3 & 3 \\ 2 & 2 & 2 \end{vmatrix} = 6 \begin{vmatrix} 0 & 0 & 0 \\ 0 & 0 & 0 \\ 1 & 1 & 1 \end{vmatrix}$

4 a) Prove Theorem 3 of Section 4.1 for $n = 2$. (*Hint:* Compare

$$\begin{vmatrix} a_{11} & a_{12} \\ a_{21} & a_{22} \end{vmatrix} \quad \text{with} \quad \begin{vmatrix} a_{11}+ca_{21} & a_{12}+ca_{22} \\ a_{21} & a_{22} \end{vmatrix}.)$$

  b) Prove, for $n = 2$, that if two rows of a matrix are the same, then the determinant is zero.

  c) Prove, for $n = 2$, that if all the entries in one row of a matrix are zeros, then the determinant is zero.

5 Use the theorems given in Section 4.1 to evaluate each of the following determinants.

a) $\begin{vmatrix} -1 & 0 & 2 \\ 0 & 0 & 0 \\ -1 & 5 & 1 \end{vmatrix}$

b) $\begin{vmatrix} 3 & 1 & 1 \\ -1 & 0 & 3 \\ 2 & 1 & 1 \end{vmatrix}$

c) $\begin{vmatrix} 2 & 1 & 3 \\ 1 & 2 & 1 \\ 4 & 0 & 0 \end{vmatrix}$

d) $\begin{vmatrix} 20 & 12 & 8 \\ 5 & 3 & 2 \\ 5 & 7 & 2 \end{vmatrix}$

e) $\begin{vmatrix} 1 & 0 & 2 \\ 1 & -3 & 0 \\ 0 & 3 & 1 \end{vmatrix}$

6 a) *Principle of Duality.* Consider Theorems 1, 2, and 3 in Section 4.1. If we replace the word "row" with the word "column," then the theorems are still true. Rewrite the three theorems with this substitution.

b) Use the theorems of part a to evaluate the determinants in Problem 5.

7 Use Cramer's rule to solve each of the following systems, if possible.

a) $2x - y = 0$
$x + y = 1$

b) $-3x + y = 3$
$-2x - y = -5$

c) $x + y = 0$
$x - y = 0$

d) $3x + y = 1$
$9x + 3y = -4$

e) $2x_1 - x_2 + x_3 = 3$
$-x_1 + 2x_2 - x_3 = 1$
$3x_1 + x_2 + 2x_3 = -1$

f) $3x + 2z = 8 - 2y$
$x - 5y + 6z = 8$
$6x - 8z = 4$

g) $x + y + 2z = 4$
$x + y - 2z = 0$
$x - y = 0$

h) $2x_1 - 3x_2 = 4$
$x_1 + x_2 - 2x_3 = 1$
$x_1 - x_2 - x_3 = 5$

i) $x + y + z = 4$
$x - y + 2z = 8$
$2x + y - z = 3$

j) $2x + 3y + z = 6$
$x - 2y + 3z = -3$
$3x + y - z = 8$

8 Given the triangle of Figure 1, show that the area of the triangle is given by

$$\frac{1}{2}\begin{vmatrix} x_1 & y_1 & 1 \\ 0 & 0 & 1 \\ x_2 & 0 & 1 \end{vmatrix}$$

Figure 1

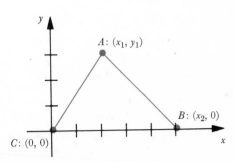

$A: (x_1, y_1)$

$B: (x_2, 0)$

$C: (0, 0)$

9 For each of the following problems, set up a linear system that will serve as a model of the situation. Then use Cramer's rule to solve the system.

a) A watch, chain, and ring together cost $225. The watch costs $50 more than the chain, and the ring costs $25 more than the watch and the chain together. What is the cost of each?

b) Twice the sum of two numbers is 30, and three times the smaller equals twice the larger. Determine the numbers.

5   Quadratic Functions

Let us return to the study of polynomial functions. A *quadratic function* is a polynomial function of degree 2. Hence, if $y = f(x)$ is a quadratic function, the standard form of f is given by $f(x) = ax^2 + bx + c$, where a, b, and c are real numbers, $a \neq 0$. A quadratic function can also be considered in set form as $\{(x,y) \mid y = ax^2 + bx + c,\ a \neq 0,\ a, b, c \in R\}$. In graphing any quadratic function, we will locate the x intercepts of f; that is, we will identify those real number values of x for which $f(x) = 0$. For example, since $x = -7$ and $x = -1$ satisfy the equation $x^2 + 8x + 7 = 0$, the graph of $f(x) = x^2 + 8x + 7$ intersects the x axis at points $(-7, 0)$ and $(-1, 0)$; $x = -2$ satisfies the equation $x^2 + 4x + 4 = 0$ so that the graph of $f(x) = x^2 + 4x + 4$ intersects the x axis at the point $(-2, 0)$. Since there is no real number x that satisfies the equation $x^2 + 1 = 0$, the graph of $f(x) = x^2 + 1$ does not intersect the x axis. In order to determine the x intercepts, it is necessary to solve equations of the form $ax^2 + bx + c = 0$.

5.1   Quadratic Equations

An equation that is equivalent to an equation of the form $ax^2 + bx + c = 0$, where a, b, and c are real numbers, $a \neq 0$, is called a *quadratic equation in x*. For example, $3x^2 = 7x + 3$ and $7x^2 = 5$ are quadratic equations because the first can be expressed as $3x^2 + (-7)x + (-3) = 0$, and the second as $7x^2 + 0x + (-5) = 0$.

The *solution* or *solution set* of a quadratic equation is the set of all possible roots. A *root* of an equation in one unknown is a value which, when substituted for the unknown, yields a true statement.

Quadratic equations can be solved by the following three methods.

1   FACTOR METHOD

We know from the algebra of real numbers that certain quadratic expressions can be represented in *factored form*, that is, as the product of two linear expressions containing real numbers. This factorization can be used to solve quadratic equations by applying the following rule:

If $ab = 0$ then $a = 0$ or $b = 0$

For example, the equation $x^2 - 5x + 4 = 0$ may be solved in the following manner:

$$\{x \mid x^2 - 5x + 4 = 0\} = \{x \mid (x-1)(x-4) = 0\}$$
$$= \{x \mid x-1 = 0 \text{ or } x-4 = 0\}$$
$$= \{x \mid x-1 = 0\} \cup \{x \mid x-4 = 0\}$$
$$= \{1, 4\}$$

EXAMPLE

Solve $2x^2 - 5x + 2 = 0$ by the factor method.

SOLUTION

$$
\begin{aligned}
\{x \mid 2x^2 - 5x + 2 = 0\} &= \{x \mid (2x-1)(x-2) = 0\} \\
&= \{x \mid 2x-1 = 0 \text{ or } x-2 = 0\} \\
&= \{x \mid 2x-1 = 0\} \cup \{x \mid x-2 = 0\} \\
&= \{\tfrac{1}{2}\} \cup \{2\} \\
&= \{\tfrac{1}{2}, 2\}
\end{aligned}
$$

2 COMPLETING-THE-SQUARE METHOD

Suppose that we are to solve a quadratic equation that is not readily factorable. Let us say, for example, that we are to solve $3x^2 - 2x - 2 = 0$. This quadratic equation can be solved by a process known as *completing the square*, which proceeds as follows.

First, "isolate" the x terms of

$$3x^2 - 2x - 2 = 0$$

to get

$$3x^2 - 2x = 2$$

Next, change the resulting equation to an equivalent equation that has 1 as the coefficient of the x^2 term to get

$$x^2 - \tfrac{2}{3}x = \tfrac{2}{3}$$

Finally, make the left-hand side a "perfect square" by adding the appropriate number. In order to form a perfect square on the left side, take one-half the coefficient of x, square it, and then add the result to both sides of the equation to get

$$
\begin{aligned}
x^2 - \tfrac{2}{3}x + (\tfrac{1}{3})^2 &= \tfrac{2}{3} + (\tfrac{1}{3})^2 \\
x^2 - \tfrac{2}{3}x + \tfrac{1}{9} &= \tfrac{7}{9}
\end{aligned}
$$

or

$$(x - \tfrac{1}{3})^2 = \tfrac{7}{9}$$

This implies, then, that

$$x - \tfrac{1}{3} = \sqrt{\tfrac{7}{9}} \quad \text{or} \quad x - \tfrac{1}{3} = -\sqrt{\tfrac{7}{9}}$$

Hence

$$x = \frac{1}{3} + \frac{\sqrt{7}}{3} \quad \text{or} \quad x = \frac{1}{3} - \frac{\sqrt{7}}{3}$$

are the solutions of $3x^2 - 2x - 2 = 0$, and the solution set is

$$\{(1+\sqrt{7})/3, \ (1-\sqrt{7})/3\}.$$

EXAMPLE

Solve $x^2 + 4x + 2 = 0$ by the method of completing the square.

SOLUTION

$$x^2 + 4x + 2 = 0$$

so that

$$x^2 + 4x = -2$$

Now, if the square of one-half the coefficient of the x term, namely $[\frac{1}{2} \cdot 4]^2 = 4$, is added to each side of the equation, we have

$$x^2 + 4x + [\tfrac{1}{2}(4)]^2 = -2 + [\tfrac{1}{2}(4)]^2$$

so that

$$x^2 + 4x + 4 = -2 + 4$$

or

$$x^2 + 4x + 4 = 2$$

The left-hand side of the equation is the square of $(x+2)$; thus the equation can be written as

$$(x+2)^2 = 2$$

which implies that

$$x + 2 = \pm\sqrt{2}$$

so that

$$x = -2 + \sqrt{2} \qquad \text{or} \qquad x = -2 - \sqrt{2}$$

That is, the solution set is $\{-2-\sqrt{2}, \ -2+\sqrt{2}\}$.

Instead of repeating the process of completing the square for each quadratic equation that is not easily factorable, we can generalize the method of completing the square to arrive at a formula that enables us to solve any quadratic equation.

3 THEOREM QUADRATIC FORMULA

If $ax^2 + bx + c = 0$, with a, b, and c real numbers, $a \neq 0$, then

$$x = \frac{-b \pm \sqrt{b^2 - 4ac}}{2a}$$

are the roots of the equation.

PROOF. (Follow the method of completing the square given above to understand this proof.)

$$ax^2 + bx + c = 0 \qquad a \neq 0$$

$$ax^2 + bx = -c$$

$$x^2 + \frac{bx}{a} = \frac{-c}{a}$$

$$x^2 + \frac{bx}{a} + \left(\frac{b}{2a}\right)^2 = \frac{-c}{a} + \left(\frac{b}{2a}\right)^2$$

$$\left(x + \frac{b}{2a}\right)^2 = \frac{b^2}{4a^2} + \frac{-c}{a} = \frac{b^2 - 4ac}{4a^2}$$

$$\left| x + \frac{b}{2a} \right| = \sqrt{\frac{b^2 - 4ac}{4a^2}}$$

Hence

$$x + \frac{b}{2a} = \sqrt{\frac{b^2 - 4ac}{4a^2}} \qquad \text{or} \qquad x + \frac{b}{2a} = -\sqrt{\frac{b^2 - 4ac}{4a^2}}$$

From this we get

$$x = \frac{-b}{2a} + \frac{\sqrt{b^2 - 4ac}}{2a} \qquad \text{or} \qquad x = \frac{-b}{2a} - \frac{\sqrt{b^2 - 4ac}}{2a}$$

However,

$$x = \frac{-b \pm \sqrt{b^2 - 4ac}}{2a}$$

is usually used as an abbreviated way of writing these two possible roots.

The number $b^2 - 4ac$ is called the *discriminant*. The discriminant of a quadratic equation indicates the type of roots: If it is zero, there is only one real double root. (Why?) If it is positive, there are two distinct real roots. (Why?) If it is negative, there are no real roots. (Why?)

EXAMPLES

1 Solve $2x^2 - 8x + 3 = 0$ by the quadratic formula.

SOLUTION. Here $a = 2$, $b = -8$, and $c = 3$. Since the discriminant is 40, this equation has two distinct real roots. The solution set is

$$\{x \mid 2x^2 - 8x + 3 = 0\} = \left\{ x \mid x = \frac{-(-8) \pm \sqrt{64 - 4(2)(3)}}{2(2)} \right\} = \left\{ x \mid x = \frac{8 \pm \sqrt{40}}{4} \right\}$$

$$= \left\{ x \mid x = \frac{8 \pm 2\sqrt{10}}{4} \right\} = \left\{ x \mid x = \frac{2(4 \pm \sqrt{10})}{4} \right\}$$

$$= \left\{ x \mid x = \frac{4 \pm \sqrt{10}}{2} \right\} = \left\{ 2 + \frac{\sqrt{10}}{2}, \ 2 - \frac{\sqrt{10}}{2} \right\}$$

2 Solve $3x^2 - 8x + 5 = 0$ by the quadratic formula.

SOLUTION. $a = 3$, $b = -8$, $c = 5$. The discriminant is 4, and so the solution consists of two roots. The solution set is

$$\{x \mid 3x^2 - 8x + 5 = 0\} = \left\{ x \mid x = \frac{-(-8) \pm \sqrt{64 - 4(3)(5)}}{2(3)} \right\}$$

$$= \left\{ x \mid x = \frac{8 \pm \sqrt{4}}{6} \right\} = \{1, \tfrac{5}{3}\}$$

5.2 Properties of Quadratic Functions

The graph of any quadratic function $f(x) = ax^2 + bx + c$ will have the same general shape as the graph of $f(x) = x^2$ (Figure 1), although the location of the graph will vary, depending upon the specific values of a, b, and c.

Figure 1

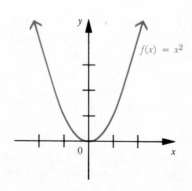

Such graphs are called *parabolas*. (A detailed discussion of parabolas appears in Chapter 9, Section 6.) Specific examples of quadratic functions are shown in Figure 2. Notice that the parabola opens upward and has a low point or *minimum point* when the coefficient of the x^2 term is positive (Figures 2*a* and *c*). The parabola opens downward and has a high point or *maximum point* when the coefficient of the x^2 term is negative (Figures 2*b* and *d*). This property, together with the location of intercepts, simplifies graphing all quadratic functions.

Figure 2

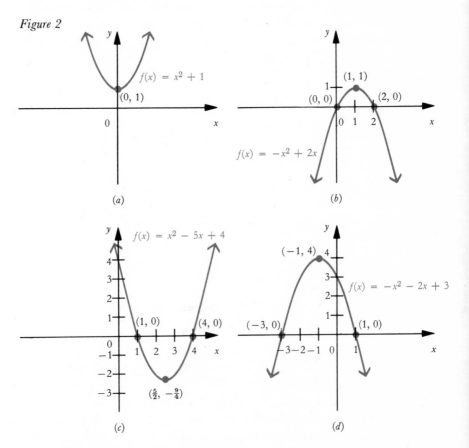

(a)

$f(x) = x^2 + 1$

(0, 1)

(b)

$f(x) = -x^2 + 2x$

(1, 1)

(0, 0)

(2, 0)

(c)

$f(x) = x^2 - 5x + 4$

(1, 0)

(4, 0)

$\left(\frac{5}{2}, -\frac{9}{4}\right)$

(d)

(-1, 4)

$f(x) = -x^2 - 2x + 3$

(-3, 0)

(1, 0)

For example, in order to graph $y = f(x) = 3x^2 - 5x + 2$, we notice that the solution of the quadratic equation $3x^2 - 5x + 2 = 0$ is $\frac{2}{3}$ or 1, so the x intercepts of f are the points $(\frac{2}{3}, 0)$ and $(1, 0)$. The y intercept can be obtained by setting $x = 0$ to get the point $(0, 2)$. The domain of f is the set of real numbers R. The range of the function f can be described easily if we determine the "extreme point." This can be done as follows: First, factor out 3 from the two terms involving x in the given quadratic function $y = 3x^2 - 5x + 2$ to get

$$y = 3\left(x^2 - \tfrac{5}{3}x\right) + 2$$

Next, complete the square of the expression $x^2 - \frac{5}{3}x$ to obtain

$$y = 3\left(x^2 - \tfrac{5}{3}x + \tfrac{25}{36}\right) + 2 - \tfrac{25}{12}$$

so that

$$y = 3\left(x - \tfrac{5}{6}\right)^2 - \tfrac{1}{12}$$

Since $3\left(x - \frac{5}{6}\right)^2 \geq 0$, it follows that $y \geq -\frac{1}{12}$. This means that $f(x)$ is always greater than or equal to $-\frac{1}{12}$, so $-\frac{1}{12}$ is a minimum value and the parabola will open upward. Notice that the minimum value for y occurs when $x - \frac{5}{6} = 0$, that is, when $x = \frac{5}{6}$. The graph of f (Figure 3) displays the minimum point $\left(\frac{5}{6}, -\frac{1}{12}\right)$. Thus the range of f is the set $\left[-\frac{1}{12}, \infty\right)$.

Figure 3

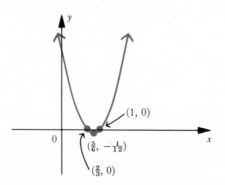

The next theorem generalizes the process used in this example and provides us with a simple way of locating the *extreme points* of parabolas.

THEOREM 1

The quadratic function $y = f(x) = ax^2 + bx + c$ has an extreme point $(-b/2a, f(-b/2a))$. If $a > 0$, the extreme point is a minimum point and the parabola opens upward. If $a < 0$, the extreme point is a maximum and the parabola opens downward.

PROOF. First we factor out a from the two terms involving x in the given quadratic function $y = ax^2 + bx + c$ to get

$$y = a\left(x^2 + \frac{b}{a}x\right) + c$$

Next we complete the square of the expression

$$x^2 + \frac{b}{a}x$$

to obtain

$$y = a\left(x^2 + \frac{b}{a}x + \frac{b^2}{4a^2}\right) + c - \frac{b^2}{4a}$$

so that

$$y = a\left(x + \frac{b}{2a}\right)^2 + c - \frac{b^2}{4a}$$

If $a > 0$, then $a\left(x + \frac{b}{2a}\right)^2 \geq 0$ since $\left(x + \frac{b}{2a}\right)^2 \geq 0$. Hence

(1) $\quad y = \left[a\left(x + \frac{b}{2a}\right)^2\right] + c - \frac{b^2}{4a} \geq c - \frac{b^2}{4a}$

That is, y is always larger than or equal to $c - b^2/4a$ if $a > 0$.

Similarly, if $a < 0$, then

(2) $\quad y \leq c - \frac{b^2}{4a}$ $\qquad$ (see Problem 4a)

That is, y is always less than or equal to $c - b^2/4a$ if $a < 0$.

Now, if we substitute $x = -b/2a$ into the given quadratic function $y = ax^2 + bx + c$, we get $y = a(b^2/4a^2) + b(-b/2a) + c = c - b^2/4a$, which is the same as the right-hand expressions in both (1) and (2). Consequently, if $a > 0$, $f(-b/2a) = c - b^2/4a$ is the *minimum* value of y; the graph of f, a *parabola*, opens upward as in Figure 4a, and its range is $[f(-b/2a), \infty)$. If $a < 0$, $f(-b/2a)$ is the *maximum* value of y; the graph of f, a parabola, opens downward as in Figure 4b, and its range is $(-\infty, f(-b/2a)]$. In either case, the value of the function $f(x) = ax^2 + bx + c$ at $x = -b/2a$ gives the *extreme* value of $y = f(x)$. Thus the *extreme point* of the graph of a quadratic function $y = f(x)$ is $(-b/2a, f(-b/2a))$ (Figure 4).

Figure 4

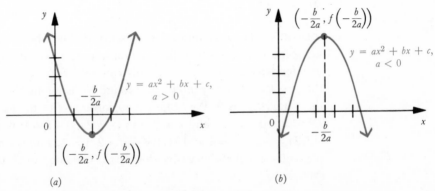

(a)

(b)

Locating the extreme point, together with the x and y intercepts simplifies sketching the graphs of quadratic functions. For example, consider the quadratic function $y = x^2 - 3x + 2$. The y intercept can be obtained by setting $x = 0$ to get the point $(0, 2)$. The x intercept occurs when $y = 0$, and it is found by solving the quadratic equation $0 = x^2 - 3x + 2$ to get the points $(1, 0)$ and $(2, 0)$. The extreme point, which occurs when $x = -b/2a = -(-3)/2 = 3/2$, is the point $(\frac{3}{2}, -\frac{1}{4})$ since $f(\frac{3}{2}) = (\frac{3}{2})^2 - 3(\frac{3}{2}) + 2 = \frac{9}{4} - \frac{9}{2} + 2 = -\frac{1}{4}$. It is a minimum since the coefficient of x^2 is positive (Figure 5).

Figure 5

x	y
0	2
$\frac{3}{2}$	$-\frac{1}{4}$
1	0
2	0

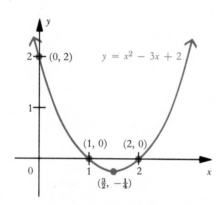

The range of the function f is $[-\frac{1}{4}, \infty)$.

EXAMPLES

Find the domain, the x and y intercepts, the extreme point, and the range of each of the following quadratic functions. Also sketch the graph.

1 $f(x) = -x^2 + 2x$

SOLUTION.   The domain of f is the set of real numbers. In locating the x intercepts of the graph of the function f, let $f(x) = 0$, and then find values of x for which $-x^2 + 2x = 0$. This can be written as $-x(x-2) = 0$, so that $x = 0$ or 2. The y intercept is found by assigning the value 0 to x, so the y intercept is 0. We obtain the coordinates of the extreme point by substituting $a = -1$ and $b = 2$ into $(-b/2a, f(-b/2a))$ to get $(1, 1)$ since $-b/2a = -2/-2 = 1$ and $f(1) = -1^2 + 2 \cdot 1 = 1$. This is a maximum since the coefficient of x^2 is negative (Figure 6). The range of f is the set $(-\infty, 1]$. (Why?)

Figure 6

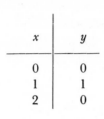

x	y
0	0
1	1
2	0

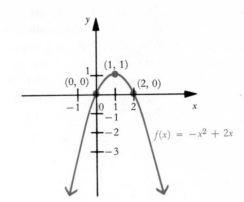

2 $f(x) = x^2 - 7x + 6$

SOLUTION. The domain of f is the set of real numbers. The x intercepts of f are the values of x such that $x^2 - 7x + 6 = 0$; that is, $(x-6)(x-1) = 0$, and $x = 1$ or 6. The y intercept of f is $f(0) = 6$. To obtain the coordinates of the extreme point we substitute $a = 1$ and $b = -7$ into $(-b/2a, f(-b/2a))$ to get $(\frac{7}{2}, -\frac{25}{4})$ since

$$f(\tfrac{7}{2}) = (\tfrac{7}{2})^2 - 7(\tfrac{7}{2}) + 6 = \tfrac{49}{4} - \tfrac{49}{2} + 6 = -\tfrac{25}{4}$$

This is a minimum point (why?) (Figure 7). The range of f is the set $[-\frac{25}{4}, \infty)$.

Figure 7

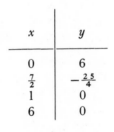

x	y
0	6
$\frac{7}{2}$	$-\frac{25}{4}$
1	0
6	0

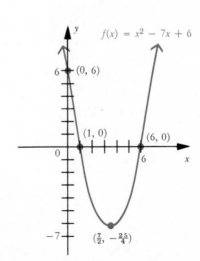

3 The path of a missile is parabolic, and it is represented by $P(t) = -t^2 + t + 1$, where t represents time. Determine the value of t at which the missile reaches its highest point, and the value of t at which the missile strikes the ground. Graph the path.

SOLUTION. $t = (-1/-2) = \frac{1}{2}$ gives the value of t at which the missile reaches its maximum height of $P(\frac{1}{2}) = \frac{5}{4}$. The missile strikes the ground when $P(t) = 0 = -t^2 + t + 1$, $P(t) = 0$ when $t = (-1 \pm \sqrt{5})/(-2)$. Since t represents time, $t \geq 0$, so $t = (-1 - \sqrt{5})/(-2) = (1 + \sqrt{5})/2$. The graph is given in Figure 8.

Figure 8

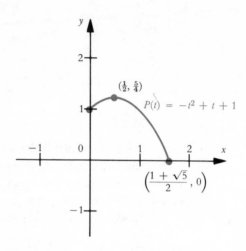

$$\left(\frac{1 + \sqrt{5}}{2}, 0\right)$$

5.3   Quadratic Inequalities

Any inequality that can be expressed either in the form $ax^2 + bx + c > 0$ or in the form $ax^2 + bx + c < 0$ is a *quadratic inequality*. For example, $x^2 < 2x - 2$ can be written as $x^2 - 2x + 2 < 0$, and $3 - x^2 < 2x$ can be written as $-x^2 - 2x + 3 < 0$.

The graphs of quadratic functions can be used to solve quadratic inequalities. For example, consider the graph of the quadratic function $y = f(x) = x^2 + 5x + 6$ shown in Figure 9. Observe that the graph of f

Figure 9

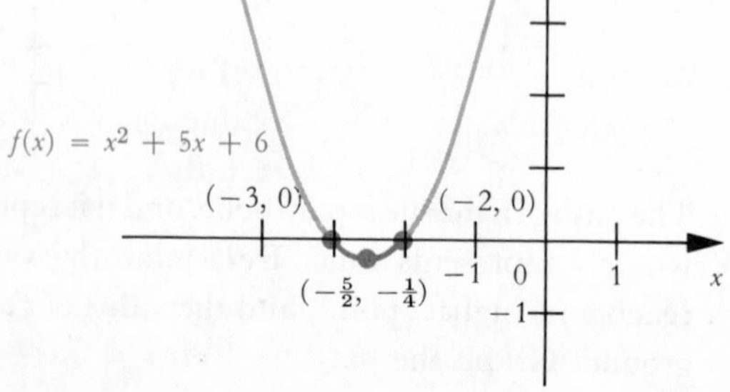

intersects the x axis at the two points $(-3, 0)$ and $(-2, 0)$. Notice also that if $x < -3$ or if $x > -2$, then $f(x) > 0$; if $-3 < x < -2$, then $f(x) < 0$. Hence

$$\{x \mid f(x) = x^2 + 5x + 6 > 0\} = \{x \mid x < -3\} \cup \{x \mid x > -2\}$$
$$= (-\infty, -3) \cup (-2, \infty)$$

and

$$\{x \mid f(x) = x^2 + 5x + 6 < 0\} = \{x \mid -3 < x < -2\} = (-3, -2)$$

EXAMPLES

1   Use the graph of $f(x) = 2x^2 + x - 6$ to find the values of x for which $2x^2 + x - 6 < 0$.

SOLUTION.   The graph of f is given in Figure 10. The inequality is satisfied by all values of x of the graph of f which have ordinates below the x axis, that is, by all x such that $f(x) < 0$. Thus

$$\{x \mid 2x^2 + x - 6 < 0\} = \{x \mid f(x) < 0\}$$
$$= \{x \mid -2 < x < \tfrac{3}{2}\}$$
$$= (-2, \tfrac{3}{2})$$

Figure 10

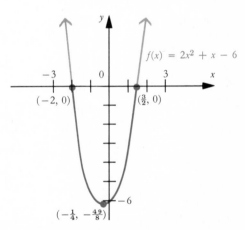

$$f(x) = 2x^2 + x - 6$$

$(-2, 0)$ $(\tfrac{3}{2}, 0)$

$(-\tfrac{1}{4}, -\tfrac{49}{8})$

2   Use the graph of $f(x) = x^2 - 2x - 3$ to find the values of x for which $x^2 - 2x - 3 > 0$.

SOLUTION.   The graph of $f(x) = x^2 - 2x - 3$ is given in Figure 11. The inequality is satisfied by all values of x for which the graph of f lies above the x axis, that is, by all x such that $f(x) > 0$. Hence

$$\{x \mid x^2 - 2x - 3 > 0\} = \{x \mid x < -1 \text{ or } x > 3\}$$
$$= \{x \mid x < -1\} \cup \{x \mid 3 < x\}$$
$$= (-\infty, -1) \cup (3, \infty)$$

Figure 11

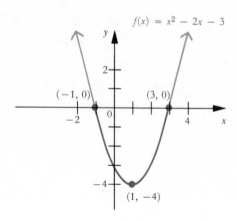

$f(x) = x^2 - 2x - 3$

3   Find the values of x for which $x^2 < -1$.

SOLUTION.   The inequality $x^2 < -1$ is equivalent to $x^2 + 1 < 0$. Let us examine the graph of the associated function $f(x) = x^2 + 1$ (Figure 12). Since $x^2 + 1 > 0$ for all real numbers, $\{x \mid x^2 + 1 < 0\} = \varnothing$. That is, $x^2 < -1$ has no solution.

Figure 12

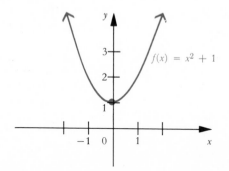

$f(x) = x^2 + 1$

We have seen how to solve quadratic inequalities by using the graphs of the associated functions. Although this method is easy to follow, another algebraic method, which we shall refer to as the *cut-point method*, can be used. The *cut points* of a quadratic inequality are merely the x intercepts of the associated quadratic function.

For example, the cut points of the inequality $x^2 + 2x - 15 < 0$ are the x intercepts of the function $f(x) = x^2 + 2x - 15$, $(-5, 0)$ and $(3, 0)$. These points divide the number line into three intervals, $(-\infty, -5)$, $(-5, 3)$, and $(3, \infty)$ (Figure 13). Notice that the function is either always positive or

Figure 13

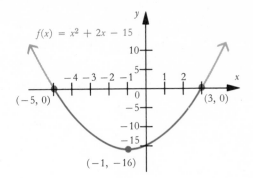

$f(x) = x^2 + 2x - 15$

$(-5, 0)$ $(3, 0)$

$(-1, -16)$

always negative on each of the three intervals; hence $\{x \mid x^2 + 2x - 15 < 0\} = (-5, 3)$ (Figure 14).

Figure 14

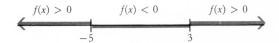

$f(x) > 0$ $f(x) < 0$ $f(x) > 0$

-5 3

Since the graphs of quadratic functions are parabolas, it follows that there are only three possible cases in determining cut points: either there is no cut point (Figure 15*a*), there is one cut point (Figure 15*b*), or there are two cut points (Figure 15*c*).

In order to determine the intervals on which $f(x)$ preserves its sign, it is enough to know the cut points; hence the examples of Figures 15*a*, *b*, and *c*, result in the situations given in Figure 16.

Figure 15

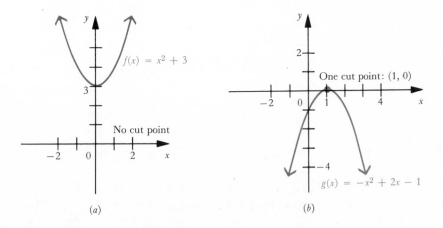

$f(x) = x^2 + 3$

No cut point

One cut point: $(1, 0)$

$g(x) = -x^2 + 2x - 1$

(*a*)   (*b*)

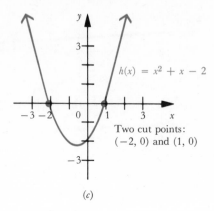

$h(x) = x^2 + x - 2$

Two cut points:
$(-2, 0)$ and $(1, 0)$

(c)

In general, once the cut points are determined, it is enough to test one point in each of the intervals, since the sign of the value of the associated function does not change for any value in each of these intervals. This happens because quadratic functions are continuous.

Figure 16

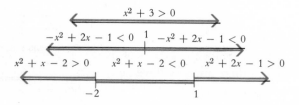

EXAMPLES

1   Use cut points to solve $x^2 + 5x + 6 < 0$.

SOLUTION.   The cut points of $x^2 + 5x + 6 < 0$ are $(-3, 0)$ and $(-2, 0)$, since the graph intersects the x axis at the points $(-3, 0)$ and $(-2, 0)$. Hence the cut points determine three intervals: $(-\infty, -3)$, $(-3, -2)$, and $(-2, \infty)$ (Figure 17).

Figure 17

-3 -2

Since $-4 \in (-\infty, -3)$ does not satisfy $x^2 + 5x + 6 < 0$, the interval $(-\infty, -3)$ is not part of the solution set. Similarly, since $0 \in (-2, \infty)$

does not satisfy the given inequality, $(-2, \infty)$ is not part of the solution set. However, $-\frac{5}{2} \in (-3, -2)$ does satisfy the inequality, so the solution of $\{x \mid x^2 + 5x + 6 < 0\}$ is the interval $(-3, -2)$ (Figure 18).

Figure 18

$$-3 \qquad -2$$

2 Use cut points to solve $x^2 - 2x - 3 > 0$.

SOLUTION. Since $\{x \mid x^2 - 2x - 3 = 0\} = \{x \mid (x+1)(x-3) = 0\} = \{-1, 3\}$, the cut points are -1 and 3 (Figure 19).

Figure 19

$$-1 \qquad\qquad 3$$

We now test the three intervals determined by the cut points—$(-\infty, -1)$, $(-1, 3)$, and $(3, \infty)$—merely by testing one member within each of the three intervals.

For $x = -2$, $(-2)^2 - 2(-2) - 3 = 5 > 0$. Hence $(-\infty, -1)$ is part of the solution set. Similarly, for $x = 0$, we get $-3 > 0$, so the interval $(-1, 3)$ is not part of the solution. Finally, for $x = 4$, we have $5 > 0$, so $(3, \infty)$ is part of the solution set. Hence

$$\{x \mid x^2 - 2x - 3 > 0\} = (-\infty, -1) \cup (3, \infty) \qquad \text{(Figure 20)}$$

Figure 20

$$-1 \qquad\qquad 3$$

PROBLEM SET 4

1 Solve each of the following equations by the factor method.

a) $3x^2 - 7x = 0$

b) $2x^2 - 19x - 33 = 0$

c) $5(x+25) = 6x^2$

d) $x(x-2) = 9 - 2x$

e) $x^2 - 6x = -8$

f) $4x^2 + 11x + 6 = 0$

g) $4x^2 - 16 = 0$

h) $18x^2 + 61x - 7 = 0$

2 Solve each of the following equations using the method of completing the square.

a) $3x^2 - 8x + 2 = 0$

b) $x^2 - 12x + 35 = 0$

c) $x^2 + 3x - 1 = 0$

d) $x^2 - 3x + 2 = 0$

e) $3x^2 - 7x - 3 = 0$ f) $x^2 + 18x + 12 = 0$

g) $x^2 - 13x + 3 = 0$ h) $x^2 + 2x - 3 = 0$

3   Solve each of the following quadratic equations using the quadratic formula. Indicate the discriminant of each equation.

a) $2x^2 - 5x + 1 = 0$ b) $2x^2 + x - 1 = 0$

c) $20 = 12x - x^2$ d) $12x^2 + 29x - 11 = 0$

e) $x^2 - x + 1 = 0$ f) $5 + x = 6x^2$

g) $x^2 - 18x + 56 = 0$ h) $32x^2 - 4x + 21 = 0$

4   a) Prove that if $a < 0$, then $y \leq c - b^2/4a$ for $y = ax^2 + bx + c$.

b) Explain why quadratic functions do not have inverses.

5   For each of the following quadratic functions, determine the domain, the range, the extreme point, the x and y intercepts, and graph the function.

a) $f(x) = 2x^2 - 4$ b) $f(x) = x^2 + 2x$

c) $f(x) = (x + 2)^2$ d) $f(x) = -(x + 1)^2$

e) $f(x) = -x^2 - 1$ f) $f(x) = 2x^2 - 3x$

g) $f(x) = -2x^2 + 3x - 1$ h) $f(x) = x^2 + 3x - 4$

i) $f(x) = 3x^2 - x + 3$ j) $f(x) = -x^2 + x + 5$

6   a) Graph on the same coordinate system $y = ax^2$, where $a \in \{-10, -5, -2, -1, 0, 1, 2, 5, 10\}$. Compare the graphs to the different values of a. What do you notice?

b) Sketch the graph of each of the following functions on the same coordinate system.

$$\begin{array}{ccc} f(x) = x^2 - 2 & & f(x) = x^2 - 1 \\ f(x) = x^2 + 1 & \text{and} & f(x) = x^2 + 2 \end{array}$$

In general, how do graphs of functions of the form $f(x) = x^2 + k$, k a constant, compare?

7   Graph each of the following quadratic functions, and then use the graph to solve the inequality. Write the solution in interval form, and then show the set on the real line.

a) $y = -2x^2 + 5x - 3$; $-2x^2 + 5x - 3 \leq 0$

b) $y = 2x^2 - x - 1$; $2x^2 - x - 1 > 0$

c) $y = (x - 1)^2$; $x^2 - 2x < -1$

d) $y = x^2 + 6x + 8$; $-x^2 - 6x - 8 \geq 0$

e) $y = -x^2 - 2$; $x^2 + 2 < 0$

f) $y = 2x^2 - 7x + 6$; $2x^2 - 7x + 6 > 0$

g) $y = 3x^2 - 5x + 1$; $3x^2 - 5x + 1 < 0$

h) $y = \frac{1}{3}x^2 - \frac{1}{2}x - \frac{3}{2}$; $\frac{1}{3}x^2 - \frac{1}{2}x - \frac{3}{2} < 0$

8 A projectile is fired from a balloon in such a way that it is h feet above the ground t seconds after the firing. If $h = -16t^2 + 96t + 256$, find
  a) h when $t = 0$
  b) The maximum height reached by the projectile
  c) The graph of the function

9 Use the cut method to solve each of the following inequalities:
  a) $2x^2 + 5x - 3 < 0$ b) $3x^2 - 7x + 2 > 0$
  c) $x^2 - 4x + 4 \leq 0$ d) $2x^2 + 3x - 9 < 0$
  e) $x^2 - 7x + 6 \geq 0$ f) $x^2 + 4 > 0$
  g) $x^2 + 2x - 3 \leq 0$ h) $-x^2 + 16 > 0$

10 A rectangular field is adjacent to a river and is to have fencing on three sides, the side on the river requiring no fencing. Assume also that 50 yards of fencing is used.
  a) Construct a quadratic function in which the area A of the field is a function of the width x.
  b) Sketch the graph of the function of part a.
  c) What is the maximum point? What is the largest area that can be enclosed?

6 Polynomial Functions of Degree Greater Than 2

We have seen that polynomial functions are functions expressible in the form $f(x) = a_n x^n + a_{n-1} x^{n-1} + \cdots + a_1 x + a_0$, where the coefficients are real numbers, $a_n \neq 0$, and n is a nonnegative integer or $f(x) = 0$.

Let us now investigate the graphs of polynomial functions of degree greater than 2. In graphing a polynomial function, it is helpful to determine the x intercepts, that is, the points where the graph intersects the x axis. For example, the x intercepts for $f(x) = x^3 - x^2$ are $(0,0)$ and $(1,0)$; they are determined by solving the equation $0 = x^3 - x^2$. Clearly, the x intercepts of a function occur at those values of x which, when substituted into the function, give 0 as the corresponding y. These values of x are called the *zeros* of the function. Hence $x = 0$ and $x = 1$ are the zeros of the polynomial function $f(x) = x^3 - x^2$.

The zeros of the polynomial function $f(x) = 2x^2 + x - 1$ can be determined by solving $0 = 2x^2 + x - 1$ to get $x = -1$ and $x = \frac{1}{2}$; the zeros of $f(x) = 4x^3 - x^2$ are $x = 0$ and $x = \frac{1}{4}$; the function $f(x) = x^2 + 1$ has no real zeros. (Why?)

The graphing of a polynomial function also requires plotting a few points. This means, of course, that specific members of the domain must be substituted into the function so as to determine corresponding range members. For certain situations the computation of the range members is not difficult. For example, if $f(x) = x^3 + 1$, $f(-1) = 0$, $f(0) = 1$, $f(1) = 2$, and $f(2) = 9$ can easily be determined by substitution. In other situations the computation becomes rather tedious. For example, if $f(x) = x^5 - 3x^4 + 7x - 8$, the determination of $f(3), f(-2)$, and $f(7)$ becomes more involved. We will see how a shorthand division process, called *synthetic division*, can be used to simplify computations of the latter type. However, before examining synthetic division, it is necessary to comment briefly on the division of polynomials.

6.1 Division of Polynomials

The polynomial $x^3 + 2x^2 - 2x + 3$ is divisible by $x + 3$, since

$$x^3 + 2x^2 - 2x + 3 = (x^2 - x + 1)(x + 3)$$

whereas the division of $x^2 + 3x + 7$ by $x + 1$ gives a quotient of $x + 2$ and a remainder of 5, so that $x^2 + 3x + 7 = (x + 2)(x + 1) + 5$. In either case, the division of polynomials is based on the following property, called the *division algorithm*, which is stated as follows:

6.2 Division Algorithm

If $f(x)$ and $D(x)$ are polynomials such that the degree of $f(x)$ is greater than or equal to the degree of $D(x)$ with $D(x) \neq 0$, then there exist unique polynomials $Q(x)$ and $R(x)$ such that $f(x) = D(x) \cdot Q(x) + R(x)$, where the degree of $R(x)$ is less than or equal to the degree of $D(x)$ or $R(x) = 0$. $D(x)$ is called the *divisor*, $f(x)$ the *dividend*, $Q(x)$ the *quotient*, and $R(x)$ the *remainder*.

EXAMPLE

Suppose that $f(x) = 3x^3 - x^2 + 2x - 1$, and $D(x) = x - 3$. Find $Q(x)$ and $R(x)$ such that $3x^3 - x^2 + 2x - 1 = Q(x)(x - 3) + R(x)$, where $R(x)$ satisfies the division algorithm.

SOLUTION.   We arrange this division in the following manner:

$$
\begin{array}{r}
3x^2 + 8x\ + 26 \\
x - 3 \overline{\smash{\big)}\ 3x^3 - x^2 + 2x - 1} \\
\underline{\pm 3x^3 \mp 9x^2} \\
8x^2 + 2x \\
\underline{\pm 8x^2 \mp 24x} \\
26x - 1 \\
\underline{\pm 26x \mp 78} \\
77
\end{array}
$$

Hence $Q(x) = 3x^2 + 8x + 26$, and $R(x) = 77$.

6.3   Synthetic Division

Synthetic division is merely a *shorthand method* for performing the division of any polynomial by a polynomial of the form $x - r$. In the example above, the division can be displayed as follows:

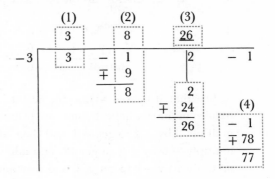

Here the variables are implied rather than explicitly indicated and the $\mp$ indicates the change of sign necessary in subtracting. This latter form could also be rearranged as

$$
\begin{array}{c|cccc}
 & (1) & (2) & (3) & (4) \\
-3 & 3 & -1 & 2 & -1 \\
 & & \mp 9 & \mp 24 & \mp 78 \\
\hline
 & 3 & 8 & 26 & 77
\end{array}
$$

Note that in this latter representation, the coefficients of the quotient appear in the last row, together with the remainder in the last position. Finally, the last form can be abbreviated as

$$
\begin{array}{cc|cccc}
 & & (1) & (2) & (3) & (4) & \\
\text{Abbreviation} \rightarrow & 3 & 3 & -1 & 2 & -1 & \leftarrow \text{Coefficients of} \\
\text{for divisor} & & & & & & \text{dividend} \\
 & & & 9 & 24 & 78 & \\
\hline
\text{Coefficients} \longrightarrow & & 3 & 8 & 26 & \boxed{77} & \leftarrow \text{Remainder} \\
\text{of quotient} & & & & & &
\end{array}
$$

Since the sign of the number to the left of the bar (-3 in this example) is changed, it is not necessary to perform subtraction as we do in the long division method. All we need do is to add algebraically. Hence $3x^3 - x^2 + 2x - 1 = (x - 3)(3x^2 + 8x + 26) + 77$. It is the latter form of the division that is called *synthetic division*.

EXAMPLES

1   Find the quotient $Q(x)$ and the remainder $R(x)$ using synthetic division if the dividend is $f(x) = 3x^3 - 8x + 1$ and the divisor is $D(x) = x + 2$.

SOLUTION.   First, write the divisor $x + 2$ in the form $x - r$ as $x - (-2)$. Then use -2 as the "divisor" in the synthetic division as follows:

$$
\begin{array}{r|rrrr}
-2 & 3 & 0 & -8 & 1 \\
   &   & -6 & 12 & -8 \\
\hline
   & 3 & -6 & 4 & -7 \\
\end{array}
$$

Notice that 0 is used as the coefficient of the x^2 term. Also notice the pattern in the synthetic division: we bring down the first coefficient, 3, and multiply 3 by -2 to get -6, and then add 0 to -6 to get -6. Next we multiply -6 by -2 to get 12, and add -8 to 12 to get 4. Finally, we multiply 4 by -2 to get -8, and then add 1 to -8 to get the remainder -7. Consequently, $3x^3 - 8x + 1 = (3x^2 - 6x + 4)(x + 2) + (-7)$, so that $Q(x) = 3x^2 - 6x + 4$ and $R(x) = -7$.

2   Use synthetic division to divide $f(x)$ by $x - r$. Then use the result of the division to determine $f(r)$ if $f(x) = x^5 - 6x^2 - 4x^3 - 9$ and $r = -3$.

SOLUTION

$$
\begin{array}{r|rrrrrr}
-3 & 1 & 0 & -4 & -6 & 0 & -9 \\
   &   & -3 & 9 & -15 & 63 & -189 \\
\hline
   & 1 & -3 & 5 & -21 & 63 & -198 \\
\end{array}
$$

Hence

$$
\begin{aligned}
f(x) &= x^5 - 6x^2 - 4x^3 - 9 \\
     &= (x^4 - 3x^3 + 5x^2 - 21x + 63)(x + 3) + (-198)
\end{aligned}
$$

so $f(-3) = 0 + (-198) = -198$. Notice that the function value $f(-3)$ is the same as the remainder we get when dividing $f(x)$ by $x - (-3)$.

This example suggests two theorems.

THEOREM 1   REMAINDER THEOREM

If a polynomial $f(x)$ of degree $n > 0$ is divided by $x - r$, the remainder R is a constant and is equal to the value of the polynomial when r is substituted for x; that is, $f(r) = R$.

PROOF.   Let $Q(x)$ be the quotient, so that, by the division algorithm, $f(x) = (x - r)Q(x) + R(x)$. Since the remainder $R(x)$ is of degree less than the divisor $x - r$, it must be constant and we will denote it as R, The equation $f(x) = (x - r)Q(x) + R$ holds for all x, and if we set $x = r$, we find that $f(r) = (r - r)Q(r) + R = 0 \cdot Q(r) + R = R$, so that $f(r) = R$.

THEOREM 2 COROLLARY*: FACTOR THEOREM

If $f(r) = 0$, then $x - r$ is a factor of the polynomial $f(x)$ of degree $n > 0$; conversely, if $x - r$ is a factor, r is a zero.

PROOF. Assume that $f(x) = (x - r)Q(x) + R$. If $f(r) = 0$, then, by Theorem 1, $R = 0$ and $f(x) = Q(x)(x - r)$. Conversely, if $x - r$ is a factor of $f(x)$, then $f(x) = Q(x)(x - r)$, so $f(r) = Q(r) \cdot 0 = 0$.

Theorem 1 tells us what was hinted in the second example above. The value of a polynomial $f(r)$ is the same as the remainder of the polynomial we get when dividing by $x - r$. But this type of operation can be performed by synthetic division. Hence the values of a polynomial function can be determined by synthetic division. Of course, in most cases it is just as simple to calculate the value directly. For example, $f(3)$ for $f(x) = x^5 - 3x^4 + 7x - 8$ can be determined as follows:

$$
\begin{array}{r|rrrrrr}
3 & 1 & -3 & 0 & 0 & 7 & -8 \\
  &   & 3 & 0 & 0 & 0 & 21 \\
\hline
  & 1 & 0 & 0 & 0 & 7 & 13 \\
\end{array}
$$

so that $f(3) = 13$.

EXAMPLES

1 Use synthetic division to find the quotient $Q(x)$ and the remainder R if $f(x) = 3x^5 - 5x^3 + 1$ is divided by $x - 2$. Use your result to determine $f(2)$.

SOLUTION

$$
\begin{array}{r|rrrrrr}
2 & 3 & 0 & -5 & 0 & 0 & 1 \\
  &   & 6 & 12 & 14 & 28 & 56 \\
\hline
  & 3 & 6 & 7 & 14 & 28 & 57 \\
\end{array}
$$

Hence $Q(x) = 3x^4 + 6x^3 + 7x^2 + 14x + 28$; also, by the remainder theorem, $f(2) = 57$.

2 Find the remainder if $f(x) = x^{1001} + 3$ is divided by $x + 1$.

SOLUTION

$$f(-1) = (-1)^{1001} + 3 = -1 + 3 = 2$$

Hence, by the remainder theorem, the remainder if $f(x) = x^{1001} + 3$ is divided by $x + 1$ is 2.

3 Show that $x + 2$ is a factor of $f(x) = x^4 - 2x^2 + 3x - 2$ and find $Q(x)$.

---

* A corollary is a theorem that follows directly from another theorem.

SOLUTION

$$
\begin{array}{r|rrrrr}
-2 & 1 & 0 & -2 & 3 & -2 \\
   &   & -2 & 4 & -4 & 2 \\
\hline
   & 1 & -2 & 2 & -1 & 0
\end{array}
$$

Hence $f(-2) = 0$, and $Q(x) = x^3 - 2x^2 + 2x - 1$, so

$$
f(x) = (x+2)(x^3 - 2x^2 + 2x - 1)
$$

4   Find the value of k if $x - 3$ is a factor of $f(x) = 3x^3 - 4x^2 - kx - 33$.

SOLUTION

$$
\begin{array}{r|rrrr}
3 & 3 & -4 & -k & -33 \\
  &   & 9 & 15 & 3(-k+15) \\
\hline
  & 3 & 5 & -k+15 & -3k+12
\end{array}
$$

Since $x - 3$ is a factor of $f(x) = 3x^3 - 4x^2 - kx - 33$, it follows from the factor theorem that $-3k + 12 = 0$. Hence $k = 4$.

6.4   Graphs of Polynomial Functions of Degree Greater Than 2

The general study of polynomial functions of degree greater than 2 is more difficult than the study of polynomial functions of degree less than or equal to 2. It can be shown (in calculus) that polynomial functions are *continuous functions*, and we use this result in graphing polynomial functions. For example, we can graph $f(x) = 1, f(x) = x, f(x) = x^2, f(x) = x^3$, and $f(x) = x^4$ by plotting a few points and assuming continuity between these points to sketch the graph (Figure 1).

Figure 1

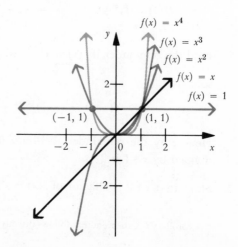

Consider $f(x) = x^3 + \frac{3}{2}x^2 - 6x - 2$. First, we can prepare a table of values of x and $f(x)$. Next the points can be located. From the table we observe that the points $(x, f(x))$ to be located are $(-4, -18)$, $(-3, 2.5)$, $(-2, 8)$, $(-1, 4.5)$, $(0, -2)$, $(1, -5.5)$, and $(2, 0)$ (Figure 2).

Figure 2

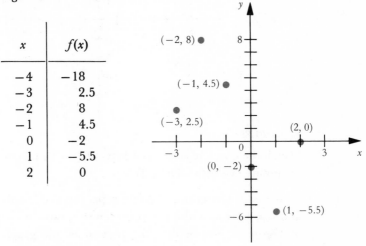

x	$f(x)$
-4	-18
-3	2.5
-2	8
-1	4.5
0	-2
1	-5.5
2	0

Since polynomial functions are continuous, if $a < b$ and $f(a) < 0 < f(b)$, then there is a zero, say c, such that $a < c < b$ and $f(c) = 0$ (Figure 3). The problem, of course, is that there may be more than one such zero between a and b.

Figure 3

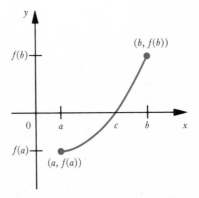

The question is whether the points we have already plotted for $f(x) = x^3 + \frac{3}{2}x^2 - 6x - 2$ are sufficient to give us a fairly accurate sketch of the graph. Are there hidden peaks not shown thus far? We are *not* in a position to answer this question, but if we plot more points in between those already located, we can get a rough sketch of the graph (Figure 4).

Figure 4

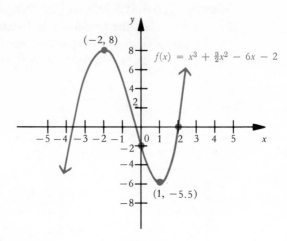

$f(x) = x^3 + \frac{3}{2}x^2 - 6x - 2$

(-2, 8)

(1, -5.5)

EXAMPLES

1 Sketch the graph of $f(x) = (x+1)(x-2)(x-3)$. Use the graph to solve the inequality $(x+1)(x-2)(x-3) < 0$.

SOLUTION.   The graph of f intercepts the x axis only at $x = -1$, $x = 2$, and $x = 3$. Some additional points on the graph of f are given in the table of Figure 5. Notice that the portion of the graph (Figure 5) below the x axis suggests the values of x that satisfy the inequality. Thus the solution set of the inequality is given by $\{x \mid (x+1)(x-2)(x-3) < 0\} = (-\infty, -1) \cup (2, 3)$.

Figure 5

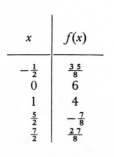

x	$f(x)$
$-\frac{1}{2}$	$\frac{35}{8}$
0	6
1	4
$\frac{5}{2}$	$-\frac{7}{8}$
$\frac{7}{2}$	$\frac{27}{8}$

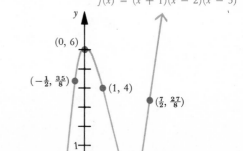

$f(x) = (x + 1)(x - 2)(x - 3)$

(0, 6)

$(-\frac{1}{2}, \frac{35}{8})$ (1, 4)

$(\frac{7}{2}, \frac{27}{8})$

$(\frac{5}{2}, -\frac{7}{8})$

2 Graph $f(x) = (x-1)^2(x-2)(x-3)^3$. Use the graph to solve the inequality $(x-1)^2(x-2)(x-3)^3 > 0$.

SOLUTION.   The graph of f intercepts the x axis only at $x = 1$, $x = 2$, and $x = 3$, the zeros of $f(x)$. Additional points on the graph of $f(x)$ are given in the table of Figure 6. Notice that the portion of the graph (Figure 6) above the x axis suggests the values of x which satisfy the inequality, so $\{x \mid (x-1)^2(x-2)(x-3)^3 > 0\}$ is the set $(-\infty, 1) \cup (1, 2) \cup (3, \infty)$.

Figure 6

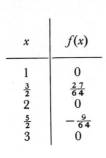

x	$f(x)$
1	0
$\frac{3}{2}$	$\frac{27}{64}$
2	0
$\frac{5}{2}$	$-\frac{9}{64}$
3	0

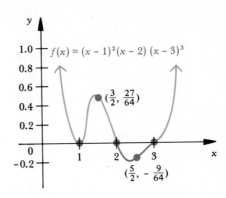

6.5   Rational Zeros

So far we have seen how the zeros of polynomial equations are used to determine the x intercepts of the graphs of polynomial functions. If the polynomial is of first or second degree it is easy to determine the zeros, but as the degree increases, the problem of determining the zeros becomes much more difficult. The general method to be taken up here enables us to determine zeros that are rational numbers of polynomial functions with integral coefficients. Before stating the result, it is necessary to recall one of the basic properties of the system of integers:

> *An integer can be written in one and only one way as a product of prime numbers.*

For example, $24 = 2^3 \cdot 3$, and there are no other prime factors of 24.

THEOREM 1   RATIONAL ROOT THEOREM

If $f(x) = a_n x^n + a_{n-1} x^{n-1} + \cdots + a_1 x + a_0$, and if the coefficients are *integers* and p/q is a rational root in lowest terms, then p is a divisor of a_0 and q is a divisor of a_n.

PROOF.   (This proof is given for completeness sake, not because the student is expected to know it.) Since $a_n(p/q)^n + a_{n-1}(p/q)^{n-1} + \cdots + a_1(p/q) + a_0 = 0$, it follows, after multiplying both sides by q^n, that

$$a_n p^n + a_{n-1} p^{n-1}q + \cdots + a_1 pq^{n-1} + a_0 q^n = 0$$

Thus

$$(1) \quad a_n p^n + a_{n-1} p^{n-1}q + \cdots + a_1 pq^{n-1} = -a_0 q^n$$

or, equivalently,

$$(2) \quad a_{n-1} p^{n-1} q + \cdots + a_1 p q^{n-1} + a_0 q^n = -a_n p^n$$

Since both sides of equation (1) are integers, p is a divisor of the left side and, therefore, also of the right side. But p and q have no common factors, since p/q is in lowest terms. Hence every prime factor of p must be a factor of a_0 and the first part of the proof is finished. Similarly, in (2), q is a factor of the left side, and hence of the right side. As before, q has no factors in common with p, so q must be a divisor of a_n.

EXAMPLES

1 Find all the possible rational roots of $f(x) = x^3 + 2x^2 - 4x - 8 = 0$.

SOLUTION.   Assume that p/q is a rational root of $f(x) = 0$. By the rational root theorem, p is a divisor of $a_0 = -8$ and q is a divisor of $a_n = 1$, so p can assume any of the values 1, -1, 2, -2, 4, -4, 8, or -8 and q can assume values 1 or -1; therefore, the possible rational roots p/q are given by 1, -1, 2, -2, 4, -4, 8, or -8. After testing these possible rational roots by synthetic division, we find that $x = 2$ and $x = -2$ are roots; in fact, $x^3 + 2x^2 - 4x - 8 = (x+2)^2 (x-2)$, so $x = 2$ and $x = -2$ are the only roots.

2 Use the rational root theorem to prove that $\sqrt{2}$ is an irrational number.

PROOF. $x^2 - 2 = 0$ is a polynomial equation whose root is $\sqrt{2}$. Let p/q be a possible rational root of $x^2 - 2 = 0$, so $p = 1$, -1, 2, or -2, and $q = 1$ or -1; therefore, by the rational root theorem, 1, -1, 2, or -2 are the only possible rational roots. Since $\sqrt{2}$ is not one of the possible rational roots, $\sqrt{2}$ is not a rational number.

PROBLEM SET 5

1 Find all the zeros of each of the following polynomial functions.
   a) $f(x) = -3x + 2$
   b) $f(x) = 3x^2 + x - 2$
   c) $f(x) = (x-1)(x^2 - 3x + 2)$
   d) $f(x) = (x-1)(x-2)(x-5)$
   e) $f(x) = x^4 - 16$
   f) $f(x) = (x-1)^2 - 9$
   g) $f(x) = x^3 - 9x$
   h) $f(x) = x^4 - x^3$

2 Use long division to perform each of the following divisions.
   a) $5x^3 - 2x^2 + 3x - 4$ by $x - 3$
   b) $2x^4 + 3x^3 - 5x^2 + 2x - 1$ by $x + 1$
   c) $5x^5 - 3x^4 + 2x^3 + x^2 - 7x + 3$ by $x - 2$
   d) $2x^4 - 3x^3 + 5x^2 + 6x - 3$ by $x + 2$
   e) $-4x^6 - 5x^3 + 3x^2 + x + 7$ by $x - 1$
   f) $2x^4 + 3x^3 - 3x^2 + x - 1$ by $x + 4$

3 Use synthetic division to perform the divisions in Problem 2.

4 Use synthetic division to find $Q(x)$ and $f(r)$ so that
$f(x) = (x-r) Q(x) + f(r)$.

    a) $f(x) = 3x^3 + 6x^2 - 10x + 7$ and $r = 2$
    b) $f(x) = 3x^3 + 4x^2 - 7x + 16$ and $r = -1$
    c) $f(x) = 2x^3 - 5x^2 + 5x + 11$ and $r = \frac{1}{2}$
    d) $f(x) = -2x^4 + 3x^3 + 5x - 13$ and $r = 3$
    e) $f(x) = -3x^4 - 3x^3 + 3x^2 + 2x - 4$ and $r = -2$

5 If $f(x) = x^3 + 2x^2 - 13x + 10$, use synthetic division to determine $f(-5)$, $f(-4), f(-3), f(-1), f(0), f(1), f(2), f(3), f(4)$, and $f(5)$. What are the factors of $f(x)$?

6 a) If $f(x) = 2x^3 - 6x^2 + x + k$, find k so that $f(3) = -2$.
    b) Find k so that $x - 2$ is a factor of $f(x) = 3x^3 + 4x^2 + kx - 20$.

7 a) Show that $x - 1$ is a factor of $f(x) = 14x^{99} - 65x^{56} + 51$.
    b) Show that $x + 4$ is a factor of $f(x) = 2x^2 + 3x - 20$.

8 For what values of n, where n is a positive integer, is each of the following true?
    a) $x^n + a^n$ is divisible by $x + a$.
    b) $x^n + a^n$ is divisible by $x - a$.
    c) $x^n - a^n$ is divisible by $x + a$.
    d) $x^n - a^n$ is divisible by $x - a$.

9 Sketch the graph of each of the following polynomial functions. Also determine the x and y intercepts.
    a) $f(x) = x^3 - 3x^2 + 4$
    b) $f(x) = x^3 - 2x^2 - 5x + 6$
    c) $f(x) = x(x-1)(x+2)$
    d) $f(x) = 2x^3 - 3x^2 - 3x + 2$
    e) $f(x) = (2x+1)^4$

10 Sketch the graph of the following polynomial functions and then use the graph to solve the inequality. Write the solution in interval form.
    a) $f(x) = x(x-1)(x+2); \; x(x-1)(x+2) > 0$
    b) $f(x) = (x-2)^2(x+1); \; (x-2)^2(x+1) < 0$
    c) $f(x) = (x-1)(x+1)(x+2); \; (x-1)(x+1)(x+2) < 0$
    d) $f(x) = (x-1)^3(x+1)^2; \; (x-1)^3(x+1)^2 > 0$
    e) $f(x) = (x-1)^2x^3(x+1); \; (x-1)^2x^3(x+1) < 0$
    f) $f(x) = (x+1)^3x^2(x-1); \; (x+1)^3x^2(x-1) > 0$

11 Write down all rational numbers that might be roots of the following polynomial functions. Use synthetic division and the remainder theorem to test the possibilities to determine which of them are roots.
    a) $f(x) = 3x^3 - 7x^2 + 8x - 2$

b) $f(x) = x^3 + 2x - 12$

c) $f(x) = 5x^3 - 12x^2 + 17x - 10$

d) $f(x) = x^3 - x^2 - 14x + 24$

12 Use the rational root theorem to prove that $\sqrt{3}$ is an irrational number.

7 Rational Functions

In this section we will study functions that are quotients of polynomials. Such functions are called *rational functions*. For example, the functions $k(x) = (x+1)/(x-2)$, $h(x) = 5/x$, and $g(x) = (3x^3+5x)/(x^2+2)$ are rational functions. It should be noted that the domain of a rational function $T(x) = f(x)/g(x)$ does not contain the zeros of the polynomial function g since division by zero is meaningless.

If the real number a is a zero of both f and g in the rational function $T(x) = f(x)/g(x)$, that is, if $f(a) = 0$ and $g(a) = 0$ for $T(x) = f(x)/g(x)$, it follows from the factor theorem that

$$T(x) = f(x)/g(x) = \frac{f_1(x)(x-a)}{g_1(x)(x-a)}$$

$$= \frac{f_1(x)}{g_1(x)} \qquad x \neq a$$

For example, the rational function

$$T(x) = (x^2 - 4)/(x - 2) = ((x+2)(x-2))/(x-2)$$
$$= x + 2 \qquad \text{for } x \neq 2$$

Note that $T(2)$ is not defined. The graph in Figure 1 displays the fact that 2 is not in the domain of T.

Figure 1

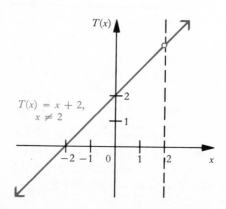

$T(x) = x + 2,$
$x \neq 2$

Next let us consider an example of a rational function $T(x) = f(x)/g(x)$, where $g(a) = 0$ and $f(a) \neq 0$, to see what happens to the graph near a. Suppose that we have $T(x) = (2x+1)/(x-1)$. T has a y intercept at $(0, -1)$ and an x intercept at $(-\frac{1}{2}, 0)$. The domain of T is $\{x \mid x \neq 1\}$. Upon examining the behavior of the values in Table 1 we see that as x gets closer to 1 "from the right," the corresponding values of $T(x)$ become very large.

This situation results from the fact that the denominator of the fraction, $x-1$, is becoming very close to zero, while the numerator of the fraction,

Table 1

x	2	$1\frac{1}{2}$	$1\frac{1}{4}$	$1\frac{1}{8}$	$1\frac{1}{16}$	$\cdots$	$1\frac{1}{1024}$
$T(x)$	5	8	14	26	50	$\cdots$	3074

$2x+1$, is getting closer to 3 in value. Table 2 displays the fact that as x gets closer to 1 "from the left," the corresponding values of $T(x)$ that are negative numbers become very large in absolute value.

Table 2

x	$\frac{1}{2}$	$\frac{3}{4}$	$\frac{7}{8}$	$\frac{9}{10}$	$\cdots$	$\frac{1999}{2002}$
$T(x)$	-4	-10	-22	-28	$\cdots$	-2000

The graph of $T(x) = (2x+1)/(x-1)$ shown in Figure 2 displays the behavior of $T(x)$ near $x = 1$. Notice that there is no point on the graph corresponding to $x = 1$. Furthermore, the values of $T(x)$ become larger in

Figure 2

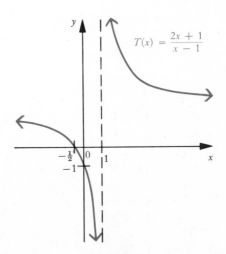

absolute value as x gets closer to 1. This situation is described by saying that the graph of $T(x) = (2x+1)/(x-1)$ is getting closer to the line $x = 1$ "asymptotically," and the line $x = 1$ is called a *vertical asymptote*.

It is important to compare $S(x) = (x^2-4)/(x-2)$ and $T(x) = (2x+1)/(x-1)$. Each of these two functions is undefined when the denominator is zero. For $S(x) = (x^2-4)/(x-2)$, the restriction on the denominator results in a graph with "one missing point" since both the numerator and denominator are simultaneously zero when $x = 2$ (Figure 1). On the other hand, for $T(x) = (2x+1)/(x-1)$, the restriction on the denominator results in a graph that not only is undefined when the denominator is zero, but also becomes infinitely larger in absolute value as x gets closer to the number that causes the denominator to be zero (Figure 2). The latter situation occurs because only the denominator is zero when $x = 1$, whereas the numerator is not zero at $x = 1$. $x = 2$ is not an asymptote for $S(x) = (x^2-4)/(x-2)$, whereas $x = 1$ is an asymptote for $T(x) = (2x+1)/(x-1)$.

In general, $T(x) = f(x)/g(x)$ has a *vertical asymptote* at $x = a$ if $g(a) = 0$ and if $|f(x)/g(x)|$ becomes infinitely large as x gets closer to a. This situation occurs when $g(a) = 0$ and $f(a) \neq 0$.

EXAMPLES

For each of the following rational functions find the domain, the x and y intercepts, the asymptotes, and sketch the graph.

1 $f(x) = \dfrac{4}{2x-3}$

SOLUTION.   The domain of f is $\{x \mid x \neq \frac{3}{2}\}$. There is no x intercept and the y intercept is $(0, -\frac{4}{3})$. A vertical asymptote occurs when $2x-3 = 0$, that is, when $x = \frac{3}{2}$. The graph is given in Figure 3.

Figure 3

x	$f(x)$
-2	$-\frac{4}{7}$
-1	$-\frac{4}{5}$
0	$-\frac{4}{3}$
1	-4
$1\frac{1}{4}$	-8
$1\frac{2}{5}$	-20
$1\frac{49}{100}$	-200
$1\frac{99}{200}$	-400

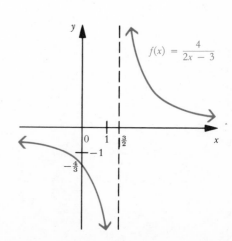

Notice that as x gets closer to $\frac{3}{2}$, $|f(x)|$ becomes very large. More precisely, as x gets closer to $\frac{3}{2}$ "from the left," the values of $f(x)$ become "very large" negative numbers, and as x gets closer to $\frac{3}{2}$ "from the right," the values of $f(x)$ become very large positive numbers.

2 $\quad f(x) = \dfrac{x - \frac{1}{2}}{x^2 - 1}$

SOLUTION. The domain of f is $\{x \mid x \neq 1, x \neq -1\}$. The graph of f intersects the y axis at $(0, \frac{1}{2})$ and the x axis at $(\frac{1}{2}, 0)$; $x = 1$ and $x = -1$ are vertical asymptotes. (Why?)

Let us examine the behavior of f near the asymptotes.

i As x gets closer to -1 from the left (for $x < -1$), then $x - \frac{1}{2} < 0$ and $x^2 - 1 > 0$, and $x^2 - 1$ is close to 0. Thus $f(x)$ is becoming negatively large.

ii As x gets closer to -1 from the right (for $x > -1$), then $x - \frac{1}{2} < 0$ and $x^2 - 1 < 0$, and $x^2 - 1$ is close to 0. Thus $f(x)$ is becoming positively large.

iii As x gets closer to 1 from the right (for $x > 1$), then $x - \frac{1}{2} > 0$ and $x^2 - 1 > 0$, and $x^2 - 1$ is close to 0. Thus $f(x)$ is becoming positively large.

iv As x gets closer to 1 from the left (for $x < 1$), then $x - \frac{1}{2} > 0$ and $x^2 - 1 < 0$, and $x^2 - 1$ is close to 0. Thus $f(x)$ is becoming negatively large.

The graph of f is given in Figure 4.

Figure 4

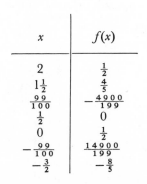

x	$f(x)$
2	$\frac{1}{2}$
$1\frac{1}{2}$	$\frac{4}{5}$
$\frac{99}{100}$	$-\frac{4900}{199}$
$\frac{1}{2}$	0
0	$\frac{1}{2}$
$-\frac{99}{100}$	$\frac{14900}{199}$
$-\frac{3}{2}$	$-\frac{8}{5}$

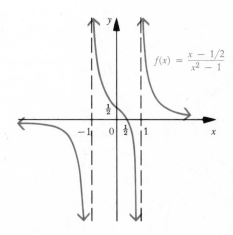

$$f(x) = \frac{x - 1/2}{x^2 - 1}$$

3 $\quad f(x) = \dfrac{5}{x^2 + 4}$

SOLUTION. The domain of f is the set $\{x \mid x \in R\}$. The y intercept of f is $(0, \frac{5}{4})$. There is no vertical asymptote since $x^2 + 4 = 0$ has no real root. $f(x) > 0$ for all $x \in R$. The graph of f is symmetric with respect to the y axis because $f(-x) = f(x)$ (Figure 5).

Figure 5

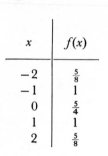

x	$f(x)$
-2	$\frac{5}{8}$
-1	1
0	$\frac{5}{4}$
1	1
2	$\frac{5}{8}$

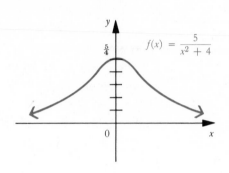

$$4 \quad f(x) = \frac{3x}{x-1}$$

SOLUTION. The domain of f is $\{x \mid x \neq 1\}$. $(0,0)$ is the x and y intercept. $x = 1$ is a vertical asymptote. Using the division algorithm, we can rewrite f as $f(x) = 3 + 3/(x-1)$. Now as x becomes larger in absolute value, that is, as x becomes very large in the positive sense or very small in the negative sense, the rational expression $3/(x-1)$ gets closer to zero so that if $|x|$ is very large $f(x) = 3 + 3/(x-1)$ gets very close to 3. In other words, as x becomes very large (or very small) $3/(x-1)$ approaches zero so that $f(x)$ approaches 3 "asymptotically." The line $y = 3$ is a horizontal asymptote. The graph of f displays this behavior (Figure 6).

Figure 6

x	$f(x)$
-2	2
$-\frac{3}{2}$	$\frac{9}{5}$
-1	$\frac{3}{2}$
0	0
$\frac{1}{2}$	-3
$\frac{3}{2}$	9

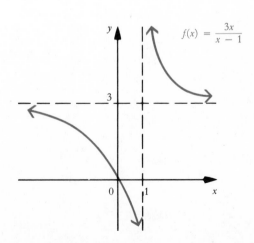

Notice that in this example the degree of the numerator is at least as large as the degree of the denominator. (In examples 1, 2, and 3, the degree of the denominator was greater than the degree of the numerator.) This comparison of the degrees of the numerator and denominator is what prompts the use of the division algorithm here.

5 $f(x) = \dfrac{2x^2 + 1}{2x^2 - 3x}$

SOLUTION.   The domain of f is $\{x \mid x \neq 0,\ x \neq \frac{3}{2}\}$. There is no x intercept and no y intercept; $x = 0$ and $x = \frac{3}{2}$ are vertical asymptotes. Since the degree of the numerator is the same as the degree of the denominator we will use the division algorithm (as we did in Example 4) to get

$$f(x) = \frac{2x^2 + 1}{2x^2 - 3x} = 1 + \frac{3x + 1}{2x^2 - 3x}$$

Then, upon rewriting,

$$\frac{3x + 1}{2x^2 - 3x} \quad \text{as} \quad \frac{\dfrac{3}{x} + \dfrac{1}{x^2}}{2 - \dfrac{3}{x}}$$

by dividing both the numerator and denominator by x^2, we notice that as $|x|$ becomes very large $(3x+1)/(2x^2 - 3x)$ approaches 0 so that $f(x)$ approaches 1 asymptotically; that is, $y = 1$ is a horizontal asymptote (Figure 7).

Figure 7

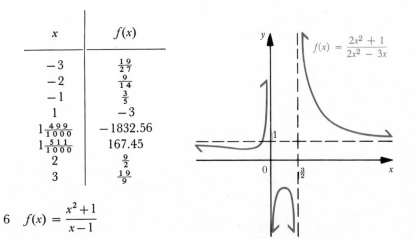

x	$f(x)$
-3	$\frac{19}{27}$
-2	$\frac{9}{14}$
-1	$\frac{3}{5}$
1	-3
$1\frac{499}{1000}$	-1832.56
$1\frac{511}{1000}$	167.45
2	$\frac{9}{2}$
3	$\frac{19}{9}$

$$f(x) = \frac{2x^2 + 1}{2x^2 - 3x}$$

6 $f(x) = \dfrac{x^2 + 1}{x - 1}$

SOLUTION.   The domain of f is $\{x \mid x \neq 1\}$; $(0, -1)$ is the y intercept and $x = 1$ is a vertical asymptote. Since the degree of the numerator is greater than the degree of the denominator, we will use the division algorithm to rewrite f as

$$f(x) = \frac{x^2 + 1}{x - 1} = x + 1 + \frac{2}{x - 1}$$

As $|x|$ becomes very large, $2/(x - 1)$ approaches 0, so $f(x)$ approaches the linear expression $x + 1$ asymptotically (Figure 8).

Figure 8

x	$f(x)$
-2	$-\frac{5}{3}$
-1	-1
0	-1
$\frac{1}{2}$	$-\frac{5}{2}$
$\frac{999}{1000}$	-1998.001
$\frac{1001}{1000}$	2002.001

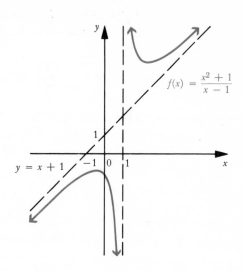

PROBLEM SET 6

For each of the following rational functions, determine the domain, the asymptotes, the x and y intercepts, and sketch the graph. Use the division algorithm to examine those functions in which the degree of the numerator is greater than or equal to the degree of the denominator.

1 $f(x) = \dfrac{9x^2 - 4}{3x - 2}$

2 $f(x) = \dfrac{x + 3}{x^2 + x - 6}$

3 $f(x) = \dfrac{5}{3x + 2}$

4 $f(x) = \dfrac{6x}{x^2 - 4}$

5 $f(x) = \dfrac{8}{3x^2}$

6 $f(x) = \dfrac{x^2+1}{x^2-1}$

7 $f(x) = \dfrac{4}{x^2-9}$

8 $f(x) = \dfrac{x^2}{(x-1)(x-3)}$

9 $f(x) = \dfrac{2x^2+3}{x-1}$

10 $f(x) = \dfrac{x(x-1)}{(x^2-1)(x-3)}$

11 $f(x) = \dfrac{3(x-2)}{x^2-3x+2}$

12 $f(x) = \dfrac{x(x^2+1)}{3(x^2+3)}$

13 $f(x) = \dfrac{x^3}{x^3+1}$

14 $f(x) = \dfrac{x^3-2x^2-11x+12}{x^2-5x+4}$

15 $f(x) = \dfrac{4x^3-5x+1}{x^2-1}$

REVIEW PROBLEM SET

1 Find the degree and the coefficients for each of the following polynomial functions.
a) $f(x) = 4x^3+2x+6$
b) $f(x) = 2x^3+5x^2+7x-3$
c) $f(x) = 7x^5-4x^4+3x^3+2x^2-5x-7$
d) $f(x) = 2x^7-3$

2 Let f be a linear function. Give conditions so that each of the following equations are true for all real numbers.
a) $3f(x) = f(3x)$ b) $f(x+7) = f(x)+f(7)$
c) $f(3x+4) = 3f(x)+f(4)$ d) $f(3) = 4$ and $f(5) = 6$

3   Identify the slope and the x and y intercepts for each of the following functions where $y = f(x)$ and sketch the line in each case. Is f increasing or decreasing? Find f^{-1} if it exists and graph it.

    a) $y = \frac{3}{5}x + \frac{7}{3}$ b) $x/3 + y/5 = 1$

    c) $7x + 5y - 13 = 0$ d) $y = 5$

    e) $x = -3 - y$ f) $y - 2 = -\frac{5}{3}(x - 1)$

4   Show that if f and g are linear functions, then

    a) $f - g$ is a linear function   b) $f \circ g$ is a linear function

5   Find the equation of each of the lines and sketch the line in each case.

    a)   Slope $m = 3$; containing the point $(4, 5)$

    b)   Slope $m = 0$; containing the point $(-2, 3)$

    c)   Containing the points $(3, -2)$ and $(5, 6)$

    d)   Containing the points $(0, 2)$ and $(3, 0)$

6   Assume that the slope of a line $f(x) = mx + b$ is negative and that $x_1 < x_2$. Is it true that $f(x_1) > f(x_2)$? Prove your assertion. What if the slope is positive? Zero?

7   a)   Is the line containing $(3, 5)$ and $(1, 9)$ parallel to the line containing $(5, 8)$ and $(11, -4)$? Explain.

    b)   Express by an equation the condition that the line containing (a, b) and $(-5, 0)$ is perpendicular to the line containing (a, b) and $(5, 0)$, where a and b are constants.

    c)   Three vertices of a rectangle are $(9, 3)$, $(5, 9)$, and $(-7, 1)$. Find its fourth vertex.

    d)   Two opposite vertices of a rectangle are $(6, 2)$ and $(-5, 4)$. Two sides of the rectangle are parallel to the line $8x - 6y + 5 = 0$. Find the equations of the four sides and the coordinates of the other two vertices.

8   A projectile fired upward attains a velocity of v feet per second after t seconds of flight, and the relationship between the two numbers v and t is linear. If the projectile is fired at a velocity of 200 feet per second and reaches a velocity of 90 feet per second after 2 seconds of flight, express v as a function of t.

9   Solve the following systems of linear equations by using the elimination method. In each step of the process show the corresponding matrix form of the system.

    a) $4x - y = -4$ b) $2x + 3y = 9$

         $6x + y = 15$ $5x - 2y = 5$

    c) $3x - y + 2z = 1$ d) $2x + 4y - 3z = -1$

         $6x - 2y + 4z = 2$ $3x + 2y - 5z = 7$

         $9x - 3y + 6z = 3$ $5x - 9y + 2z = 7$

10 Solve each of the following for x.

a) $\begin{vmatrix} x & 0 & 0 \\ 3 & x & 2 \\ 0 & 4 & 1 \end{vmatrix} = 5$

b) $\begin{vmatrix} 1 & x^2 & x \\ 1 & 0 & 2 \\ 4 & 3 & 1 \end{vmatrix} = 28$

c) $\begin{vmatrix} -2 & 1 & x \\ 1 & x+1 & -2 \\ x-6 & 3 & -1 \end{vmatrix} = 0$

11 Use Cramer's rule, if possible, to solve each of the systems in Problem 9.

12 Show that the determinant

$$\begin{vmatrix} 1 & x & x^2 \\ 1 & y & y^2 \\ 1 & z & z^2 \end{vmatrix}$$

can be factored as $(y-z)(z-x)(x-y)$.

13 Solve the following quadratic equations.
a) $(x+1)^2 = 7$
b) $x^2+2x-3 = 0$
c) $-2x^2+x+1 = 0$
d) $x^2-x-1 = 0$

14 Find a value of k such that:
a) The roots of $3x^2-5x+4k = 0$ are real and equal.
b) The roots of $3x^2-2x-k = 0$ are real.

15 Graph each of the following quadratic functions by locating the x and y intercepts and the extreme point. Then use the graph to determine the range of f and to solve the given associated inequality.
a) $f(x) = 6x^2-5x-4; \; 6x^2-5x-4 \leq 0$
b) $f(x) = -3+10x-8x^2; \; -3+10x-8x^2 < 0$
c) $f(x) = (7x-3)(1-2x); \; (7x-3)(1-2x) > 0$
d) $f(x) = -x^2+2x-2; \; -x^2+2x-2 < 0$

16 A projectile is fired from a balloon in such a way that it is h feet above the ground t seconds after the firing where $h = 88t-16t^2$.
a) Find t when $h = 0$.
b) Find the maximum height reached by the projectile.

17 Use synthetic division to determine $Q(x)$ and $f(r)$, so that
$f(x) = (x-r)Q(x)+f(r)$.
a) $f(x) = 3x^3+5x^2+7x-3, \; r = 2$
b) $f(x) = 5x^4-2x^3+11x^2+5x+36, \; r = 1$
c) $f(x) = 2x^3+4x^2+3x-18, \; r = -1$
d) $f(x) = x^7-5, \; r = 2$

18   Solve each of the following inequalities.

a) $(x+2)(x+3)(x-6) < 0$ b) $(x-3)(x-2)(x+4) > 0$

c) $(x-2)(x-3)(x-2)(x+6) > 0$

19   Let $f(x) = x^5 + x^4 - 3x^3 - 5x^2 + x + 7$. Use synthetic division to find each of the following values.

a) $f(2)$ b) $f(3)$ c) $f(-1)$ d) $f(-2)$

20   Use the rational root theorem to prove that $\sqrt{7}$ is an irrational number.

21   Find all the rational zeros of each of the following functions.

a) $\{(x, f(x)) \mid f(x) = x^3 - 4x^2 + x + 6\}$

b) $\{(x, f(x)) \mid f(x) = 48x^4 - 52x^3 + 13x - 3\}$

c) $\{(x, f(x)) \mid f(x) = 2x^4 + 5x^3 - 9x^2 - 15x + 9\}$

22   Explain why the graphs of $f(x) = x - 4$ and $f(x) = (x^2 - 16)/(x+4)$ are different.

23   Determine the domain, the x and y intercepts, the asymptotes, and sketch the graph of each of the following rational functions.

a) $f(x) = \dfrac{2}{x-5}$ b) $f(x) = \dfrac{x}{x+3}$

c) $f(x) = \dfrac{x-2}{x^2-9}$ d) $f(x) = \dfrac{x^2}{x+1}$

e) $f(x) = \dfrac{x^2+2}{x^2+3}$

CHAPTER 4

Exponential and Logarithmic Functions

1 Introduction

This chapter begins with a brief review of the properties of exponents. Next we approach the study of exponential functions by answering the questions: What are the domain and range? What are the graphs? Are the functions even or odd? Are the functions increasing or decreasing? Are the functions continuous? Do the exponential functions have inverses? The answer to the last question will lead to the development of logarithmic functions and the properties of logarithms. Also, mathematical induction will be introduced so that some of the properties of exponents and the binomial theorem can be proved. Finally, summation notation and geometric series will be surveyed.

2 Properties of Exponents

Initially, exponents can be considered to be shorthand notations for indicating repeated multiplication of the same factor. For example, a^4 represents $a \cdot a \cdot a \cdot a$, $7^2 = 7 \cdot 7 = 49$, and $(x+y)^3 = (x+y)(x+y)(x+y)$.

In general, if n is a positive integer and a is a real number, then a^n, the nth *power of a*, represents the product of n factors each equal to a, that is,

$$a^n = \overbrace{a \cdot a \cdot a \cdot a \cdots a}^{n \text{ factors of } a}$$

a is called the *base* and n is called the *exponent* of a^n.

The basic laws of exponents are given in the following theorem without proof.

THEOREM 1

Suppose that m and n are positive integers and assume that a and b are real numbers. Then

i $a^n \cdot a^m = a^{n+m}$
ii $(a^m)^n = a^{mn}$
iii $(ab)^n = a^n b^n$

iv $\quad \dfrac{a^m}{a^n} = \begin{cases} a^{m-n} & \text{if } m > n \\[2mm] \dfrac{1}{a^{n-m}} & \text{if } n > m \\[2mm] 1 & \text{if } n = m \end{cases}$ $\qquad$ provided that $a \neq 0$

The meaning of exponents can be extended to include the negative integers together with zero so that the properties of Theorem 1 continue to hold.

DEFINITION 1 a^0 AND a^{-n}

If a is a real number different from 0, then a^0 is defined as $a^0 = 1$ and a^{-n} is defined as $a^{-n} = 1/a^n$, where n is a positive integer.

EXAMPLES

Write each of the following expressions in a "simplified" form that has only positive integral exponents.

1 $\quad \dfrac{3ab^{-2}}{c^3 d^{-4}}$

SOLUTION

$$\frac{3ab^{-2}}{c^3 d^{-4}} = \frac{3a(1/b^2)}{c^3(1/d^4)} = \frac{3a}{b^2} \cdot \frac{d^4}{c^3} = \frac{3ad^4}{b^2 c^3}$$

2 $\quad \dfrac{\left(x^{k-4n}\right)^k}{x^{k(k-5n)} x^{n(k-6n)}}$

SOLUTION

$$\frac{\left(x^{k-4n}\right)^k}{x^{k(k-5n)} x^{n(k-6n)}} = \frac{x^{k^2-4nk}}{\left(x^{k^2-5nk}\right)\left(x^{nk-6n^2}\right)} \qquad \text{(Theorem 1, ii)}$$

$$= \frac{x^{k^2-4nk}}{x^{k^2-4nk-6n^2}} \qquad \text{(Theorem 1, i)}$$

$$= x^{6n^2} \qquad \text{(Theorem 1, iv)}$$

3 $\quad \dfrac{\left(2^{4x+2}\right)\left(8^3\right)}{\left(2^{4x}\right)\left(2^{10}\right) - \left(2^{4x+7}\right)\left(2^6\right)}$

SOLUTION

$$\frac{\left(2^{4x+2}\right)\left(8^3\right)}{\left(2^{4x}\right)\left(2^{10}\right) - \left(2^{4x+7}\right)\left(2^6\right)} = \frac{\left(2^{4x+2}\right)\left(2^3\right)^3}{2^{4x+10} - 2^{4x+13}} \qquad \text{(Theorem 1, i)}$$

$$= \frac{\left(2^{4x+2}\right)\left(2^9\right)}{2^{4x+10} - 2^{4x+13}} \qquad \text{(Theorem 1, ii)}$$

$$= \frac{2^{4x+11}}{2^{4x+10} - 2^{4x+13}} \qquad \text{(Theorem 1, i)}$$

$$= \frac{2^{4x+11}}{2^{4x+10}(1 - 2^3)} \qquad \text{(why?)}$$

$$= \frac{2}{1 - 8} \qquad \text{(Theorem 1, iv)}$$

$$= -\frac{2}{7}$$

Exponents can also be extended to include rational numbers. Rational number exponents are defined so that the properties of Theorem 1 continue to hold. We begin by defining rational exponents of the form $1/n$.

DEFINITION 2 PRINCIPAL ROOT

If a is a real number and n is a positive integer, then $a^{1/n}$, called the nth *principal root* of a, is defined to be the number x which satisfies $x^n = a$. If $a > 0$, $a^{1/n} > 0$; if $a < 0$ and n is odd, $a^{1/n} < 0$; if $a < 0$ and n is even, $a^{1/n}$ is not defined. $a^{1/n}$ can also be expressed in the form $\sqrt[n]{a}$; the latter form is called a *radical* with *index n*.

For example,

$$4^{1/2} = 2 \qquad \text{since} \qquad 2^2 = 4$$
$$(-8)^{1/3} = -2 \qquad \text{since} \qquad (-2)^3 = -8$$

and

$$(-9)^{1/2} \qquad \text{is not defined}$$

Note that $\sqrt{4} = 2$, not -2, because of Definition 2.

Definition 2, together with Theorem 1, can be used to motivate the definition of exponents that are rational numbers, for if p/q is a positive rational number, then the rule $(a^n)^m = a^{nm}$ suggests that

$$a^{p/q} = (a^{1/q})^p = (\sqrt[q]{a})^p$$

or, equivalently,

$$a^{p/q} = (a^p)^{1/q} = (\sqrt[q]{a^p})$$

so that we have the following definition.

DEFINITION 3 RATIONAL NUMBER EXPONENT

If p/q is a positive rational number and if a is a real number, then $a^{p/q} = (a^p)^{1/q} = (a^{1/q})^p$, where a and q must satisfy the conditions of Definition 2.

If p/q is a negative rational number and $a \neq 0$, $a^{p/q} = 1/a^{-p/q}$ and the first part of the definition is applicable to $a^{-p/q}$, since $-p/q$ is a positive rational number.

For example,

$$32^{2/5} = (32^{1/5})^2 = (\sqrt[5]{32})^2 = 2^2 = 4$$

and

$$32^{-2/5} = \frac{1}{32^{2/5}} = \frac{1}{4}$$

Care must be taken in applying Theorem 1 to rational exponents. For example, $(-1)^{1/3} = -1$; however, $(-1)^{1/3} = (-1)^{2/6} = [(-1)^2]^{1/6} = 1^{1/6} = 1$ is not true because $(-1)^{1/6}$ is not defined.

EXAMPLES

1 Simplify each of the following expressions.

a) $\sqrt{125}$ b) $\sqrt{(\frac{4}{25})^3}$ c) $\sqrt[3]{\sqrt{9}+\sqrt{25}}$

d) $\left(\frac{x^{n^2+1}}{x^{1-n}}\right)^{1/(n+1)}$ n is a positive integer

SOLUTION

a) $\sqrt{125} = 125^{1/2} = (5\cdot25)^{1/2} = 5^{1/2}\cdot25^{1/2} = 5^{1/2}\cdot5 = \sqrt{5}\cdot5 = 5\sqrt{5}$

b) $\sqrt{(\frac{4}{25})^3} = (\frac{4}{25})^{3/2} = \frac{4^{3/2}}{25^{3/2}} = \frac{(4^{1/2})^3}{(25^{1/2})^3} = \frac{2^3}{5^3} = \frac{8}{125}$

c) $\sqrt[3]{\sqrt{9}+\sqrt{25}} = (9^{1/2}+25^{1/2})^{1/3} = (3+5)^{1/3} = 8^{1/3} = 2$

d) $\left(\frac{x^{n^2+1}}{x^{1-n}}\right)^{1/(n+1)} = (x^{n^2+n})^{1/(n+1)} = (x^{n(n+1)})^{1/(n+1)} = x^n$

2 Express each of the numbers without radicals in the denominator. (This process is often referred to as *rationalizing the denominator*.)

a) $1/\sqrt{3}$ b) $1/(2+\sqrt{3})$ c) $\sqrt{2}/(1+\sqrt{2}+\sqrt{3})$

SOLUTION

a) $\frac{1}{\sqrt{3}} = \frac{1}{\sqrt{3}}\cdot\frac{\sqrt{3}}{\sqrt{3}} = \frac{\sqrt{3}}{3}$

b) $\dfrac{1}{2+\sqrt{3}} = \dfrac{1}{2+\sqrt{3}} \cdot \dfrac{2-\sqrt{3}}{2-\sqrt{3}} = 2 - \sqrt{3}$

c) $\dfrac{\sqrt{2}}{1+\sqrt{2}+\sqrt{3}} = \dfrac{\sqrt{2}}{(1+\sqrt{2})+\sqrt{3}} \cdot \dfrac{(1+\sqrt{2})-\sqrt{3}}{(1+\sqrt{2})-\sqrt{3}}$

$= \dfrac{\sqrt{2}+2-\sqrt{6}}{(1+\sqrt{2})^2 - 3} = \dfrac{\sqrt{2}+2-\sqrt{6}}{1+2\sqrt{2}+2-3}$

$= \dfrac{\sqrt{2}+2-\sqrt{6}}{2\sqrt{2}} \cdot \dfrac{\sqrt{2}}{\sqrt{2}} = \dfrac{2+2\sqrt{2}-2\sqrt{3}}{4}$

$= \dfrac{1+\sqrt{2}-\sqrt{3}}{2}$

3   For what value of x does $27^x = 81$ hold?

SOLUTION.   Since $27 = 3^3$, $27^x = (3^3)^x = 3^{3x}$. Also $81 = 3^4$, so that $3^{3x} = 3^4$. This latter equation holds if $3x = 4$, that is, if $x = \frac{4}{3}$ and $\{\frac{4}{3}\}$ is the solution set.

4   Solve $2^{2x+2} - 9(2^x) + 2 = 0$.

SOLUTION. $2^{2x+2} - 9(2^x) + 2 = 0$ can be written $(2^x)^2 2^2 - 9(2^x) + 2 = 0$. If we let $y = 2^x$, we get $4y^2 - 9y + 2 = 0$, that is, $(4y-1)(y-2) = 0$, so that $y = \frac{1}{4}$ or $y = 2$, and $2^x = \frac{1}{4}$ or $2^x = 2$. Hence $x = -2$ or $x = 1$ and the solution set is $\{-2,1\}$.

It can be shown, by methods beyond the scope of this text, that the properties of exponents can be extended to include irrational numbers, so that Theorem 1 is true for all real number exponents if the further restrictions that $a > 0$ and $b > 0$ are imposed. Furthermore, it can be shown that if $a > 0$, then $a^x > 0$ for any real number x. The latter fact will be needed for investigating the properties of exponential functions.

PROBLEM SET 1

1   Simplify each of the following expressions and justify each step by using Theorem 1 and the definitions of Section 2.

a) $(-5)^2$

b) $(3^{-2})(3^7)$

c) $(3^4)(3^{-5})$

d) $5^6/-5^8$

e) $(\frac{1}{3})^2 (\frac{1}{3})^4 (3^{-5})$

f) $[(3^{-2})(3^{-1})]^{-2}$

g) $16^{-2}[(2^{-1})(2)(2^5)]^4$

h) $[(5^{-1})(5^2 \cdot 5^{-3})]^{-1}$

i) $\dfrac{3^{-1}}{(1+\frac{2}{3}-\frac{1}{3})^3}$

j) $\dfrac{(x^2)(x^3)(x^{-1})(x^{-2})^3}{x^2}$

k) $\dfrac{(x^{-1})(x^2)(x^6)^{-3}}{x^{-4}}$

l) $\dfrac{32^{4/5}(2^{n+1})^4}{8^3\cdot(16)^{n-1}}$

m) $\dfrac{9^2(27)^{2-n}}{81^{-n}(27)^3(3^{n-1})}$

2 Prove each of the following statements if a and b are real numbers and m and n are positive integers. Also, indicate what conditions a, b, m, and n must satisfy in order for the expressions to be defined (see Definition 2).

a) $(\sqrt[n]{a})^n = a$

b) $\sqrt[n]{ab} = \sqrt[n]{a}\sqrt[n]{b}$

c) $\sqrt[n]{\dfrac{a}{b}} = \dfrac{\sqrt[n]{a}}{\sqrt[n]{b}}$

d) $\sqrt[m]{\sqrt[n]{a}} = \sqrt[mn]{a}$

3 Simplify each of the following expressions.

a) $\sqrt{3}\cdot\sqrt[3]{27}\cdot\sqrt[4]{9}$

b) $(\sqrt[4]{x^3})(\sqrt[5]{x^2})$

c) $\sqrt[4]{x^2}\sqrt{x\sqrt[3]{x}}$

d) $(\sqrt{2})(\sqrt[3]{3})(\sqrt[4]{4})$

e) $\sqrt{9+\sqrt{27}}$

f) $[\sqrt[3]{x}(16x^{-5/4}/x^{4/3})]^{1/2}$

g) $\sqrt[n]{\dfrac{4^n\cdot6}{4^{2n+1}+2^{4n+1}}}$

h) $\sqrt[n]{\dfrac{9(3^{4n+1}+9^{2n})}{(3^{2n+2})(4)}}$

4 a) Explain why it is *not* true in general that if x is a real number, then $\sqrt{x^2}=x$. Give examples. How should $\sqrt{x^2}$ be defined?

b) Is $\sqrt{x^2}+\sqrt{y^2} = |x| + |y|$? Give two examples.

c) Is $\sqrt{x^2+y^2} = |x+y|$? Give two examples.

5 Rationalize the denominator in each of the following expressions.

a) $\dfrac{2}{\sqrt{2}}$

b) $\dfrac{5}{\sqrt{3}-\sqrt{2}}$

c) $\dfrac{5}{\sqrt{x+1}}$

d) $\dfrac{2\sqrt{3}+\sqrt{5}}{\sqrt{3}}$

e) $\dfrac{\sqrt{x+1}+\sqrt{x-1}}{\sqrt{x+1}-\sqrt{x-1}}$

f) $\dfrac{\sqrt{3}+1}{\sqrt{2}+\sqrt{3}+\sqrt{5}}$

6 Evaluate -3^2 and $(-3)^2$. In general, what is the difference between $-a^x$ and $(-a)^x$? Are the two numbers ever the same? If one of the two is defined, is it necessarily true that the other is also defined? Give examples to support your assertions.

7 Solve each of the following equations for x.

a) $8^x = 32$

b) $6^{2x} = 216$

c) $2^{x+4x} = \frac{1}{8}$

d) $2^{3-x}(4^{2x-1}) = 16$

e) $(3^{4x-3})^{-2} = (27)^{-x-8}$

f) $(5^3)^{2x-6} = (5^{-2})^{4-x}$

g) $(5^{2x})(25) = (125)^{x-1}$

h) $2^{2x+2}+2^{x+2} = 3$

8 Criticize the following "proof" that $-1 = 1$.

Proof

$$\frac{-1}{1} = \frac{1}{-1} \qquad \text{(both equal } -1)$$

$$\sqrt{\frac{-1}{1}} = \sqrt{\frac{1}{-1}} \qquad \text{(if } a = b, \text{ then } \sqrt{a} = \sqrt{b})$$

$$\frac{\sqrt{-1}}{\sqrt{1}} = \frac{\sqrt{1}}{\sqrt{-1}} \qquad \begin{array}{l}\text{(quotient property of radicals – see} \\ \text{Problem Set 1, Problem 2c)}\end{array}$$

$$\sqrt{-1} \cdot \sqrt{-1} = \sqrt{1} \cdot \sqrt{1} \qquad \begin{array}{l}\text{(multiply each side by the} \\ \text{common denominator)}\end{array}$$

$$\therefore \ -1 = 1 \qquad (\sqrt{a} \cdot \sqrt{a} = a)$$

9 Indicate the domain and the range of each of the following functions. Indicate whether the function is even or odd and graph it. Is the function increasing or decreasing?

a) $f(x) = x^{1/2}$ b) $f(x) = \sqrt{1-x}$

c) $f(x) = \sqrt{x^2 + 1}$ d) $f(x) = \sqrt{-x}$

e) $f(x) = \sqrt[4]{x}$

3 Exponential Functions and Their Properties

Suppose that a biologist grows a colony of a certain kind of bacteria. As part of his investigation he wishes to discover how the number of bacteria changes with time. He discovers that the time required for the number of bacteria to triple does not depend on the number of bacteria that are initially present. To be specific, assume that on a given day there are x_0 bacteria present and the number triples each day. There are $3x_0$ present after 1 day, $3(3x_0) = 3^2 x_0$ after 2 days, $3(3^2 x_0) = 3^3 x_0$ after 3 days, and so on, so that after n days there are $3^n x_0$ bacteria present. This phenomena of bacteria growth can be represented by the function $f(t) = 3^t x_0$, where t represents the number of days after the experiment begins and x_0 represents the number of bacteria initially present.

Now, although it is true that the actual experiment supplies data only when t is a positive integer, we can assume that the function $f(t) = 3^t x_0$ is continuous, so that $f(t)$ indicates the number of bacteria present after t days, where t is any positive number. For example, after 2 days, 4 hours, there would be $3^{2\frac{1}{6}} x_0 = 3^{13/6} x_0$ bacteria present.

It is our purpose here to investigate the properties of functions of the type $f(t) = 3^t x_0$.

DEFINITION   EXPONENTIAL FUNCTION

If b is a positive number, then the function $f(x) = b^x$ is called an *exponential function* with *base b*.

For example, $f(x) = 3^x$ is an exponential function with base 3; $f(t) = (\frac{1}{3})^t$ is an exponential function with base $\frac{1}{3}$; $f(x) = 2^{-x} = (2^{-1})^x = (\frac{1}{2})^x$ is an exponential function with base $\frac{1}{2}$. As usual, we take the domain of the function to be as much as possible of the real numbers. We know from Section 2 that the domain includes the positive and negative integers, and indeed all the rational numbers. We assume here that b^x may be adequately defined for all irrationals as well (Example 1 displays an intuitive reason for this). Hence the domain of an exponential function is the entire set of real numbers. For $b \neq 1$, the range is the set of positive real numbers, and Example 1 justifies this claim.

EXAMPLES

1   Discuss the properties of $f(x) = 3^x$.

SOLUTION. $f(x) = 3^x$ has base 3. The domain of $f(x) = 3^x$, as with all exponential functions, is the set of real numbers R, and the range is the set of positive real numbers. It can be shown by the more advanced techniques of calculus that all exponential functions are continuous over the entire real line, and this property is used to graph the function (Figure 1).

Figure 1

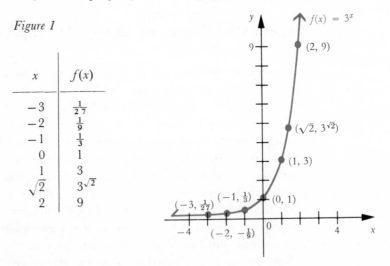

x	$f(x)$
-3	$\frac{1}{27}$
-2	$\frac{1}{9}$
-1	$\frac{1}{3}$
0	1
1	3
$\sqrt{2}$	$3^{\sqrt{2}}$
2	9

The graph of $f(x) = 3^x$ shows that f is an increasing function, so the function has an inverse. Notice that if $x = \sqrt{2}$, then the definition of the exponential function does not apply, since $\sqrt{2}$ is an irrational number. However, the continuity of the exponential function permits us to assume that the point $(\sqrt{2}, 3^{\sqrt{2}})$ is on the graph of $f(x) = 3^x$. Also, the x axis is a horizontal asymptote.

2   Discuss the properties of $f(x) = (\frac{1}{4})^x$.

SOLUTION. $f(x) = (\frac{1}{4})^x$ has base $\frac{1}{4}$; the domain is the set of real numbers R, and the range is the set of positive real numbers. The graph is given in Figure 2. Here again we use the continuity of exponential functions. $f(x) = (\frac{1}{4})^x$ is decreasing, and, consequently, it has an inverse. Also, the x axis is a horizontal asymptote

Figure 2

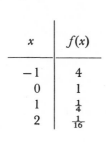

x	$f(x)$
-1	4
0	1
1	$\frac{1}{4}$
2	$\frac{1}{16}$

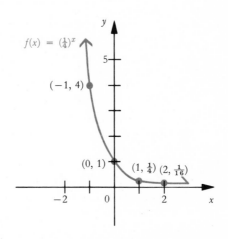

3   Let $y = f(x)$ be an exponential function with base b. Show that $f(x_1 + x_2) = f(x_1) \cdot f(x_2)$ for any two real numbers x_1 and x_2.

SOLUTION.   Since $y = f(x)$ is an exponential function with base b, $f(x) = b^x$. By the rules of exponents, $b^{x_1 + x_2} = b^{x_1} \cdot b^{x_2}$, so $f(x_1 + x_2) = f(x_1) \cdot f(x_2)$.

4   If the graph of an exponential function contains the point $(2, 25)$, what is the base?

SOLUTION.   Since the function is exponential, it is of the form $f(x) = b^x$. $(2, 25)$ on the graph implies that $25 = b^2$, so $b = 5$ is the base. Note that although -5 satisfies $b^2 = 25$, -5 is not the base because of the definition of the exponential function.

PROBLEM SET 2

1   Discuss the properties of each of the following functions. Indicate the domain and range. Graph the function. Is the function increasing or decreasing? Does the function have an inverse?

a) $f(x) = 2^x$ b) $f(x) = 1^x$
c) $f(x) = 3^{x+1}$ d) $f(x) = \sqrt{2^x}$
e) $f(x) = (\frac{4}{7})^x$ f) $f(x) = 2^{-x}$
g) $f(x) = (\frac{1}{5})^{-x}$ h) $f(x) = (0.1)^x$
i) $f(x) = (5)(3^x)$ j) $f(x) = -4^x$

2 Use the graph of $f(x) = 3^x$ to approximate each of the following numbers.
   a) $3^{1/2}$ b) $3^{2\frac{1}{4}}$ c) 3^{π}
   d) $3^{\sqrt{2}}$ e) $3^{-1.2}$

3 What is the base of the exponential function $f(x) = b^x$ whose graphs contain the following point?
   a) $(2, 9)$ b) $(2, 16)$
   c) $(3, 27)$ d) $(5, 3125)$
   e) $(0, 1)$ f) $(\frac{1}{2}, \sqrt{10})$

4 Graph $f(x) = 2^x$ and $f(x) = 2^{-x}$ on the same coordinate system. Are the two graphs symmetric? Explain. Can you generalize this situation?

5 a) Given exponential function $f(x) = b^x$, indicate whether f is increasing or decreasing for $b = 1$, for $0 < b < 1$, and for $b > 1$. Give examples to support your conclusions.
   b) Can an exponential function f be even?—Odd? Explain.

6 Graph $f(x) = 5^x$ and $f(x) = -5^x$ on the same coordinate system. Compare the functional properties (see Problem 1). Can you generalize your results so that you can use the properties of $f(x) = b^x$ to determine the properties of $f(x) = -b^x$?

7 Let $f(x) = 3^x$ and $g(x) = 5^x$. Find
   a) $(f \circ g)(2)$ b) $(g \circ f)(2)$
   c) $(f \circ g)(x)$ d) $(g \circ f)(x)$
   e) $(g \circ f)(0)$

8 Let $f(x) = 3^x + 3^{-x}$ and $g(x) = 3^x - 3^{-x}$. Find
   a) $f + g$ b) $f \cdot g$
   c) f/g d) $[f(x)]^2 - [g(x)]^2$
   e) $f(x^2) - g(x^2)$

9 Suppose that a biologist grows a colony of a certain kind of bacteria. Assume that in an experiment, $N = x_0 3^n$ represents the number of bacteria present at the end of n days, where x_0 is the number of bacteria that are initially present. Suppose that there are 333,000 bacteria present at the end of 2 days.
   a) Find the number of bacteria present at the end of 4 days.
   b) What is the number of days at the end of which there are 111,000 bacteria present?

10 A department store's annual profit, y, from the sale of a certain toy is given by the equation $y = 8000 + 30,000 (2^{-0.4x})$, where y is in dollars and x denotes the number of years the toy has been on the market.
   a) Calculate the store's annual profit for $x = 1, 2, 3, 5, 10,$ and 15.
   b) Use the results of part a to sketch the graph of this function.

11 A survey by a tire company showed the proportion y of tires still usable after having been driven for x miles is given by $y = 2^{-0.003x}$.
   a) Calculate y for $x = 1000, 2000, 3000, 6000, 9000$, and $20,000$.
   b) Use the results of part a to sketch the graph of this function.

4 Logarithmic Functions and Their Properties

The exponential function $f(x) = b^x$ is either an increasing or decreasing function if $b \neq 1$. We know (Chapter 2, Section 5) that such a function is one-to-one and consequently has an inverse. This inverse function is called the *logarithmic function*. More formally, we define the logarithmic function as follows.

DEFINITION   LOGARITHMIC FUNCTION
The function $f^{-1}(x) = \log_b x$, the *logarithmic function with base b*, is the inverse function of the exponential function $f(x) = b^x$, where $b \neq 1$.
   Using set notation, $f(x) = b^x$ can be written as

$$f = \{(x,y) \mid y = b^x\}$$

The inverse f^{-1} is formed from f merely by "reversing" the roles of the members of the ordered pairs of f to get

$$f^{-1} = \{(y,x) \mid y = b^x\}$$

By convention, the symbol x is usually used to represent the domain variable; hence f^{-1} is written as $f^{-1} = \{(x,y) \mid x = b^y\}$. But we have defined f^{-1} as $f^{-1}(x) = \log_b x$, so we also have

$$f^{-1} = \{(x,y) \mid y = \log_b x\}$$

Now, if the latter two representations of f^{-1} are compared, it becomes apparent that

$$y = \log_b x \quad \text{is equivalent to} \quad x = b^y$$

For example,

$$3 = \log_2 8 \quad \text{is equivalent to} \quad 8 = 2^3$$

and

$$x = \log_3 64 \quad \text{is equivalent to} \quad 64 = 3^x$$

EXAMPLES

1   Write each of the following exponential statements as an equivalent logarithmic statement.

a) $3^2 = 9$ b) $10^0 = 1$ c) $\sqrt[3]{27} = 3$

SOLUTION.   By the definition of logarithmic functions we know that $b^y = x$ is equivalent to $y = \log_b x$; hence

a) $3^2 = 9$ is equivalent to $2 = \log_3 9$
b) $10^0 = 1$ is equivalent to $0 = \log_{10} 1$
c) $\sqrt[3]{27} = 27^{1/3} = 3$ is equivalent to $\tfrac{1}{3} = \log_{27} 3$

2   Write each of the following logarithmic statements as an equivalent exponential statement.

a) $\log_{10} 10 = 1$ b) $\log_{1/2} 4 = -2$
c) $\log_{1/2} \tfrac{1}{4} = 2$

SOLUTION.   By the definition of logarithmic functions we know that $y = \log_b x$ is equivalent to $b^y = x$; hence

a) $\log_{10} 10 = 1$ is equivalent to $10^1 = 10$
b) $\log_{1/2} 4 = -2$ is equivalent to $(\tfrac{1}{2})^{-2} = 4$
c) $\log_{1/2} \tfrac{1}{4} = 2$ is equivalent to $(\tfrac{1}{2})^2 = \tfrac{1}{4}$

3   Evaluate $\log_b b$ and $\log_b 1$.

SOLUTION.   If $\log_b b = t$, then $b^t = b$, so $t = 1$. Hence $\log_b b = 1$. Also, $\log_b 1 = t$ implies that $b^t = 1$, so $t = 0$, that is, $\log_b 1 = 0$.

4   Evaluate

$$\frac{\log_{49} 7 - \log_8 64}{\log_9 27 + \log_{10} 100}$$

SOLUTION.   Since $\log_{49} 7 = \tfrac{1}{2}$, $\log_8 64 = 2$, $\log_9 27 = \tfrac{3}{2}$, and $\log_{10} 100 = 2$ (why?), it follows that

$$\frac{\log_{49} 7 - \log_8 64}{\log_9 27 + \log_{10} 100} = \frac{\tfrac{1}{2} - 2}{\tfrac{3}{2} + 2} = \frac{-3}{7}$$

Since the graph of the inverse f^{-1} is a reflection of the graph of f across the line $y = x$, the graph of $f^{-1}(x) = \log_b x$ is a reflection of the graph of $f(x) = b^x$, $b \neq 1$, across the line $y = x$. For example, $g(x) = \log_2 x$ is the inverse of $f(x) = 2^x$ and the graph of $g(x) = \log_2 x$ is the reflection of the graph of $f(x) = 2^x$ (Figure 1). Hence the domain of $f(x) = 2^x$ is the range of $g(x) = \log_2 x$ and the range of $f(x) = 2^x$ is the domain of $g(x) = \log_2 x$, that is, the domain of $g(x) = \log_2 x$ is the set of all positive real numbers and the range is R. The function $g(x) = \log_2 x$ is continuous and increasing; and the y axis is a vertical asymptote of the graph of g.

Figure 1

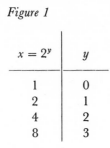

$x = 2^y$	y
1	0
2	1
4	2
8	3

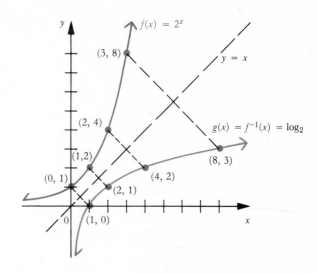

In general, the domain of f^{-1} is the range of f, and the range of f^{-1} is the domain of f. Consequently, since the domain of the exponential function $f(x) = b^x$ is the set of all real numbers, the range of $g(x) = \log_b x$ is also the set of all real numbers; and, since the range of $f(x) = b^x$ is the set of all positive real numbers whenever $b \neq 1$, the domain of $g(x) = \log_b x$ is also the set of all positive real numbers. For example, the domain of $f(x) = \log_5 x$ is the set of positive real numbers, whereas the domain of $f(x) = \log_5(-x)$ is the set of negative real numbers since $-(-x)$ is positive.

EXAMPLE

Find the domain of $f(x) = \log_8(1 - 2x)$.

SOLUTION.   Since the logarithm is defined only for positive real numbers, $1 - 2x > 0$ or $x < \frac{1}{2}$. Hence the domain of $f(x) = \log_8(1 - 2x)$ is $\{x \mid x < \frac{1}{2}\} = (-\infty, \frac{1}{2})$.

A more direct way to graph a logarithmic function $y = \log_b x$ other than to reflect the graph of $y = b^x$ is to use the equivalent equation $x = b^y$ to locate some points on the graph. Thus to graph $y = \log_3 x$ we can use the equivalent equation $x = 3^y$. For $y = 0$, $x = 3^0 = 1$; for $y = 1$, $x = 3^1 = 3$; for $y = 2$, $x = 3^2 = 9$; and so on. One should note that we are reversing the usual technique in finding points on the graph. Here we have selected a value of $y = f(x)$ first and then determined the corresponding value of x (Figure 2).

Figure 2

$x = 3^y$	y
$\frac{1}{9}$	-2
$\frac{1}{3}$	-1
1	0
3	1
9	2
27	3

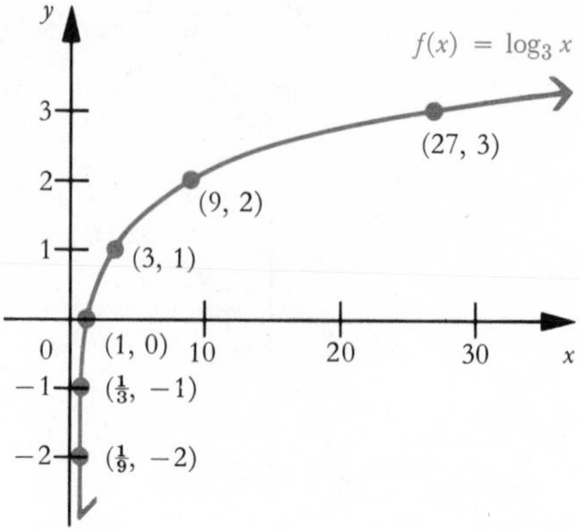
y
f(x) = log3 x
3
2
1
0
−1
−2
(27, 3)
(9, 2)
(3, 1)
(1, 0)
(1/3, −1)
(1/9, −2)
10
20
30
x

EXAMPLES

Discuss the properties of the following functions and sketch the graph.

1 $f(x) = \log_4 x$

SOLUTION. The domain of f is the set of positive real numbers, and the range is the set of real numbers R. Notice that $y = f(x) = \log_4 x$ is equivalent to $4^y = x$; the table of values can be determined by the latter exponential equation. The graph of $f(x) = \log_4 x$ (Figure 3) shows that $f(x) = \log_4 x$ is an increasing, continuous function and the y axis is a vertical asymptote.

Figure 3

$x = 4^y$	y
$\frac{1}{4}$	-1
1	0
4	1
16	2
64	3

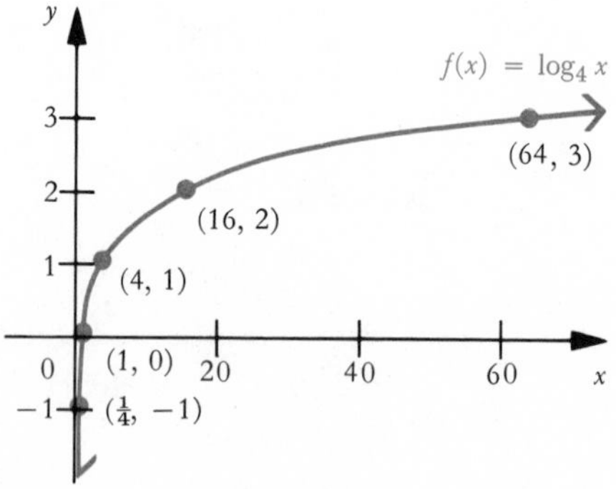
y
f(x) = log4 x
3
2
1
0
−1
(64, 3)
(16, 2)
(4, 1)
(1, 0)
(1/4, −1)
20
40
60
x

2 $f(x) = \log_{1/4} x$

SOLUTION. The domain of f is the set of all positive real numbers and the range is R. The graph can be determined by considering $y = f(x) = \log_{1/4} x$ as $(\frac{1}{4})^y = x$ to locate some points (Figure 4). $f(x) = \log_{1/4} x$ is continuous and decreasing, and the y axis is a vertical asymptote.

Figure 4

$x = (\tfrac{1}{4})^y$	y
$\tfrac{1}{2}$	$\tfrac{1}{2}$
1	0
2	$-\tfrac{1}{2}$

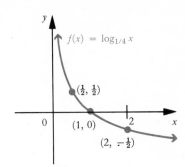

PROBLEM SET 3

1 a) Write each of the following exponential statements as an equivalent logarithmic statement.

   i) $5^3 = 125$ ii) $4^{-2} = \tfrac{1}{16}$

   iii) $\sqrt[5]{32} = 2$ iv) $(\tfrac{1}{3})^{-2} = 9$

   v) $\sqrt{9} = 3$ vi) $(\tfrac{1}{6})^2 = \tfrac{1}{36}$

   b) Write each of the following logarithmic statements as an equivalent exponential statement.

   i) $\log_9 81 = 2$ ii) $\log_{10} 0.0001 = -4$

   iii) $\log_{1/3} 9 = -2$ iv) $\log_{10} \tfrac{1}{10} = -1$

   v) $\log_{\sqrt{16}} 2 = \tfrac{1}{2}$ vi) $\log_{36} 216 = \tfrac{3}{2}$

2 Simplify each of the following expressions.

   a) $\dfrac{\log_2 \sqrt{\tfrac{1}{8}} - \log_2 16^{1/3}}{\log_8 64 \cdot \log_3 3^{1/6}}$ b) $\dfrac{\log_7 7^{4.2} + \log_6 6^{-0.6}}{\log_{25} 5^{3/2} + \log_{0.1} 10}$

   c) $\dfrac{\log_3 243}{\log_2 \sqrt[4]{64} + \log_e e^{-10}}$ d) $\dfrac{\log_\pi \pi^2 + \log_e e^3}{\log_4 16 + \log_2 32}$

3 Discuss the properties of each of the following functions. Indicate the domain and range. Graph the function. Is the function increasing or decreasing? What is the inverse?

   a) $f(x) = \log_6 x$ b) $f(x) = \log_5 x$

   c) $f(x) = \log_{1/2} x$ d) $f(x) = \log_{1/3} x$

   e) $f(x) = \log_\pi x$

4 Use the graph of $f(x) = \log_{10} x$ to approximate each of the following values.

   a) $\log_{10} \tfrac{1}{2}$ b) $\log_{10} 2$

   c) $\log_{10} \pi$ d) $\log_{10} \sqrt{2}$

   e) $\log_{10} 0.1$

5 Graph $f(x) = \log_b x$, where $b = 2, 3, 4$ and 5 on the same coordinate system. How do the graphs compare?

6 Graph $f(x) = \log_2 x$ and $g(x) = 2^x$ on the same coordinate system. Explain the symmetry with respect to the line $y = x$.

7 Compare the functional properties of $f(x) = \log_3 x$ to the functional properties of $f(x) = \log_{1/3} x$. How are they the same? How are they different? In general, how do the logarithmic functions with base less than 1 compare to the logarithmic functions with base greater than 1?

8 Let $f(x) = 10^x$ and $g(x) = \log_{10} x$. Find $f \circ g$ and $g \circ f$. Simplify your answer by using the fact that $f(x) = \log_{10} x$ is equivalent to $10^{f(x)} = x$.

9 What is the domain of each of the following functions?
a) $f(x) = \log_{10} x^2$ b) $f(x) = \log_{10}|x+1|$
c) $f(x) = \log_{10}(-x)$ d) $f(x) = \log_{10}(5x-1)$
e) $f(x) = \log_{10}(x^2 - 4)$

10 Graph $y = \log_{10} x$.
a) For what values of x is $y < 0$?
b) For what values of x is $y > 0$?
c) For what value of x is $y = 0$?
d) If $x_1 < x_2$, how does $\log_{10} x_1$ compare to $\log_{10} x_2$?

5 Properties of Logarithms

Since $y = \log_b x$ is equivalent to $b^y = x$, it follows that the properties of logarithms can be derived from the properties of exponents (see Theorem 1 of Section 2).

THEOREM 1

Suppose that M, N, and b are positive real numbers, $b \neq 1$, and that r is any real number. Then
i $b^{\log_b x} = x$
ii $\log_b MN = \log_b M + \log_b N$
iii $\log_b N^r = r \log_b N$
iv $\log_b(M/N) = \log_b M - \log_b N$

PROOF OF i. Let $u = \log_b x$; then, by the definition of the logarithmic function $b^u = x$, so that by substituting $\log_b x$ for u, we have

$$b^{\log_b x} = x$$

PROOF OF ii. $M = b^{\log_b M}$ and $N = b^{\log_b N}$ by Property i. Hence

$$MN = b^{\log_b M} \cdot b^{\log_b N}$$

and, by applying the properties of exponents, we have

$$MN = b^{\log_b M + \log_b N}$$

so that

$$\log_b MN = \log_b M + \log_b N$$

PROOF OF iii

$$N = b^{\log_b N}$$

so that

$$N^r = (b^{\log_b N})^r$$

and, by the properties of exponents,

$$N^r = b^{r \log_b N}$$

Thus

$$\log_b N^r = r \log_b N$$

PROOF OF iv. Since M/N can be written as MN^{-1}, we have

$$\log_b (M/N) = \log_b (MN^{-1}) = \log_b M + \log_b N^{-1}$$

so that, by Property iii,

$$\log_b (M/N) = \log_b M + (-1) \log_b N$$

or

$$\log_b (M/N) = \log_b M - \log_b N$$

In order to solve logarithmic equations, we will make use of the following property: $\log_b M = \log_b N$ if and only if $M = N$. This result follows from the fact that the logarithmic function is a one-to-one function.

EXAMPLES

1 Let $f(x) = \log_b x$, $f(2) = 0.35$, $f(3) = 0.55$, and $f(5) = 0.82$. Use Theorem 1 to find each of the following values.

a) $f(\tfrac{2}{3})$ b) $f(2^3)$ c) $f(2)/f(3)$
d) $[f(2)]^3$ e) $f(24)$ f) $f(\sqrt{\tfrac{2}{3}})$
g) $f(60/b)$ h) $f(0.6)$

SOLUTIONS

a) $f(\tfrac{2}{3}) = \log_b (\tfrac{2}{3}) = \log_b 2 - \log_b 3 = 0.35 - 0.55 = -0.20$
b) $f(2^3) = \log_b 2^3 = 3 \log_b 2 = 3(0.35) = 1.05$
c) $f(2)/f(3) = 0.35/0.55 = \tfrac{7}{11}$
d) $[f(2)]^3 = (\log_b 2)^3 = (0.35)^3 = 0.042875$
e) $f(24) = \log_b 24 = \log_b (2^3 \cdot 3) = \log_b 2^3 + \log_b 3$
   $= 3 \log_b 2 + \log_b 3 = 3(0.35) + 0.55 = 1.60$

f) $f(\sqrt{\tfrac{2}{3}}) = \log_b (\tfrac{2}{3})^{1/2} = \tfrac{1}{2} \log_b (\tfrac{2}{3}) = \tfrac{1}{2}(\log_b 2 - \log_b 3)$
$\qquad = \tfrac{1}{2}(0.35 - 0.55) = -0.10$

g) $f(60/b) = \log_b (60/b) = \log_b 60 - \log_b b$
$\qquad = \log_b (2^2 \cdot 3 \cdot 5) - \log_b b$
$\qquad = 2 \log_b 2 + \log_b 3 + \log_b 5 - \log_b b$
$\qquad = 2(0.35) + 0.55 + 0.82 - 1$
$\qquad = 0.70 + 0.55 + 0.82 - 1 = 1.07$

h) $f(0.6) = \log_b (0.6) = \log_b (\tfrac{3}{5}) = \log_b 3 - \log_b 5$
$\qquad = 0.55 - 0.82 = -0.27$

2 Solve for x.

a) $\log_4 (3x - 2) = 2$ b) $5^{\log_5 x} = 3$

SOLUTION

a) $\log_4 (3x - 2) = 2$ is equivalent to $4^2 = 3x - 2$, so that $16 = 3x - 2$ or $18 = 3x$, from which it follows that $x = 6$ and the solution set is $\{6\}$.

b) Using Theorem 1, part i, that is, $b^{\log_b x} = b$, we have $x = 3$.

3 Solve for x if $\log_3 (x + 1) + \log_3 (x + 3) = 1$.

SOLUTION. $\log_3 (x + 1) + \log_3 (x + 3) = \log_3 [(x + 1)(x + 3)] = 1$.   (Why?) Hence $x^2 + 4x + 3 = 3$. (Why?) Now $\{x \mid x^2 + 4x + 3 = 3\} = \{x \mid x^2 + 4x = 0\} = \{0, -4\}$; however, -4 does not satisfy the original equation because the logarithm of a negative number is not defined. $x = 0$ is the only solution and the solution set is $\{0\}$.

4 Solve for x if $\log_4 (x + 3) - \log_4 x = 1$.

SOLUTION. $\log_4 (x + 3) - \log_4 x = \log_4 [(x + 3)/x] = 1$; therefore, $(x + 3)/x = 4^1 = 4$, from which it follows that $x = 1$ and the solution set is $\{1\}$.

5 Use the properties of logarithms to write the following expressions as a single logarithmic expression.

a) $\log_2 \tfrac{13}{5} + 2 \log_2 \tfrac{5}{2} - \log_2 \tfrac{169}{8}$ b) $\log_a (c^2 - cd) - \log_a (2c - 2d)$

SOLUTION

a) Using Theorem 1, we have
$$\log_2 \tfrac{13}{5} + 2 \log_2 \tfrac{5}{2} - \log_2 \tfrac{169}{8} = \log_2 \tfrac{13}{5} + \log_2 (\tfrac{5}{2})^2 - \log_2 \tfrac{169}{8}$$
$$= \log_2 \tfrac{13}{5} + \log_2 \tfrac{25}{4} - \log_2 \tfrac{169}{8}$$
$$= \log_2 (\tfrac{13}{5} \cdot \tfrac{25}{4}) - \log_2 \tfrac{169}{8}$$
$$= \log_2 \frac{\tfrac{13}{5} \cdot \tfrac{25}{4}}{\tfrac{169}{8}} = \log_2 \tfrac{10}{13}$$

b) $\log_a (c^2 - cd) - \log_a (2c - 2d) = \log_a \dfrac{c^2 - cd}{2c - 2d} = \log_a \dfrac{c(c - d)}{2(c - d)} = \log_a \dfrac{c}{2}$

PROBLEM SET 4

1 Let $f(x) = \log_{10} x$. If $f(2) = 0.3010$ and $f(3) = 0.4771$, find

a) $f(4)$ b) $f(18)$

c) $f(1000)$ d) $f(\sqrt[5]{2})$

e) $f(3000)$ f) $f(5)$

g) $f(60)$ h) $f(0.5)$

i) $f(\frac{1}{3})$ j) $f(x^0)$, $x \neq 0$

2 Use $x_1 = 10{,}000$, $x_2 = 10$, $b = 10$, and $p = 3$ to show that the following statements are false.

a) $\log_b (x_1/x_2) = (\log_b x_1)/(\log_b x_2)$

b) $(\log_b x_1)/(\log_b x_2) = \log_b x_1 - \log_b x_2$

c) $\log_b x_1 \cdot \log_b x_2 = \log_b x_1 + \log_b x_2$

d) $\log_b (x_1 x_2) = \log_b x_1 \cdot \log_b x_2$

e) $\log_b (x_1^p) = (\log_b x_1)^p$

f) $(\log_b x_1)^p = p \log_b x_1$

3 Solve each of the following equations.

a) $\log_{10} x = 5$ b) $\log_2 2^4 = x$

c) $\log_4 x = \frac{3}{2}$ d) $\log_x 4 = \frac{2}{5}$

e) $\log_3 3^x = 1$ f) $x^{\log_x 2} = 2$

g) $9^{\log_x 7} = 7$ h) $\log_7 (2x - 7) = 2$

i) $\log_5 (5x - 1) = -2$

                             j) $\log_8 \sqrt{\dfrac{3x+4}{x}} = 0$

4 Prove that $\log_b (xyz) = \log_b x + \log_b y + \log_b z$, where b, x, y, and z are positive numbers, with $b \neq 1$. (*Hint*: Use Theorem 1, ii.)

5 Solve each of the following equations.

a) $\log_2 (x^2 - 9) - \log_2 (x + 3) = 2$

b) $\log_{10} (x + 1) - \log_{10} x = 1$

c) $\log_4 x + \log_4 (6x + 10) = 1$

d) $\log_3 x + \log_3 (x - 6) = \log_3 7$

e) $\log_{10} (x^2 - 144) - \log_{10} (x + 12) = 1$

f) $\dfrac{\log_{10} (7x - 12)}{\log_{10} x} = 2$

6 Let $\log_b 2 = A$, $\log_b 3 = B$, and $\log_b 5 = C$. Express $\log_b (0.012)$ in terms of A, B, and C.

7 Use Theorem 1 to simplify each of the following expressions.

a) $\log_5 \frac{5}{7} + \log_5 \frac{49}{25}$

b) $\log_2 \frac{32}{11} + \log_2 \frac{121}{16} - \log_2 \frac{4}{5}$

c) $\log_c (a^2 - ab) - \log_c (7a - 7b)$

d) $\log_7 \left(\dfrac{1}{4} - \dfrac{1}{x^2} \right) - \log_7 \left(\dfrac{1}{2} - \dfrac{1}{x} \right)$

e) $\log_x\left(a + \dfrac{a}{b}\right) - \log_x\left(c + \dfrac{c}{b}\right)$

f) $\log_a \dfrac{a}{\sqrt[3]{x}} - \log_a \dfrac{\sqrt[3]{x}}{a}$

g) $3\log_e \dfrac{a^2b}{c^2} + 2\log_e \dfrac{bc^2}{a^4} + 2\log_e \dfrac{abc}{2}$

6 Computation of Logarithms

The two logarithmic bases used most often are base 10 and base e, where e, an irrational number, is approximately equal to 2.718. We will see how base 10 can be used to simplify certain computations, such as determining approximations to roots of numbers, and to solve certain types of equations, such as $2^x = 3$ and $5^{x-2} = 8$, equations that cannot be solved by the usual methods of algebra.

6.1 Logarithms—Base 10

Base 10 is called the common *base*. By convention, we usually do not write the index 10 when using logarithmic notation so that $\log x$ is the abbreviated way of writing $\log_{10} x$. For certain values of x it is easy to determine $\log x$. For example, $\log 10 = 1$ (why?) and $\log 100 = \log 10^2 = 2$ (why?); in fact, $\log 10^n = n$ (why?).

For other values of x it is *not* easy to determine $\log x$. Let us examine two such values to see how far we could get in computing $\log x$.

Suppose that $x = 5340$. Clearly, we could represent x as $5.34 \cdot 10^3$, so

$$\begin{aligned}\log 5340 &= \log(5.34 \cdot 10^3) \\ &= \log 5.34 + \log 10^3 \\ &= \log 5.34 + 3\end{aligned}$$

Hence the problem has been reduced to finding $\log 5.34$.

Suppose that $x = 0.000234$. Then

$$\begin{aligned}\log 0.000234 &= \log(2.34 \cdot 10^{-4}) \\ &= \log 2.34 + \log(10^{-4}) \\ &= \log 2.34 + (-4)\end{aligned}$$

Here the problem is reduced to finding $\log 2.34$. In both cases, the computation has been reduced to the computation of the logarithm of a number between 1 and 10. The procedure that is suggested by the two examples can be generalized.

Any positive number x can be represented as $x = s \cdot 10^n$, where $1 \le s < 10$ and n is an integer. This form is called the *scientific notation* for x; n is called the *characteristic* of x.

But $\log x = \log s \cdot 10^n = \log s + \log 10^n = (\log s) + n$, so $\log x = (\log s) + n$. The latter form is called the *standard form* of $\log x$ and $\log s$ is called the *mantissa*. Note that since $y = \log x$ is increasing (see Problem Set 3, Problem 10d) and since $1 \le s < 10$, we have the fact that $\log 1 \le \log s < \log 10$, that is, $0 \le \log s < 1$. In other words, the mantissa is always a number between 0 and 1, possibly equal to 0.

Hence the task of determining $\log x$ is reduced to determining $\log s$, where s is always between 1 and 10; but $\log s$ can be determined from the *common log table* (Table I in Appendix A), which gives us *approximations* to the values of the logarithms.

EXAMPLES

In each of the following parts, express the number in scientific notation, and then determine the common logarithm. Indicate the characteristic and the mantissa.

1 53,900.

SOLUTION. $53,900 = 5.39 \cdot 10^4$, so that

$$\begin{aligned}
\log 53,900 &= \log 5.39 + 4 \\
&= 0.7316 + 4 \\
&= 4.7316
\end{aligned}$$

Hence $10^{4.7316} = 53,900$. The mantissa is 0.7316 and the characteristic is 4.

2 0.0035.

SOLUTION. $0.0035 = 3.5 \cdot 10^{-3}$, so that

$$\begin{aligned}
\log 0.0035 &= \log 3.5 - 3 \\
&= 0.5441 - 3 \\
&= -2.4559
\end{aligned}$$

Hence $10^{-2.4559} = 0.0035$. The mantissa is 0.5441 and the characteristic is -3. Note that although $\log 0.0035 = -2.4559$, we say that the mantissa of $\log 0.0035$ is 0.5441 and not 0.4559 or even -0.4559. This is because in applications, the equivalent form, $0.5441 - 3$, is much more convenient to work with.

The process of determining logarithm values can be "reversed." Given a number r, the task is to determine the value of x such that $\log x = r$. This number x is called the *antilogarithm* of r and is sometimes written as $x = \text{antilog}\, r$.

As with finding the logarithm, it is easy to determine x for some values of r, but not so easy for others. For example, if $\log x = -3$, $x = 10^{-3} = 0.001$; if $\log x = 5$, $x = 10^5 = 100,000$; however, if $\log x = 4.4969$, the value of x is not so easy to determine.

The antilog 4.4964, or the solution of $\log x = 4.4969$, can be determined by reversing the process of determining the logarithm. First, we write $\log x = 4.4969$ in standard form, that is, as the sum of a number between 0 and 1 (mantissa) and an integer (characteristic).

$$\begin{aligned} \log x &= 4.4969 \\ &= 0.4969 + 4 \end{aligned}$$

Second, use Table I to find a value s such that $\log s = 0.4969$. Here $s = 3.14$. Hence

$$\begin{aligned} \log x &= 4.4969 \\ &= 0.4969 + 4 \\ &= \log 3.14 + 4 \\ &= \log 3.14 + \log 10^4 \\ &= \log(3.14 \cdot 10^4) \end{aligned}$$

so $x = 31,400$.

EXAMPLES

1   Solve $\log x = 2.7210$.

SOLUTION

$$\begin{aligned} \log x &= 0.7210 + 2 \\ &= \log s + 2 \\ &= \log 5.26 + 2 \\ &= \log(5.26 \cdot 10^2) \end{aligned}$$

so that $x = 526$.

2   Find $10^{-2.0804}$.

SOLUTION.   Let $x = 10^{-2.0804}$, so

$$\begin{aligned} \log x &= -2.0804 \\ &= (-2.0804 + 3) - 3 \\ &= 0.9196 - 3 \\ &= \log 8.31 - 3 \\ &= \log(8.31 \cdot 10^{-3}) \end{aligned}$$

Hence $x = 0.00831$.

6.2 Interpolation

The logarithms and antilogarithms that we computed in Section 6.1 were special in the sense that we were able to find the necessary numbers in Table I. But what if this were not the case? Suppose, for example, that we wanted to find $\log 1.234$ or suppose that we wanted to find the antilog 0.2217. We would not be able to find $\log 1.234$ or mantissa 0.2217 in Table I. This problem can be resolved by using an approximation method called *linear interpolation*.

Let us investigate the meaning of linear interpolation by determining $\log 1.234$:

$$\log 1.24 = 0.0934$$

and

$$\log 1.23 = 0.0899$$

Next let us examine that portion of the graph of $y = \log x$, where $x \in [1.23, 1.24]$ (Figure 1). $\log 1.234$ is the length of the ordinate associated with abscissa 1.234, so $\log 1.234 = 0.0899 + \overline{d}$, where $\overline{d}$ is the "distance" between $\log 1.234$ and 0.0899. $\overline{d}$ is the number that we will approximate. First, we "replace" the arc of the log graph with a line segment (Figure 2). Next, we assume that $\overline{d}$ is approximately the same as d in Figure 2. (The amount of "curvature" of $y = \log x$ has been exaggerated in Figure 2 for illustrative purposes.)

Figure 1

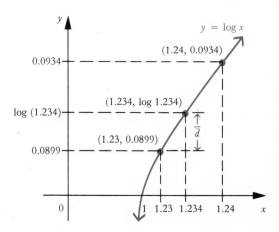

Finally, d can be determined by using the proportionality of the sides of the similar right triangles that have been formed. Thus

$$\frac{d}{0.0035} = \frac{0.004}{0.01}$$

Figure 2

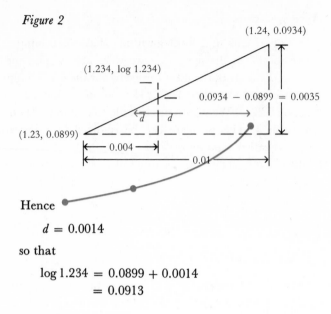

Hence

$$d = 0.0014$$

so that

$$\log 1.234 = 0.0899 + 0.0014$$
$$= 0.0913$$

EXAMPLE

Find x so that $\log x = 0.2217$, by linear interpolation.

SOLUTION.   From Table I we find

$$\log 1.67 = 0.2227$$

and

$$\log 1.66 = 0.2201$$

As before, we examine the graph of $y = \log x$ (Figure 3) for $x \in [1.66, 1.67]$.

Figure 3

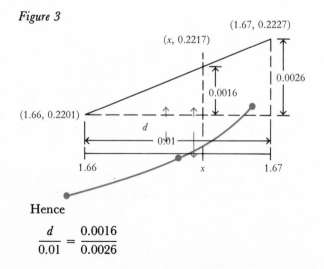

Hence

$$\frac{d}{0.01} = \frac{0.0016}{0.0026}$$

That is, d is approximately 0.006, so that

$$x = 1.66 + 0.006$$
$$= 1.666$$

Essentially, then, linear interpolation is an approximation method that replaces an arc of a curve with a straight-line segment; the accuracy of this method of approximation depends on the "straightness" of the curve between the end points.

The two examples above can be organized in a manner that displays the mechanics involved in linear interpolation as follows.

EXAMPLES

1 Find log 1.234.

SOLUTION

$$0.01 \begin{bmatrix} \log 1.24 & = 0.0934 \\ 0.004 \begin{bmatrix} \log 1.234 = & ? \\ \log 1.23 & = 0.0899 \end{bmatrix} d \end{bmatrix} 0.0035$$

First,

$$\frac{d}{0.0035} = \frac{0.004}{0.01}$$

so that $d = 0.0014$. Hence

$$\log 1.234 = 0.0899 + d$$
$$= 0.0899 + 0.0014$$
$$= 0.0913$$

2 Solve $\log x = 0.2217$.

SOLUTION

$$0.01 \begin{bmatrix} \log 1.67 = 0.2227 \\ d \begin{bmatrix} \log x & = 0.2217 \\ \log 1.66 = 0.2201 \end{bmatrix} 0.0016 \end{bmatrix} 0.0026$$

First,

$$\frac{d}{0.01} = \frac{0.0016}{0.0026}$$

so that $d = 0.006$. Hence

$$x = 1.66 + d$$
$$= 1.66 + 0.006$$
$$= 1.666$$

3   Use logarithms to approximate
$$\frac{(134)^5 (0.35)^8}{(49)^3}$$

SOLUTION.   Let
$$x = \frac{(134)^5 (0.35)^8}{(49)^3}$$

Then
$$\log x = 5 \log 134 + 8 \log 0.35 - 3 \log 49 \qquad \text{(Why?)}$$

But
$$5 \log 134 = 5(2.1271) = 10.6355$$
$$8 \log 0.35 = 8(0.5441 - 1) = 4.3528 - 8$$

and
$$3 \log 49 = 3(1.6902) = 5.0706$$

so that
$$\log x = 10.6355 + (4.3528 - 8) - 5.0706$$

or
$$x = \text{antilog } 1.9177 = 82.74 \text{ (approximately)}$$

4   Use logarithms to approximate $\sqrt[5]{17}$.

SOLUTION.   Assume that $x = \sqrt[5]{17}$. Then $x = 17^{1/5}$, so that

$$\begin{aligned}
\log x &= \tfrac{1}{5} \log 17 \qquad \text{(why?)} \\
&= \tfrac{1}{5}(\log 1.7 + 1) \\
&= \tfrac{1}{5}(0.2304 + 1) \\
&= 0.2461
\end{aligned}$$

Next we can determine x, that is, $\sqrt[5]{17}$, by finding the antilog 0.2461, which, after interpolation, gives $x = 1.762$ (approximately).

5   Use logarithms to solve $5^x = 7$.

SOLUTION.   Since $5^x = 7$, we have $\log 5^x = \log 7$, so that $x \log 5 = \log 7$. Hence

$$x = \frac{\log 7}{\log 5} = \frac{0.8451}{0.699} = \frac{2817}{233}$$

6   If P dollars represents the amount invested at an annual interest rate r, then the amount A_n accumulated in n years when interest is compounded t times a year is given by

$$A_n = P\left(1 + \frac{r}{t}\right)^{nt}$$

Suppose that \$1000 is placed at a yearly interest rate of 6 percent compounded every 4 months.

a) How much money is accumulated after 4 years?

b) In how many years would the money double at this rate?

SOLUTION. Here $P = 1000$, $r = 0.06$, and $t = 3$, so

$$A_n = 1000\left(1 + \frac{0.06}{3}\right)^{3n} \quad \text{or} \quad A_n = 1000(1.02)^{3n}$$

a) If $n = 4$, $A_4 = 1000(1.02)^{12}$. Now we can use logarithms to get

$$\begin{aligned}
\log A_4 &= \log[1000(1.02)^{12}] \\
&= \log 1000 + \log(1.02)^{12} \\
&= 3 + 12 \log 1.02 \\
&= 3 + 12(0.0086) \\
&= 3.1032
\end{aligned}$$

Finally, we determine the antilogarithm:

$$\begin{aligned}
\log A_4 &= 0.1032 + 3 \\
&= \log s + 3 \\
&= \log 1.268 + 3
\end{aligned}$$

so that

$$A_4 = \$1268 \quad \text{(approximately)}$$

b) We must solve

$$2000 = 1000(1.02)^{3n} \quad \text{or} \quad (1.02)^{3n} = 2$$

so that

$$\log(1.02)^{3n} = \log 2 \quad \text{or} \quad 3n \log 1.02 = \log 2$$

Hence

$$n = \frac{\log 2}{3 \log 1.02} = 11.7 \text{ years} \quad \text{(approximately)}$$

6.3 Logarithms—Base e

Base e occurs in many of the formulas that are used in applied mathematics. Base e is called the *natural base*, and *the natural logarithm*, $\log_e x$, *is usually written* $\ln x$. In order to compute natural logarithms, we can use Table II of Appendix A, together with the general properties of logarithms. In Table I of common logarithms it was necessary to write down only the values of $\log_{10} x$ for $1 \leq x < 10$. Logarithms of other numbers were obtained by using the characteristics, the logarithms of the integral powers of 10. Table II of natural logarithms, $\ln x$, are also written for $1 \leqq x < 10$ as in Table I, but now it is more difficult to find logarithms of other numbers, because the logarithms of the integer powers of 10 are not themselves integers. For example, $\ln 3.28 = 1.1878$ can be found directly in Table II whereas $\ln 32.8$ can be found as follows:

$$\ln 32.8 = \ln [(3.28)(10)] = \ln 3.28 + \ln 10 = \ln 3.28 + \ln (2)(5) \quad (why?)$$
$$= \ln 3.28 + \ln 2 + \ln 5 = 1.1878 + 0.6931 + 1.6094 = 3.4903$$

EXAMPLES

1 Use Table II to determine the following values.
   a) $\ln 7.47$ b) $\ln 17.4$

   SOLUTION

   a) Using Table II, we have $\ln 7.47 = 2.0109$.
   b) 17.4 can be written as 1.74×10, so that

   $$\ln 17.4 = \ln 1.74 + \ln 10 = \ln 1.74 + \ln (2)(5) = \ln 1.74 + \ln 2 + \ln 5$$
   $$= 0.5539 + 0.6931 + 1.6094 = 2.8564$$

2 Use natural logarithms to solve $e^{-3t} = 0.5$.

   SOLUTION.   After taking the natural logarithm of both sides of the equation, we get $\ln e^{-3t} = \ln 0.5$, so $-3t \ln e = \ln 0.5$; that is, $-3t = \ln 0.5$, from which it follows, by Table II, that

   $$t = \frac{-0.6931}{-3} = \frac{6931}{30000}$$

3 Suppose that the number of bacteria N present after t hours is given by $N = Pe^{2t}$, where P represents the number of bacteria initially present. How long will it take for the number of bacteria to triple?

   SOLUTION.   We want $N = 3P$, so $3P = Pe^{2t}$, that is, $3 = e^{2t}$, from which it follows that $\ln 3 = \ln e^{2t}$, and $2t \ln e = \ln 3$ or, equivalently, $t = (\ln 3)/2 = 1.0986/2 = 0.55$ hour (approximately).

4 *Radium decomposition.* Suppose that the amount of radium R present after t years is given by $R = Pe^{-kt}$, where P represents the number of grams initially present and k is a constant. Find k if 30 percent of the radium disappears in 100 years.

   SOLUTION.   When $t = 100$, $R = P - 0.3P = 0.7P$. Hence $0.7P = Pe^{-100k}$, or, equivalently, $e^{-100k} = 0.7$, so $-100k \ln e = \ln 0.7$; that is,

   $$k = \frac{\ln 0.7}{-100} = \frac{-0.3566}{-100} = 0.0036 \quad \text{(approximately)}$$

   The next theorem enables us to relate natural and common logarithms.

THEOREM 1

$\log_b x = (\log_a x)/(\log_a b)$, where a and b are positive real numbers different from 1.

PROOF. From Theorem 1, i, of Section 5 we know that $b^{\log_b x} = x$, so that $\log_a (b^{\log_b x}) = \log_a x$. Hence $\log_b x \cdot \log_a b = \log_a x$, from which it follows that

$$\log_b x = \frac{\log_a x}{\log_a b}$$

EXAMPLES

1 Express $\ln x$ in terms of common logarithms.

SOLUTION

$$\ln x = \log_e x = \frac{\log x}{\log e}$$

2 Express $\log_2 10$ in terms of common logarithms.

SOLUTION

$$\log_2 10 = \frac{\log 10}{\log 2} = \frac{1}{\log 2}$$

3 Show that $\log_a b \cdot \log_b a = 1$ for a and b positive numbers different from 1.

SOLUTION

$$\log_a b = \frac{\log_b b}{\log_b a} = \frac{1}{\log_b a}$$

so that

$$\log_a b \cdot \log_b a = 1$$

4 Show that $\ln (\ln x) = \ln (\log x) - \ln (\log e)$.

SOLUTION. We know that $\ln x = (\log x)/(\log e)$, so that

$$\ln (\ln x) = \ln \left(\frac{\log x}{\log e} \right) = \ln (\log x) - \ln (\log e)$$

PROBLEM SET 5

1 Express each of the following numbers in scientific notation, and then compute the common logarithm. Interpolate if necessary. Indicate the mantissa and characteristic.

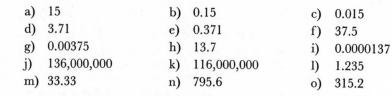

2 Graph $y = \sqrt{x}$. Use linear interpolation, together with the fact that $\sqrt{1} = 1$ and $\sqrt{4} = 2$, to approximate $\sqrt{2}$. Square your result to see how "close" your approximation is to $\sqrt{2}$.

3 Find the approximate value of the antilogarithm. Interpolate if necessary.
a) $\log x = 0.9138$
b) $\log x = 1.7419$
c) $\log x = -0.5933$
d) $\log x = -2.4639$
e) $\log x = 3.9025$
f) $\ln x = 0.1989$
g) $\ln x = 2.1365$
h) $\log x = 3.1461$
i) $\log x = -1.5050$
j) $\log x = -2.5401$
k) $\log x = 2.9977$
l) $\log x = 0.7446$

4 Find each of the following values.
a) $10^{-0.1614}$
b) $10^{1.5428}$
c) $10^{2.8432}$
d) $10^{-2.4857}$

5 Use logarithms to approximate each of the following values.
a) 3.11×3.41
b) $\dfrac{(3.87)^2 (1.326)}{2.11}$
c) $(0.515)^3 (4.7)^5 / (97)^2$
d) $\sqrt[3]{731}$
e) $\dfrac{(45.17)^3 \sqrt{31.1}}{\sqrt[5]{65.31}}$
f) $\sqrt[7]{32.46}$

6 Use common logarithms to approximate each of the following values.
a) $\log_2 5$
b) $\log_{1/10} 10$
c) $\ln 3$
d) $\log_5 \left(\frac{1}{3}\right)$
e) $\log_{1/3} \left(\frac{4}{3}\right)$

7 Use logarithms to solve each of the following equations.
a) $2^x = 7$
b) $e^x = 5$
c) $17^{2x-1} = 4^{-x}$
d) $(0.2)^{x+1} = 0.5$
e) $7^{3x-1} = 5$
f) $5^{3x-5} = 6^{2x}$

8 Suppose that $1500.00 is put into a savings plan that yields a yearly interest rate of $5\frac{1}{2}$ percent. How much money is accumulated after five years, and in how many years would the money triple if the money is compounded?
a) annually
b) semiannually
c) quarterly
d) monthly

9 Suppose that $500.00 is placed at a yearly interest rate of 2 percent compounded annually.
a) How much money is accumulated after 7 years?
b) In how many years would the money double at this rate?

10 Assume that it is known that the number of bacteria N present after t minutes is given by $N = 1000e^{0.25t}$.
a) How many bacteria are present after $\frac{1}{2}$ hour?
b) How long will it take to have 50,000 bacteria?

11 The half-life of radium is defined to be the time required for a given amount of radium to decrease by one-half. Assume that the amount of radium R present after t years is given by $R = Pe^{-kt}$, where P represents the number of grams initially present and k is a constant.
a) Find k if the half-life is 50 years.
b) Find the half-life if $k = 0.005$.

12 If interest is compounded continuously, that is, the number of conversions within any year becomes "infinite," the continuously compounded interest formula

$$A = Pe^{nr}$$

is used, where P represents the principal, r represents the rate of interest per year expressed as a decimal, A represents the total accumulation after n years, and e is the natural base. Use this formula to answer the following questions. Suppose that \$1000 is placed at a yearly interest rate of 6 percent compounded continuously?
a) How much money is accumulated after 4 years?
b) In how many years would the \$1000 double at this rate?

7 Mathematical Induction

In deductive reasoning the process usually begins with a basic assumption called the *hypothesis*. Next, established axioms, definitions, and theorems are used to "deduce" from the hypothesis a true statement. This latter statement is used in turn to deduce another true statement and so on until a sequence of true statements has been organized that concludes with the statement to be proved.

The use of deductive reasoning process is not always apparent when proving general mathematical statements. Let us consider an example. Suppose we were given the task to prove or disprove the following assertion:

$$1 + 2 + 3 + \cdots + n = \frac{n(n+1)}{2} \qquad \text{for any positive integer } n$$

Observe that if $n = 1$, then $1 = 1(1+1)/2$, which is true; if $n = 2$, then $1+2 = 2(2+1)/2$, which is also true; if $n = 3$, then $1+2+3 = 3(3+1)/2$, which is again true.

These tested values of n can only give us an *impression* that the equation is true. Simply testing these values is not adequate to establish a formal proof of the equation for all possible values of n.

Let us consider another example. Suppose we were to prove or disprove the following statement:

$$1 + 4 + 7 + \cdots + (3n-2) = \frac{n(3n-1)}{2} \qquad \text{for any positive integer } n$$

We can begin to test the equation for specific values of n as we did in the first example. If $n = 1$, then the statement becomes $1 = 1(3 \cdot 1 - 1)/2$, which is true; if $n = 2$, then we have $1 + 4 = 2(3 \cdot 2 - 1)/2$, which is also true; if $n = 3$, then we have $1 + 4 + 7 = 3(3 \cdot 3 - 1)/2$, which is true. Again, however, our testing process is not enough to give us a generalized proof.

In order to establish a formal proof of each of the above statements, we need a method of proof that makes use of the *principle of mathematical induction*.

7.1 Principle of Mathematical Induction

Suppose that $S_1, S_2, S_3, \ldots$ is a sequence of assertions; that is, suppose that for each positive integer n we have a corresponding assertion S_n. Assume that the following two conditions hold:

i S_1 is true.
ii For each fixed positive integer k, S_k implies S_{k+1}.

Then it follows that every assertion $S_1, S_2, S_3, \ldots$ is true, that is, S_n is true for all positive integers.

Let us see how the principle can be applied to prove the statement on the bottom of page 235.

S_n is the assertion

$$1 + 2 + 3 + \cdots + n = \tfrac{1}{2}n(n+1)$$

So far as we know, S_n may be true for certain values of n and false for other values of n. In order to show that S_n is, in fact, true for all values of n, we need to verify the following conditions:

i S_1 is true.
ii If S_k is true, then S_{k+1} is also true, where k is a fixed positive integer.

Condition i can be verified by direct computation, for S_1 is the assertion which states that $1 = (\tfrac{1}{2})(1)(1+1)$, which is clearly true.

To prove condition (2) we must show that S_k implies S_{k+1}; that is, we must show that if S_k is assumed to be true, then S_{k+1} must be true. To this end, assume that S_k is true; that is, assume that

$$1 + 2 + 3 + \cdots + k = (\tfrac{1}{2})k(k+1)$$

Since S_k is a true equation, we can add $(k+1)$ to both sides of this equation, to get

$$1 + 2 + 3 + \cdots + k + (k+1) = (\tfrac{1}{2})k(k+1) + (k+1)$$
$$= (\tfrac{1}{2})(k+1)[(k+1) + 1]$$

But the latter assertion is precisely S_{k+1}.

Hence, we have proved condition ii. Thus, by the principle of mathematical induction, we conclude that S_n is true for any positive integer n; therefore, $1 + 2 + 3 + \cdots + n = (\tfrac{1}{2})n(n+1)$ for any positive integer n.

Let us investigate other examples to see how the principle of mathematical induction is especially valuable in proving certain statements in algebra which are to hold for all positive integers.

EXAMPLES

1 Use mathematical induction to prove that for any positive integer n,

$$1 + 4 + 7 + \cdots + (3n-2) = \frac{n(3n-1)}{2}$$

PROOF. Since the principle of mathematical induction will be used, we must prove the two conditions where S_n is the equation

$$1 + 4 + 7 + \cdots + (3n-2) = \frac{n(3n-1)}{2}$$

i S_1 becomes $1 = 1(3 \cdot 1 - 1)/2$, which is true.
ii Assume that S_k is true; that is, assume that for any fixed positive integer k,

$$1 + 4 + 7 + \cdots + (3k-2) = \frac{k(3k-1)}{2}$$

We are to prove that S_{k+1} is true; that is, we must prove that for each positive integer k,

$$1 + 4 + 7 + \cdots + (3k-2) + (3k+1) = \frac{(k+1)(3k+2)}{2}$$

Since S_k is a true equation, we can add $(3k+1)$ to each side of this equation, so that

$$1 + 4 + 7 + \cdots + (3k-2) + (3k+1) = \frac{k(3k-1)}{2} + (3k+1)$$

$$= \frac{3k^2 + 5k + 2}{2}$$

$$= \frac{(k+1)(3k+2)}{2}$$

Hence S_{k+1} is true and we have proved condition ii. Thus we conclude that

$$1 + 4 + 7 + \cdots + (3n-2) = \frac{n(3n-1)}{2}$$

is true for any positive integer n.

2   Use mathematical induction to prove that for any positive integer n,

$$1^2 + 2^2 + 3^2 + \cdots + n^2 = (\tfrac{1}{6})n(n+1)(2n+1)$$

PROOF.   Let S_n represent the equation

$$1^2 + 2^2 + 3^2 + \cdots + n^2 = (\tfrac{1}{6})n(n+1)(2n+1)$$

Using the principle of mathematical induction, we have

i S_1 becomes $1^2 = (\tfrac{1}{6})(1)(1+1)(2\cdot 1+1) = (\tfrac{1}{6})(1)(2)(3) = 1$, which is true.

ii   Assume that S_k is true; that is, assume that for any fixed positive integer k,

$$1^2 + 2^2 + 3^2 + \cdots + k^2 = (\tfrac{1}{6})k(k+1)(2k+1)$$

We must now prove that S_{k+1} is true where S_{k+1} is the equation

$$1^2 + 2^2 + 3^2 + \cdots + k^2 + (k+1)^2 = (\tfrac{1}{6})(k+1)(k+2)(2k+3)$$

Adding $(k+1)^2$ to both sides of equation S_k we have

$$1^2 + 2^2 + 3^2 + \cdots + k^2 + (k+1)^2 = (\tfrac{1}{6})k(k+1)(2k+1) + (k+1)^2$$

$$= (k+1)[(\tfrac{1}{6})k(2k+1) + (k+1)]$$

$$= (k+1)\frac{2k^2 + k + 6k + 6}{6}$$

$$= (k+1)\frac{2k^2 + 7k + 6}{6}$$

$$= (\tfrac{1}{6})(k+1)(k+2)(2k+3)$$

Since the last equation is S_{k+1}, then S_{k+1} is true if S_k is true. Thus, by the principle of mathematical induction, we conclude that S_n is true for any positive integer n. That is, for any positive integer n

$$1^2 + 2^2 + 3^2 + \cdots + n^2 = (\tfrac{1}{6})n(n+1)(2n+1)$$

3   Use mathematical induction to prove that $x^n - y^n$ is divisible by $(x-y)$ for any positive integer n whenever $x \neq y$.

PROOF.   Using the principle of mathematical induction, we must prove the following two conditions.

i   For $n = 1$, $x^n - y^n = x - y$, which is divisible by $x - y$, so the assertion S_1 is true.

ii   Assume that the assertion S_k is true; that is, for any fixed positive integer k, assume that $x^k - y^k$ is divisible by $(x-y)$. We are to show that S_{k+1} is true; that is, we must show that $x^{k+1} - y^{k+1}$ is divisible by $(x-y)$. Now

$$x^{k+1} - y^{k+1} = x^{k+1} - xy^k + xy^k - y^{k+1}$$
$$= x(x^k - y^k) + y^k(x-y)$$

so the right-hand side of the above equation is divisible by $(x-y)$; hence the left-hand side, $x^{k+1} - y^{k+1}$, is also divisible by $(x-y)$ (why?); therefore, the assertion S_{k+1} is true and the proof is complete.

7.2   Binomial Expansions

The principle of mathematical induction is used to prove a theorem that provides a technique other than repeated multiplication for expanding positive integral powers of binomial expressions, such as $(x+3)^{10}$ and $(2x-y)^8$. This theorem is called the *binomial theorem*.

Before investigating this theorem, let us consider some examples of expanding the binomial $a+b$ to integral powers.

$$(a+b)^1 = a + b$$
$$(a+b)^2 = a^2 + 2ab + b^2$$
$$(a+b)^3 = a^3 + 3a^2b + 3ab^2 + b^3$$
$$(a+b)^4 = a^4 + 4a^3b + 6a^2b^2 + 4ab^3 + b^4$$
$$(a+b)^5 = a^5 + 5a^4b + 10a^3b^2 + 10a^2b^3 + 5ab^4 + b^5$$

Notice that the following pattern holds for the a and b terms in the expansion of $(a+b)^n$, where n is a positive integer.

i   There are $n+1$ terms; the "first term" is a^n; the "last term" is b^n.

ii   The power of a decreases by 1 for each term, and the power of b increases by 1 for each term. In any case, the sum of the exponents of a and b is n for each term.

The pattern for the coefficients is easier to detect if the following notation is used. $0!$ (zero factorial) is defined as $0! = 1$. If n is a positive integer, $n!$ (n factorial) is defined as $n! = n(n-1)(n-2) \cdots 3 \cdot 2 \cdot 1$. For example,

$$5! = 5 \cdot 4 \cdot 3 \cdot 2 \cdot 1 = 120$$

and

$$7! = 7 \cdot 6 \cdot 5! = (42)(120) = 5040$$

Now, if $0 \leq k \leq n$, where k and n are integers, then the binomial coefficient, $\binom{n}{k}$, is given by

$$\binom{n}{k} = \frac{n!}{k!(n-k)!}$$

For example,

$$\binom{5}{3} = \frac{5!}{3!(5-3)!} = \frac{5!}{3!2!} = 10$$

$$\binom{17}{14} = \frac{17!}{14!(17-14)!} = \frac{17!}{14!3!} = 680$$

$$\binom{5}{2} = \frac{5!}{2!(5-2)!} = \frac{5!}{2!3!} = 10$$

The binomial theorem formalizes the patterns that occur in binomial expansions.

THEOREM 1   BINOMIAL THEOREM

Let a and b be real numbers and let n be a positive integer; then

$$(a+b)^n = \binom{n}{0}a^n + \binom{n}{1}a^{n-1}b + \cdots + \binom{n}{k}a^{n-k}b^k + \cdots + \binom{n}{n}b^n$$

PROOF.   Use the principle of mathematical induction. Let S_1 be the statement: $(a+b)^1 = (a^1 + b^1)$.

i S_1 is certainly true. (Why?)

ii   We must show that if S_n is true, then S_{n+1} is also true. [Notice that we are using n instead of k for (ii).]

To this end assume that S_n is true; that is, assume that

$$(a+b)^n = \binom{n}{0}a^n + \binom{n}{1}a^{n-1}b + \cdots + \binom{n}{k}a^{n-k}b^k + \cdots + \binom{n}{n}b^n$$

for n a positive integer. After multiplying both sides by $(a+b)$, we obtain

$$(a+b)^n(a+b) = (a+b)\left[\binom{n}{0}a^n + \binom{n}{1}a^{n-1}b + \cdots \right.$$

$$\left. + \binom{n}{k}a^{n-k}b^k + \cdots + \binom{n}{n}b^n\right]$$

$$= \binom{n}{0}(a^{n+1}+a^nb) + \binom{n}{1}(a^nb+a^{n-1}b^2) + \cdots$$

$$+ \binom{n}{k}(a^{n-k+1}b^k+a^{n-k}b^{k+1}) + \cdots$$

$$+ \binom{n}{n}(ab^n+b^{n+1})$$

$$= \binom{n}{0}a^{n+1} + \left[\binom{n}{0}+\binom{n}{1}\right]a^nb$$

$$+ \left[\binom{n}{1}+\binom{n}{2}\right]a^{n-1}b^2 + \cdots$$

$$+ \left[\binom{n}{k-1}+\binom{n}{k}\right]a^{n+1-k}b^k + \cdots + \binom{n}{n}b^{n+1}$$

But

$$\binom{n}{k-1}+\binom{n}{k} = \frac{n!}{(k-1)!(n-k+1)!} + \frac{n!}{k!(n-k)!}$$

$$= \frac{n!k + n!(n-k+1)}{k!(n-k+1)!}$$

$$= \frac{n!(n+1)}{k!(n+1-k)!}$$

$$= \frac{(n+1)!}{k!(n+1-k)!}$$

$$= \binom{n+1}{k}$$

so that

$$(a+b)^{n+1} = \binom{n}{0}a^{n+1} + \binom{n+1}{1}a^nb + \cdots$$

$$+ \binom{n+1}{k}a^{n+1-k}b^k + \cdots + \binom{n+1}{n+1}b^{n+1}$$

But the latter assertion is precisely S_{n+1}, and the proof is complete.

EXAMPLES

Use the binomial theorem to solve each of the following problems.

1 Expand $(x+y)^5$.

SOLUTION. By the binomial theorem,

$$(x+y)^5 = \binom{5}{0}x^5 + \binom{5}{1}x^4y + \binom{5}{2}x^3y^2 + \binom{5}{3}x^2y^3$$

$$+ \binom{5}{4}xy^4 + \binom{5}{5}y^5$$

$$= x^5 + \frac{5!}{1!4!}x^4y + \frac{5!}{2!3!}x^3y^2 + \frac{5!}{3!2!}x^2y^3 + \frac{5!}{4!1!}xy^4 + y^5$$

$$= x^5 + 5x^4y + 10x^3y^2 + 10x^2y^3 + 5xy^4 + y^5$$

2 Expand $(x-3)^4$.

SOLUTION

$$(x-3)^4 = (x+(-3))^4$$

$$= \binom{4}{0}x^4 + \binom{4}{1}x^3(-3) + \binom{4}{2}x^2(-3)^2$$

$$+ \binom{4}{3}x(-3)^3 + \binom{4}{4}(-3)^4$$

$$= x^4 + 4x^3(-3) + 6x^2(-3)^2 + 4x(-3)^3 + (-3)^4$$

$$= x^4 - 12x^3 + 54x^2 - 108x + 81$$

Notice the "symmetry" of the values of the coefficients of the binomial expansions.

$$(a+b)^2 = 1a^2 + 2ab + 1b^2$$

$$(a+b)^3 = 1a^3 + 3a^2b + 3ab^2 + 1b^3$$

and

$$(a+b)^4 = 1a^4 + 4a^3b + 6a^2b^2 + 4ab^3 + 1b^4$$

In general,

$$(a+b)^n = \binom{n}{0}a^n + \binom{n}{1}a^{n-1}b + \binom{n}{2}a^{n-2}b^2 + \cdots + \binom{n}{n-2}a^2b^{n-2} + \binom{n}{n-1}ab^{n-1} + \binom{n}{n}b^n$$

where

$$\binom{n}{k} = \binom{n}{n-k}$$

since

$$\binom{n}{k} = \frac{n!}{k!(n-k)!}$$

and

$$\binom{n}{n-k} = \frac{n!}{(n-k)!(n-(n-k))!} = \frac{n!}{(n-k)!\,k!}$$

3   Find the expansion of $\left(3x^2 - \frac{1}{2}\sqrt{y}\right)^4$.

SOLUTION

$$\left(3x^2 - \tfrac{1}{2}\sqrt{y}\right)^4 = (3x^2)^4 - 4(3x^2)^3\left(\tfrac{1}{2}\sqrt{y}\right) + 6(3x^2)^2\left(\tfrac{1}{2}\sqrt{y}\right)^2$$
$$- 4(3x^2)\left(\tfrac{1}{2}\sqrt{y}\right)^3 + \left(\tfrac{1}{2}\sqrt{y}\right)^4$$
$$= 81x^8 - 54x^6\sqrt{y} + \tfrac{27}{2}x^4 y - \tfrac{3}{2}x^2 y^{3/2} + \tfrac{1}{16}y^2$$

4   Find the sixth term of the expansion of $(2x - y^2)^8$.

SOLUTION.   The sixth term is

$$\binom{8}{5}(2x)^3(-y^2)^5 = -\binom{8}{5}8x^3 y^{10}$$
$$= -\frac{8\cdot7\cdot6}{3\cdot2\cdot1}8x^3 y^{10}$$
$$= -448x^3 y^{10}$$

PROBLEM SET 6

1   Prove the following identities by using the principle of mathematical induction.
   a) $1+3+5+\cdots+(2n-1) = n^2$
   b) $1^3 + 2^3 + 3^3 + \cdots + n^3 = [n^2(n+1)^2]/4$
   c) $2+4+6+\cdots+2n = n^2 + n$
   d) $1^2 + 3^2 + 5^2 + \cdots + (2n-1)^2 = \tfrac{1}{3}n(2n-1)(2n+1)$
   e) $1\cdot2 + 2\cdot3 + 3\cdot4 + \cdots + n(n+1) = \tfrac{1}{3}n(n+1)(n+2)$

2   Prove the following identities by using mathematical induction.

   a) $\dfrac{1}{1\cdot2} + \dfrac{1}{2\cdot3} + \dfrac{1}{3\cdot4} + \cdots + \dfrac{1}{n(n+1)} = \dfrac{n}{n+1}$

   b) $4 + 4^2 + 4^3 + \cdots + 4^n = \tfrac{4}{3}(4^n - 1)$

3   Prove that $a^{2n} - b^{2n}$ is divisible by $a+b$, where a and b are real numbers and n is a positive integer.

4   Prove that $n(n^2+5)$ is divisible by 6 for any positive integer n.

5   Prove that for any number $x \neq 1$,

$$x^0 + x^1 + x^2 + \cdots + x^n = \frac{1-x^{n+1}}{1-x}$$

6   Use induction to prove that $a+(a+d)+(a+2d)+\cdots+[a+(n-1)d] = \frac{1}{2}n[2a+(n-1)d]$ for any positive integer n.

7   Evaluate each of the following expressions.

a) $\dbinom{15}{10}$ b) $\dbinom{n}{n}$ c) $\dbinom{6}{2}$

d) $\dbinom{15}{5}$ e) $\dbinom{n}{0}$ f) $\dbinom{6}{5}$

g) $\dbinom{15}{3}$ h) $\dbinom{n}{1}$ i) $\dbinom{6}{4}$

8   a)   Show that

$$\dbinom{n}{r} + \dbinom{n}{r+1} = \dbinom{n+1}{r+1}$$

b)   Let $a = x = 1$ in the expansion of $(a+x)^n$, and find the sum

$$\dbinom{n}{0} + \dbinom{n}{1} + \dbinom{n}{2} + \cdots + \dbinom{n}{n}$$

9   Find the first four terms of the following expansions.
a) $(x^2+2a)^{10}$ b) $(2a-1/b)^6$
c) $(\sqrt{x/2}+2y)^7$ d) $(1/a+x/2)^{11}$
e) $(a^{3/2}-2x^2)^8$

10 Use the binomial theorem to expand each of the following expressions.
a) $(2z+x)^5$ b) $(x-3)^4$
c) $(y^2-2x)^4$ d) $(1/a+x/2)^3$

11 Find the indicated term for each of the following expressions by using the binomial theorem.
a) $(x^2/2+a)^{15}$, 4th term   b) $(y^2-2z)^{10}$, 6th term
c) $(2x^2-a^2/3)^9$, 7th term   d) $(x+\sqrt{a})^{12}$, middle term
e) $(a+x^2/3)^9$, term containing x^{12}
f) $(2\sqrt{y}-x/2)^{10}$, term containing y^4

8 Finite Sums and Series

We shall discover in this section an abbreviated way, called sigma notation, to write finite sums of a recursive nature such as those sums produced by the binomial theorem. Motivated by the notion of finite sums, we will survey geometric series that are, in a sense, special types of "infinite sums."

8.1 Finite Sums

$\sum_{k=1}^{n} a_k$ is called *sigma notation* and is an abbreviated way of writing the finite sum $a_1 + a_2 + \cdots + a_n$, that is, $\sum_{k=1}^{n} a_k = a_1 + a_2 + \cdots + a_n$. Here a_k represents a function where k, called the *index*, represents the domain variable. For example,

$$\sum_{k=1}^{4} \frac{1}{k} = 1 + \tfrac{1}{2} + \tfrac{1}{3} + \tfrac{1}{4} = \tfrac{25}{12} \qquad \text{with } a_k = \frac{1}{k}$$

EXAMPLES

Evaluate each of the following finite sums in 1, 2, and 3.

1 $\displaystyle\sum_{k=1}^{3} (4k^2 - 3k)$

SOLUTION. $a_k = 4k^2 - 3k$. To find the indicated sum, we substitute the integers 1, 2, and 3 for k in succession, then add the resulting numbers. Thus

$$\sum_{k=1}^{3} (4k^2 - 3k) = (4(1^2) - 3(1)) + (4(2^2) - 3(2)) + (4(3^2) - 3(3))$$
$$= 1 + 10 + 27 = 38$$

2 $\displaystyle\sum_{k=3}^{6} k(k-2)$

SOLUTION. $a_k = k(k-2)$. Notice here that the index begins with 3. To find the indicated sum, we substitute the integers 3, 4, 5, and 6 in succession, then add the resulting numbers. Thus

$$\sum_{k=3}^{6} k(k-2) = [3(3-2)] + [4(4-2)] + [5(5-2)] + [6(6-2)]$$
$$= 3 + 8 + 15 + 24 = 50$$

3 $\displaystyle\sum_{=2}^{5} \frac{k-1}{k+1}$

SOLUTION.   Here $a_k = (k-1)/(k+1)$, so

$$\sum_{k=2}^{5} \frac{k-1}{k+1} = \left(\frac{2-1}{2+1}\right) + \left(\frac{3-1}{3+1}\right) + \left(\frac{4-1}{4+1}\right) + \left(\frac{5-1}{5+1}\right)$$

$$= \tfrac{1}{3} + \tfrac{2}{4} + \tfrac{3}{5} + \tfrac{4}{6} = \tfrac{21}{10}$$

4   Write each of the following finite sums in sigma notation.

a) $1 + \tfrac{1}{2} + \tfrac{1}{4} + \tfrac{1}{8} + \tfrac{1}{16}$

b)   The binomial theorem formula:

$$(a+b)^n = \binom{n}{0}a^n + \binom{n}{1}a^{n-1}b + \cdots + \binom{n}{k}a^{n-k}b^k + \cdots + \binom{n}{n}b^n$$

SOLUTION

a) $1 + \tfrac{1}{2} + \tfrac{1}{4} + \tfrac{1}{8} + \tfrac{1}{16} = (\tfrac{1}{2})^0 + (\tfrac{1}{2})^1 + (\tfrac{1}{2})^2 + (\tfrac{1}{2})^3 + (\tfrac{1}{2})^4$

$$= \sum_{k=0}^{4} (\tfrac{1}{2})^k$$

or, equivalently,

$$1 + \tfrac{1}{2} + \tfrac{1}{4} + \tfrac{1}{8} + \tfrac{1}{16} = \sum_{k=1}^{5} (\tfrac{1}{2})^{k-1} \qquad (\text{why?})$$

b) $\binom{n}{0}a^n + \binom{n}{1}a^{n-1}b + \cdots + \binom{n}{k}a^{n-k}b^k + \cdots + \binom{n}{n}b^n$

$$= \binom{n}{0}a^n b^0 + \binom{n}{1}a^{n-1}b^1 + \cdots + \binom{n}{k}a^{n-k}b^k + \cdots + \binom{n}{n}a^{n-n}b^n$$

$$= \sum_{k=0}^{n} \binom{n}{k}a^{n-k}b^k$$

so that

$$(a+b)^n = \sum_{k=0}^{n} \binom{n}{k}a^{n-k}b^n$$

8.2   Geometric Series

If a_n determines a sequence (see Chapter 2, Section 3.3) and $s_n = \sum_{k=1}^{n} a_k = a_1 + a_2 + \cdots + a_n$, then the sequence determined by s_n is called an *infinite series* and it is usually written as $\sum_{k=1}^{\infty} a_k$. We will be concerned with determining the "sums" of geometric series.

Geometric series are series of the form $\sum_{k=1}^{\infty} ar^{k-1}$, where a is a constant and $|r| < 1$. For example,

$$\sum_{k=1}^{\infty} (\tfrac{1}{2})^{k-1} = 1 + \tfrac{1}{2} + \tfrac{1}{4} + \tfrac{1}{8} + \cdots + (\tfrac{1}{2})^k + \cdots$$

is a geometric series in which $a = 1$ and $r = \tfrac{1}{2}$.

The notion of an "infinite sum" is defined as follows.

Suppose that $\sum_{k=1}^{\infty} a_k$ is an infinite series.

First, *partial sums* are formed:

$$s_1 = a_1$$

$$s_2 = a_1 + a_2$$

$$s_3 = a_1 + a_2 + a_3$$

$$\cdots\cdots\cdots\cdots\cdots\cdots\cdots\cdots$$

$$s_n = a_1 + a_2 + \cdots + a_n = \sum_{k=1}^{n} a_k$$

Then the sum S of $\sum_{k=1}^{\infty} a_k$, written $S = \sum_{k=1}^{\infty} a_k$, is defined to be the "limit value" that s_n approaches as "n approaches infinity," if the limit value is finite.

Let us use this notion to determine the formula for the sum of a geometric series

$$\sum_{k=1}^{\infty} ar^{k-1} \qquad |r| < 1$$

Here $a_k = ar^{k-1}$, so

$$s_1 = a$$

$$s_2 = a + ar$$

(1) $\qquad s_3 = a + ar + ar^2$

$$\cdots\cdots\cdots\cdots\cdots\cdots\cdots\cdots\cdots$$

$$s_n = a + ar + ar^2 + ar^3 + \cdots + ar^{n-1}$$

Upon multiplying both sides of equation (1) by r, we get

(2) $\qquad rs_n = ar + ar^2 + ar^3 + \cdots + ar^n$

Next subtract Equation (2) from (1) to get

$$s_n - rs_n = a - ar^n$$

so that

$$(1-r)s_n = a - ar^n$$

Hence

$$s_n = \frac{a - ar^n}{1 - r} \qquad \text{where } |r| < 1$$

However, the latter equation can be written as

$$s_n = \frac{a}{1 - r} - \frac{ar^n}{1 - r}$$

Intuitively, we can see that as n becomes increasingly large (as n approaches infinity) r^n approaches 0 since $|r| < 1$. [Consider what happens for example to the values of $(\frac{1}{2})^n$ as n becomes larger and larger by examining the graph of $f(n) = (\frac{1}{2})^n$.]

Consequently, s_n approaches $a/(1 - r)$ as n approaches infinity, so that

$$\sum_{k=1}^{\infty} ar^{k-1} = \frac{a}{1 - r} \qquad \text{where } |r| < 1$$

For example,

$$\sum_{k=1}^{\infty} (\tfrac{1}{2})^{k-1} = \frac{1}{1 - \frac{1}{2}} = 2$$

The next example shows that geometric series have an interesting application in connection with the repeating decimals that were introduced in Chapter 1.

EXAMPLE

Use geometric series to find the rational number that corresponds to the following decimals.

a) $0.33\overline{3}$

b) $0.24\overline{2424}$

SOLUTION

a)   From the expression $0.33\overline{3}$ we obtain the geometric series

$$0.33\overline{3} = \frac{3}{10} + \frac{3}{100} + \frac{3}{1000} + \frac{3}{10,000} + \cdots + \frac{3}{10^n} + \cdots$$

$$= \frac{3}{10} + \frac{3}{10}\left(\frac{1}{10}\right) + \frac{3}{10}\left(\frac{1}{10}\right)^2 + \frac{3}{10}\left(\frac{1}{10}\right)^3 + \cdots$$

$$+ \frac{3}{10}\left(\frac{1}{10}\right)^{n-1} + \cdots$$

$$= \sum_{k=1}^{\infty} \frac{3}{10}\left(\frac{1}{10}\right)^{k-1}$$

Here $a = \frac{3}{10}$ and $r = \frac{1}{10}$, so the sum is given by

$$0.33\overline{3} = \frac{\frac{3}{10}}{1 - \frac{1}{10}} = \frac{1}{3}$$

b) $0.24242\overline{4}$ can be written as

$$0.2424\overline{24} = \frac{24}{100} + \frac{24}{10,000} + \frac{24}{1,000,000} + \cdots$$

$$= \sum_{k=1}^{\infty} \left(\frac{24}{100}\right)\left(\frac{1}{100}\right)^{k-1}$$

We have a geometric series in which $a = \frac{24}{100}$ and $r = \frac{1}{100}$. Hence

$$0.2424\overline{24} = \frac{\frac{24}{100}}{1 - \frac{1}{100}} = \frac{24}{99} = \frac{8}{33}$$

PROBLEM SET 7

1 Find the numerical values of each of the following finite sums.

a) $\sum\limits_{k=1}^{5} k$

b) $\sum\limits_{k=0}^{4} 3^{2k}$

c) $\sum\limits_{k=1}^{3} (2k+1)$

d) $\sum\limits_{k=1}^{4} k^k$

e) $\sum\limits_{k=1}^{5} \frac{1}{k(k+1)}$

f) $\sum\limits_{k=0}^{4} \frac{2^k}{(k+1)}$

g) $\sum\limits_{k=2}^{5} 2^{k-2}$

h) $\sum\limits_{k=1}^{5} (3k^2 - 5k + 1)$

i) $\sum\limits_{k=1}^{100} 5$

j) $\sum\limits_{k=1}^{4} \frac{3}{k}$

2 Use the principle of mathematical induction to prove each of the following formulas.

a) $\sum\limits_{k=1}^{n} k = \frac{n(n+1)}{2}$

b) $\sum\limits_{k=1}^{n} (2k-1) = n^2$

c) $\sum\limits_{k=1}^{n} k^2 = \frac{n^3}{3} + \frac{n^2}{2} + \frac{n}{6}$

3 Express each of the following finite sums in sigma notation.
a) $1 + 4 + 7 + 10 + 13$
b) $\frac{1}{2} + \frac{1}{4} + \frac{1}{8} + \frac{1}{16} + \frac{1}{32}$
c) $\frac{3}{5} + \frac{9}{25} + \frac{27}{125} + \frac{81}{625}$
d) $\frac{1}{6} + \frac{2}{11} + \frac{3}{16} + \frac{4}{21}$

4 Determine whether each of the following statements is true or false. Give the reason.

a) $\sum_{k=0}^{100} k^3 = \sum_{k=1}^{100} k^3$

b) $\sum_{k=0}^{100} 2 = 200$

c) $\sum_{k=0}^{100} (k+2) = \left(\sum_{k=0}^{100} k\right) + 2$

d) $\sum_{k=0}^{99} (k+1)^2 = \sum_{k=1}^{100} k^2$

e) $\sum_{k=0}^{100} k^2 = \left(\sum_{k=0}^{100} k\right)^2$

5 Use geometric series to find the rational number that is represented by each of the following decimal numbers.

a) $0.32\overline{3232}$

b) $0.049\overline{999}$

c) $0.4646\overline{46}$

d) $0.072072\overline{072}$

e) $3.561561\overline{561}$

f) $32.421842184\overline{2184218}$

6 Find the sum of each of the following geometric series.

a) $\frac{1}{3}+\frac{1}{9}+\frac{1}{27}+\frac{1}{81}+\cdots+\left(\frac{1}{3}\right)^n+\cdots$

b) $\frac{2}{3}+\frac{4}{9}+\frac{8}{27}+\cdots+\left(\frac{2}{3}\right)^n+\cdots$

c) $\frac{1}{5}+\frac{1}{25}+\frac{1}{125}+\cdots+\left(\frac{1}{5}\right)^n+\cdots$

d) $\frac{9}{8}+\frac{9}{64}+\frac{9}{512}+\cdots+9\left(\frac{1}{8}\right)^n+\cdots$

REVIEW PROBLEM SET

1 Simplify each of the following expressions.

a) $a^{n+4}a^n$

b) $2^n \cdot 4^{n-1} \cdot 8^{n+1}$

c) $x^{n+2} \cdot x^{-n+2}$

d) $(x^{-1})^3$

e) $(4x/y)\left(\sqrt{y^2/4x}\right)$

f) $\sqrt[3]{\frac{1}{2}x^2} \cdot \sqrt[4]{\frac{1}{3}x^3}$

2 Simplify each of the following expressions.

a) $\dfrac{\log_2 \frac{1}{8} - \log_2 16}{\log_8 64 \cdot \log_3 3}$

b) $\dfrac{\log_3 \sqrt{243}\sqrt{81^3}\sqrt{3}}{\log_2 \sqrt[4]{64} + \log_e e^{-10}}$

c) $\dfrac{\log_7 49 + \log_6 216}{\log_{25} 5 + \log_{0.1} 10}$

d) $\dfrac{\log_\pi \pi^4 + \log_e e^5}{\log_4 64 + \log_2 32}$

3 Determine x in each of the following equations.

a) $6^{\log_6 5} + 7^{\log_7 6} = 3^{\log_3 x}$

b) $9^{\log_x 3} = 3$

c) $\log_x 3^4 = 4$

d) $\log_7 7^4 = x$

e) $\log_3 3^x = 4$

4 Sketch the graph of $f(x) = (\sqrt{3})^x$ and find the domain and the range of f. Use the graph to give an approximation of $(\sqrt{3})^{1/2}$.

5 Sketch the graph of each of the following functions, and find the domain and range of each function. Also indicate whether the function is increasing or decreasing.

a) $f(x) = (\frac{1}{2})^x$

b) $f(x) = 2^x + 2^{-x}$

c) $f(x) = 4^x$

d) $f(x) = 2^x - 2^{-x}$

e) $f(x) = 3^{x^2}$

6 Find the domains and ranges of f and g, where $f(x) = 5^x$ and $g(x) = \log_5 x$. Also, verify that

$$(f \circ g)(x) = x \quad \text{and} \quad (g \circ f)(x) = x$$

7 Solve each of the following equations.

a) $2^x = 8^{x-1}$

b) $4^{-x} = 8^{x+2}$

c) $3^{2x+2} - 18(3^x) + 9 = 0$

d) $2^{2x+4} - 7(2^{x+2}) + 6 = 0$

8 a) If $\log_b x = 3$, find $\log_{1/b} x$.

b) If $\log_b x = 7$, find $\log_b (1/x)$.

c) Show that $\log_{1/b} x = \log_b (1/x)$.

9 Let $\log_a 2 = 0.69$, $\log_a 3 = 1.10$, $\log_a 5 = 1.62$, and $\log_a 7 = 1.94$. Find each of the following values.

a) $\log_a (3^5 \cdot 3^7)$

b) $\log_a (\sqrt[5]{5^3 \cdot 7^4})$

c) $\log_a \sqrt[3]{16}$

d) $\sqrt[3]{\log_a 16}$

e) $\log_a (2^4/3^4)$

f) $\log_a (\sqrt[3]{\frac{3}{2}})$

g) $\log_a (60/a)$

h) $\log_a \frac{25}{27}$

10 Explain why $f(x) = \log_b x$ is a one-to-one function, where $b \neq 1$, $b > 0$.

11 Find the domain and range of each of the following functions, and then determine the inverse.

a) $f(x) = 5^x$

b) $f(x) = \log_5 x$

c) $f(x) = \log_5 (3x) + \log_5 x$

d) $f(x) = \log_5 (5x + 3)$

12 Solve each of the following equations.

a) $\log_5 (x^2 - 4) = 0$

b) $\log_5 (2x - 1) + \log_5 (2x + 1) = 1$

c) $\log_{1/2} (4x^2 - 1) - \log_{1/2} (2x + 1) = 1$

13 Combine each of the following expressions into a single term.

a) $\log \frac{11}{5} + \log \frac{14}{3} - \log \frac{22}{15}$

b) $\log \frac{6}{7} - \log \frac{27}{4} + \log \frac{21}{16}$

c) $\log x^3 - \log (2/x^4) + \log x^3 + \log (2/x)$

14 Let $\log_b 2 = A$, $\log_b 3 = B$, and $\log_b 5 = C$. Express $\log_b (0.006)$ in terms of A, B, and C.

15 Sketch the graph of each of the following functions.

a) $g(x) = \log_{1/2} x$

b) $g(x) = \log_2 (x + 1)$

c) $f(x) = (\log_2 x)^2$

16   Let $f(x) = \log_{16} x$. Find the domain and the range of f; also find each of the following values.

a) $f(256)$ b) $f(64)$

c) $f(32)$ d) $f(\sqrt[5]{2})$

17   Use $f(x) = \log x$ to solve the equation $2f(x) = 2f(100) - 3$.

18   If the graph of a logarithmic function contains the point $(1000, 3)$, what is the base?

19   Solve the equation $\log(\log x) = 0$.

20   Solve each of the following equations.

a) $(\frac{1}{2})^x = 8$ b) $(\frac{2}{3})^x = \frac{1}{5}$

c) $(2.06)^x = 300$ d) $(1.03)^x = 3$

e) $3^x = 7$ f) $10^x = 6$

21   If $S = [a(1 - r^n)/(1 - r)]$, solve for n if $S = 9$, $a = 25$, and $r = \frac{3}{4}$.

22   Let $f(x) = cb^x$, $f(0) = 7$, and $f(1) = 14$. Find b and c.

23   Find the composition $f \circ g$ of each of the following pairs of functions.

a) $f(x) = 5^x$ and $g(x) = x^2 + 1$

b) $f(x) = \log_3 x$ and $g(x) = x$

c) $f(x) = x^2 + 3x - 2$ and $g(x) = 2^x$

d) $f(x) = \log_2 x$ and $g(x) = x^2 + x + 1$

24   Use mathematical induction to prove each of the following equations.

a) $2 + 2^2 + 2^3 + \cdots + 2^n = 2(2^n - 1)$

b) $1 - \frac{1}{2} + \frac{1}{4} + \cdots + (-1)^{n-1}(\frac{1}{2})^{n-1} = \frac{2}{3}[1 - (-\frac{1}{2})^n]$

c) $1 \cdot 3 + 2 \cdot 4 + 3 \cdot 5 + \cdots + n(n+2) = \frac{1}{6}n(n+1)(2n+7)$

25   Use the binomial theorem to expand each of the following expressions.

a) $(3x + y)^4$ b) $(3x + \sqrt{x})^5$

c) $(2x + 1/y)^3$ d) $(x - 1/x)^8$

e) $(3y + 1/3\sqrt{y})^6$ f) $(x^3 - 1/\sqrt{x})^9$

26   a)   What amount must have been deposited 20 years ago to amount to \$10,000 now at the yearly interest rate of 3 percent compounded continuously?

b)   Find the accumulated amount from an initial \$1000 after 4 years if the yearly interest rate is 5 percent per year compounded continuously.

27   In the study of probability theory, the normal distribution curve is described by the equation $y = ae^{-cx^2}$. Find the value of c if $a = 2.2$ when $x = 0.5$ and $y = 0.345$.

28   The charge on a condenser is described by $Q = CE(1 - e^{-t/CR})$, where Q is the charge in coulombs, C is the capacity of the condenser in farads, E

is the applied voltage, R is the resistance in ohms, and t is the time in seconds after the voltage is applied. Find Q if $C = 7 \times 10^{-6}$ farads, $R = 600$ ohms, $E = 130$ volts, and $t = 0.025$ seconds.

29 Find the numerical values of each of the following finite sums.

a) $\displaystyle\sum_{k=1}^{5} k(2k-1)$

b) $\displaystyle\sum_{k=1}^{4} 2k^2(k-3)$

c) $\displaystyle\sum_{k=2}^{6} (k+1)(k+2)$

d) $\displaystyle\sum_{k=5}^{10} (2k-1)^2$

e) $\displaystyle\sum_{k=1}^{6} 3^{k+1}$

f) $\displaystyle\sum_{k=4}^{7} \frac{1}{k(k-3)}$

30 Find the sum of each of the following geometric series.

a) $\frac{1}{10}+\frac{1}{100}+\frac{1}{1000}+\cdots+(\frac{1}{10})^n+\cdots$

b) $\frac{3}{5}+\frac{9}{25}+\frac{27}{125}+\cdots+(\frac{3}{5})^n+\cdots$

c) $\frac{3}{4}+\frac{9}{16}+\frac{27}{64}+\cdots+(\frac{3}{4})^n+\cdots$

d) $(\frac{2}{3})^2+(\frac{2}{3})^4+(\frac{2}{3})^6+\cdots+(\frac{2}{3})^{2n}+\cdots$

Circular Functions

1 Introduction

The word "trigonometry" stems from the Greek words for "triangle measurement," and the subject was originally developed to solve geometric problems involving triangles. Today, however, many of the important applications of trigonometry do not involve triangles specifically. The study of periodic phenomena such as sound waves, alternating electric current, and business cycles is based on the principles developed in trigonometry, not from the triangle viewpoint but from the functional approach. The functional approach is concerned with the properties and applications of "circular functions."

Here our study of the circular functions will begin by constructing a function called the "wrapping function."

2 The Wrapping Function—A Periodic Function

The *unit circle* is the circle of radius 1 with center $(0,0)$, whose equation is $x^2 + y^2 = 1$ (see Chapter 1, Section 5.4) (Figure 1). Let us see how it is possible to associate each real number with a point on the unit circle.

Figure 1

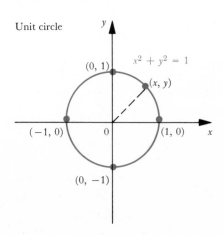

First, associate the set of real numbers with the points on the number line (see Chapter 1, Section 2.2) (assume the same scale unit as the one on the unit circle):

Next, the number line can be "wrapped" around the unit circle using the following scheme. The number 0 on the number line is associated with the point $(1, 0)$ on the unit circle; the positive part of the real line is "wrapped around" the circumference of the circle in the counterclockwise sense; the negative part of the real line is "wrapped around" the circumference of the unit circle in the clockwise sense. Hence, if t is any real number, the point on the unit circle associated with t is determined by moving $|t|$ units along the circumference of the unit circle, beginning at the point $(1, 0)$—in the clockwise sense if $t < 0$ or in the counterclockwise sense if $t > 0$ (Figure 2).

Figure 2

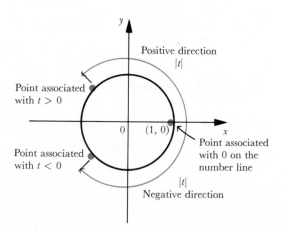

Since the circumference C of a circle is given by $C = 2\pi r$, where r is the radius, the circumference of the unit circle is 2π. Hence it can be seen (Figure 3) that using this wrapping scheme, the point on the unit circle associated with the real number $\pi/2$ is $(0, 1)$; the point associated with π is $(-1, 0)$; the point associated with 2π is $(1, 0)$; the point associated with $-\pi$ is $(-1, 0)$.

The scheme that has been described here suggests a function, called the *wrapping function P*, which associates each real number t with a point (x, y) on the unit circle (Figure 4).

Figure 3

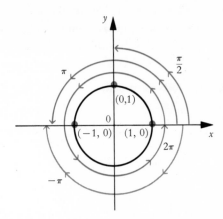

Hence the example illustrated in Figure 3 can be rewritten as

$$P\left(\frac{\pi}{2}\right) = (0,1) \qquad P(\pi) = (-1,0)$$

$$P(-\pi) = (-1,0) \qquad P(2\pi) = (1,0)$$

Figure 4

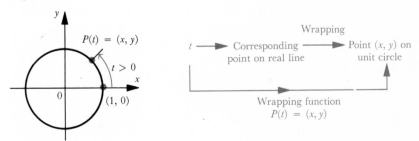

Since the circumference of the unit circle is 2π, we have $P(t) = P(t+2\pi)$ for any real number t; hence the wrapping function associates more than one real number with the same point on the unit circle (Figure 5). For example, $P(0) = P(2\pi) = P(4\pi) = P(6\pi) = (1,0)$. In any case, the function values of P repeat every 2π units.

Figure 5

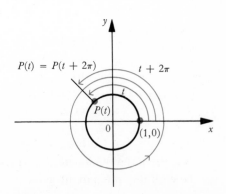

This is our first example of a function which has the property that it repeats in value at regular intervals (in this case, in intervals of 2π). A function that has this property is called a *periodic function*.

2.1 Periodic Functions

A function f is said to be a *periodic function with period a*, where a is a nonzero constant if, for all x in the domain of f, $x + a$ is also in the domain of f, and

$$f(x+a) = f(x)$$

The smallest *positive* period of f is called the *fundamental period of f*.

If f is a periodic function of period a, we can use the definition repeatedly to get

$$
\begin{aligned}
f(x) &= f(x+a) = f[(x+a)+a] \\
&= f(x+2a) = f[(x+2a)+a] \\
&= f(x+3a) \qquad \text{(and so on)}
\end{aligned}
$$

For example, the wrapping function P is a periodic function of period 2π since $P(t) = P(t+2\pi)$ for any real number t. Hence

$$P(t) = P(t+2\pi) = P(t+4\pi) = P(t+6\pi)$$

In general, if f is a periodic function of period a, we have $f(x) = f(x+na)$ for any integer n.

THEOREM 1

If f is a periodic function with period a, then $f(x+na) = f(x)$, where n is any positive integer.

PROOF. We use mathematical induction to prove this result. Let S_n be the statement that $f(x+na) = f(x)$.

i S_1 is certainly true, since $f(x+a) = f(x)$ by the definition of a periodic function.

ii We must show that if S_n is true, then S_{n+1} is also true. To this end, assume that S_n is true; that is, assume that $f(x+na) = f(x)$; then

$$
\begin{aligned}
f[x+(n+1)a] &= f[(x+na)+a] \\
&= f(x+na) \qquad \text{(why?)} \\
&= f(x) \qquad\quad \text{(why?)}
\end{aligned}
$$

But the latter statement is precisely S_{n+1}; hence, by induction, S_n is true for any positive integer n.

Theorem 1 can also be proved when n is a negative integer (see Problem 4). Notice, for example, that if f is a periodic function of period a, then

$$f(x-a) = f[(x-a)+a] = f(x),$$
$$f(x-2a) = f[(x-2a)+a] = f(x-a) = f(x),$$
$$f(x-3a) = f[(x-3a)+a] = f(x-2a) = f(x)$$

and so on. Combining Theorem 1 with the result of Problem 4, we have $f(x+na) = f(x)$ for any integer n.

EXAMPLES

1 Suppose that f is a periodic function with fundamental period 3 and domain R. Assume also that the function values of f are known for $[0,3) \subset R$; that is, assume that $S = \{f(x) \mid x \in [0,3)\}$ is determined. Use Theorem 1 to express each of the following values as values in S.
   a) $f(4)$ b) $f(11)$ c) $f(-3)$
   d) $f(7.5)$ e) $f(-7.5)$

SOLUTION. Since f is a function of period 3, it follows from Theorem 1 that $f(x+n \cdot 3) = f(x)$ for any positive integer n. Consequently,
   a) $f(4) = f(1+3) = f(1)$
   b) $f(11) = f(2+3 \cdot 3) = f(2)$
   c) $f(-3) = f(-3+3) = f(0)$
   d) $f(7.5) = f(1.5+2 \cdot 3) = f(1.5)$
   e) $f(-7.5) = f(-7.5+3 \cdot 3) = f(1.5)$

This example illustrates that if f is a periodic function of fundamental period a, we need only know the values of f on $[0,a)$ in order to determine any value of f (Figure 6).

Figure 6

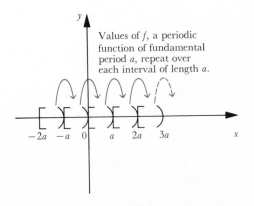

Values of f, a periodic function of fundamental period a, repeat over each interval of length a.

$-2a$ $-a$ 0 a $2a$ $3a$ x

2   If a function f is periodic, then f is not one-to-one. In other words, no periodic function has an inverse (see Chapter 2, Section 5, Theorem 1).

PROOF.   Suppose that f is a periodic function that has fundamental period a.

Since $a \neq 0$, $x+a \neq x$, so that $(x, f(x))$ and $(x+a, f(x+a))$ are two different members of function f such that $f(x) = f(x+a)$; hence f is not one-to-one (Figure 7).

Figure 7

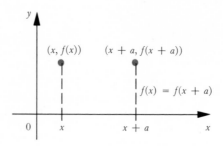

3   Assume that f and g are periodic functions both with fundamental period a. Show that each of the following functions are periodic functions of period a.

a) $f+g$ b) $f \circ f$

SOLUTION.   Since f and g both have fundamental period a, it follows that $f(x+a) = f(x)$ and $g(x+a) = g(x)$.

a)   Thus $(f+g)(x+a) = f(x+a) + g(x+a)$
$$= f(x) + g(x)$$
$$= (f+g)(x)$$

so that $f+g$ is also a periodic function of period a.

b) $(f \circ f)(x+a) = f[f(x+a)]$
$$= f[f(x)]$$
$$= (f \circ f)(x)$$

so that $f \circ f$ is also a periodic function of period a.

2.2   Special Values of $P(t)$

Returning to the wrapping function, we know that $P(t) = P(t+2\pi)$. By examining the unit circle, it can be "seen" that 2π is the *fundamental* period of the wrapping function P. Also, the symmetry of the circle enables us to conclude that if $P(t) = (x,y)$, then $P(-t) = (x, -y)$ (Figure 8). These two properties of P, together with some elementary facts from geometry, can be used to evaluate P for special values.

Figure 8

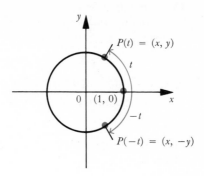

EXAMPLES

1   Determine $P(\pi/4)$, $P(3\pi/4)$, $P(5\pi/4)$, and $P(7\pi/4)$.

SOLUTION.   Let $P(\pi/4) = (x,y)$. Since $P(\pi/4) = P(\frac{1}{2}\cdot\pi/2)$, $P(\pi/4)$ is the midpoint of the arc joining the points $(1,0)$ and $(0,1)$ on the unit circle (Figure 9).

Figure 9

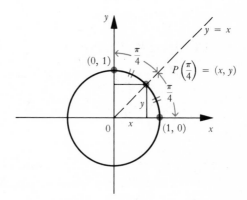

Using the symmetry of the circle, we have $x = y$. Since $x^2 + y^2 = 1$, it follows that $x^2 + x^2 = 2x^2 = 1$; that is,

$$x = \frac{1}{\sqrt{2}} = \frac{\sqrt{2}}{2} \qquad \text{or} \qquad x = \frac{-1}{\sqrt{2}} = \frac{-\sqrt{2}}{2}$$

But x and y are positive in quadrant I; hence

$$x = y = \frac{1}{\sqrt{2}} = \frac{\sqrt{2}}{2}$$

Consequently,

$$P\left(\frac{\pi}{4}\right) = \left(\frac{\sqrt{2}}{2}, \frac{\sqrt{2}}{2}\right)$$

Now the values of $P(3\pi/4)$, $P(5\pi/4)$ and $P(7\pi/4)$ can be determined by using the symmetry of the circle as shown in Figure 10.

Figure 10

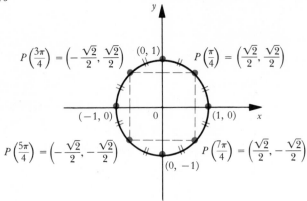

2   Determine $P(\pi/3)$, $P(2\pi/3)$, $P(4\pi/3)$, and $P(5\pi/3)$.

SOLUTION.   Let $P(\pi/3) = (x,y)$; then, using the symmetry of the circle,

$$P\left(\frac{2\pi}{3}\right) = (-x,y) \qquad \text{(Figure 11)}$$

Figure 11

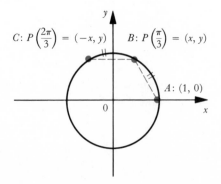

Since the lengths of arcs $\widehat{AB}$ and $\widehat{BC}$ are equal (each is of length $\pi/3$), it follows from geometry that the chords $\overline{AB}$ and $\overline{BC}$ are equal, so that, using the distance formula, we get

$$\overline{AB} = \overline{BC}$$

$$\overline{AB}^2 = \overline{BC}^2$$

(1) $(x-1)^2 + y^2 = 4x^2$

But, since $x^2 + y^2 = 1$, $y^2 = 1 - x^2$. By substitution into equation (1), we have

$$(x-1)^2 + (1-x^2) = 4x^2$$

so that

$$4x^2 + 2x - 2 = 0$$

or, equivalently,

$$2(2x-1)(x+1) = 0 \qquad \text{from which it follows that } x = \tfrac{1}{2}$$

[Notice that $x = -1$ is also a root, but since the point $P(\pi/3)$ is in quadrant I, we choose only $x = \tfrac{1}{2}$.] Hence

$$y^2 = 1 - \tfrac{1}{4} = \tfrac{3}{4}$$

so that

$$y = \frac{\sqrt{3}}{2} \qquad \text{(why?)}$$

and

$$P\left(\frac{\pi}{3}\right) = \left(\frac{1}{2}, \frac{\sqrt{3}}{2}\right)$$

Finally, symmetry can be used to determine the values of $P(2\pi/3)$, $P(4\pi/3)$ and $P(5\pi/3)$ as shown in Figure 12.

Figure 12

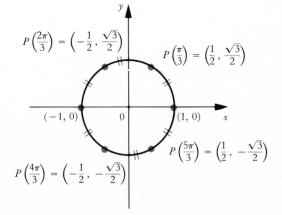

$P\left(\dfrac{2\pi}{3}\right) = \left(-\dfrac{1}{2}, \dfrac{\sqrt{3}}{2}\right)$

$P\left(\dfrac{\pi}{3}\right) = \left(\dfrac{1}{2}, \dfrac{\sqrt{3}}{2}\right)$

$(-1, 0)$ 0 $(1, 0)$

$P\left(\dfrac{5\pi}{3}\right) = \left(\dfrac{1}{2}, -\dfrac{\sqrt{3}}{2}\right)$

$P\left(\dfrac{4\pi}{3}\right) = \left(-\dfrac{1}{2}, -\dfrac{\sqrt{3}}{2}\right)$

3   Find each of the following values.

a) $P\left(\dfrac{9\pi}{4}\right)$ b) $P\left(\dfrac{32\pi}{3}\right)$ c) $P\left(\dfrac{11\pi}{2}\right)$

SOLUTION.   Using the fact that P is of period 2π and using the results of the preceding examples we have

a) $P\left(\dfrac{9\pi}{4}\right) = P\left(\dfrac{\pi}{4} + 2\pi\right) = P\left(\dfrac{\pi}{4}\right) = \left(\dfrac{\sqrt{2}}{2}, \dfrac{\sqrt{2}}{2}\right)$ (Figure 13)

b) $P\left(\dfrac{32\pi}{3}\right) = P\left(\dfrac{2\pi}{3} + 5\cdot2\pi\right) = P\left(\dfrac{2\pi}{3}\right) = \left(-\dfrac{1}{2}, \dfrac{\sqrt{3}}{2}\right)$ (Figure 13)

c) $P\left(\dfrac{11\pi}{2}\right) = P\left(\dfrac{3\pi}{2} + 2\cdot2\pi\right) = P\left(\dfrac{3\pi}{2}\right) = (0, -1)$ (Figure 13)

Figure 13

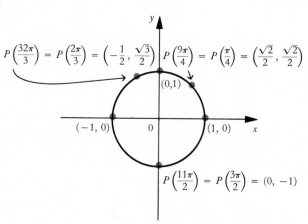

$$P\left(\dfrac{32\pi}{3}\right) = P\left(\dfrac{2\pi}{3}\right) = \left(-\dfrac{1}{2}, \dfrac{\sqrt{3}}{2}\right) \qquad P\left(\dfrac{9\pi}{4}\right) = P\left(\dfrac{\pi}{4}\right) = \left(\dfrac{\sqrt{2}}{2}, \dfrac{\sqrt{2}}{2}\right)$$

$(0,1)$

$(-1, 0)$ 0 $(1, 0)$

$$P\left(\dfrac{11\pi}{2}\right) = P\left(\dfrac{3\pi}{2}\right) = (0, -1)$$

4   Determine each of the following values.

a) $P\left(\dfrac{-5\pi}{4}\right)$ b) $P\left(\dfrac{-2\pi}{3}\right)$

SOLUTION

a) $P\left(\dfrac{-5\pi}{4}\right) = P\left(\dfrac{-5\pi}{4} + 2\pi\right) = P\left(\dfrac{3\pi}{4}\right)$

$= \left(-\dfrac{\sqrt{2}}{2}, \dfrac{\sqrt{2}}{2}\right)$ (See Figure 10, p. 262)

b) $P\left(\dfrac{-2\pi}{3}\right) = P\left(\dfrac{-2\pi}{3} + 2\pi\right) = P\left(\dfrac{4\pi}{3}\right)$

$= \left(\dfrac{-1}{2}, \dfrac{-\sqrt{3}}{2}\right)$ (See Figure 12, p. 263)

PROBLEM SET 1

1 Find each of the following values.
a) $P(0)$ b) $P(-2\pi)$ c) $P(3\pi/2)$
d) $P(-\pi/2)$ e) $P(3\pi)$ f) $P(5\pi/2)$

2 a) Explain why 2π is the *fundamental period* of the wrapping function P.
b) Explain why *any* multiple of 2π is also a period of the wrapping function P.

3 Suppose that $y = f(x)$, a periodic function of fundamental period 2, has been graphed for $x \in [1, 3)$ in Figure 14. Graph $y = f(x)$ for $x \in [-3, 7]$.

Figure 14

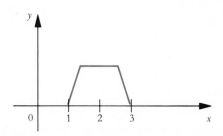

4 Show that if f is a periodic function of period a, then

$$f(x + na) = f(x) \qquad \text{where } n \text{ is a negative integer}$$

5 Let $y = f(x)$ be a periodic function of fundamental period 6. Express each of the following values in terms of values at $x \in [0, 6)$.
a) $f(7)$ b) $f(10)$
c) $f(51)$ d) $f(-\tfrac{1}{2})$
e) $f(-7.15)$

6 Let f and g be periodic functions of period a. Which of the following functions must be periodic?
a) $f \cdot g$ b) $f - g$ c) $f \circ g$

7 Let f be a periodic function of fundamental period π; find two values of x, where $0 \le x \le \pi$, such that
a) $f(15\pi) = f(x)$ b) $f(-\pi) = f(x)$

8 Suppose that f is a periodic odd function of fundamental period 2 such that $f(\tfrac{1}{2}) = 3$. Determine each of the following values.
a) $f(\tfrac{9}{2})$ b) $f(\tfrac{7}{2})$ c) $f(9) + f(-7)$

9 Let P be the wrapping function. Determine the quadrant in which each of the following points lie.
a) $P(0.8)$ b) $P(\tfrac{29}{3})$ c) $P(6)$
d) $P(3.6)$ e) $P(4.5)$ f) $P(-13)$
g) $P(-\tfrac{31}{3})$ h) $P(-3.9)$ i) $P(-7.8)$

10 Determine $P(\pi/6)$, $P(5\pi/6)$, $P(7\pi/6)$, and $P(11\pi/6)$. [*Hint:* If $P(\pi/6) = (x,y)$, then $P(-\pi/6) = (x, -y)$. Notice that $\overset{\frown}{BC} = \overset{\frown}{BA}$, and then use the methods of Example 2 on page 262 (Figure 15).]

Figure 15

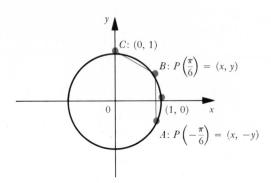

11 Use the properties of the wrapping function P, together with the information in the examples and Problem 10, to determine each of the following values.

a) $P(5\pi)$ b) $P(9\pi/2)$ c) $P(-7\pi/2)$
d) $P(-\pi/4)$ e) $P(13\pi/4)$ f) $P(17\pi/6)$
g) $P(-22\pi/3)$ h) $P(-11\pi/4)$ i) $P(59\pi/6)$
j) $P(43\pi/6)$ k) $P(71\pi/2)$ l) $P(103\pi/3)$

3   Circular Functions—Sine and Cosine

For each real number t, the wrapping function P associates an ordered pair of real numbers (x, y) on the unit circle, so that $P: t \rightarrow (x, y)$. Since it is difficult to develop the properties of a function whose range is a set of ordered pairs rather than a set of single numbers, we shall use the wrapping function P to construct two other functions.

DEFINITION   CIRCULAR FUNCTIONS

The *circular functions*—the *cosine* and *sine*—are defined in the following manner.

$$\text{cosine} = \{(t, x) \mid P(t) = (x,y)\}$$

and

$$\text{sine} = \{(t,y) \mid P(t) = (x,y)\}$$

Using functional notation, the cosine of t, abbreviated $\cos t$, and the sine of t, abbreviated $\sin t$, are given by

$$\cos t = x \quad \text{and} \quad \sin t = y \quad \text{where } P(t) = (x,y)$$

In a certain sense, the cosine and sine functions are constructed by "separating" the ordered pairs in the range of the function P (Figure 1).

Figure 1

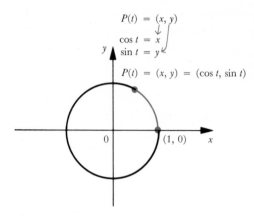

$P(t) = (x, y)$
$\cos t = x$
$\sin t = y$

$P(t) = (x, y) = (\cos t, \sin t)$

$(1, 0)$

For example, since $P(0) = (1,0)$, it follows from the definition of circular function that $\cos 0 = 1$ and $\sin 0 = 0$.

EXAMPLES

1   Find the values of $\cos t$ and $\sin t$ if the coordinates of $P(t)$ are given by $(1/\sqrt{10}, 3/\sqrt{10})$.

SOLUTION.   Since $P(t) = (1/\sqrt{10}, 3/\sqrt{10})$ it follows from the definition of the circular functions that $\cos t = 1/\sqrt{10} = \sqrt{10}/10$ and $\sin t = 3/\sqrt{10} = 3\sqrt{10}/10$.

2   Use the results of Section 2 together with the definition of the circular functions to determine $\cos t$ and $\sin t$ if

a) $t = \pi/2$ b) $t = -\pi/3$ c) $t = 7\pi/4$

SOLUTION

a)   Since $P(\pi/2) = (0, 1)$, $\cos(\pi/2) = 0$ and $\sin(\pi/2) = 1$.
b)   Since $P(-\pi/3) = (\frac{1}{2}, -\sqrt{3}/2)$, $\cos(-\pi/3) = \frac{1}{2}$
     and $\sin(-\pi/3) = -\sqrt{3}/2$.
c)   Since $P(7\pi/4) = (\sqrt{2}/2, -\sqrt{2}/2)$, $\cos(7\pi/4) = \sqrt{2}/2$
     and $\sin(7\pi/4) = -\sqrt{2}/2$.

3   Suppose that $P(t) = (2a, a)$ is a point in quadrant I. Find a, $\cos t$, and $\sin t$.

SOLUTION.   Since $(2a, a)$ is a point on the unit circle, it must satisfy the equation of the circle $x^2 + y^2 = 1$. Thus $(2a)^2 + a^2 = 1$, so that $5a^2 = 1$ or

$a = \pm 1/\sqrt{5}$. However, the point $P(t)$ is in quadrant I where the coordinates are positive so that $a = 1/\sqrt{5}$. By the definition of the circular functions $\cos t = 2a = 2/\sqrt{5} = 2\sqrt{5}/5$ and $\sin t = a = 1/\sqrt{5} = \sqrt{5}/5$.

3.1 Properties of the Cosine and Sine

We have seen that if $P(t) = (x,y)$, then $\cos t = x$ and $\sin t = y$. Now the properties of P can be used to determine some elementary properties of the cosine and sine function; in this sense we sometimes say that the cosine and sine "inherit" certain properties from the wrapping function P.

1   The Domain and Range

Since $P(t) = (x,y)$ is defined for *any* real number t, $\cos t$ and $\sin t$ are defined for any real number; that is, R is the domain for both the cosine and sine functions. Because of the fact that any (x,y) on the unit circle satisfies $-1 \le x \le 1$ and $-1 \le y \le 1$, $\cos t = x$ has range $[-1,1]$ and $\sin t = y$ has range $[-1,1]$ (Figure 2).

Figure 2

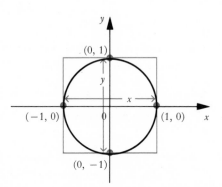

2   Signs of the Values

Also, the signs of the values of the cosine and sine functions depend on the quadrant in which the point $P(t)$ is located (Figure 3).

Figure 3

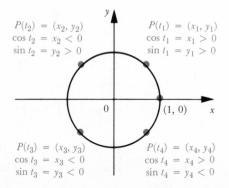

3 Periodicity

Finally, the cosine and sine are periodic functions with period 2π; for $P(t) = (x,y) = P(t+2\pi)$ implies that $\cos t = x = \cos(t+2\pi)$ and $\sin t = y = \sin(t+2\pi)$. Hence, if we can determine the values of the $\cos t$ and $\sin t$ for $t \in [0, 2\pi)$, we can use these values to determine $\cos t$ and $\sin t$ for *any real number* t. We will use this property later when we evaluate the functions, and again when we graph them.

EXAMPLES

Use the results of Section 2, together with the periodicity of the functions, to determine each of the following values.

1 $\cos(25\pi/6)$ and $\sin(25\pi/6)$

SOLUTION

$$\cos\frac{25\pi}{6} = \cos\left(\frac{\pi}{6} + 2\cdot 2\pi\right) = \cos\frac{\pi}{6} = \frac{\sqrt{3}}{2}$$

and

$$\sin\frac{25\pi}{6} = \sin\left(\frac{\pi}{6} + 2\cdot 2\pi\right) = \sin\frac{\pi}{6} = \frac{1}{2} \qquad \text{(Figure 4)}$$

Figure 4

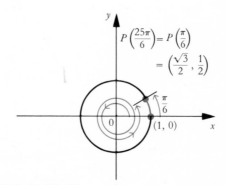

2 $\cos(-27\pi/4)$ and $\sin(-27\pi/4)$

SOLUTION

$$\cos\left(\frac{-27\pi}{4}\right) = \cos\left(\frac{-27\pi}{4} + 4\cdot 2\pi\right) = \cos\left(\frac{5\pi}{4}\right) = \frac{-\sqrt{2}}{2}$$

and

$$\sin\left(\frac{-27\pi}{4}\right) = \sin\left(\frac{-27\pi}{4} + 4\cdot 2\pi\right) = \sin\left(\frac{5\pi}{4}\right) = \frac{-\sqrt{2}}{2}$$

(See Figure 10, page 262.)

3 Use the fact that the sine function is a periodic function of period 2π to determine the period of each of the following functions.

a) $f(t) = \sin 2t$ b) $f(t) = \sin \dfrac{t}{5}$

SOLUTION

a) $f(t) = \sin 2t$
$$= \sin (2t + 2\pi) \qquad \text{(the sine has period } 2\pi)$$
$$= \sin [2(t + \pi)]$$
$$= f(t + \pi)$$
so that f is a periodic function of period π.

b) $f(t) = \sin \dfrac{t}{5}$

$$= \sin \left(\frac{t}{5} + 2\pi\right) \qquad \text{(why?)}$$

$$= \sin [\tfrac{1}{5}(t + 10\pi)]$$

$$= f(t + 10\pi)$$

so that f is a periodic function of period 10π.

By using the fact that x and y are the coordinates of $P(t)$, we can prove more properties of the cosine and sine functions.

THEOREM 1

For any real number t,

i $\cos^2 t + \sin^2 t = 1$
ii $\cos (-t) = \cos t$ (the cosine is an even function)
iii $\sin (-t) = -\sin t$ (the sine is an odd function)

PROOF OF i. Assume that $P(t) = (x,y)$, so that $\cos t = x$ and $\sin t = y$, where $x^2 + y^2 = 1$. By substituting $\cos t$ for x and $\sin t$ for y, we have $(\cos t)^2 + (\sin t)^2 = 1$ [the power notation for $(\cos t)^n$ is $\cos^n t$ and for $(\sin t)^n$ is $\sin^n t$], so that

$$\cos^2 t + \sin^2 t = 1$$

PROOF OF ii AND iii. Although we shall illustrate this result for a value of t which displays $P(t)$ in quadrant I, it is important to remember that the argument can be used for any real number t. Assume that $P(t) = (x,y)$. Then, by definition, $P(t) = (x,y) = (\cos t, \sin t)$ and $P(-t) = (x, -y) = (\cos (-t), \sin (-t))$ (Figure 5). Hence, by the symmetry of the circle, $\cos t = x = \cos (-t)$, and $\sin (-t) = -y = -\sin t$.

Figure 5

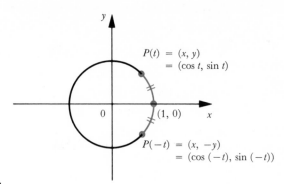

$P(t) = (x, y)$
$\qquad = (\cos t,\ \sin t)$

0

$(1, 0)$

x

$P(-t) = (x, -y)$
$\qquad\ \ = (\cos(-t),\ \sin(-t))$

EXAMPLES

1   Use the property of the signs of the circular functions, together with the identity $\cos^2 t + \sin^2 t = 1$, to solve the following problems.
   a)   If $\sin t = \frac{3}{10}$ and $0 < t < \pi/2$, find $\cos t$.
   b)   If $\sin t = \frac{3}{5}$ and $\pi/2 < t < \pi$, find $\cos t$.
   c)   If $\cos t = 1/\sqrt{2}$ and $3\pi/2 < t < 2\pi$, find $\sin t$.

SOLUTION

   a)   Using $\cos^2 t + \sin^2 t = 1$, we have $\cos^2 t = 1 - \sin^2 t$, so that $\cos t = \pm\sqrt{1 - \sin^2 t}$. Since $P(t)$ lies in quadrant I (Figure 6a), $\cos t > 0$, and

$$\cos t = \sqrt{1 - \sin^2 t} = \sqrt{1 - \left(\frac{3}{10}\right)^2} = \sqrt{1 - \frac{9}{100}} = \sqrt{\frac{91}{100}} = \frac{\sqrt{91}}{10}$$

   b) $\cos t = \pm\sqrt{1 - \sin^2 t}$ but, since $P(t)$ is in quadrant II (Figure 6b), where $\cos t < 0$, we use

$$\cos t = -\sqrt{1 - \sin^2 t} = -\sqrt{1 - \left(\frac{3}{5}\right)^2} = -\sqrt{1 - \frac{9}{25}} = -\frac{4}{5}$$

   c)   Since $\sin t < 0$ in quadrant IV (Figure 6c), we have

$$\sin t = -\sqrt{1 - \cos^2 t} = -\sqrt{1 - \left(\frac{1}{\sqrt{2}}\right)^2} = -\sqrt{1 - \frac{1}{2}} = -\frac{\sqrt{2}}{2}$$

Figure 6

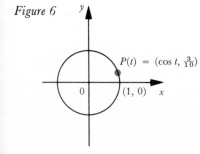

$P(t) = (\cos t,\ \frac{3}{10})$

$(1, 0)$

0

x

(a)

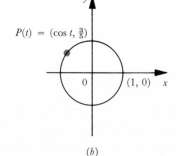

$P(t) = (\cos t,\ \frac{3}{5})$

0

$(1, 0)$

x

(b)

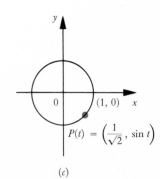

0

$(1, 0)$

x

$P(t) = \left(\frac{1}{\sqrt{2}},\ \sin t\right)$

(c)

2   Assume that $f(t) = \sin t$ and $g(t) = \cos t$ to evaluate
   a) $f(-5\pi)$ b) $g(-10\pi/3)$

SOLUTION

a) $f(-5\pi) = \sin(-5\pi)$
$= -\sin(5\pi)$ (the sine is an odd function)
$= -\sin(\pi + 2 \cdot 2\pi)$ (the sine has period 2π)
$= -\sin \pi$
$= 0$ since $P(\pi) = (-1, 0)$

b) $g\left(\dfrac{-10\pi}{3}\right) = \cos\left(\dfrac{-10\pi}{3}\right)$

$= \cos\left(\dfrac{10\pi}{3}\right)$ (the cosine is an even function)

$= \cos\left(\dfrac{4\pi}{3} + 2\pi\right)$ (why?)

$= \cos\left(\dfrac{4\pi}{3}\right)$

$= -\dfrac{1}{2}$ since $P\left(\dfrac{4\pi}{3}\right) = \left(-\dfrac{1}{2}, \dfrac{-\sqrt{3}}{2}\right)$

3   For what values of t (if any) does $\sin t = \cos t$ if $0 < t < 2\pi$?

SOLUTION.   Using Theorem 1,

$$1 = \sin^2 t + \cos^2 t = \sin^2 t + \sin^2 t = 2 \sin^2 t$$

so that

$$\sin t = \pm\frac{\sqrt{2}}{2} \quad \text{and} \quad \cos t = \pm\frac{\sqrt{2}}{2}$$

Hence it follows that $\sin t = \cos t$ if $t = \pi/4$ or if $t = 5\pi/4$ (Figure 7).

Figure 7

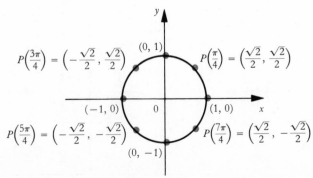

PROBLEM SET 2

1 Use the information given in Figure 8 to complete the table below.

Figure 8

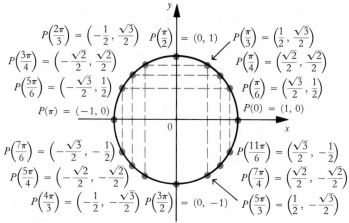

$$P\left(\frac{2\pi}{3}\right) = \left(-\frac{1}{2}, \frac{\sqrt{3}}{2}\right) \quad P\left(\frac{\pi}{2}\right) = (0, 1) \quad P\left(\frac{\pi}{3}\right) = \left(\frac{1}{2}, \frac{\sqrt{3}}{2}\right)$$

$$P\left(\frac{3\pi}{4}\right) = \left(-\frac{\sqrt{2}}{2}, \frac{\sqrt{2}}{2}\right) \qquad P\left(\frac{\pi}{4}\right) = \left(\frac{\sqrt{2}}{2}, \frac{\sqrt{2}}{2}\right)$$

$$P\left(\frac{5\pi}{6}\right) = \left(-\frac{\sqrt{3}}{2}, \frac{1}{2}\right) \qquad P\left(\frac{\pi}{6}\right) = \left(\frac{\sqrt{3}}{2}, \frac{1}{2}\right)$$

$$P(\pi) = (-1, 0) \qquad\qquad P(0) = (1, 0)$$

$$P\left(\frac{7\pi}{6}\right) = \left(-\frac{\sqrt{3}}{2}, -\frac{1}{2}\right) \qquad P\left(\frac{11\pi}{6}\right) = \left(\frac{\sqrt{3}}{2}, -\frac{1}{2}\right)$$

$$P\left(\frac{5\pi}{4}\right) = \left(-\frac{\sqrt{2}}{2}, -\frac{\sqrt{2}}{2}\right) \qquad P\left(\frac{7\pi}{4}\right) = \left(\frac{\sqrt{2}}{2}, -\frac{\sqrt{2}}{2}\right)$$

$$P\left(\frac{4\pi}{3}\right) = \left(-\frac{1}{2}, -\frac{\sqrt{3}}{2}\right) \quad P\left(\frac{3\pi}{2}\right) = (0, -1) \quad P\left(\frac{5\pi}{3}\right) = \left(\frac{1}{2}, -\frac{\sqrt{3}}{2}\right)$$

t	$\cos t$	$\sin t$	t	$\cos t$	$\sin t$
0	1	0	π	-1	0
$\dfrac{\pi}{6}$	$\dfrac{\sqrt{3}}{2}$	$\dfrac{1}{2}$	$\dfrac{7\pi}{6}$	$-\dfrac{\sqrt{3}}{2}$	$-\dfrac{1}{2}$
$\dfrac{\pi}{4}$	$\dfrac{\sqrt{2}}{2}$	$\dfrac{\sqrt{2}}{2}$	$\dfrac{5\pi}{4}$	$-\dfrac{\sqrt{2}}{2}$	$-\dfrac{\sqrt{2}}{2}$
$\dfrac{\pi}{3}$	$\dfrac{1}{2}$	$\dfrac{\sqrt{3}}{2}$	$\dfrac{4\pi}{3}$	$-\dfrac{1}{2}$	$-\dfrac{\sqrt{3}}{2}$
$\dfrac{\pi}{2}$	0	1	$\dfrac{3\pi}{2}$	0	-1
$\dfrac{2\pi}{3}$	$-\dfrac{1}{2}$	$\dfrac{\sqrt{3}}{2}$	$\dfrac{5\pi}{3}$	$\dfrac{1}{2}$	$-\dfrac{\sqrt{3}}{2}$
$\dfrac{3\pi}{4}$	$-\dfrac{\sqrt{2}}{2}$	$\dfrac{\sqrt{2}}{2}$	$\dfrac{7\pi}{4}$	$\dfrac{\sqrt{2}}{2}$	$-\dfrac{\sqrt{2}}{2}$
$\dfrac{5\pi}{6}$	$-\dfrac{\sqrt{3}}{2}$	$\dfrac{1}{2}$	$\dfrac{11\pi}{6}$	$\dfrac{\sqrt{3}}{2}$	$-\dfrac{1}{2}$

2 a) Explain why $P(t) = (\cos t, \sin t)$.

b) Explain why $P(-t) = (\cos t, -\sin t)$.

3 a) Find a, $\cos t$, and $\sin t$ if $P(t) = (3/a, 4/a)$ is in quadrant I.

b) Find $\sin t$ if $\cos t = \frac{5}{13}$ and $P(t)$ is in quadrant IV.

c) Find $\cos t$ if $\sin t = -\frac{5}{6}$ and $P_t(t)$ is in quadrant IV.

d) Find $\sin t$ if $\cos t = -\frac{1}{3}$ and $P(t)$ is in quadrant III.

e) Explain why $\sin t > \cos t$ for any $t \in (\pi/2, \pi)$.

f) Express $\cos^2 t - \sin^2 t$ in terms of a and b if $\cos t = 2ab/(a^2 + b^2)$.

4 a) Is it possible to solve the equation $\sin t = 5$? Explain.

b) Explain why the circular functions are not invertible.

5 Use the fact that the cosine and sine are periodic functions of fundamental period of 2π to find the period of each of the following functions.

a) $f(t) = \sin 4t$

b) $f(t) = \cos 3t$

c) $f(t) = \cos(t/2)$

d) $f(t) = \sin(2t + 1)$

6 Define the function $Q(t) = (\sin t)/(\cos t)$, where $\cos t \neq 0$. Show that Q is an odd function. What can you say about the symmetry of the graph of Q? Define $R(t) = 1/\cos t$. Show that R is an even function.

7 Use the periodicity of the circular functions, Figure 8 in Problem 1, and the fact that the cosine is even and the sine is odd to find $\cos t$ and $\sin t$ for each of the following values of t.

a) $13\pi/6$

b) $-5\pi/4$

c) -5π

d) $-4\pi/3$

e) $71\pi/3$

f) $41\pi/2$

g) $59\pi/6$

h) $-31\pi/6$

i) $67\pi/4$

j) $107\pi/6$

8 Explain the difference between $g \circ f$ and $f \circ g$ if $f(t) = \sin t$ and $g(t) = 2t$.

9 Use $f(t) = \cos t$ and $g(t) = 3t$ to answer each of the following parts.

a) Form $g \circ f$, $f \circ g$, and $f \circ f$.

b) Indicate the domain and range of $g \circ f$ and $f \circ g$.

c) Indicate which of the functions of part a is even or odd.

10 Show that

$$\begin{vmatrix} \cos t & \sin t \\ -\sin t & \cos t \end{vmatrix} = 1$$

4 Evaluation of Sine and Cosine

Since the cosine and sine functions are periodic functions of period 2π, the evaluation of either the cosine or sine at any given real number t can always be reduced to evaluating the function at a number in $[0, 2\pi)$ (see Section 2). Let us examine a few examples in order to clarify this important fact.

EXAMPLE

In each of the following parts use the fact that the cosine is an even function and the sine is an odd function, together with the fact that the cosine and sine are periodic functions of period 2π, to reduce the evaluation of the given function at the given number to an evaluation of the same function at a number in $[0, 2\pi)$. Use 3.14 as the approximate value of π.

a) $\cos(-7)$ b) $\cos 81$ c) $\sin(-15)$

SOLUTION

a) $\cos(-7) = \cos 7$

$= \cos(0.72 + 6.28)$

$= \cos 0.72$ (Figure 1a)

b) $\cos 81 = \cos[5.64 + 12(6.28)]$

$= \cos 5.64$ (Figure 1b)

c) $\sin(-15) = -\sin(15)$

$= -\sin[2.44 + 2(6.28)]$

$= -\sin 2.44$ (Figure 1c)

An alternate method proceeds as follows.

$\sin(-15) = \sin[-15 + 3(6.28)]$

$= \sin 3.84$ (Figure 1c)

Figure 1

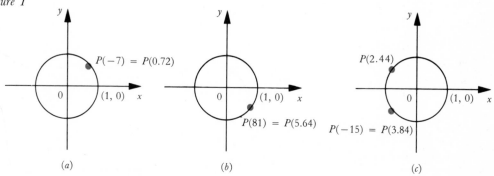

(a)   (b)   (c)

4.1 Reference Numbers

Next, by considering various cases, it can be shown that the evaluation of either the cosine or sine function at $t \in [0, 2\pi)$ can be reduced to an evaluation of the function at 0, $\pi/2$, π, $3\pi/2$ or at some number in interval $(0, \pi/2)$ with a possible adjustment in sign.

CASE 1. $t \in \{0, \pi/2, \pi, 3\pi/2\}$. For these special values, it is easy to evaluate the cosine and sine, since $P(0) = (1, 0)$, $P(\pi/2) = (0, 1)$, $P(\pi) = (-1, 0)$, and $P(3\pi/2) = (0, -1)$ (Figure 2).

Figure 2

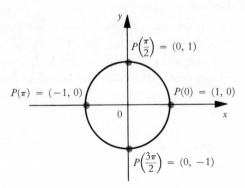

CASE 2. $t \in (3\pi/2, 2\pi)$ or $P(t)$ *is in quadrant IV* (Figure 3). First, the "shortest" arc length between $P(t)$ and the x axis is determined. This arc length is called the *reference number*, and it is denoted by t_R. Here $t_R = 2\pi - t$ (Figure 3).

By the symmetry of the circle, it follows that

$$\cos t = x = \cos t_R \qquad \text{and} \qquad \sin t = -y = -\sin t_R$$

Figure 3

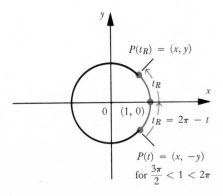

Hence the evaluation of the cosine and sine at $t \in (3\pi/2, 2\pi)$ has been reduced to an evaluation at the reference number t_R with the sign adjustment for the sine function. In other words, $\cos t$ and $\cos t_R$ are exactly the same, whereas $\sin t$ and $\sin t_R$ agree in absolute value (why?) but are opposite in sign (Figure 3). For example, for $t = 11\pi/6$, $t_R = 2\pi - (11\pi/6) = \pi/6$ so that

$$\cos \frac{11\pi}{6} = \cos \frac{\pi}{6} \qquad \text{and} \qquad \sin \frac{11\pi}{6} = -\sin \frac{\pi}{6}$$

Finally, we will show that $t_R \in (0, \pi/2)$.

Since $t \in (3\pi/2, 2\pi)$, $3\pi/2 < t < 2\pi$, from which it follows that

$$\frac{-3\pi}{2} > -t > -2\pi \qquad \text{(why?)}$$

$$2\pi - \frac{3\pi}{2} > 2\pi - t > 0 \qquad \text{(why?)}$$

$$0 < 2\pi - t < \frac{\pi}{2}$$

so that

$$0 < t_R < \frac{\pi}{2}$$

CASE 3. $t \in (\pi, 3\pi/2)$ or $P(t)$ *is in quadrant III* (Figure 4). The reference number is determined as before; that is, it is the shortest arc length between $P(t)$ and the x axis. Here $t_R = t - \pi$ (Figure 4).

Figure 4

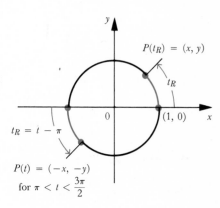

By the symmetry of the circle, we have, for $\pi < t < 3\pi/2$,

$$\cos t = -x = -\cos t_R \qquad \text{and} \qquad \sin t = -y = -\sin t_R$$

For example, for $t = 4\pi/3$, $t_R = (4\pi/3) - \pi = \pi/3$ so that

$$\cos \frac{4\pi}{3} = -\cos \frac{\pi}{3} \qquad \text{and} \qquad \sin \frac{4\pi}{3} = -\sin \frac{\pi}{3}$$

Again, $t_R \in (0, \pi/2)$, for $\pi < t < 3\pi/2$ implies that

$$\pi - \pi < t - \pi < \frac{3\pi}{2} - \pi \quad \text{so that} \quad 0 < t - \pi < \frac{\pi}{2}$$

That is,

$$0 < t_R < \frac{\pi}{2}$$

Hence the evaluation at $t \in (\pi, 3\pi/2)$ has been reduced again to an evaluation at $t_R \in (0, \pi/2)$ with the proper adjustment in sign.

CASE 4. $t \in (\pi/2, \pi)$ or $P(t)$ *is in quadrant II* (Figure 5). As before, we determine the reference number as the shortest arc length between $P(t)$ and the x axis. Here $t_R = \pi - t$ (Figure 5).

Figure 5

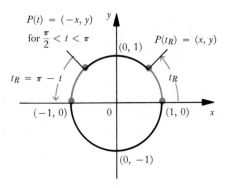

We have, by symmetry,

$$\cos t = -\cos t_R \quad \text{and} \quad \sin t = \sin t_R$$

For example, for $t = 3\pi/4$, $t_R = \pi - (3\pi/4) = \pi/4$, so that

$$\cos \frac{3\pi}{4} = -\cos \frac{\pi}{4} \quad \text{and} \quad \sin \frac{3\pi}{4} = \sin \frac{\pi}{4}$$

Again $t_R \in (0, \pi/2)$ (see Problem 2).

CASE 5. $t \in (0, \pi/2)$ or $P(t)$ *is in quadrant I.* Here $t_R = t$ (Figure 6), so that

$$\cos t = \cos t_R \quad \text{and} \quad \sin t = \sin t_R$$

Figure 6

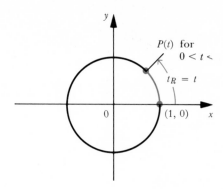

Clearly,

$$t_R \in \left(0, \frac{\pi}{2}\right)$$

EXAMPLES

In each of the following parts, reduce the evaluation of the given function to an evaluation of the same function at a reference number, that is, a number t_R such that $t_R \in (0, \pi/2)$ (use $\pi = 3.14$).

1 $\sin 2.5$ and $\cos 2.5$

SOLUTION. The reference number is $t_R = 3.14 - 2.50 = 0.64$ (Figure 7). Since $P(2.5)$ is in quadrant II, where the sine is positive and the cosine is negative,

$$\sin 2.5 = \sin 0.64 \qquad \text{and} \qquad \cos 2.5 = -\cos 0.64 \qquad \text{(Figure 7)}$$

Figure 7

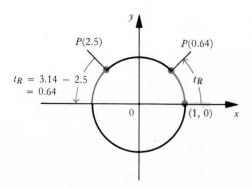

2    $\sin 6$ and $\cos 6$

SOLUTION.   Since $t_R = 6.28 - 6 = 0.28$, and since $P(6)$ is in quadrant IV (Figure 8),

$$\sin 6 = -\sin 0.28 \qquad \text{and} \qquad \cos 6 = \cos 0.28$$

Figure 8

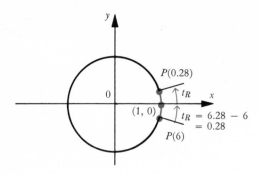

3    $\cos 29.12$ and $\sin 29.12$

SOLUTION.   The periodicity of the cosine and sine can be used to get

$$\cos 29.12 = \cos [4 + 4(6.28)] = \cos 4$$

and

$$\sin 29.12 = \sin [4 + 4(6.28)] = \sin 4$$

But the reference number t_R for 4 is

$$t_R = 4 - 3.14 = 0.86$$

Since the cosine and sine are both negative in quadrant III (Figure 9),

$$\cos 4 = -\cos 0.86 \qquad \text{and} \qquad \sin 4 = -\sin 0.86$$

Figure 9

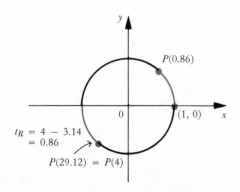

4.2   Tables

The evaluation of the cosine and sine functions can always be reduced to an evaluation at a reference number t, where $t \in (0, \pi/2)$, with the proper adjustment in sign, or at 0, $\pi/2$, π, or $3\pi/2$. The only question that remains is how we can determine the values of $\cos t$ and $\sin t$ for $0 < t < \pi/2$. For certain values, such as $\pi/6$, $\pi/4$, and $\pi/3$, the question has been answered (see Problem Set 2, Problem 1). For other values, there is a table (see Table III of Appendix A) which has been constructed using some more advanced techniques of mathematics. As with the logarithmic tables, this table gives *approximations* to the values. Whenever Table III is used, the use of the equal sign will be understood to mean "approximately equal to."

EXAMPLE

Use Table III and the reference number to determine each of the following values (use $\pi = 3.14$).
a) $\cos 0.65$ and $\sin 0.65$
b) $\cos 5.68$ and $\sin 5.68$
c) $\cos 2.14$ and $\sin 2.14$

SOLUTION

a)   Since $P(0.65)$ is in quadrant I, the reference number is $t_R = 0.65$ (Figure 10a). The cosine and sine are both positive in quadrant I, so that we get directly from Table III,

$$\cos 0.65 = 0.7961 \qquad \text{and} \qquad \sin 0.65 = 0.6052$$

b) $P(5.68)$ is in quadrant IV and the reference number is $t_R = 6.28 - 5.68 = 0.60$ (Figure 10b). Since the cosine is positive and the sine is negative in quadrant IV, we have

$$\cos 5.68 = \cos 0.60 = 0.8253$$

and

$$\sin 5.68 = -\sin 0.60 = -0.5646$$

Figure 10

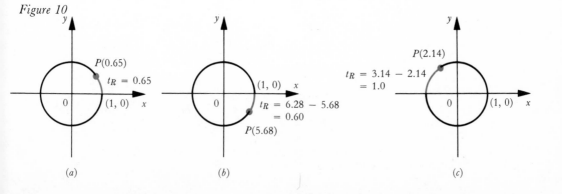

(a)   (b)   (c)

c) $P(2.14)$ is in quadrant II and the reference number is $t_R = 3.14 - 2.14 = 1.0$ (Figure 10c). In quadrant II the cosine is negative and the sine is positive, so that

$$\cos 2.14 = -\cos 1 = -0.5403 \qquad \text{and} \qquad \sin 2.14 = \sin 1 = 0.8415$$

It is obvious that Table III does not contain entries for all values between 0 and $\pi/2$. (For example, $\sin 1.514$ and $\sin 0.005$ cannot be found in Table III.) Fortunately, as we shall see in the examples below, *linear interpolation* (see Chapter 4, Section 6.2) can be used to give some reasonably accurate approximations.

Let us summarize the steps for finding $T(t)$, where $T(t)$ is either $\cos t$ or $\sin t$.

1   Locate the quadrant in which $P(t)$ lies.

2   Use the fact that the cosine is even and the sine is odd, together with the fact that the cosine and sine have period 2π, to reduce the evaluation of $T(t)$ to an evaluation at a number in $[0, 2\pi)$.

3   Determine the reference number t_R associated with t.

4   Use Table III and linear interpolation if necessary to find $T(t_R) = |T(t)|$.

5   Finally, adjust the sign of $T(t)$ according to the quadrant in which $P(t)$ lies.

EXAMPLES

Use Table III and linear interpolation to find each of the following values (use $\pi = 3.14$).

1   $\sin 1.514$

SOLUTION. $P(1.514)$ is in quadrant I, so that the reference number is 1.514 (Figure 11). Table III, together with linear interpolation, can be used to find $\sin 1.514$ as follows.

$$0.01 \left[0.006 \begin{bmatrix} \sin 1.51 & = 0.9982 \\ \sin 1.514 = & ? \\ \sin 1.52 & = 0.9987 \end{bmatrix} d \right] 0.0005$$

$$\frac{0.006}{0.01} = \frac{d}{0.0005}$$

so that

$$d = 0.0003$$

Figure 11

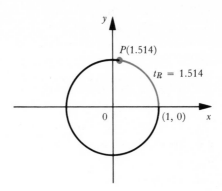

Hence

$$\sin 1.514 = 0.9987 - 0.0003$$
$$= 0.9984$$

2 $\sin(-2.955)$

SOLUTION. Since the sine is an odd function, $\sin(-2.955) = -\sin 2.955$. But $\sin 2.955 = \sin 0.185$ (Figure 12), since the sine is positive in quadrant II and since the reference number for 2.955 is 0.185. Thus $\sin(-2.955) = -\sin 0.185$.

Figure 12

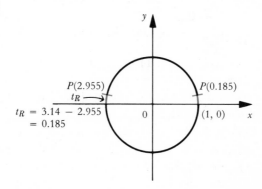

Using linear interpolation, $\sin 0.185$ can be approximated as follows.

$$0.01\left[0.005\begin{bmatrix}\sin 0.18 & = 0.1790 \\ \sin 0.185 = & ? \\ \sin 0.19 & = 0.1889\end{bmatrix}d\right]0.0099$$

$$\frac{0.005}{0.01} = \frac{d}{0.0099}$$

Hence

$$d = 0.005$$

and

$$\sin 0.185 = 0.1889 - 0.005 = 0.1839$$

so that

$$\sin(-2.955) = -0.1839$$

PROBLEM SET 3

1 Sketch $P(t)$ and determine the reference number for each of the following values of t (use $\pi = 3.14$).

a) 3   b) 1   c) 5.8   d) 2

e) 14.3   f) −5   g) −10   h) 6.12

2 Assume that $t \in (\pi/2, \pi)$; that is, $\pi/2 < t < \pi$. Prove that if $t_R = \pi - t$, then $0 < t_R < \pi/2$.

3 Use Table III, together with the fact that the cosine is an even function and the sine is an odd function, to compute $\cos t$ and $\sin t$ for each of the following values of t (use $\pi = 3.14$).

a) 0.7   b) 1.38   c) 0.5   d) −0.35   e) −0.01

4 a) Follow the path of the point $P(t)$ around the unit circle as t increases from 0 to $\pi/2$. What is the behavior of $\cos t$? Next, examine Table III to see the behavior of $\cos t$ as t increases from 0 to 1.57. What do you notice?

  b) Answer part a) for $\sin t$.

5 Perform the five steps on page 282 to find each of the following values (use $\pi = 3.14$).

a) $\sin 3.82$ b) $\cos(-10)$

c) $\sin 7.25$ d) $\cos 8$

e) $\sin(-2.9)$ f) $\cos(-47)$

g) $\sin 6.195$ h) $\cos 1.111$

i) $\cos 1.588$ j) $\sin(-4.232)$

k) $\cos(-12.407)$ l) $\cos 0.333$

6 Follow the path of $P(t)$ on the unit circle to determine the behavior of $\cos t$ and $\sin t$ if:

a) t increases from $\pi/2$ to π

b) t increases from π to $3\pi/2$

c) t increases from $3\pi/2$ to 2π

Do the table values show these behaviors? Explain.

5   Graphs of the Sine and Cosine Functions

In this section we shall investigate the graphs of

$$f(x) = \sin x \quad \text{and} \quad g(x) = \cos x$$

Hereafter we shall ordinarily use x instead of t to denote a real number. In this usage, of course, x can be thought of as representing an arc length along the unit circle, as t did.

Since $f(x) = \sin x$ is a function of period 2π, it is sufficient to restrict our attention to the values of x where $x \in [0, 2\pi)$. Then the graph of $f(x) = \sin x$ for $x \in R$ can be extended as far as we like by repeating the graph, since the graph on interval $[0, 2\pi)$ repeats every 2π units (Figure 1). The part of the graph which occurs in interval $[0, 2\pi)$ is called a *cycle* (or *wavelength*) of the curve.

Figure 1

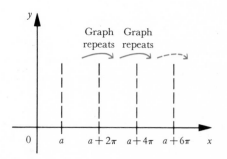

We also know that the range of $f(x) = \sin x$ is $[-1, 1]$, so that the graph is "restricted" to a horizontal strip between $y = 1$ and $y = -1$. Combining this restriction with the periodicity, it is enough to graph $f(x) = \sin x$ in the region of Figure 2 in order to establish the pattern that repeats every 2π units.

Figure 2

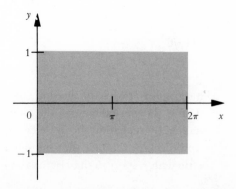

Since we know that $f(x) = \sin x$ is positive for $x \in (0, \pi)$ and negative for $x \in (\pi, 2\pi)$ (see Section 3.1, Figure 3), we can further restrict the region (Figure 3).

Figure 3

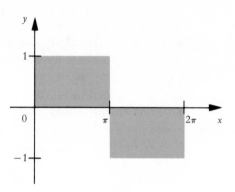

Finally, some known ordered pairs $(x, \sin x)$ can be used, together with the fact that f is increasing in the interval $(0, \pi/2)$, decreasing in the interval $(\pi/2, 3\pi/2)$, and again is increasing in the interval $(3\pi/2, 2\pi)$ (see Problem Set 3, Problem 6) to obtain the graph on $[0, 2\pi)$ contained in the shaded region (Figure 4).

Figure 4

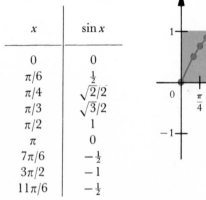

x	$\sin x$
0	0
$\pi/6$	$\frac{1}{2}$
$\pi/4$	$\sqrt{2}/2$
$\pi/3$	$\sqrt{3}/2$
$\pi/2$	1
π	0
$7\pi/6$	$-\frac{1}{2}$
$3\pi/2$	-1
$11\pi/6$	$-\frac{1}{2}$

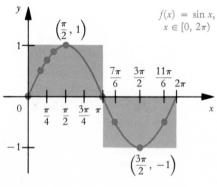

This graph is a cycle of the sine curve (see Problem 7).

Using the periodicity of the sine, the graph of $f(x) = \sin x$ on R is obtained by repeating the cycle every 2π units (Figure 5). Notice that the graph displays the fact that the sine is an odd function (its graph is symmetric with respect to the origin) and it does not have an inverse (why?).

Figure 5

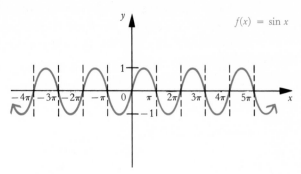

$f(x) = \sin x$

The graph of $f(x) = \cos x$ can be determined in the same way that the graph of the sine function was. Since the cosine is a function of period 2π, it is enough to graph it on $[0, 2\pi)$. As with the sine, the range of the cosine is $[-1, 1]$, so we need only consider the graph in the region of Figure 6.

Figure 6

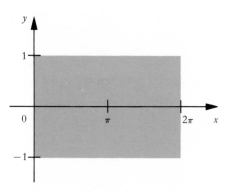

We can further restrict the region that contains the graph by using the fact that $f(x) = \cos x$ is positive for $x \in (0, \pi/2) \cup (3\pi/2, 2\pi)$ and negative for $x \in (\pi/2, 3\pi/2)$ (see Section 3.1, Figure 3) (Figure 7).

Figure 7

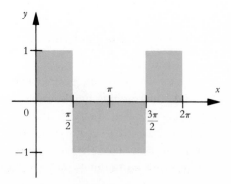

Finally, by using some functional values of the cosine, together with the known behavior of the cosine (see Problem Set 3, Problem 6), we get the graph on $[0, 2\pi)$, a *cycle* of the cosine curve contained in the shaded region (Figure 8).

Figure 8

x	$\cos x$
0	1
$\pi/3$	$\frac{1}{2}$
$\pi/2$	0
π	-1
$4\pi/3$	$-\frac{1}{2}$
$3\pi/2$	0

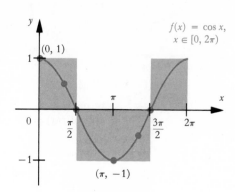

By repeating the cycle every 2π units, we get the graph of $f(x) = \cos x$ on R (Figure 9). Notice that the graph displays the fact that $f(x) = \cos x$ does not have an inverse and that it is an even function since the graph is symmetric with respect to the y axis.

Figure 9

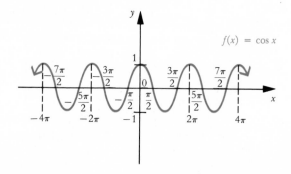

5.1   Graphs of $f(x) = a \sin(kx + b)$

We already sketched the graph of $f(x) = \sin x$ on R (Figure 5). It is worthwhile to investigate the geometric effect on this graph of the constants a, k, and b in graphing functions of the form $f(x) = a \sin(kx + b)$.

1   Amplitude

First, let us suppose that $k = 1$ and $b = 0$, so that the function takes the form $f(x) = a \sin x$. Since the sine graph $y = \sin x$ contains all points of the form $(x, \sin x)$, the graph of $f(x) = a \sin x$ can be determined from the graph of $f(x) = \sin x$ by multiplying the ordinates by a:

x	$\sin x$	$a \sin x$
$\vdots$	$\vdots$	$\vdots$
x_0	y_0	ay_0
$\vdots$	$\vdots$	$\vdots$

This can be illustrated by the following examples.

EXAMPLES

In each of the following examples, use the graph of $y = \sin x$ to graph the given function on $[0, 2\pi)$. Explain the relationship between the geometric change of the graph of $y = \sin x$ and the algebraic change of multiplying $\sin x$ by a constant. Also, find the range of f.

1 $f(x) = 3 \sin x$

SOLUTION. First, we graph one cycle of $y = \sin x$ (Figure 10a). Now the graph of $f(x) = 3 \sin x$ is obtained by locating the points $(x, 3 \sin x)$ on the graph. First, select a value for x, evaluate $\sin x$, and then multiply this latter value by 3:

$$x \rightarrow \sin x \rightarrow 3 \sin x$$

Geometrically, this means that the ordinate of $(x, \sin x)$ is "stretched" by a multiple of 3 to obtain $(x, 3 \sin x)$ (Figure 10b).

Figure 10

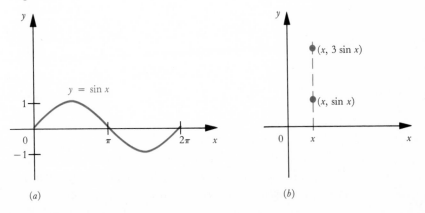

(a)   (b)

The overall geometric effect on the graph of $y = \sin x$, then, is a "vertical stretching" of the sine curve by a multiple of 3 (Figure 11). The range of f is the set $[-3, 3]$.

Figure 11

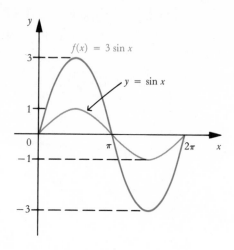

2   $f(x) = -\frac{1}{5}\sin x$

SOLUTION.   In this case each ordinate of $y = \sin x$ is multiplied by $-\frac{1}{5}$ in order to obtain the graph of $f(x) = -\frac{1}{5}\sin x$. This multiplication has the geometric effect of reflecting the graph of $y = \sin x$ across the x axis (because of the negative value) and also of "shrinking vertically" the sine curve by a multiple of $\frac{1}{5}$ (Figure 12). The range of f is the set $[-\frac{1}{5}, \frac{1}{5}]$.

Figure 12

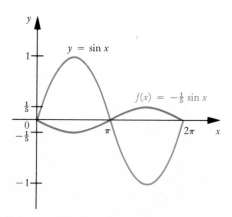

In each of the examples above notice that the constant multiple determines the range of the function. This constant multiple is used to compute the *amplitude* of the function. The *amplitude* of a function $y = f(x)$ is defined to be $\frac{1}{2}(M - m)$, where M is the largest value of $f(x)$ and m is the smallest value of $f(x)$. Hence the amplitude of $f(x) = 3\sin x$ is $\frac{1}{2}[3 - (-3)] = 3$; the amplitude of $f(x) = -\frac{1}{5}\sin x$ is $\frac{1}{2}[\frac{1}{5} - (-\frac{1}{5})] = \frac{1}{5}$; and the amplitude of $f(x) = -3\sin x$ is $\frac{1}{2}[3 - (-3)] = 3$.

In general, if $f(x) = a \sin x$, the amplitude is $|a|$ and the range is the set $[-|a|, |a|]$. (Why?)

2 Phase Shift

Next, let us consider an example in which $a = 1$, that is, $f(x) = \sin(kx + b)$, by reexamining the graph of $y = \sin x$. We know that when x varies from 0 to 2π, one cycle of the sine function graph is generated. This fact can also be expressed in the following manner:

$f(x) = \sin[g(x)]$ generates one cycle of the sine graph if $g(x)$ varies from 0 to 2π

Suppose, for example, that $g(x) = 3x - 1$. Then

$$f(x) = \sin(3x - 1)$$

generates one cycle of the sine graph if

$3x - 1$ varies from 0 to 2π $(0 \le 3x - 1 < 2\pi)$

that is, if

$3x$ varies from 1 to $2\pi + 1$ $(1 \le 3x < 2\pi + 1)$

that is, if

x varies from $\dfrac{1}{3}$ to $\dfrac{2\pi}{3} + \dfrac{1}{3}$ $\left(\dfrac{1}{3} \le x < \dfrac{2\pi}{3} + \dfrac{1}{3}\right)$

Hence if x varies from $\frac{1}{3}$ to $2\pi/3 + \frac{1}{3}$, then $3x - 1$ varies from 0 to 2π, so that, in turn, $f(x) = \sin(3x - 1)$ generates one cycle of the sine wave (Figure 13). Using the periodicity of the sine, we get the graph of $f(x) = \sin(3x - 1)$ on R (Figure 14). Notice here that the sine wave "begins" at

Figure 13

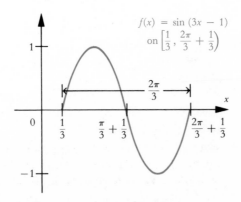

$\frac{1}{3}$ (this is called the *phase shift*), and that a cycle covers an interval of length $2\pi/3$ [$2\pi/3$ is the period of $f(x) = \sin(3x-1)$, and, by contrast, the period of $y = \sin x$ is 2π].

Figure 14

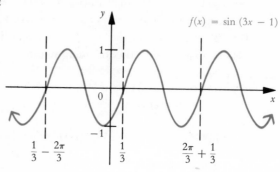

$f(x) = \sin(3x-1)$

EXAMPLES

Use the graph of $y = \sin x$ to graph one cycle of each of the following functions. Indicate the amplitude, period, and phase shift.

1   $f(x) = \sin \frac{1}{2}x$

SOLUTION.   The amplitude of f is 1. $f(x) = \sin \frac{1}{2}x$ generates one cycle if

$\frac{1}{2}x$ varies from 0 to 2π $(0 \leq \frac{1}{2}x < 2\pi)$

that is, if

x varies from 0 to 4π $(0 \leq x < 4\pi)$

Consequently, the period is 4π and the phase shift is 0 (Figure 15).

Figure 15

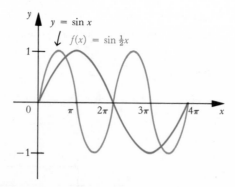

$y = \sin x$

$f(x) = \sin \frac{1}{2}x$

2   $f(x) = 3\sin(\frac{1}{2}x+2)$

SOLUTION.   Here the amplitude is 3. f generates a cycle of the sine graph if

$\frac{1}{2}x + 2$ varies from 0 to 2π $(0 \leq \frac{1}{2}x+2 < 2\pi)$

that is, if

$\frac{1}{2}x$ varies from -2 to $2\pi - 2$ $(-2 \leqq \frac{1}{2}x < 2\pi - 2)$

that is, when

x varies from -4 to $4\pi - 4$ $(-4 \leqq x < 4\pi - 4)$

Hence the period is 4π and the phase shift is -4 (Figure 16).

Figure 16

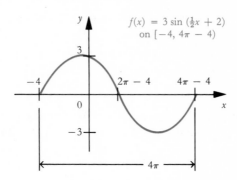

$f(x) = 3 \sin (\frac{1}{2}x + 2)$
on $[-4, 4\pi - 4)$

3    $f(x) = \sin(-x+1)$

SOLUTION. $f(x) = \sin(-x+1) = \sin[-(x-1)] = -\sin(x-1)$, since the sine function is an odd function. First, we graph $y = \sin(x-1)$. $y = \sin(x-1)$ generates a sine cycle if

$x - 1$ varies from 0 to 2π $(0 \leqq x-1 < 2\pi)$

that is, when

x varies from 1 to $2\pi + 1$ $(1 \leqq x < 2\pi + 1)$ (Figure 17)

Figure 17

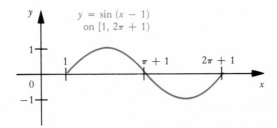

$y = \sin(x - 1)$
on $[1, 2\pi + 1)$

Finally, we obtain the graph of $f(x) = -\sin(x-1)$ from the graph of $y = \sin(x-1)$ by reflecting the latter graph across the x axis (Figure 18). The period is 2π; the phase shift is 1; the amplitude is 1.

Figure 18

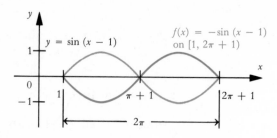

5.2   Graphs of $f(x) = a \cos(kx + b)$

We can use the graph of $y = \cos x$ to obtain the graphs of functions of the form $f(x) = a \cos(kx + b)$ by following the same approach as that which was taken for the sine function. A few examples will clarify the procedure.

EXAMPLES

Use the graph of $y = \cos x$ to obtain the amplitude, period, phase shift, and one cycle of the graph of each of the following functions.

1   $f(x) = 4 \cos x$

SOLUTION.   The graph of $f(x) = 4 \cos x$ has the same "horizontal be-havior" as $y = \cos x$, although its vertical position differs by a multiple of 4 (Figure 19). The amplitude of $f(x) = 4 \cos x$ is 4, the period is 2π, and the phase shift is 0.

Figure 19

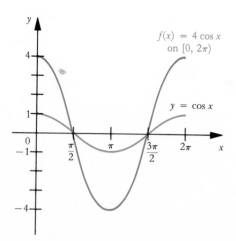

2   $f(x) = \cos(x/2)$

SOLUTION.   This function generates one cycle of the cosine curve when $x/2$ varies from 0 to 2π ($0 \le x/2 < 2\pi$), that is, when x varies from 0 to 4π ($0 \le x < 4\pi$) (Figure 20). The period is 4π, the amplitude is 1, and the phase shift is 0.

Figure 20

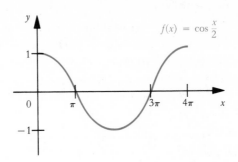

3   $f(x) = -3 \cos(x/8 + \pi/2)$

SOLUTION.   First, we will investigate $y = \cos(x/8 + \pi/2)$. This function generates a cosine cycle when $x/8 + \pi/2$ varies from 0 to 2π ($0 \leq x/8 + \pi/2 < 2\pi$), that is, when $x/8$ varies from $-\pi/2$ to $2\pi - \pi/2$ ($-\pi/2 \leq x/8 < 2\pi - \pi/2$), that is, when x varies from -4π to $16\pi - 4\pi$ ($-4\pi \leq x < 16\pi - 4\pi$) (Figure 21). Multiplication of $y = \cos(x/8 + \pi/2)$ by -3 merely stretches the

Figure 21

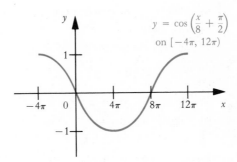

graph by a multiple by 3, and then reflects this latter graph across the x axis (why?) (Figure 22). The amplitude is 3, the period is 16π, and the phase shift is -4π.

Figure 22

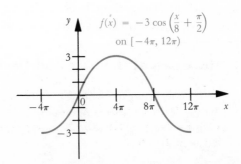

PROBLEM SET 4

1 Use the sine and cosine functions to graph a cycle of each of the following functions. Indicate the period, amplitude, and phase shift.

  a) $f(x) = 10 \sin x$

  b) $f(x) = \pi \cos x$

  c) $f(x) = \frac{1}{3} \sin (x/5)$

  d) $f(x) = 3 \cos (-2x)$

  e) $f(x) = \sin (\pi x)$

  f) $f(x) = -2 \sin 5x$

  g) $f(x) = \sin 10x$

  h) $f(x) = 4 \sin (-x/2)$

2 How does the graph of $f(x) = \cos x$ display the fact that the cosine is an even function?

3 Follow the directions of problem one for each of the following functions.

  a) $f(x) = 2 \cos (x + \pi)$

  b) $f(x) = 3 \sin (x + \pi)$

  c) $f(x) = -\frac{1}{3} \sin (x - 1)$

  d) $f(x) = 3 \cos (x - \pi/6)$

  e) $f(x) = \frac{1}{2} \cos (-x + \pi/12)$

  f) $f(x) = 5 \sin (x/4 - \pi/2)$

  g) $f(x) = 4 \sin (x/3 + \pi/6)$

  h) $f(x) = \pi \cos (x/\pi - 1/\pi)$

  i) $f(x) = 2 \cos (3 - 2x)$

  j) $f(x) = 2 \sin (2x/3 + \pi/4)$

4 Consider $f(x) = \sin x - \sin 3x$.

  a) Sketch the graphs of $g(x) = \sin x$ and $h(x) = \sin 3x$ on the same set of axes.

  b) Select appropriate points along the x axis and subtract graphically the ordinates $g(x) - h(x)$ of these points to obtain a sketch of the graph of $f(x) = g(x) - h(x)$.

5 Assume that $f(x) = a \sin (kx + b)$ where a and k are nonzero constants and b is a constant.

  a) Show that the amplitude is $|a|$.

  b) Show that the period is $2\pi/|k|$.

  c) Show that the phase shift is $-b/k$.

6 Explain why functions of the form $f(x) = a \sin (kx + b)$ and of the form $f(x) = a \cos (kx + b)$ do not have inverses.

7 We know that if t varies from 0 to 2π, $P(t)$ transverses the unit circle once, and $\sin t$ varies through all its range values. We know that $f(x) = \sin x$ generates one cycle of the sine curve when x varies from 0 to 2π ($0 \leq x < 2\pi$). What happens to $\sin t$ when t recycles the unit circle, that is, varies from 2π to 4π ($2\pi \leq t < 4\pi$)? What happens to $f(x) = \sin x$ when x varies from 2π to 4π? Draw sketches to display these variations. Can you see why the word "cycle" is used to indicate a sine wave? Explain.

6 Inverses of the Sine and Cosine

We know from our general study of functions that a function f has an inverse only when f is one-to-one. Geometrically, f^{-1} exists if the graph of $y = f(x)$ intersects any horizontal line in at most one point (see Chapter 2, Section 5).

Since the sine and cosine functions are periodic functions, it follows that they are not one-to-one (see Example 2 on page 260). For example, $\sin 0 = \sin 2\pi$ and $\cos 0 = \cos 2\pi$; or, using the horizontal line test, we can "see" from the graphs that the functions are not one-to-one, and, consequently, do not have inverses.

Although the sine and cosine functions do not have inverses, it is possible to construct an invertible function from each of the two functions by making suitable restrictions on the domains of the functions.

We shall begin by constructing the function $f(x) = \operatorname{Sin} x$ (notice that a capital letter S is used to distinguish this function from the sine function) by restricting the domain of the sine function to the interval $[-\pi/2, \pi/2]$. The range is $[-1, 1]$ (Figure 1) (why?). Since $f(x) = \operatorname{Sin} x$ is increasing in the interval $[-\pi/2, \pi/2]$, it has an inverse. This inverse is called the Arcsine function.

Figure 1

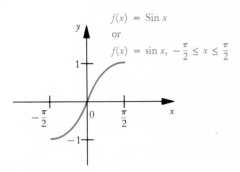

$f(x) = \operatorname{Sin} x$

or

$f(x) = \sin x, \ -\dfrac{\pi}{2} \le x \le \dfrac{\pi}{2}$

1 DEFINITION ARCSINE FUNCTION

The *Arcsine function*, denoted by $y = \operatorname{Arcsin} x$ or by $y = \sin^{-1} x$, is the inverse of the function $f(x) = \operatorname{Sin} x$.

$y = \operatorname{Arcsin} x$ is equivalent to $x = \sin y$ with $-\pi/2 \le y \le \pi/2$.

For example,

$$\frac{\pi}{3} = \operatorname{Arcsin} \frac{\sqrt{3}}{2} \qquad \text{is equivalent to} \qquad \frac{\sqrt{3}}{2} = \sin \frac{\pi}{3}$$

$$\frac{\pi}{4} = \operatorname{Arcsin} \frac{1}{\sqrt{2}} \qquad \text{is equivalent to} \qquad \frac{1}{\sqrt{2}} = \sin \frac{\pi}{4}$$

$$0 = \sin^{-1} 0 \qquad \text{is equivalent to} \qquad 0 = \sin 0$$

Since the inverse of a function is formed by interchanging the numbers in the ordered pairs, the domain and range of $f(x) = \operatorname{Sin} x$ become, respectively, the range and the domain of the inverse function. Consequently, the domain of $y = \operatorname{Arcsin} x$ is $[-1, 1]$ and the range is $[-\pi/2, \pi/2]$.

It should be noted here that even though we write $\sin^2 x = (\sin x)^2$, we never use $\sin^{-1} x = (\sin x)^{-1}$, since $(\sin x)^{-1}$ is $1/\sin x$, which is not the same as Arcsin x. The problem arises because we are using "-1" ambiguously: it denotes both reciprocals and inverse functions.

Geometrically, the graph of the inverse sine function $f^{-1}(x) = $ Arcsin x can be obtained by "reflecting" the graph of the function $f(x) = $ Sin x about the line $y = x$ (Figure 2).

Figure 2

x	$\sin^{-1} x$
-1	$-\pi/2$
$-\frac{1}{2}$	$-\pi/6$
0	0
$\frac{1}{2}$	$\pi/6$
1	$\pi/2$

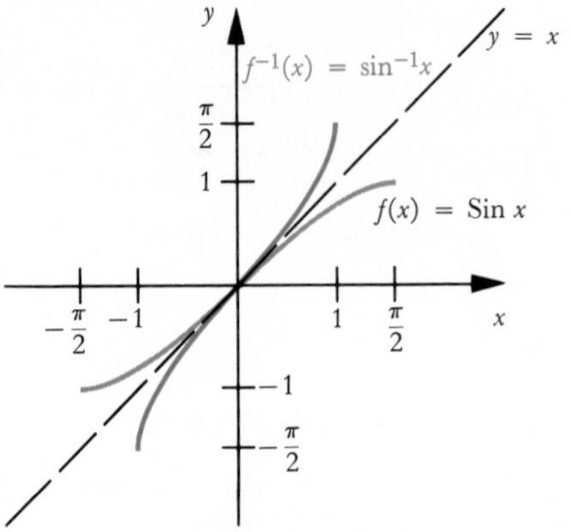

The inverse of the cosine function can be derived in the same way as the inverse sine function was derived. First, we can construct $g(x) = $ Cos x (notice the use of capital C) from $y = \cos x$ by restricting the domain to the interval $[0, \pi]$. The range of $g(x) = $ Cos x is $[-1, 1]$ (Figure 3).

Figure 3

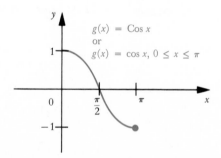

Since $g(x) = $ Cos x is a *decreasing* function it has an inverse called the Arccosine function.

2 DEFINITION ARCCOSINE FUNCTION

The *Arccosine function*, denoted by $y = $ Arccos x or by $y = \cos^{-1} x$, is the inverse of the function $g(x) = $ Cos x.

$y = $ Arccosin x is equivalent to $x = \cos y$, with $0 \leq y \leq \pi$.

For example,

$$\pi = \text{Arccos}(-1) \qquad \text{is equivalent to} \qquad -1 = \cos \pi$$

$$\frac{3\pi}{4} = \cos^{-1}\left(-\frac{1}{\sqrt{2}}\right) \qquad \text{is equivalent to} \qquad -\frac{1}{\sqrt{2}} = \cos \frac{3\pi}{4}$$

$$\frac{\pi}{2} = \cos^{-1} 0 \qquad \text{is equivalent to} \qquad 0 = \cos \frac{\pi}{2}$$

The domain of the Arccosine function is $[-1, 1]$ and the range is $[0, \pi]$. The graph of $g^{-1}(x) = \cos^{-1} x$ can be obtained by reflecting the graph of $g(x) = \text{Cos}\, x$ across the line $y = x$ (Figure 4).

Figure 4

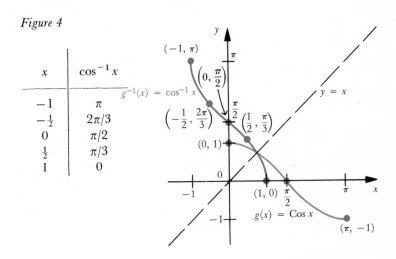

x	$\cos^{-1} x$
-1	π
$-\frac{1}{2}$	$2\pi/3$
0	$\pi/2$
$\frac{1}{2}$	$\pi/3$
1	0

EXAMPLES

1 Find each of the following values.

a) Arcsin $\frac{1}{2}$

b) $\cos^{-1}\left(-\frac{\sqrt{3}}{2}\right)$

c) $\sin^{-1}(-0.8016)$

d) $\cos^{-1}(0.2675)$

SOLUTION

a) Let Arcsin $\frac{1}{2} = t$; then $\sin t = \frac{1}{2}$, where $t \in [-\pi/2, \pi/2]$, so that $t = \pi/6$.

b) Let $\cos^{-1}(-\sqrt{3}/2) = t$; then $\cos t = -\sqrt{3}/2$, where $t \in [0, \pi]$ so that $t = 5\pi/6$.

c) Let $\sin^{-1}(0.8016) = t$; then $\sin t = 0.8016$, where $t \in [-\pi/2, \pi/2]$. From Table III we find that $t = 0.93$, so that $\sin^{-1}(0.8016) = 0.93$. Consequently, $\sin^{-1}(-0.8016) = -0.93$.

d) Let $\cos^{-1}(0.2675) = t$; then $\cos t = 0.2675$, where $t \in [0, \pi]$. From Table III we find that $t = 1.30$, so that $\cos^{-1}(0.2675) = 1.30$.

2   Evaluate each of the following.

a) $\cos\left[\sin^{-1}(-\sqrt{3}/2)\right]$ b) $\sin\left[\cos^{-1}\left(\tfrac{3}{5}\right)\right]$

SOLUTION

a) Since $\sin^{-1}(-\sqrt{3}/2) = -\pi/3$ (why?),
$$\cos\left[\sin^{-1}(-\sqrt{3}/2)\right] = \cos(-\pi/3) = \cos(\pi/3) = \tfrac{1}{2}.$$

b) Let $y = \cos^{-1}\left(\tfrac{3}{5}\right)$ so that $\cos y = \tfrac{3}{5}$, where $0 \leq y \leq \pi$. Using the identity $\sin^2 y + \cos^2 y = 1$ we have $\sin y = \pm\sqrt{1 - \cos^2 y}$. Since $0 \leq y \leq \pi$ where the sine is positive, we use $\sin y = \sqrt{1 - \cos^2 y}$. Hence
$$\sin\left[\cos^{-1}\left(\tfrac{3}{5}\right)\right] = \sin y = \sqrt{1 - \cos^2 y} = \sqrt{1 - \left(\tfrac{3}{5}\right)^2} = \sqrt{\tfrac{16}{25}} = \tfrac{4}{5}$$

3   Show that $\sin^{-1} x = \cos^{-1}\sqrt{1 - x^2}$ for $0 \leq x \leq 1$.

PROOF.   Let $t = \sin^{-1} x$ so that $\sin t = x$, where $t \in [-\pi/2, \pi/2]$. Since the cosine is positive in quadrants I and IV, we have $\cos t = \sqrt{1 - \sin^2 t}$. Substituting x for $\sin t$ yields $\cos t = \sqrt{1 - x^2}$ or $t = \cos^{-1}\sqrt{1 - x^2}$, so that, by substitution, $\sin^{-1} x = \cos^{-1}\sqrt{1 - x^2}$.

4   Graph $f(x) = \tfrac{1}{2}\sin^{-1} x$.

SOLUTION.   The graph of $f(x) = \tfrac{1}{2}\sin^{-1} x$ can be obtained from the graph of $y = \sin^{-1} x$ (see Figure 2). The ordinates of f will be one-half those of $y = \sin^{-1} x$ for corresponding values of x (Figure 5).

Figure 5

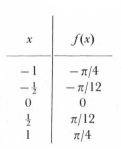

x	$f(x)$
-1	$-\pi/4$
$-\tfrac{1}{2}$	$-\pi/12$
0	0
$\tfrac{1}{2}$	$\pi/12$
1	$\pi/4$

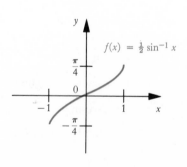

5   Express x in terms of y for each of the following equations.

a) $y = \cos^{-1} 5x$ b) $y = 5 + 3\cos^{-1} 5x$

SOLUTION

a) $y = \cos^{-1} 5x$ is equivalent to $5x = \cos y$ so that $x = \tfrac{1}{5}\cos y$.

b) $y = 5 + 3 \cos^{-1} 5x$ can be written as $(y-5)/3 = \cos^{-1} 5x$. The latter equation is equivalent to $5x = \cos[(y-5)/3]$, so that $x = \frac{1}{5} \cos[(y-5)/3]$.

PROBLEM SET 5

1 Determine each of the following values. Use Table III for parts k–n.

a) $\sin^{-1}(-\frac{1}{2})$ b) $\sin^{-1} 0$

c) $\sin^{-1}(\sqrt{2}/2)$ d) $\cos^{-1} 0$

e) $\cos^{-1} \frac{1}{2}$ f) $\cos^{-1}(-\sqrt{3}/2)$

g) $\sin^{-1}(-\sqrt{2}/2)$ h) $\cos^{-1}(-1)$

i) $\sin^{-1}(-1)$ j) $\sin^{-1}[\sin(\pi/2)]$

k) $\sin^{-1} 0.2182$ l) $\cos^{-1} 0.8628$

m) $\sin^{-1}(-0.7771)$ n) $\cos^{-1}(-0.2035)$

2 For what values of x, if any, is each of the following equations true?

a) $\sin^{-1}(\sin x) = x$ b) $\sin(\sin^{-1} x) = x$

c) $\cos^{-1}(\cos x) = x$ d) $\cos(\cos^{-1} x) = x$

3 Find the exact value of each of the following expressions.

a) $\sin^{-1}(\cos \pi)$ b) $\cos^{-1}[\sin(\pi/6)]$

c) $\cos[\sin^{-1}(\sqrt{2}/2)]$ d) $\sin^{-1}[\sin(-\pi/4)]$

e) $\sin^{-1}[\sin(-3\pi/2)]$ f) $\sin[\cos^{-1}(-1)]$

g) $\sin(\cos^{-1} \frac{3}{4})$ h) $\cos^{-1}[\cos(-\frac{7}{25})]$

i) $\cos(\cos^{-1} \frac{7}{12})$ j) $\cos[\sin^{-1}(-\frac{35}{37})]$

4 Show that the following statements are true for $x \in [0, 1]$.

a) $\sin[\sin^{-1}(-x)] = -x$ b) $\cos(\sin^{-1} x) = \sqrt{1-x^2}$

c) $\sin(\cos^{-1} x) = \sqrt{1-x^2}$

5 Express x in terms of y for each of the following.

a) $y = \cos^{-1} 2x$ b) $y = 2 \sin^{-1} 3x$

c) $y = 4 - 2 \sin^{-1} 4x$ d) $y = \cos^{-1}(2x+1)$

e) $3y = 1 + 4 \cos^{-1}(x/2)$

6 Define $f(x) = \sin x$, $x \in [\pi/2, 3\pi/2]$. Graph the function. What is the range? Explain why f^{-1} exists. Describe f^{-1}.

7 Sketch the graph of each of the following functions.

a) $f(x) = \sin^{-1} \frac{1}{2} x$ b) $f(x) = 2 \cos^{-1} \frac{1}{2} x$

c) $f(x) = \cos^{-1} 3x$ d) $f(x) = \sin^{-1} x + \cos^{-1} x$

7 Other Circular Functions

The sine and cosine functions can be used to define four other functions.

**DEFINITION   TANGENT, COTANGENT, SECANT,
AND COSECANT FUNCTIONS**

The tangent, cotangent, secant, and cosecant functions are defined as follows:

$$\text{tangent} = \left\{ (x, y) \mid y = \frac{\sin x}{\cos x}, \cos x \neq 0 \right\}$$

$$\text{cotangent} = \left\{ (x, y) \mid y = \frac{\cos x}{\sin x}, \sin x \neq 0 \right\}$$

$$\text{secant} = \left\{ (x, y) \mid y = \frac{1}{\cos x}, \cos x \neq 0 \right\}$$

$$\text{cosecant} = \left\{ (x, y) \mid y = \frac{1}{\sin x}, \sin x \neq 0 \right\}$$

or, more briefly,

$$\tan x = \frac{\sin x}{\cos x} \qquad \cos x \neq 0$$

$$\cot x = \frac{\cos x}{\sin x} \qquad \sin x \neq 0$$

$$\sec x = \frac{1}{\cos x} \qquad \cos x \neq 0$$

$$\csc x = \frac{1}{\sin x} \qquad \sin x \neq 0$$

7.1   Properties of the Tangent, Cotangent, Secant, and Cosecant

Since this definition expresses each of the four functions in terms of the cosine and/or sine, many of the functional properties of the four functions can be determined by using the known functional properties of the sine and cosine functions.

1   The Domains

Since $\tan x = \sin x/\cos x$, the domain of the tangent is given by $D_{\tan} = \{x \mid \cos x \neq 0\}$. However, $\cos x = 0$ for $x = \pi/2 + n\pi$, $n \in I$ (see Figure 9, page 288). Hence

$$D_{\tan} = \left\{ x \mid x \neq \frac{\pi}{2} + n\pi, n \in I \right\}$$

Since the domain of the secant is the same as the domain of the tangent (why?), the domain of the secant is given by

$$D_{\sec} = \left\{ x \mid x \neq \frac{\pi}{2} + n\pi, n \in I \right\}$$

The domain of the cotangent is given by

$$D_{\cot} = \{x \mid \sin x \neq 0\}$$

but, since $\sin x = 0$ for $x = n\pi$, $n \in I$ (see Figure 5, page 287), we have

$$D_{\cot} = \{x \mid x \neq n\pi, \, n \in I\}$$

Since the domain of the cosecant is the same as the domain of the cotangent, the domain of the cosecant is given by

$$D_{\csc} = \{x \mid x \neq n\pi, \, n \in I\}$$

2 The Ranges

The ranges of the secant and cosecant are not difficult to determine. We know that $\sec x = 1/\cos x$ and that $|\cos x| \leq 1$. Hence

$$|\sec x| = \left| \frac{1}{\cos x} \right| = \frac{1}{|\cos x|}$$

but

$$|\cos x| \leq 1 \quad \text{implies that} \quad 1 \leq \frac{1}{|\cos x|} \qquad \text{whenever } \cos x \neq 0$$

so that

$$1 \leq |\sec x|$$

That is, the range of the secant is given by

$$R_{\sec} = \{y \mid |y| \geq 1\} = \{y \mid y \leq -1\} \cup \{y \mid y \geq 1\}$$
$$= (-\infty, -1] \cup [1, \infty)$$

Similarly, we can use the fact that $\csc x = 1/\sin x$ and $|\sin x| \leq 1$ to conclude that the range of the cosecant is given by

$$R_{\csc} = \{y \mid |y| \geq 1\} = \{y \mid y \leq -1\} \cup \{y \mid y \geq 1\}$$
$$= (-\infty, -1] \cup [1, \infty)$$

The ranges of the tangent and cotangent are not so easy to determine. We will begin by examining the behavior of $f(x) = \tan x$ for $x \in [0, \pi/2)$. Clearly, as x increases from 0 to $\pi/2$, $y = \cos x$ decreases from 1 to 0 (Figure 1a), and $y = \sin x$ increases from 0 to 1 (Figure 1b), and $\tan x = \sin x/\cos x$ is composed of a ratio of two functions with the function in the denominator decreasing to 0 and the one in the numerator increasing to 1.

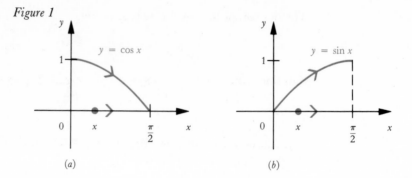

Figure 1

(a)   (b)

From the behavior of the sine and cosine function and by examining values in Table 1 below (from Table III of Appendix A), we can reasonably conclude that $\tan x$ increases from 0 through all possible real numbers to "infinity." By a similar argument it can be shown that the tangent assumes all possible negative real values for $x \in (-\pi/2, 0]$. Hence the range of the tangent is R. It can also be shown that the range of the cotangent is R (see Problem 4).

Table 1

x	$\sin x$	$\cos x$	$\tan x = \dfrac{\sin x}{\cos x}$
0	0	1	0
0.11	0.1098	0.9940	0.1104
0.44	0.4259	0.9048	0.4708
0.89	0.7771	0.6294	1.235
1.01	0.8468	0.5319	1.592
1.50	0.9975	0.0707	14.101
1.55	0.9998	0.0208	48.078
1.56	0.9999	0.0108	92.62

3   Signs of the Values

It is easy to determine the signs of the tangent, cotangent, secant, and cosecant by using what we know about the signs of the sine and cosine. Table 2, which is self-explanatory, indicates the signs of the function values.

The next theorem uses the properties of the sine and cosine functions to determine the periodicity of the tangent, cotangent, secant, and cosecant functions and to identify which of the four functions is even and which is odd

Table 2

$P(x)$ in quadrant:	I	II	III	IV
$\cos x$	$+$	$-$	$-$	$+$
$\sin x$	$+$	$+$	$-$	$-$
$\tan x = \dfrac{\sin x}{\cos x}$	$\dfrac{+}{+} = +$	$\dfrac{+}{-} = -$	$\dfrac{-}{-} = +$	$\dfrac{-}{+} = -$
$\cot x = \dfrac{\cos x}{\sin x}$	$\dfrac{+}{+} = +$	$\dfrac{-}{+} = -$	$\dfrac{-}{-} = +$	$\dfrac{+}{-} = -$
$\csc x = \dfrac{1}{\sin x}$	$\dfrac{1}{+} = +$	$\dfrac{1}{+} = +$	$\dfrac{1}{-} = -$	$\dfrac{1}{-} = -$
$\sec x = \dfrac{1}{\cos x}$	$\dfrac{1}{+} = +$	$\dfrac{1}{-} = -$	$\dfrac{1}{-} = -$	$\dfrac{1}{+} = +$

THEOREM 1

i   The tangent, cotangent, secant, and cosecant are periodic functions of period 2π. (In Chapter 6, pages 353 and 357, we will see that the *fundamental* period of the tangent and cotangent is actually π.)

ii   The secant is an even function.

iii   The tangent, cotangent, and cosecant are odd functions.

PROOF

i $\tan(x+2\pi) = \dfrac{\sin(x+2\pi)}{\cos(x+2\pi)} = \dfrac{\sin x}{\cos x} = \tan x$

and

$\sec(x+2\pi) = \dfrac{1}{\cos(x+2\pi)} = \dfrac{1}{\cos x} = \sec x$

(For the cotangent and the cosecant see Problem 2a.)

ii $\sec(-x) = \dfrac{1}{\cos(-x)} = \dfrac{1}{\cos x} = \sec x$

and

iii $\tan(-x) = \dfrac{\sin(-x)}{\cos(-x)} = \dfrac{-\sin x}{\cos x} = -\tan x$

(For the cotangent and the cosecant see Problem 2b.)

4   Evaluation of Circular Functions

Since all four of these functions are periodic functions of period 2π and since these functions are either even or odd, their evaluation at any given number in the domain follows the same procedure as that of the sine and cosine: first, the reference number; next, the given function at the reference number using Table III, together with interpolation if necessary; finally, the evaluation made by adjusting the sign according to the quadrant (Table 2, page 305). Here again Table III gives us only *approximate* values.

EXAMPLES

1   Use the definition of the other circular functions to evaluate the following.

a) $\tan\dfrac{\pi}{4}$ b) $\sec\dfrac{5\pi}{6}$

c) $\tan\left(-\dfrac{11\pi}{6}\right)$ d) $\cot\left(-\dfrac{4\pi}{3}\right)$

SOLUTION

a) $\tan\dfrac{\pi}{4} = \dfrac{\sin(\pi/4)}{\cos(\pi/4)} = \dfrac{1/\sqrt{2}}{1/\sqrt{2}} = 1$

b) $\sec\dfrac{5\pi}{6} = \dfrac{1}{\cos(5\pi/6)} = \dfrac{1}{-\sqrt{3}/2} = -\dfrac{2}{\sqrt{3}} = -\dfrac{2\sqrt{3}}{3}$

c) $\tan\left(-\dfrac{11\pi}{6}\right) = \dfrac{\sin(-11\pi/6)}{\cos(-11\pi/6)} = \dfrac{\sin(\pi/6)}{\cos(\pi/6)} = \dfrac{\frac{1}{2}}{\sqrt{3}/2} = \dfrac{1}{\sqrt{3}} = \dfrac{\sqrt{3}}{3}$

d) $\cot\left(-\dfrac{4\pi}{3}\right) = \dfrac{\cos(-4\pi/3)}{\sin(-4\pi/3)} = \dfrac{-\cos(\pi/3)}{\sin(\pi/3)} = \dfrac{-\frac{1}{2}}{\sqrt{3}/2} = -\dfrac{1}{\sqrt{3}} = -\dfrac{\sqrt{3}}{3}$

2   Evaluate the other five circular functions at t if $\sin t = -\frac{3}{5}$ and $P(t)$ is in quadrant III.

SOLUTION.   Since $\sin^2 t + \cos^2 t = 1$ and $P(t)$ is in quadrant III where the cosine is negative, we have

$$\cos t = -\sqrt{1-\sin^2 t}$$
$$= -\sqrt{1-(-\tfrac{3}{5})^2}$$
$$= -\sqrt{1-\tfrac{9}{25}}$$
$$= -\sqrt{\tfrac{16}{25}}$$
$$= -\tfrac{4}{5}$$

Next the definition of the other circular functions can be used to find the other four values.

$$\tan t = \frac{\sin t}{\cos t} = \frac{-\frac{3}{5}}{-\frac{4}{5}} = \frac{3}{4}$$

$$\cot t = \frac{\cos t}{\sin t} = \frac{-\frac{4}{5}}{-\frac{3}{5}} = \frac{4}{3}$$

$$\sec t = \frac{1}{\cos t} = \frac{1}{-\frac{4}{5}} = -\frac{5}{4}$$

$$\csc t = \frac{1}{\sin t} = \frac{1}{-\frac{3}{5}} = -\frac{5}{3}$$

3   Use the periodicity of the tangent function to simplify $\tan(t - 2\pi)$.

SOLUTION

$$\begin{aligned}
\tan(t - 2\pi) &= \tan[-(2\pi - t)] \\
&= -\tan(2\pi - t) \\
&= -\tan[(2\pi) + (-t)] \\
&= -\tan(-t) \\
&= -(-\tan t) \\
&= \tan t
\end{aligned}$$

4   Use Table III to evaluate each of the six circular functions for $t = 4.8$ (use $\pi = 3.14$).

SOLUTION.   The reference number for 4.8 is $t_R = 1.48$, and $P(4.8)$ is in quadrant IV (Figure 2). Hence by using the known signs of the six func-

Figure 2

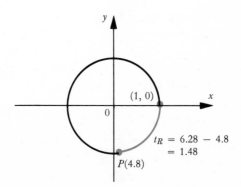

tions, together with Table III, we get

$$\sin 4.8 = -\sin 1.48 = -0.9959$$

$$\cos 4.8 = \cos 1.48 = 0.0907$$

$$\tan 4.8 = -\tan 1.48 = -10.983$$

$$\csc 4.8 = -\csc 1.48 = -1.004$$

$$\sec 4.8 = \sec 1.48 = 11.029$$

$$\cot 4.8 = -\cot 1.48 = -0.0910$$

5   Use Table III and linear interpolation to evaluate $\tan 3.048$ (use $\pi = 3.14$).

SOLUTION.   Since the reference number is $t_R = 0.092$ and $P(3.048)$ is in quadrant II (Figure 3), we have $\tan 3.048 = -\tan 0.092$. Next linear

Figure 3

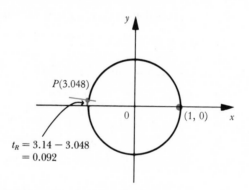

interpolation can be used to evaluate $\tan 0.092$.

$$0.01 \left[0.008 \left[\begin{array}{l} \tan 0.09 \ = 0.0902 \\ \tan 0.092 = \quad ? \\ \tan 0.10 \ = 0.1003 \end{array} \right] d \right] 0.0101$$

$$\frac{0.008}{0.01} = \frac{d}{0.0101}$$

so that

$$d = 0.0081$$

Hence

$$\tan 0.092 = 0.1003 - 0.0081 = 0.0922$$

and

$$\tan 3.048 = -0.0922$$

7.2 Graphs of the Other Circular Functions

1 The Tangent

Since $f(x) = \tan x$ is a periodic function of period 2π, it is enough to sketch the graph on $[0, 2\pi)$ in order to determine the graph on R.

First, one cycle of $g(x) = \sin x$ and one cycle of $h(x) = \cos x$ is graphed on the same coordinate system. These graphs will be called the *auxiliary graphs* (Figure 4).

Second, the points at which the $\tan x$ is undefined are indicated by drawing dashed vertical lines through the values of x at which $\cos x = 0$ (Figure 4).

Next, the points at which $\tan x = 0$ occur where $\sin x = 0$ (why?) (Figure 4).

Figure 4

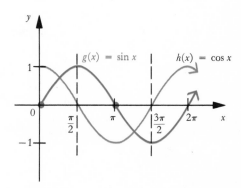

Finally, the sketch of $f(x) = \tan x$ can be determined by examining the auxiliary graphs to determine the behavior of $\tan x = \sin x / \cos x$ as x increases from 0 to 2π.

i (See Figure 5.) As x increases from 0 to $\pi/2$, $\sin x$ increases from 0 to

Figure 5

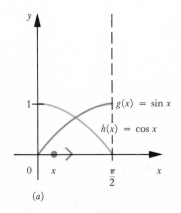

(a)

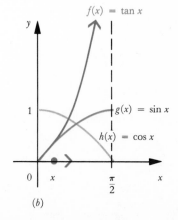

(b)

1 and $\cos x$ decreases from 1 to 0 (Figure 5a). Hence $\tan x = \sin x/\cos x$, where the numerator is increasing to 1 and the denominator is decreasing to 0; consequently, $\tan x$ is becoming increasingly large in value as x increases to $\pi/2$ (Figure 5b). However, the graph does not cross the line $x = \pi/2$. Because of this behavior, $x = \pi/2$ is a *vertical asymptote*.

ii (See Figure 6.) As x increases from $\pi/2$ to π, $\sin x$ is positive and decreasing from 1 to 0, whereas the $\cos x$ is negative and increasing in *absolute* value. Hence $\tan x = \sin x/\cos x$ is negative, with a numerator decreasing in absolute value to 0 and a denominator increasing in absolute value so that the tangent is negative and decreasing in absolute value to 0.

Figure 6

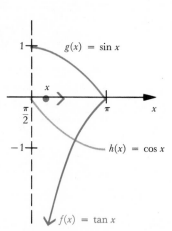

iii (See Figure 7.) As x increases from π to $3\pi/2$, $\sin x$ is negative and increasing in absolute value to 1, whereas $\cos x$ is negative and decreasing in absolute value to 0; consequently, $\tan x = \sin x/\cos x$ is positive (why?) and increasing.

Figure 7

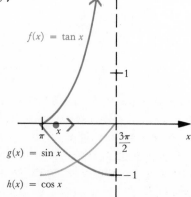

iv (See Figure 8.) Finally, as x increases from $3\pi/2$ to 2π, $\tan x$ is negative (why?) and decreasing in absolute value to 0.

Figure 8

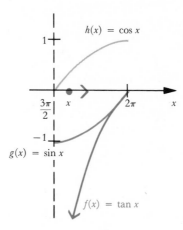

After combining these four results and locating a few points, we get the graph of $f(x) = \tan x$ on $[0, 2\pi)$ (Figure 9).

Figure 9

x	$\tan x$
$\pi/6$	0.6
$\pi/4$	1
$5\pi/4$	1
$2\pi/3$	-1.7
$7\pi/4$	-1

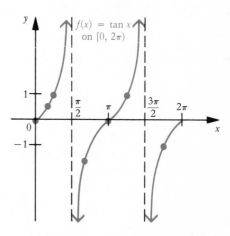

Using the fact that $f(x) = \tan x$ has a period of 2π, we get the graph on R (Figure 10).

Notice that the graph of the tangent displays some of the functional properties of the tangent. For example, the range is R; the tangent is symmetric with respect to the origin; the tangent is not continuous on R; the tangent is not one-to-one. Also, the graph displays the fact that the fundamental period of the tangent function is π, a fact that will be proved in Chapter 6, page 353.

Figure 10

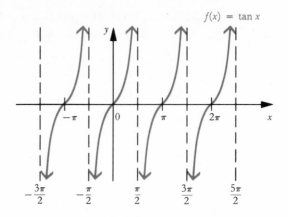

2   The Cotangent

Because the cotangent is a periodic function of period 2π, we will graph $f(x) = \cot x$ on $[0, 2\pi)$. The graphs of $g(x) = \cos x$ and $h(x) = \sin x$ can be used as auxiliary graphs to determine the graph of $\cot x = \cos x/\sin x$.

The asymptotes of $f(x) = \cot x$ occur when $\sin x = 0$; the x intercepts of $f(x) = \cot x$ occur when $\cos x = 0$ (Figure 11). Finally, by locating some points and examining the behavior of $\cos x/\sin x$ as x increases from 0 to 2π (see Problem 4), we get the graph of $f(x) = \cot x$ on $[0, 2\pi)$ (Figure 11).

Figure 11

x	$\cot x$
$\pi/6$	1.7
$\pi/4$	1
$2\pi/3$	-0.6
$5\pi/4$	1
$7\pi/4$	-1

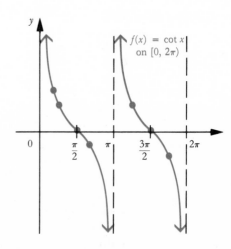

3   The Secant

Since the period of $f(x) = \sec x$ is 2π, it is enough to determine the graph on $[0, 2\pi)$. $\sec x = 1/(\cos x)$, and $\sec x$ has vertical asymptotes wherever $\cos x = 0$ (Figure 12). Finally, we can examine the behavior of $\sec x$ and sketch $f(x) = \sec x$ on $[0, 2\pi)$ by noting the behavior of $g(x) = \cos x$ as x increases from 0 to 2π (Figure 12).

Figure 12

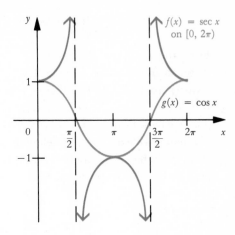

The graph of $f(x) = \sec x$ reaffirms the fact that the range of $f(x) = \sec x$ is $(-\infty, -1] \cup [1, \infty)$; the secant is not one-to-one; it is not continuous on R.

4 *The Cosecant*

For $f(x) = \csc x = 1/\sin x$, the period is 2π and the vertical asymptotes occur when $\sin x = 0$. The graph of $f(x) = \csc x$ can be determined from the auxiliary graph of $g(x) = \sin x$ (Figure 13). Here again the graph repeats every 2π units.

Figure 13

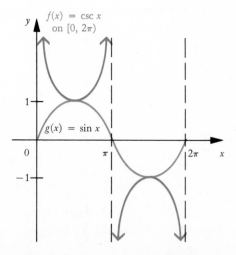

EXAMPLES

1   Use auxiliary graphs to sketch the graph of $f(x) = \tan(x-1)$.

SOLUTION

$$f(x) = \tan(x-1) = \frac{\sin(x-1)}{\cos(x-1)}$$

First, the auxiliary graphs of $g(x) = \sin(x-1)$ and $h(x) = \cos(x-1)$ are sketched. Each yields a cycle if $x-1$ varies from 0 to 2π, that is, when x varies from 1 to $2\pi+1$ (Figure 14).

Next, we locate the vertical asymptotes [where $\cos(x-1) = 0$] and the points where $\tan(x-1) = 0$ [where $\sin(x-1) = 0$] (Figure 14).

Finally, we examine the behavior of $\sin(x-1)$ and $\cos(x-1)$ to determine the behavior of $\tan(x-1)$ as x increases from 1 to $2\pi+1$. Because of periodicity, the graph repeats every 2π units.

Figure 14

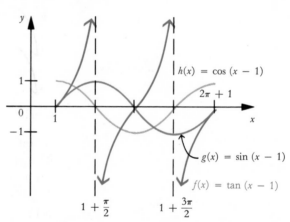

2   Use an auxiliary graph to graph one period of $f(x) = 2\sec 5x$.

SOLUTION.   First we graph $g(x) = \sec 5x = 1/\cos 5x$ by using the auxiliary graph $y = \cos x$ (Figure 15).

Figure 15

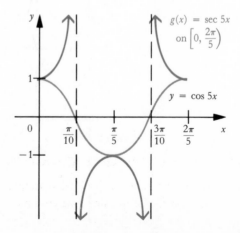

The graph of $f(x) = 2 \sec 5x$ can be obtained from the graph of $g(x) = \sec 5x$ by multiplying the ordinates of g by 2 (Figure 16).

The graph of f displays the fact that the range of $f(x) = 2 \sec 5x$ is the set $(-\infty, -2] \cup [2, \infty)$.

Figure 16

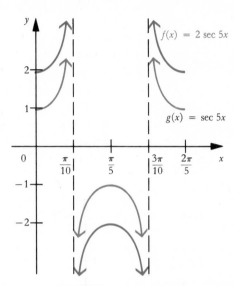

3   Use an auxiliary graph to sketch one period of $f(x) = \csc(x/4-5)$.

SOLUTION.   The auxiliary graph is that of $g(x) = \sin(x/4-5)$, since $\csc(x/4-5) = 1/[\sin(x/4-5)]$. $g(x) = \sin(x/4-5)$ generates one sine cycle when $x/4-5$ varies from 0 to 2π ($0 \leq x/4-5 < 2\pi$), that is, when x varies from 20 to $8\pi+20$ ($20 \leq x < 8\pi+20$) (Figure 17). The range of f is the set $(-\infty, -1] \cup [1, \infty)$.

Figure 17

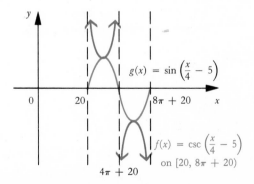

7.3   Inverses of the Other Circular Functions

In order to obtain the inverse tangent function we proceed as we did for the inverses of the sine and cosine functions.

We begin by constructing the function $f(x) = \text{Tan } x$ (notice the capital T) by restricting the domain of the tangent function to the *open* interval $(-\pi/2, \pi/2)$. The range of $f(x) = \text{Tan } x$ is the set of real numbers (Figure 18). Since the function f is an increasing function in the interval $(-\pi/2, \pi/2)$, it has an inverse that we call the *Arctangent function*.

Figure 18

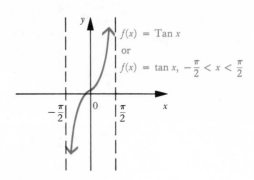

$f(x) = \text{Tan } x$

or

$f(x) = \tan x, \quad -\dfrac{\pi}{2} < x < \dfrac{\pi}{2}$

DEFINITION ARCTANGENT FUNCTION

The *Arctangent function*, denoted by $y = \text{Arctan } x$ or by $y = \tan^{-1} x$, is the inverse function of $f(x) = \text{Tan } x$. The domain of Arctangent is the set of real numbers and the range is $(-\pi/2, \pi/2)$.

$y = \tan^{-1} x$ is equivalent to $x = \tan y$ with $-\pi/2 < y < \pi/2$.

For example,

$$\frac{\pi}{4} = \tan^{-1} 1 \qquad \text{is equivalent to} \qquad 1 = \tan\frac{\pi}{4}$$

$$0 = \tan^{-1} 0 \qquad \text{is equivalent to} \qquad 0 = \tan 0$$

$$-\frac{\pi}{4} = \text{Arctan}(-1) \qquad \text{is equivalent to} \qquad -1 = \tan\left(-\frac{\pi}{4}\right)$$

The graph of $y = \text{Arctan } x$ is displayed in Figure 19 (see Problem 8).

Figure 19

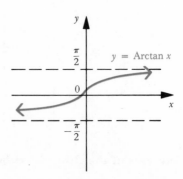

$y = \text{Arctan } x$

The derivations of the inverses for the cotangent, secant, and cosecant functions are assigned in Problems 10, 12, and 13.

PROBLEM SET 6

1 a) Use the result of Problem 1 of Problem Set 2 on page 273 to complete
       the table below.

t	$\tan t$	$\cot t$	$\sec t$	$\csc t$
0				
$\dfrac{\pi}{6}$				
$\dfrac{\pi}{4}$				
$\dfrac{\pi}{3}$				
$\dfrac{\pi}{2}$				
$\dfrac{2\pi}{3}$				
$\dfrac{3\pi}{4}$				
$\dfrac{5\pi}{6}$				
π				
$\dfrac{7\pi}{6}$				
$\dfrac{5\pi}{4}$				
$\dfrac{4\pi}{3}$				
$\dfrac{3\pi}{2}$				
$\dfrac{5\pi}{3}$				
$\dfrac{7\pi}{4}$				
$\dfrac{11\pi}{6}$				

b) Use the table to find $\tan t$, $\cot t$, $\sec t$, and $\csc t$ for each of the following values of t.

    i $11\pi/3$ ii $-\pi/3$ iii $-3\pi/4$ iv $-2\pi/3$
    v $17\pi/3$ vi $85\pi/6$ vii $71\pi/6$ viii $-11\pi/4$

2   a) Prove that the cotangent and cosecant are periodic functions of period 2π.

    b) Prove that the cotangent and cosecant are odd functions.

3   Use the five-step procedure outlined on page 282 to find $\tan t$, $\cot t$, $\sec t$, and $\csc t$ for each of the following values of t (use $\pi = 3.14$).

    a)   1.14   b) -0.53 c)   2.1   d)   9.5
    e) -10 f) -1.6 g)   3.454

4   Discuss the behavior of $f(x) = \cot x = \cos x / \sin x$ for $x \in [0, 2\pi)$ by examining the auxiliary graphs $g(x) = \cos x$ and $h(x) = \sin x$

    a)   as x increases from 0 to $\pi/2$
    b)   as x increases from $\pi/2$ to π
    c)   as x increases from π to $3\pi/2$
    d)   as x increases from $3\pi/2$ to 2π

5   Evaluate the other five circular functions at t for each of the following cases.

    a) $\cos t = \frac{5}{13}$ and $P(t)$ in quadrant IV
    b) $\sin t = \frac{1}{3}$ and $P(t)$ in quadrant II
    c) $\csc t = 5$ and $P(t)$ in quadrant I
    d) $\sec t = -2$ and $P(t)$ in quadrant III

6   Use the graphs of the six circular functions to complete the table.

x increasing from __ to __	$0 \to \dfrac{\pi}{2}$	$\dfrac{\pi}{2} \to \pi$	$\pi \to \dfrac{3\pi}{2}$	$\dfrac{3\pi}{2} \to 2\pi$
$\sin x$	Increasing	Decreasing		
$\cos x$	Decreasing			
$\tan x$	Increasing		Increasing	
$\cot x$		Decreasing		
$\sec x$				Decreasing
$\csc x$				

7   Use auxiliary graphs to graph each of the following functions.

a) $f(x) = \tan 3x$
b) $f(x) = 2 \sec x$
c) $f(x) = 3 \cot x$
d) $f(x) = \frac{1}{3} \csc x$
e) $f(x) = -8 \sec 2x$
f) $f(x) = \cot(2x - 1)$
g) $f(x) = 3 \sec(4 - x)$
h) $f(x) = 2 \tan(x/3 + 5)$

8   a) Sketch the graph of $f(x) = \tan x$, $-\pi/2 < x < \pi/2$.

b) On the same coordinate system, locate the points of the graph of $f^{-1}(x) = \text{Arctan} \, x$ for $x = -1, -\sqrt{3}, 0, 1$ and $1/\sqrt{3}$.

c) Complete the graph of $f^{-1}(x) = \text{Arctan} \, x$ by reflecting the graph of $f(x) = \tan x$, $-\pi/2 < x < \pi/2$, across $y = x$.

9   Evaluate each of the following.

a) $\tan^{-1}(\sqrt{3}/3)$
b) $\arctan(-1)$
c) $\tan^{-1}[\cos(\pi/2)]$
d) $\tan^{-1}(1.072)$

10   *The Inverse Cotangent Function*

a) Assume that $f(x) = \text{Cot} \, x$ is the function $f(x) = \cot x$, $0 < x < \pi$. Graph f. Explain why f has an inverse.

b) Define the inverse of f. Indicate the domain and the range of f^{-1}.

c) Graph f^{-1}.

11   Use the fact that $y = \text{Arccot} \, x$ (or $y = \cot^{-1} x$) is equivalent to $x = \cot y$, where $0 < y < \pi$, to evaluate the following.

a) $\cot^{-1}(\sqrt{3})$
b) $\cot^{-1}(-\sqrt{3})$
c) $\text{arccot}(1)$
d) $\cot^{-1}(0.3102)$

12   *The Inverse Cosecant Function*

a) Let $f(x) = \text{Csc} \, x$ be the function $f(x) = \csc x$, $-\pi/2 \leq x < 0$ or $0 < x \leq \pi/2$. Graph f. Explain why f has an inverse.

b) Define the inverse of f. Indicate the domain and range of f^{-1}.

c) Graph f^{-1}.

13   *The Inverse Secant Function*

a) Let $f(x) = \text{Sec} \, x$ be the function $f(x) = \sec x$, $0 \leq x < \pi/2$ or $\pi/2 < x \leq \pi$. Graph f. Explain why f has an inverse.

b) Define the inverse of f. Indicate the domain and range of f^{-1}.

c) Graph f^{-1}.

REVIEW PROBLEM SET

1   Find t in each of the following cases, where $t \in [0, 2\pi)$.

a) $P(t) = (0, 1)$
b) $P(t) = (-1, 0)$
c) $P(t) = (0, -1)$
d) $P(t) = (1, 0)$

2   Suppose that f is a periodic function with period 2π; find two values of x in the interval $[0, 4\pi)$ such that
   a) $f(x) = f(-\pi/2)$
   b) $f(x) = f(15\pi)$

3   Find the values of the six circular functions for each of the following cases.
   a) $P(t) = (\frac{1}{2}, -\sqrt{3}/2)$
   b) $P(t) = (-1/\sqrt{2}, 1/\sqrt{2})$
   c) $P(t) = (x, \frac{12}{13})$, in quadrant I.
   d) $P(t) = (-\frac{4}{5}, y)$, in quadrant III.

4   State the quadrant that $P(t)$ is in if
   a) $\sin t < 0$ and $\cos t < 0$
   b) $\sin t < 0$ and $\cos t > 0$
   c) $\tan t > 0$ and $\cos t < 0$
   d) $\sec t > 0$ and $\cot t < 0$

5   Find each of the following values without the use of tables.
   a) $\sec\left(-\dfrac{5\pi}{3}\right)$
   b) $\cot\left(-\dfrac{5\pi}{6}\right)$
   c) $\cos\dfrac{47\pi}{6}$

   d) $\sin\left(-\dfrac{89\pi}{2}\right)$
   e) $\tan\dfrac{14\pi}{3}$
   f) $\csc\dfrac{25\pi}{6}$

   g) $\sin\dfrac{125\pi}{6}$
   h) $\cos\left(-\dfrac{325\pi}{3}\right)$

6   Find $\cos t$, $\tan t$, $\cot t$, $\sec t$, and $\csc t$ in each of the following cases.
   a) If $\sin t = \frac{2}{3}$ and $P(t)$ is in quadrant II
   b) If $\sin t = -\frac{3}{8}$ and $P(t)$ is in quadrant IV
   c) If $\sin t = -\frac{1}{3}$ and $P(t)$ is in quadrant III

7   Determine each of the following values (use $\pi = 3.14$).
   a) $\cos 7.6$
   b) $\sin 13.4$
   c) $\tan 3.77$
   d) $\cot 6.37$
   e) $\sec 4.73$
   f) $\sin 5.37$
   g) $\csc 6.91$
   h) $\tan(-6.90)$
   i) $\cos(-7.5)$
   j) $\sin(-4.33)$

8   Determine each of the following values by linear interpolation (use $\pi = 3.14$).
   a) $\tan 1.245$
   b) $\cos 15.728$
   c) $\sin(-7.513)$
   d) $\cos(-0.378)$
   e) $\cot 16.191$
   f) $\sec(-11.495)$

9   Use the sine and cosine graphs to graph one cycle of each of the following functions. Indicate the amplitude, period, and the phase shift.
   a) $f(x) = \frac{1}{3}\cos x$
   b) $f(x) = \sin\frac{1}{3}x$

   c) $f(x) = 2\sin(-2x)$
   d) $f(x) = 4\sin\left(x - \dfrac{\pi}{6}\right)$

   e) $f(x) = \dfrac{1}{2}\cos\left(2x + \dfrac{\pi}{8}\right)$
   f) $f(x) = 3\sin\left(\dfrac{x}{2} - 1\right)$

10 Use the graphs of the sine and cosine functions to graph $f(x) = \sin 2x + \cos x$, $0 \leqq x < 2\pi$.

11 Graph one cycle of each of the following functions and determine the intervals where the function is increasing or decreasing.

a) $f(x) = 3 \cot x$

b) $f(x) = 4 \csc x$

c) $f(x) = 2 \tan 3x$

d) $f(x) = 3 \sec 4x$

e) $f(x) = 2 \tan\left(2x - \dfrac{\pi}{8}\right)$

12 Express x in terms of y for each of the following equations.

a) $y = \sin^{-1} 3x$

b) $y = 2 \tan^{-1}(x+3)$

c) $y = 3 + \sec^{-1} 4x$

d) $2y = \pi - \cos^{-1}(2x - 1)$

13 Find the values of each of the following expressions without the use of tables.

a) $\tan^{-1} 1$

b) $\cos^{-1}(-\tfrac{1}{2})$

c) $\sec^{-1}\sqrt{2}$

d) $\cot^{-1}(-\sqrt{3})$

e) $\csc^{-1} 2$

f) $\sin^{-1}(-1)$

14 Use Table III to find the following values (use $\pi = 3.14$).

a) $\sin^{-1} 0.4969$

b) $\tan^{-1} 0.5334$

c) $\cos^{-1}(-0.6294)$

d) $\cos^{-1} 0.8419$

e) $\sin^{-1}(-0.9959)$

f) $\tan^{-1}(-0.3203)$

15 Evaluate each of the following expressions.

a) $\sin\left[\tan^{-1}\left(\sqrt{3}/3\right)\right]$

b) $\tan\left[\sin^{-1}\left(\sqrt{3}/2\right)\right]$

c) $\tan^{-1}\left[\sin\left(\pi/2\right)\right]$

d) $\sin\left[\tan^{-1}\left(-\sqrt{3}\right)\right]$

e) $\sin\left(\tan^{-1} 1\right)$

16 Graph each of the following functions. Indicate the domain and the range.

a) $f(x) = \sin^{-1} 2x$

b) $f(x) = \cos^{-1} 3x$

c) $f(x) = \sin(\sin^{-1} x)$

d) $f(x) = \sin^{-1}(\cos x)$

e) $f(x) = \tan^{-1} 2x$

Trigonometric Functions

1 Introduction

This chapter is devoted to an introduction of the *trigonometric functions* and their relationship to the circular functions. The trigonometric functions are, in a sense, the circular functions applied to angles. The trigonometric functions are defined on angles, whereas the circular functions are defined on real numbers.

We will see that the trigonometric functions are quite useful in solving certain types of geometric problems; for example, the trigonometric functions can be used to find unknown parts of triangles.

Let us begin by reviewing some elementary plane geometry.

2 Angles

We know from plane geometry that two distinct points determine a (straight) *line* (Figure 1a). A portion of a line joining any two of its points is called a *line segment* (Figure 1b). The two points A and B are called the *end points* joining the line segment $\overline{AB}$. Any point on a line separates the line into two parts called *rays*. For example, ray $\overrightarrow{AB}$ means that the ray starts at A, goes through B, and continues indefinitely (Figure 1c).

Figure 1

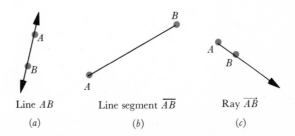

Line AB	Line segment $\overline{AB}$	Ray $\overrightarrow{AB}$
(a)	(b)	(c)

An *angle* is determined by rotating a ray about its end point, the *vertex* of the angle, from some initial position, the *initial side* of the angle, to a terminal position, the *terminal side* of the angle **(Figure 2)**. If the angle is formed by a counterclockwise rotation, the angle is said to be an angle of *positive* sense, whereas if the angle is formed by a clockwise rotation, the angle is of *negative* sense. In Figure 2 the angle determined by Q, P, and R, $\angle QPR$, is a positive angle, whereas $\angle CAB$ is negative.

Figure 2

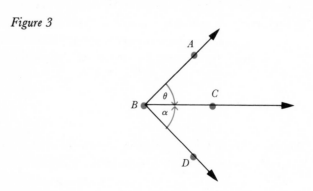

For convenience, Greek letters are often used to label angles. In Figure 3, α is the positive angle, $\angle DBC$, whereas θ is the negative angle, $\angle ABC$.

Figure 3

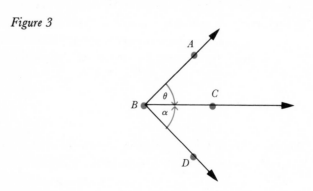

It is important to remember that an angle is determined by the initial side, the terminal side, and the rotation used to form it. For example, the angles α and β in Figure 4 have the same initial and terminal sides, yet are different angles.

Figure 4

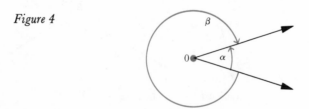

Angles are measured by using either *degrees* or *radians*. *One degree* (1°) is the measure of a positive angle which is formed by $\frac{1}{360}$ of one complete revolution **(Figure 5a)**. Furthermore, a degree can be divided into 60 equal parts called *minutes* ('); a minute can be divided into 60 equal parts called seconds ("). **On the other hand,** one *radian* (1) is the measure of a positive angle that intercepts an arc of length 1 on a circle of radius 1 **(Figure 5b).** The angle measure is positive or negative according to whether the angle is formed by a counterclockwise or clockwise rotation. Hence the radian measure of an angle is the "directed" length of its subtended arc on a circle of radius 1.

Figure 5

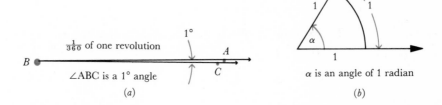

$\frac{1}{360}$ of one revolution

B

∠ABC is a 1° angle

(*a*)

α is an angle of 1 radian

(*b*)

EXAMPLES

1 Indicate both the degree measure and the radian measure of an angle α formed by
   a) 1 complete counterclockwise rotation
   b) $\frac{1}{4}$ counterclockwise rotation
   c) $\frac{1}{8}$ clockwise rotation

SOLUTION

a) Since 1° is the measure of an angle formed by $\frac{1}{360}$ of one complete revolution, the degree measure of α is $(360)(1°) = 360°$. The radian measure of α is the length of the circumference of a circle of radius 1, 2π (Figure 6).

Figure 6

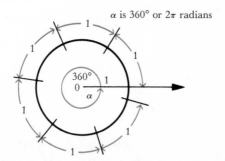

α is 360° or 2π radians

b) α is either $\frac{1}{4}(360°) = 90°$ or $\frac{1}{4}(2\pi) = \pi/2$ radians (Figure 7a).
c) $\frac{1}{8}(360°) = 45°$ and $\frac{1}{8}(2\pi) = \pi/4$; however, α is a negative angle; hence α is $-45°$ or $-\pi/4$ radians (Figure 7b).

Figure 7

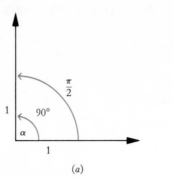 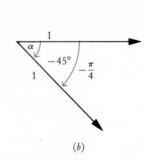

(a)   (b)

2   Express the following angles in terms of degrees, minutes, and seconds.
    a)   37.45°   b)   −84.32°

SOLUTION

a) $37.45° = 37° + 0.45°$; however, $0.45° = (0.45)(60') = 27.00'$, so that
     $37.45° = 37°27'0''$.

b) $-84.32° = -(84° + 0.32°)$; however, $0.32° = (0.32)(60') = 19.20'$.
     Also, $0.20' = (0.20)(60'') = 12''$, so that $-84.32° = -84°19'12''$.

2.1 Conversion of Angle Measures

Often an angle measure is expressed in either degrees or radians and we
need to convert its given measure to the other. For example, an angle of
1° subtends an arc of the unit circle of length $\frac{1}{360}$ of the circumference of
the circle. In determining the radian measure R of this angle, we use the
fact that R is the length of the subtended arc, together with the fact that
the circumference of the circle is 2π (Figure 8) to get $R = \frac{1}{360}(2\pi) = \pi/180$.

Figure 8

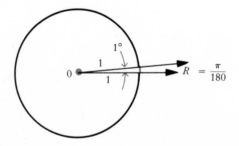

Hence 1° corresponds to $\pi/180$ radians. This means that an angle of D
degrees is $(\pi/180) D$ radians, so that in general,

$$R = \frac{\pi}{180} D$$

where R represents the radian measure of any angle and D represents the degree measure of the same angle. This relationship between R and D can also be expressed by the formula

$$D = \frac{180}{\pi} R$$

which we get by solving the first equation for D in terms of R.

EXAMPLES

1 In each case, sketch the angle with the given degree measure and find its corresponding radian measure.

a) 45°   b) 150°   c) 810°   d) −15°

SOLUTION. Using the fact that the radian measure R corresponds to $(\pi/180) D$, where D is the degree measure of the angle, we have

a) 45° corresponds to $\frac{\pi}{180}(45) = \frac{\pi}{4}$ radians   (Figure 9a)

b) 150° corresponds to $\frac{\pi}{180}(150) = \frac{5\pi}{6}$ radians   (Figure 9b)

c) 810° corresponds to $\frac{\pi}{180}(810) = \frac{9\pi}{2}$ radians   (Figure 9c)

d) −15° corresponds to $\frac{\pi}{180}(-15) = -\frac{\pi}{12}$ radians   (Figure 9d)

Figure 9

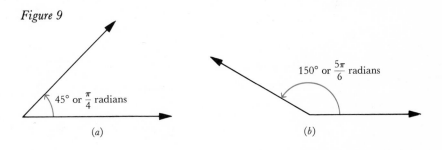

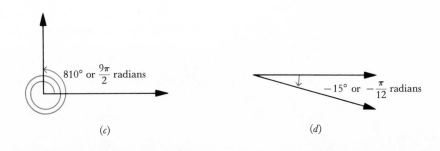

2 In each case sketch the angle with the given radian measure and find its corresponding degree measure.

a) $\dfrac{\pi}{3}$ b) $\dfrac{17\pi}{10}$ c) $\dfrac{23\pi}{10}$ d) $-\dfrac{5\pi}{12}$

SOLUTION. Using the fact that the degree measure D corresponds to $(180/\pi)\,R$, where R is the radian measure of the angle, we have

a) $\dfrac{\pi}{3}$ radians corresponds to $\dfrac{180}{\pi}\left(\dfrac{\pi}{3}\right) = 60°$ (Figure 10a)

b) $\dfrac{17\pi}{10}$ radians corresponds to $\dfrac{180}{\pi}\left(\dfrac{17\pi}{10}\right) = 306°$ (Figure 10b)

c) $\dfrac{23\pi}{10}$ radians corresponds to $\dfrac{180}{\pi}\left(\dfrac{23\pi}{10}\right) = 414°$ (Figure 10c)

d) $-\dfrac{5\pi}{12}$ radians corresponds to $\dfrac{180}{\pi}\left(-\dfrac{5\pi}{12}\right) = -75°$ (Figure 10d)

Figure 10

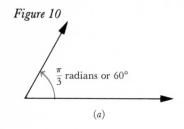

$\frac{\pi}{3}$ radians or 60°

(a)

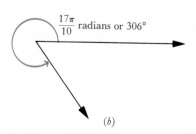

$\frac{17\pi}{10}$ radians or 306°

(b)

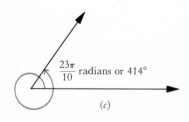

$\frac{23\pi}{10}$ radians or 414°

(c)

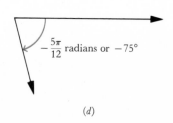

$-\frac{5\pi}{12}$ radians or −75°

(d)

2.2 Arcs and Sectors of Circles

The radian measure of a central angle of a circle can be used to determine both the length of the arc the angle subtends and the area of the sector determined by the central angle.

Suppose that θ is the central angle of a circle of radius r (Figure 11). Also assume that the angle θ has a radian measure t. (If θ is measured in degrees, we can always convert to radians.)

Figure 11

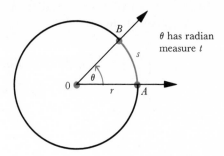

θ has radian measure t

Our task is to determine s, the length of $\widehat{AB}$. The length of the circumference of the circle is $2\pi r$; but, since θ determines an arc that is "$(t/2\pi)$th" of the circumference (why?),

$$s = \frac{t}{2\pi}(2\pi r) \qquad \text{that is} \qquad s = tr$$

Hence we have a formula for determining the length of an arc subtended by a central angle of t radians.

Next we can determine the area of sector AOB. Since the area of the circle is πr^2 and since θ determines $(t/2\pi)$th of the circle, the area A is determined by

$$A = \frac{t}{2\pi}\pi r^2 = \frac{tr^2}{2}$$

so that

$$A = \tfrac{1}{2}r^2 t$$

Using the fact that $s = tr$, we can also write the latter formula as

$$A = \tfrac{1}{2}sr$$

EXAMPLES

1 Find the arc length and the area of the sector of a circle of radius 6 inches with a central angle of measure
   a) $\pi/6$ b) $70°$

   SOLUTION

   a) $s = tr = 6(\pi/6) = \pi$ inches   and
      $A = \tfrac{1}{2}r^2 t = \tfrac{1}{2}(36)(\pi/6) = 3\pi$ square inches   (Figure 12a)

b) $s = tr$, where t is the radian measure of the angle. Since 70° corresponds to $70\pi/180 = 7\pi/18$ radians,

$$s = 6\left(\frac{7\pi}{18}\right) = \frac{7\pi}{3} \text{ inches} \qquad \text{and}$$

$$A = \frac{1}{2}r^2 t = \frac{1}{2}(36)\left(\frac{7\pi}{18}\right) = 7\pi \text{ square inches} \qquad \text{(Figure 12b)}$$

Figure 12

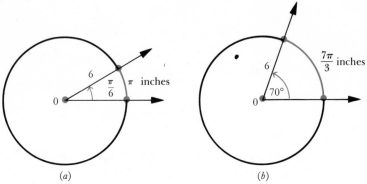

(a)   (b)

2   The tip of the minute hand of a clock travels $7\pi/10$ inches in 3 minutes. How long is the minute hand?

SOLUTION.   In 3 minutes the minute hand generates a central angle of $(3)(\frac{1}{60})(360°) = 18°$. (Why?) But an angle of 18° has a radian measure of $t = (\pi/180)(18) = \pi/10$, so that $7\pi/10 = r(\pi/10)$. Hence $r = 7$ inches (Figure 13).

Figure 13

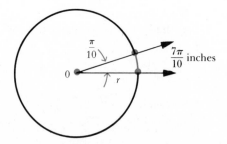

2.3   Functions Defined on Angles

An angle is in *standard position* if it is placed on a Cartesian coordinate system so that the vertex corresponds to the origin and the initial side coincides with the positive x axis. For example, an 80° angle in standard position would have its terminal side in quadrant I (Figure 14a), whereas an angle of radian measure $-3\pi/5$ in standard position has its terminal side in quadrant III (Figure 14b). [Notice that $-3\pi/5$ corresponds to $(180/\pi)(-3\pi/5) = -108°$.]

Figure 14

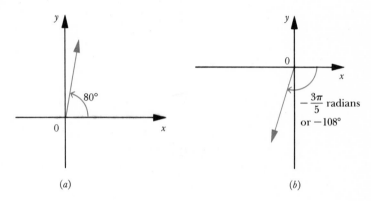

(a)   (b)

EXAMPLE

Sketch an angle in standard position whose terminal side contains the given point. Indicate by θ one of the positive angles and by ϕ one of the negative angles determined by the terminal side. Indicate the quadrant in which the terminal side of θ lies.

a) (3, 4)   b) (−4, 3)   c) (−2, −3)   d) (4, −1)

SOLUTION

a) The terminal side of any angle θ that contains the point (3, 4) lies in quadrant I (Figure 15a).

b) The terminal side of any angle θ that contains the point (−4, 3) lies in quadrant II (Figure 15b).

c) The terminal side of any angle θ that contains the point (−2, −3) lies in quadrant III (Figure 15c).

d) The terminal side of any angle θ that contains the point (4, −1) lies in quadrant IV (Figure 15d).

   Later in this section we will see how to find the measure (in degrees or radians) of these angles.

   Now, suppose that θ is a 45° angle in standard position and suppose that (x, y) is any point other than $(0, 0)$ on the terminal side of θ (Figure 16). We can compute the distance r between (x, y) and $(0, 0)$ by using the distance formula:

$$r = \sqrt{(x-0)^2 + (y-0)^2} = \sqrt{x^2 + y^2}$$

Figure 15

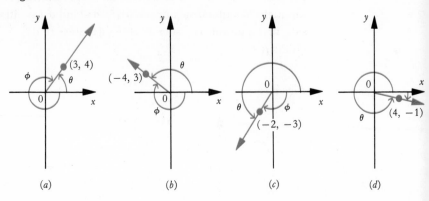

(a)   (b)   (c)   (d)

Since the triangle determined by the three points $(0,0)$, (x,y), and $(x,0)$ is isosceles (why?), we have $x = y$ and $r = \sqrt{2x^2} = \sqrt{2}x$. What we want here are functions that relate θ, r, x, and y. These functions, the *trigonometric functions*, are defined as follows.

Figure 16

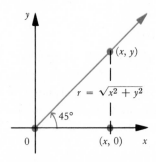

DEFINITION   TRIGONOMETRIC FUNCTIONS

Let θ be an angle in standard position and let (x, y) be any point other than $(0,0)$ on the terminal side of θ, and let r be the distance between (x, y) and $(0,0)$ (Figure 17). Notice in this illustration that the terminal side of θ lies

Figure 17

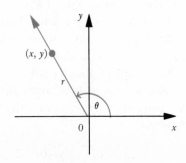

in quadrant II. In general, the quadrant in which the terminal side of θ lies depends, of course, upon the measure of the angle. The definition applies to all cases.

The six *trigonometric* functions are defined in the following manner.

$$\text{sine} \quad = \left\{ (\theta, \sin \theta) \mid \sin \theta = \frac{y}{r} \right\}$$

$$\text{cosine} \quad = \left\{ (\theta, \cos \theta) \mid \cos \theta = \frac{x}{r} \right\}$$

$$\text{tangent} \quad = \left\{ (\theta, \tan \theta) \mid \tan \theta = \frac{y}{x}, \, x \neq 0 \right\}$$

$$\text{cotangent} = \left\{ (\theta, \cot \theta) \mid \cot \theta = \frac{x}{y}, \, y \neq 0 \right\}$$

$$\text{secant} \quad = \left\{ (\theta, \sec \theta) \mid \sec \theta = \frac{r}{x}, \, x \neq 0 \right\}$$

$$\text{cosecant} \quad = \left\{ (\theta, \csc \theta) \mid \csc \theta = \frac{r}{y}, \, y \neq 0 \right\}$$

where $r = \sqrt{x^2 + y^2}$. These functions are usually written in the following abbreviated forms.

$$\sin \theta = \frac{y}{r}$$

$$\cos \theta = \frac{x}{r}$$

$$\tan \theta = \frac{y}{x} \qquad x \neq 0$$

$$\cot \theta = \frac{x}{y} \qquad y \neq 0$$

$$\sec \theta = \frac{r}{x} \qquad x \neq 0$$

$$\csc \theta = \frac{r}{y} \qquad y \neq 0$$

Returning to θ, the 45° angle, it has been noted that $x = y$ for any (x, y) on the terminal side of the angle and that $r = \sqrt{2}\,x$ (Figure 18).

Figure 18

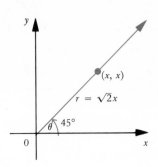

Hence, by the definition of the trigonometric functions,

$$\sin \theta = \sin 45° = \frac{x}{\sqrt{2}\,x} = \frac{1}{\sqrt{2}} = \frac{\sqrt{2}}{2}$$

$$\cos \theta = \cos 45° = \frac{x}{\sqrt{2}\,x} = \frac{1}{\sqrt{2}} = \frac{\sqrt{2}}{2}$$

$$\tan \theta = \tan 45° = \frac{x}{x} = 1$$

$$\cot \theta = \cot 45° = \frac{x}{x} = 1$$

$$\sec \theta = \sec 45° = \frac{\sqrt{2}\,x}{x} = \sqrt{2}$$

$$\csc \theta = \csc 45° = \frac{\sqrt{2}\,x}{x} = \sqrt{2}$$

It is important to understand that the values of the six trigonometric functions for the 45° angle depend only on the position of the terminal side of the angle in the sense that no matter what particular point may be selected on the terminal side to evaluate the six functions [other than $(0, 0)$ of course] the results are the same. For example, $(1, 1)$ with $r = \sqrt{2}$ will yield the same results as $(5, 5)$ with $r = 5\sqrt{2}$.

In general, if (x, y) and (x_1, y_1) are two different points in quadrant I on the terminal side of θ other than $(0, 0)$ (Figure 19), so that $r = \sqrt{x^2 + y^2}$ and $r_1 = \sqrt{x_1^2 + y_1^2}$, we have, because of the similar triangles, $\triangle OBP_1$ and and $\triangle OAP$ (Figure 19),

$$\frac{y_1}{r_1} = \frac{y}{r} \qquad \frac{x_1}{r_1} = \frac{x}{r} \qquad \frac{y_1}{x_1} = \frac{y}{x} \qquad \frac{x_1}{y_1} = \frac{x}{y} \qquad \frac{r_1}{x_1} = \frac{r}{x} \qquad \frac{r_1}{y_1} = \frac{r}{y}$$

Figure 19

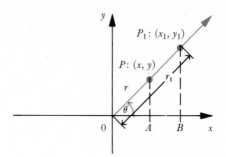

Hence the values of the trigonometric functions are the same no matter what two points are selected on the terminal side of the angle other than $(0, 0)$. Furthermore, by a similar construction, we can show that the above results are true for all angles regardless of the quadrant where the terminal sides lie (see Problem 4).

EXAMPLES

1 Evaluate each of the following.

a) $\sin 180°$ and $\cos 180°$

b) $\sin\left(-\dfrac{\pi}{2}\right)$ and $\cos\left(-\dfrac{\pi}{2}\right)$

SOLUTION. First, select points on the terminal sides of the 180° angle and the $-\pi/2$ angle after the angles are placed in standard position (Figure 20).

a) Here $(-3, 0)$ is selected on the terminal side of a 180° angle, so that $\sin 180° = 0/3 = 0$ and $\cos 180° = -3/3 = -1$.

b) The point $(0, -2)$ is selected on the terminal side of the $-\pi/2$ angle, so that $\sin(-\pi/2) = -2/2 = -1$ and $\cos(-\pi/2) = 0/2 = 0$.

Figure 20

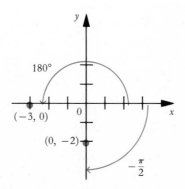

Notice that the results would be the same if any points were chosen on the terminal sides.

2 Evaluate the trigonometric functions of θ if θ has $(3, -5)$ on its terminal side. Is θ unique?

SOLUTION. Figure 21 suggests the fact that there are infinitely many different angles with this same terminal side. Hence θ is not unique.

Figure 21

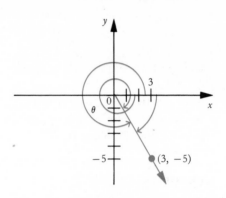

Using the fact that $r = \sqrt{3^2 + (-5)^2} = \sqrt{34}$, we have

$$\sin \theta = \frac{y}{r} = \frac{-5}{\sqrt{34}} \qquad \cos \theta = \frac{x}{r} = \frac{3}{\sqrt{34}}$$

$$\tan \theta = \frac{y}{x} = -\frac{5}{3} \qquad \cot \theta = \frac{x}{y} = -\frac{3}{5}$$

$$\sec \theta = \frac{r}{x} = \frac{\sqrt{34}}{3} \qquad \csc \theta = \frac{r}{y} = \frac{\sqrt{34}}{-5}$$

3 If $\tan \theta = -\frac{4}{3}$ and θ has its terminal side in quadrant II, find the values of the remaining trigonometric functions.

Figure 22

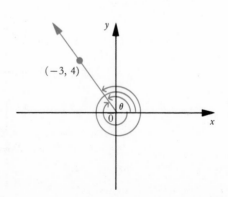

SOLUTION. $\tan\theta = y/x = -\frac{4}{3}$ and θ has its terminal side in quadrant II so that one point on the terminal side of θ is $(-3, 4)$ (Figure 22). Also $r = \sqrt{(-3)^2 + 4^2} = 5$ for this point, so by the definition of the trigonometric functions, we have

$$\sin\theta = \frac{y}{r} = \frac{4}{5} \qquad\qquad \sec\theta = \frac{r}{x} = \frac{5}{-3}$$

$$\cos\theta = \frac{x}{r} = \frac{-3}{5} \qquad\qquad \csc\theta = \frac{r}{y} = \frac{5}{4}$$

$$\cot\theta = \frac{x}{y} = \frac{-3}{4}$$

4   In what quadrant is the terminal side of θ if $\sin\theta = -\frac{7}{9}$? Evaluate $\cos\theta$.

SOLUTION. $\sin\theta = y/r = -\frac{7}{9}$, from which we can deduce the possibilities shown in Figure 23. Since $9 = \sqrt{x^2 + 49}$, $x^2 = 32$. Hence $x = \pm 4\sqrt{2}$, so $\cos\theta = 4\sqrt{2}/9$ if θ has a terminal side in quadrant IV or $\cos\theta = -4\sqrt{2}/9$ if θ has terminal side in quadrant III. Notice that there are actually infinitely many different possible values for θ; however, any such θ must have one of the two terminal sides (Figure 23).

Figure 23

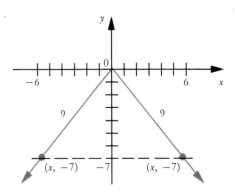

PROBLEM SET 1

1   a)   In each of the following parts sketch the angle in standard position and convert it to degree measure.

i $2\pi/3$	ii $11\pi/6$	iii $7\pi/18$	iv $121\pi/360$
v $-7\pi/12$	vi $43\pi/6$	vii $-4\pi/9$	viii $-7\pi/2$

b)   In each of the following parts, sketch the angle in standard position and convert it to radian measure.

i $40°$	ii $75°$	iii $240°$	iv $330°$
v $-95°$	vi $-220°$	vii $-444°$	viii $100°$

2 Explain why an angle of 1 radian is larger in measure than an angle of 1°.

3 Find the arc length and the area of a sector of a circle of radius r that is subtended by a central angle θ for each of the following situations.

a) $r = 7$ inches and $\theta = 3\pi/14$ b) $r = 4$ inches and $\theta = 2\pi/5$

c) $r = 6$ inches and $\theta = 50°$ d) $r = 9$ inches and $\theta = 275°$

e) $r = 7$ inches and $\theta = \pi°$

4 a) Suppose that (x, y) and (x_1, y_1) are two different points in quadrant II on the terminal side of θ other than $(0,0)$ (Figure 24). Show that

$$\frac{y_1}{r_1} = \frac{y}{r} \qquad \frac{x_1}{r_1} = \frac{x}{r} \qquad \frac{y_1}{x_1} = \frac{y}{x}$$

$$\frac{x_1}{y_1} = \frac{x}{y} \qquad \frac{r_1}{x_1} = \frac{r}{x} \qquad \frac{r_1}{y_1} = \frac{r}{y}$$

Figure 24

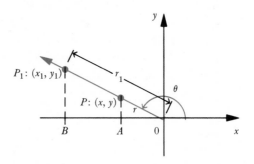

b) Show that the ratios in part a) also hold if the points are in quadrant III.

c) Show that the ratios in part a) also hold if the points are in quadrant IV.

5 Use the definition of the trigonometric functions to prove each of the following identities.

a) $\tan\theta = \sin\theta/\cos\theta$ b) $\cot\theta = \cos\theta/\sin\theta$

c) $\sec\theta = 1/\cos\theta$ d) $\csc\theta = 1/\sin\theta$

e) $\sin^2\theta + \cos^2\theta = 1$

6 Draw a 60° angle in standard position, and then use Figure 25 to evaluate the trigonometric functions of 60°. Would the value differ if the angle were 420°? −300°? $\pi/3$? Explain.

Figure 25

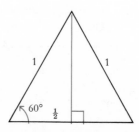

7 Determine the values of the trigonometric functions of θ in each of the following cases. Sketch any two possible angles θ in standard position in each case.

a) θ has $(-5, 0)$ on its terminal side. What is the measure of θ?

b) θ has $(-3, -4)$ on its terminal side.

c) θ has $(7, -10)$ on its terminal side.

d) θ has $(0, 1)$ on its terminal side. What is the measure of θ?

e) θ has $(-1, \sqrt{3})$ on its terminal side. What is the measure of θ?

f) $(x, -4)$ is 11 units from $(0, 0)$ and is on the terminal side of θ.

g) $(3x, x)$ is on the terminal side of θ and $x \neq 0$.

8 The length of each chain supporting the seat of a baby swing is 8 feet. When the swing is at its highest points forward and backward, the radian measure of the angle determined by these chains is $5\pi/6$. How far does the baby travel in one trip between these high points; that is, what is the length of the arc generated by one swing between the two high points?

9 Construct each of the following angles in standard position whose terminal side is in the indicated quadrant, and evaluate the other five trigonometric functions of θ.

a) $\sin \theta = \frac{15}{17}$, quadrant II

b) $\cos \theta = \sqrt{3}/2$, quadrant IV

c) $\tan \theta = -1$, quadrant IV

d) $\cot \theta = \sqrt{3}$, quadrant III

e) $\sec \theta = -\frac{13}{5}$, quadrant II

f) $\csc \theta = \sqrt{2}$, quadrant I

3 Trigonometric and Circular Functions

Suppose that t is a real number. If the wrapping function P gives $P(t) = (x, y)$, then the *circular* functions yield $\cos t = x$ and $\sin t = y$ (Figure 1). Now let θ be any angle between the positive x axis and the ray through the point $P(t)$ in Figure 1. (The argument holds regardless of what quadrant the terminal side of θ lies in.)

Figure 1

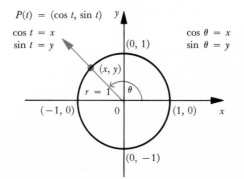

By the definition of the trigonometric functions on page 332, the *trigonometric* functions on θ yield $\cos\theta = x/r$ and $\sin\theta = y/r$, where $(x, y) \neq (0, 0)$ is a point on the terminal side of θ and $r = \sqrt{x^2 + y^2}$. But with (x, y) selected as above, we have $r = 1$, so that $\cos\theta = x/1 = x$ and $\sin\theta = y/1 = y$ (Figure 1). Thus

$$\cos\theta = \cos t \quad \text{and} \quad \sin\theta = \sin t$$

Furthermore, t equals the arc length subtended by θ. (Why?) Hence (by definition of radian measure on page 325) t equals the measure in *radians* of θ. Here $\cos\theta$ and $\sin\theta$ are the values of the trigonometric functions, whereas $\cos t$ and $\sin t$ are the values of the circular functions.

In other words, the circular sine and cosine functions can be thought of as the trigonometric sine and cosine functions defined on angles, and, conversely, the trigonometric sine and cosine functions defined on angles can be thought of as the circular sine and cosine functions defined on radian measures of the angles.

Finally, since the other four trigonometric functions can be expressed in terms of the sine and cosine (see Problem Set 1, Problem 5) in the same way the circular functions can be expressed in terms of the sine and cosine, we have

$$\tan\theta = \frac{\sin\theta}{\cos\theta} = \frac{\sin t}{\cos t} = \tan t$$

$$\cot\theta = \frac{\cos\theta}{\sin\theta} = \frac{\cos t}{\sin t} = \cot t$$

$$\sec\theta = \frac{1}{\cos\theta} = \frac{1}{\cos t} = \sec t$$

$$\csc\theta = \frac{1}{\sin\theta} = \frac{1}{\sin t} = \csc t$$

where t is the radian measure of θ. Thus the following theorem holds.

THEOREM 1

If T is any one of the six trigonometric functions and C is the corresponding circular function, then

$$T(\theta) = C(t) \quad \text{where } \theta \text{ is an angle with radian measure } t$$

Hence it follows that all the properties of the circular functions become properties of the trigonometric functions. (See Problems 2, 4, and 6.)

EXAMPLES

Use Theorem 1 to find the values of the six trigonometric functions at the given angle θ.

1 $\theta = 60°$

SOLUTION.   Since $T(\theta) = C(t)$ and $60°$ is the degree measure of an angle of $\pi/3$ radians (Figure 2), we have

Figure 2

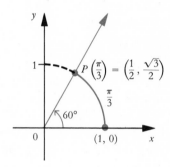

$$\sin 60° = \sin\frac{\pi}{3} = \frac{\sqrt{3}}{2} \qquad \cot 60° = \cot\frac{\pi}{3} = \frac{1}{\sqrt{3}} = \frac{\sqrt{3}}{3}$$

$$\cos 60° = \cos\frac{\pi}{3} = \frac{1}{2} \qquad \sec 60° = \sec\frac{\pi}{3} = 2$$

$$\tan 60° = \tan\frac{\pi}{3} = \sqrt{3} \qquad \csc 60° = \csc\frac{\pi}{3} = \frac{2}{\sqrt{3}} = \frac{2\sqrt{3}}{3}$$

2 $\theta = 150°$

SOLUTION.   Since $T(\theta) = C(t)$ and $150°$ is the degree measure of an angle of measure $5\pi/6$ radians (Figure 3), then

Figure 3

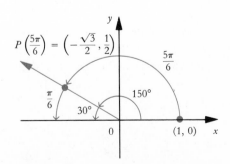

$$\sin 150° = \sin \frac{5\pi}{6} = \frac{1}{2} \qquad\qquad \cot 150° = \cot \frac{5\pi}{6} = -\sqrt{3}$$

$$\cos 150° = \cos \frac{5\pi}{6} = -\frac{\sqrt{3}}{2} \qquad \sec 150° = \sec \frac{5\pi}{6} = -\frac{2}{\sqrt{3}} = -\frac{2\sqrt{3}}{3}$$

$$\tan 150° = \tan \frac{5\pi}{6} = -\frac{1}{\sqrt{3}} = -\frac{\sqrt{3}}{3} \quad \csc 150° = \csc \frac{5\pi}{6} = 2$$

3.1 Reference Angles

Since $T(\theta) = C(t)$, where t is the radian measure of angle θ, it follows that the trigonometric functions are periodic functions of period 2π if radian measure is used or of 360° if degree measure is used (see Problem 4a) and that the evaluations of the trigonometric functions parallel the evaluations of the circular functions.

Given any angle, we can find another angle with the same terminal side as the given angle so that the "new" angle is between 0° and 360° (or 0 and 2π). If the angle is 0°, 90°, 180°, or 270° the evaluation is not difficult; if not, we can find a *reference angle* as we found a reference number. For example, the reference angle for 300° is 60°; for −200° it is 20°; and for 280° it is 80° (Figure 4). Notice that the reference angle of a given angle is an *acute angle* formed by the terminal side of the given angle and the x axis.

Figure 4

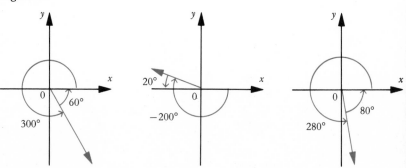

As with the circular functions, the value of the given trigonometric function T on the given angle θ agrees in absolute value with the value of the function on the reference angle θ_R; hence we can compute $T(\theta)$ as follows.

1   Locate the quadrant where the terminal side of θ lies.

2   Determine the reference angle θ_R, an acute angle formed by the terminal side of θ and the x axis.

3   Use Table IV of Appendix A if degrees are used or Table III if radians are used, and linear interpolation if necessary, to find $T(\theta_R) = |T(\theta)|$.

4   Finally, adjust the sign to get $T(\theta)$ according to which quadrant contains the terminal side of θ.

The values of the trigonometric functions of θ_R can be determined *approximately* by using Table IV. As with the circular functions, we shall assume that the table values of the trigonometric functions are approximated correct to four decimal places even though we will use the equal sign.

EXAMPLES

1   Use Table IV to determine each of the following values.
   a) $\sin 36°$ b) $\cos 65°$ c) $\tan(72°10')$
   d) $\cot(83°40')$ e) $\sec(41°30')$ f) $\csc(58°50')$

SOLUTION.   The values of the trigonometric functions can be determined directly from Table IV to get
   a) $\sin 36° = 0.5878$ b) $\cos 65° = 0.4226$
   c) $\tan(72°10') = 3.108$ d) $\cot(83°40') = 0.1110$
   e) $\sec(41°30') = 1.335$ f) $\csc(58°50') = 1.169$

2   Reduce the evaluation of the given trigonometric function to an evaluation of the same function at the reference angle θ_R. Then use Table IV and linear interpolation if necessary to determine the value.
   a) $\sin 245°$ and $\cos 245°$
   b) $\tan(320°10')$ and $\cot(320°10')$
   c) $\sec 481°$ and $\csc 481°$

SOLUTION.   To find the values of the trigonometric functions, we follow the steps outlined on page 342.
   a)   The terminal side of 245° lies in quadrant III, so the reference angle $\theta_R = 245° - 180° = 65°$. Also the sine and the cosine are negative in quadrant III (Figure 5a), so that

$$\sin 245° = -\sin 65° = -0.9063$$

and

$$\cos 245° = -\sin 65° = -0.4226$$

   b)   Since the terminal side of 320° 10' lies in quadrant IV, the reference angle $\theta_R = 360° - 320° 10' = 39° 50'$ (Figure 5b). Also, both the

Figure 5

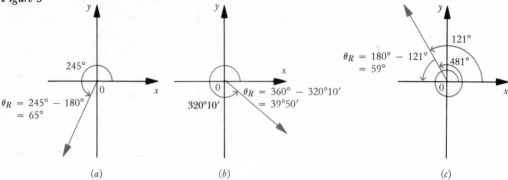

(a)   (b)   (c)

tangent and cotangent are negative in quadrant IV, so that

$$\tan 320° \, 10' = -\tan 39° \, 50' = -0.8342$$

and

$$\cot 320° \, 10' = -\cot 39° \, 50' = -1.199$$

c) Since $481° = 360° + 121°$, the terminal side of a 481° angle in standard position lies in quadrant II, and the reference angle $\theta_R = 59°$ (Figure 5c). Since the secant is negative in quadrant II, and the cosecant is positive in quadrant II, we have

$$\sec 481° = \sec 121° = -\sec 59° = -1.942$$

and

$$\csc 481° = \csc 121° = \csc 59° = 1.167$$

3    $\sin(-27° \, 43')$

SOLUTION.   The reference angle $\theta_R = 27° \, 43'$ (Figure 6); $27° \, 43'$ is between $27° \, 40'$ and $27° \, 50'$; hence, by linear interpolation,

$$10' \left[3' \begin{bmatrix} \sin 27° \, 40' = 0.4643 \\ \sin 27° \, 43' = \quad ? \\ \sin 27° \, 50' = 0.4669 \end{bmatrix} d \right] 0.0026$$

Therefore, $d/0.0026 = \frac{3}{10}$, so that $d = 0.3(0.0026) = 0.00078 = 0.0008$ (approximately). Hence $\sin 27° \, 43' = 0.4643 + 0.0008 = 0.4651$, but since $-27° \, 43'$ is in quadrant IV where the sine is negative, we have

$$\sin(-27° \, 43') = -0.4651$$

Figure 6

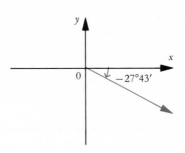

4    $\tan(200°25')$

SOLUTION.   An angle of 200°25′ in standard position has its terminal side in quadrant III. The reference angle is 20° 25′ (Figure 7). Since the tangent is positive in quadrant III, we have $\tan 200°25' = \tan 20°25'$. But $\tan 20°25'$ can be approximated by linear interpolation as follows:

$$10'\left[5'\left[\begin{array}{l}\tan 20°20' = 0.3706 \\ \tan 20°25' = \quad ? \\ \tan 20°30' = 0.3739\end{array}\right]d\right]0.0033$$

Figure 7

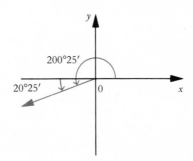

Therefore,

$$\frac{d}{0.0033} = \frac{5}{10}$$

so that

$$d = 0.0017 \text{ (approximately)}$$

Hence

$$\tan 20°25' = 0.3706 + 0.0017$$
$$= 0.3723$$

PROBLEM SET 2

1   Complete the table.

θ Measure		$\sin\theta$	$\cos\theta$	$\tan\theta$	$\cot\theta$	$\sec\theta$	$\csc\theta$
Degree	Radian						
0°							
	$\dfrac{\pi}{6}$						
45°		$\dfrac{\sqrt{2}}{2}$					
60°							
	$\dfrac{\pi}{2}$						
135°							
150°							
	π						

2   a)   Determine the domain of each of the trigonometric functions by examining the definition of the trigonometric functions. Describe the domains in terms of the degree measures of the angles. Describe the domains in terms of the radian measures of the angles. How do these domains compare to the domains of the corresponding circular functions?

   b)   Use the definition of the trigonometric functions to determine the ranges of the trigonometric functions, and then compare your results to the circular function ranges.

3   Use the fact that $T(\theta) = C(t)$ to determine each of the following values without the use of tables.

   a) $\sin(-150°)$ b) $\cos 315°$ c) $\tan 240°$
   d) $\cot 330°$ e) $\sec 300°$ f) $\csc(-240°)$
   g) $\cot(-315°)$ h) $\cos(-300°)$ i) $\csc 225°$

4   a)   Show that the trigonometric functions are periodic functions of period 360° (or 2π radians).

b) Explain why the trigonometric sine, tangent, cotangent, and cosecant functions are odd and the trigonometric cosine and secant functions are even.

5 Use the fact that $T(\theta) = C(t)$ and the results of Problem 4a to determine the following trigonometric values.

a) $\sin 390°$ b) $\cos 420°$ c) $\tan 510°$

d) $\cot(-405°)$ e) $\sec 780°$ f) $\csc(-750°)$

6 a) Assume that angle θ has a terminal side in quadrant I. Indicate the sign of the value of each of the trigonometric functions, then compare your results to the results in Chapter 5.

b) Do part a for θ in quadrant II.

c) Do part a for θ in quadrant III.

d) Do part a for θ in quadrant IV.

7 Follow the steps outlined on page 342 to determine the following trigonometric values.

a) $\cos 235°$ b) $\sin(325°10')$ c) $\tan(320°30')$

d) $\cot(462°40')$ e) $\sin 1059°$ f) $\sec(-100°)$

g) $\cos(23°45')$ h) $\tan(133°15')$ i) $\cot(331°25')$

j) $\sin(-37°35')$ k) $\sec(121°27')$ l) $\csc(167°16')$

4 Trigonometric and Circular Function Identities

In this section some of the standard formulas or identities that enable us to express the cosine or sine of a sum or difference of two numbers or two angles in terms of cosines and sines of the two numbers or the two angles will be derived. For example, suppose that $\cos(t+s)$ were to be computed using the values $\cos t$, $\cos s$, $\sin t$, and $\sin s$. The first response might be to say the $\cos(t+s) = \cos t + \cos s$, however, this is *not* the case in general. For example,

$$\cos(\pi/2 + \pi/2) = \cos \pi = -1 \neq \cos(\pi/2) + \cos(\pi/2) = 0$$

or

$$\cos(45° + 45°) = \cos 90° = 0 \neq \cos 45° + \cos 45° = \sqrt{2}$$

Notice here that although the proofs of the following theorems are given in terms of real numbers s and t, the results are also true if s and t are interpreted as angles. In other words, use s and t to represent real numbers and angles interchangeably.

THEOREM 1

If t and s are any real numbers, then

i $\cos(t-s) = \cos t \cos s + \sin t \sin s$

ii $\cos(t+s) = \cos t \cos s - \sin t \sin s$

PROOF

i   Although this identity holds for all real numbers t and s, we will assume for definiteness that $s > 0$, $t > 0$, and $t > s$ as illustrated in Figure 1. Since the arc $\overset{\frown}{P_0 P_3}$ is equal to the arc $\overset{\frown}{P_1 P_2}$ (why?), it follows from geometry that the chord $\overline{P_0 P_3}$ is equal to the chord $\overline{P_1 P_2}$. Using the distance formula, we obtain

$$\overline{P_1 P_2}^2 = (\cos t - \cos s)^2 + (\sin t - \sin s)^2$$

and

$$\overline{P_0 P_3}^2 = [\cos(t-s) - 1]^2 + [\sin(t-s) - 0]^2$$

Figure 1

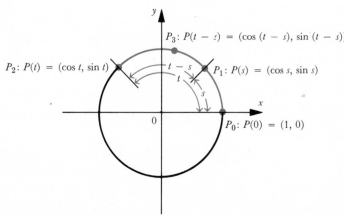

$P_3: P(t - s) = (\cos(t - s), \sin(t - s))$

$P_2: P(t) = (\cos t, \sin t)$

$P_1: P(s) = (\cos s, \sin s)$

$P_0: P(0) = (1, 0)$

Equating $\overline{P_1 P_2}^2$ with $\overline{P_0 P_3}^2$, we have

$$(\cos t - \cos s)^2 + (\sin t - \sin s)^2 = [\cos(t-s) - 1]^2 + \sin^2(t-s)$$

The left-hand side of this equation simplifies to $2 - 2(\cos t \cos s + \sin t \sin s)$, and the right-hand side simplifies to $2 - 2\cos(t-s)$ (why?), so that

$$2 - 2(\cos t \cos s + \sin t \sin s) = 2 - 2\cos(t-s)$$

Hence

$$\cos(t-s) = \cos t \cos s + \sin t \sin s \qquad \text{(why?)}$$

ii $\cos(t+s) = \cos[t-(-s)] = \cos t \cos(-s) + \sin t \sin(-s)$ because of the identity in i above; however, the latter expression can be written as

$$\cos t \cos s - \sin t \sin s$$

since

$$\cos(-s) = \cos s \qquad \text{and} \qquad \sin(-s) = -\sin s$$

Hence

$$\cos(t+s) = \cos t \cos s - \sin t \sin s$$

THEOREM 2

If t is any real number, then

i $\quad \cos\left(\dfrac{\pi}{2} - t\right) = \sin t$

ii $\quad \sin\left(\dfrac{\pi}{2} - t\right) = \cos t$

PROOF.   Apply Theorem 1, i to get

i $\quad \cos\left(\dfrac{\pi}{2} - t\right) = \cos\dfrac{\pi}{2}\cos t + \sin\dfrac{\pi}{2}\sin t$

$$= 0 + \sin t = \sin t$$

ii $\quad \sin\left(\dfrac{\pi}{2} - t\right) = \cos\left[\dfrac{\pi}{2} - \left(\dfrac{\pi}{2} - t\right)\right] \qquad \text{(Theorem 2, i)}$

$$= \cos\left(\dfrac{\pi}{2} - \dfrac{\pi}{2} + t\right)$$

$$= \cos t$$

Notice that if Theorem 2 is applied to angles measured in degrees, then the identities would be written as $\cos(90° - \theta) = \sin\theta$ and $\sin(90° - \theta) = \cos\theta$.

THEOREM 3

If t and s are any real numbers, then

i $\quad \sin(t+s) = \sin t \cos s + \cos t \sin s$

ii $\quad \sin(t-s) = \sin t \cos s - \cos t \sin s$

PROOF

i $\quad \sin(t+s) = \cos\left[\dfrac{\pi}{2} - (t+s)\right] \qquad \text{(why?)}$

$$= \cos\left[\left(\dfrac{\pi}{2} - t\right) - s\right]$$

$$= \cos\left(\dfrac{\pi}{2} - t\right)\cos s + \sin\left(\dfrac{\pi}{2} - t\right)\sin s \qquad \text{(why?)}$$

$$= \sin t \cos s + \cos t \sin s$$

ii $\sin(t-s) = \sin[t+(-s)]$

$$= \sin t \cos(-s) + \cos t \sin(-s)$$

$$= \sin t \cos s - \cos t \sin s \qquad \text{(why?)}$$

THEOREM 4

If t is any real number, then

i $\cos 2t = \cos^2 t - \sin^2 t$

$$= 2\cos^2 t - 1$$

$$= 1 - 2\sin^2 t$$

ii $\sin 2t = 2\sin t \cos t$

PROOF

i $\cos 2t = \cos(t+t)$

$$= \cos t \cos t - \sin t \sin t \qquad \text{(Theorem 1, ii)}$$

$$= \cos^2 t - \sin^2 t$$

However, since $\cos^2 t + \sin^2 t = 1$, we also have

$$\cos 2t = (1 - \sin^2 t) - \sin^2 t = 1 - 2\sin^2 t$$

or

$$\cos 2t = \cos^2 t - (1 - \cos^2 t) = 2\cos^2 t - 1$$

ii $\sin 2t = \sin(t+t)$

$$= \sin t \cos t + \cos t \sin t \qquad \text{(why?)}$$

$$= 2\sin t \cos t$$

THEOREM 5

If t is any real number, then

i $\cos^2 \dfrac{t}{2} = \dfrac{1 + \cos t}{2}$

ii $\sin^2 \dfrac{t}{2} = \dfrac{1 - \cos t}{2}$

PROOF

i   From Theorem 4, i, we have, upon substituting t for $2t$,

$$\cos t = 2\cos^2 \frac{t}{2} - 1$$

so that

$$\frac{\cos t + 1}{2} = \cos^2 \frac{t}{2}$$

ii Using Theorem 4, i, we have, upon substituting t for $2t$, $\cos t = 1 - 2\sin^2(t/2)$, so that

$$\sin^2\frac{t}{2} = \frac{1-\cos t}{2}$$

EXAMPLES

Use the identities to solve each of the following problems.

1 Prove that if t is a real number in the domains of the given functions, then
a) $\tan^2 t + 1 = \sec^2 t$ b) $\cot^2 t + 1 = \csc^2 t$

PROOF

a) $\tan^2 t + 1 = \dfrac{\sin^2 t}{\cos^2 t} + 1 = \dfrac{\sin^2 t + \cos^2 t}{\cos^2 t}$

$$= \frac{1}{\cos^2 t} = \sec^2 t$$

b) $\cot^2 t + 1 = \dfrac{\cos^2 t}{\sin^2 t} + 1 = \dfrac{\cos^2 t + \sin^2 t}{\sin^2 t}$

$$= \frac{1}{\sin^2 t} = \csc^2 t$$

2 Prove that $1 - \tan^4 t = 2\sec^2 t - \sec^4 t$ if t is in the domain of the tangent and secant.

PROOF

$$
\begin{aligned}
1 - \tan^4 t &= (1-\tan^2 t)(1+\tan^2 t) \\
&= (1-\tan^2 t)\sec^2 t \quad \text{(see Example 1a)} \\
&= [1-(\sec^2 t - 1)]\sec^2 t \\
&= (2-\sec^2 t)\sec^2 t \\
&= 2\sec^2 t - \sec^4 t
\end{aligned}
$$

3 Simplify each of the following expressions.

a) $\cos\left(\dfrac{\pi}{2} + t\right)$ b) $\sin\left(\dfrac{\pi}{2} + t\right)$

c) $\sin 27° \cos 18° + \cos 27° \sin 18°$ d) $\cos 25° \cos 35° - \sin 25° \sin 35°$

e) $\cos(s-t)\cos t - \sin(s-t)\sin t$

SOLUTION

a) Using Theorem 1, i, we have

$$\cos\left(\frac{\pi}{2} + t\right) = \cos\frac{\pi}{2}\cos t - \sin\frac{\pi}{2}\sin t = 0 - \sin t = -\sin t$$

b) Using Theorem 3, i, we have

$$\sin\left(\frac{\pi}{2}+t\right) = \sin\frac{\pi}{2}\cos t + \cos\frac{\pi}{2}\sin t = \cos t + 0 = \cos t$$

c) Using Theorem 3, i, we have

$$\sin 27° \cos 18° + \cos 27° \sin 18° = \sin(27°+18°) = \sin 45° = \frac{\sqrt{2}}{2}$$

d) Using Theorem 1, ii, we have

$$\cos 25° \cos 35° - \sin 25° \sin 35° = \cos(25°+35°) = \cos 60° = \tfrac{1}{2}$$

e) Using Theorem 1, ii, we have

$$\cos(s-t)\cos t - \sin(s-t)\sin t = \cos[(s-t)+t] = \cos t$$

4   Use the fact that $7\pi/12 = \pi/3 + \pi/4$ to evaluate $\cos(7\pi/12)$.

SOLUTION

$$\cos\frac{7\pi}{12} = \cos\left(\frac{\pi}{3}+\frac{\pi}{4}\right)$$

$$= \cos\frac{\pi}{3}\cos\frac{\pi}{4} - \sin\frac{\pi}{3}\sin\frac{\pi}{4}$$

$$= \frac{1}{2}\frac{\sqrt{2}}{2} - \frac{\sqrt{3}}{2}\frac{\sqrt{2}}{2}$$

$$= \frac{\sqrt{2}}{4} - \frac{\sqrt{6}}{4}$$

5   Given $\sin t = \tfrac{3}{5}$ where $\pi/2 < t < \pi$ and $\cos s = \tfrac{4}{5}$ where $0 < s < \pi/2$, determine each of the following values.

a) $\sin s$ b) $\cos t$ c) $\sin(t+s)$
d) $\cos(t+s)$ e) $\tan(t+s)$ f) $\sec(t+s)$

SOLUTION

a) s defines an angle with its terminal side in quadrant I so that

$$\sin s = \sqrt{1-\cos^2 s} = \sqrt{1-\tfrac{16}{25}} = \tfrac{3}{5};$$

and, since t defines an angle with the terminal side in quadrant II,

b) $\cos t = -\sqrt{1-\sin^2 t} = -\sqrt{1-\tfrac{9}{25}} = -\tfrac{4}{5}$

c) $\sin(t+s) = \sin t \cos s + \cos t \sin s$
$$= (\tfrac{3}{5})(\tfrac{4}{5}) + (-\tfrac{4}{5})(\tfrac{3}{5})$$
$$= 0$$

d) $\cos(t+s) = \cos t \cos s - \sin t \sin s = (-\frac{4}{5})(\frac{4}{5}) - (\frac{3}{5})(\frac{3}{5}) = -\frac{25}{25} = -1$

e) $\tan(t+s) = \dfrac{\sin(t+s)}{\cos(t+s)} = \dfrac{0}{-1} = 0$

f) $\sec(t+s) = \dfrac{1}{\cos(t+s)} = \dfrac{1}{-1} = -1$

6   Prove that $[\sin(t/2)+\cos(t/2)]^2 - \sin t = 1$ for any real number t.

PROOF

$$\left(\sin\frac{t}{2} + \cos\frac{t}{2}\right)^2 - \sin t = \sin^2\frac{t}{2} + 2\sin\frac{t}{2}\cos\frac{t}{2} + \cos^2\frac{t}{2} - \sin t$$

$$= 1 + 2\sin\frac{t}{2}\cos\frac{t}{2} - \sin t$$

$$= 1 + \sin t - \sin t \qquad \text{(Theorem 4, ii)}$$

$$= 1$$

7   Use the fact that $\frac{1}{2}(30°) = 15°$, together with Theorem 5, ii, to evaluate $\sin 15°$.

SOLUTION

$$\sin 15° = \sqrt{\frac{1-\cos 30°}{2}} \qquad \text{(Theorem 5, ii)}$$

$$= \frac{\sqrt{2-\sqrt{3}}}{2}$$

8   Prove each of the following identities.

a) $\cos 2t + 2\sin^2 t = 1$ b) $\dfrac{2\cos 2t}{\sin 2t - 2\sin^2 t} = \cot t + 1$

PROOF

a) $\cos 2t + 2\sin^2 t = \cos^2 t - \sin^2 t + 2\sin^2 t \qquad$ (why?)

$= \cos^2 t + \sin^2 t = 1$

b) $\dfrac{2\cos 2t}{\sin 2t - 2\sin^2 t} = \dfrac{2(\cos^2 t - \sin^2 t)}{2\sin t \cos t - 2\sin^2 t} = \dfrac{2(\cos t - \sin t)(\cos t + \sin t)}{2\sin t(\cos t - \sin t)}$

$= \dfrac{\cos t + \sin t}{\sin t} = \dfrac{\cos t}{\sin t} + 1 = \cot t + 1$

9   a) Prove that $\tan(t+s) = (\tan t + \tan s)/[1-(\tan t \tan s)]$ if t, s, and $t+s$ are in the domain of the tangent function.

b) Prove that $f(x) = \tan x$ is a periodic function of fundamental period π.

PROOF

a) $\quad \tan(t+s) = \dfrac{\sin(t+s)}{\cos(t+s)}$

$\qquad\qquad = \dfrac{\sin t \cos s + \sin s \cos t}{\cos t \cos s - \sin t \sin s}$

Dividing both the numerator and the denominator of the right-hand side by $\cos s \cos t$, we get

$$\tan(t+s) = \frac{(\sin t \cos s)/(\cos t \cos s) + (\sin s \cos t)/(\cos s \cos t)}{(\cos t \cos s)/(\cos t \cos s) - (\sin t \sin s)/(\cos t \cos s)}$$

so that, after simplifying the above expression, we have

$$\tan(t+s) = \frac{\tan t + \tan s}{1 - \tan t \tan s}$$

b) From the result of part a, we have

$$\tan(x+\pi) = \frac{\tan x + \tan \pi}{1 - \tan x \tan \pi} = \frac{\tan x + 0}{1 - 0} = \tan x$$

It can be shown that π is the *smallest* positive period of the tangent function, hence π is the fundamental period of $f(x) = \tan x$.

10   Find the exact value of $\sin[2 \cos^{-1}(-\tfrac{3}{5})]$.

SOLUTION

$\sin[2 \cos^{-1}(-\tfrac{3}{5})]$

$\quad = 2 \sin[\cos^{-1}(-\tfrac{3}{5})] \cos[\cos^{-1}(-\tfrac{3}{5})]$ (Theorem 4, ii)

$\quad = 2\sqrt{1-(-\tfrac{3}{5})^2}\cdot(-\tfrac{3}{5})$

$\quad = 2\sqrt{\tfrac{16}{25}}\cdot(-\tfrac{3}{5}) = 2(\tfrac{4}{5})(-\tfrac{3}{5}) = -\tfrac{24}{25}$

PROBLEM SET 3

1   Given $\tan t = -\tfrac{5}{12}$, find $\sin t$ and $\cos t$. Are the answers unique? (*Hint:* Use Example 1a to find $\sec t$ first.)

2   Use the identities to simplify each of the following expressions.
   a) $(\cos t + \sin^2 t \sec t)/\sec t$ b) $(\sec t \csc t)/(\tan t + \cot t)$
   c) $\cos^6 t + \sin^6 t$ d) $(\tan t + \cot t)/(\sec t \csc t)$

3   Express each of the following functions in terms of the sine and/or cosine.
   a) $(\sec t + \csc t)^2 \tan t$ b) $(\csc t + \cot t)^2$
   c) $1/(1+\tan^2 t)$ d) $1/(\sec^2 x) - 1/(\tan^2 x)$
   e) $\tan^2 x$ f) $(1-\sin^2 x)/\cos x$
   g) $\cot^2 t + 1$ h) $[\sin(t/2)+\cos(t/2)]^2$

4   Give two specific examples to prove that each of the following equations does *not* always hold.
   a) $\cos(t+s) = \cos t + \cos s$
   b) $\sin(t-s) = \sin t - \sin s$

5   Express each of the following expressions as a trigonometric function of θ.
   a) $\sin(90° - \theta)$ b) $\cos(\pi - \theta)$
   c) $\tan(270° - \theta)$ d) $\tan(3\pi/2 + \theta)$
   e) $\cos(180° + \theta)$ f) $\sec(90° - \theta)$
   g) $\csc(90° - \theta)$ h) $\cot(\pi/2 - \theta)$
   i) $\cos(90° - \theta)$ j) $\sin(270° + \theta)$
   k) $\sin(10\pi - \theta)$ l) $\cos(7\pi/2 - \theta)$

6   Use identities to express each of the following expressions as a single function evaluated at some angle.
   a) $\sin 33° \cos 27° + \cos 33° \sin 27°$ b) $\sin 35° \cos 25° + \cos 35° \sin 25°$
   c) $\cos 55° \cos 25° - \sin 55° \sin 25°$ d) $\sin 2\theta \cos \theta + \cos 2\theta \sin \theta$

   e) $2 \sin 40° \cos 40°$ f) $\dfrac{\tan 32° + \tan 43°}{1 - \tan 32° \tan 43°}$

7   a)   Use the identities to find the exact value of each of the following expressions. (*Hint:* $5\pi/12 = \pi/4 + \pi/6$.)

   i) $\sin \dfrac{5\pi}{12}$ ii) $\cos \dfrac{5\pi}{12}$ iii) $\tan \dfrac{5\pi}{12}$

   b)   Use the fact that $255° = 225° + 30°$ together with the identities to find the exact values of the following expressions.
   i) $\sin 255°$ ii) $\cos 255°$ iii) $\cot 255°$

8   Prove that $\sin(t+s) \cdot \sin(t-s) = \sin^2 t - \sin^2 s$.

9   a)   If $\sin t = \frac{1}{2}$ and $\cos s = -\sqrt{3}/2$, evaluate each of the following expressions where $\pi/2 < t < \pi$ and $\pi < s < 3\pi/2$.
   i $\sin(\pi + t)$ ii $\sin(s - t)$
   iii $\sin 2t$ iv $\cos(s + t)$
   v $\cos(4\pi - t)$ vi $\sin(t/2)$

   b)   If $\sin t = -\frac{12}{13}$ and $\cos s = \frac{4}{5}$, evaluate each of the following expressions where $\pi < t < 3\pi/2$ and $3\pi/2 < s < 2\pi$.
   i $\sin(s - t)$ ii $\cos(s - t)$
   iii $\tan(s - t)$ iv $\cot(s - t)$
   v $\sec(s - t)$ vi $\csc(s - t)$

10   a)   Prove each of the following identities.
   i $\sin(u+v) + \sin(u-v) = 2 \sin u \cos v$
   ii $\sin(u+v) - \sin(u-v) = 2 \cos u \sin v$

    iii $\cos(u-v)+\cos(u+v)=2\cos u\cos v$

    iv $\cos(u-v)-\cos(u+v)=2\sin u\sin v$

b) Use the results of part a to prove each of the following identities. [*Hint:* Using part a, let $s=u+v$ and $t=u-v$; therefore, $u=(s+t)/2$ and $v=(s-t)/2$. Substitute for u and v in terms of s and t.]

    i $\sin s+\sin t=2\sin[(s+t)/2]\cos[(s-t)/2]$

    ii $\sin s-\sin t=2\cos[(s+t)/2]\sin[(s-t)/2]$

    iii $\cos s+\cos t=2\cos[(s+t)/2]\cos[(s-t)/2]$

    iv $\cos t-\cos s=2\sin[(s+t)/2]\sin[(s-t)/2]$

11 a) Suppose that $\tan t=\frac{3}{4}$; find $\sin 2t$ and $\cos 2t$.

     b) If $\csc t=\frac{4}{3}$, find the values of $\tan 2t$ and $\cot 2t$.

12 Suppose that $\sin t=\frac{3}{5}$, where $0<t<\pi/2$. Evaluate each of the following expressions.

  a) $\sin 2t$ b) $\cos 2t$

  c) $\sin 3t$ d) $\cos 3t$

13 Find the exact value of each of the following expressions by using Theorem 5.

  a) $\sin(\pi/8)$ b) $\cos(\pi/12)$

  c) $\sin(5\pi/12)$ d) $\cos(5\pi/8)$.

  e) $\sin(7\pi/12)$

14 a) Use the identity for $\tan(t+s)$ (see Example 9), together with the fact that the tangent is an odd function, to prove that

$$\tan(t-s)=\frac{\tan t-\tan s}{1+\tan t\tan s}$$

     b) Use the identity for $\tan(t+s)$ to prove that

$$\tan 2t=\frac{2\tan t}{1-\tan^2 t}$$

15 Prove each of the following identities.

  a) $\cot t+\tan t=\sec t\csc t$ b) $\dfrac{\sin 3t}{\sin t}=2\cos 2t+1$

  c) $\sec^4 t-\sec^2 t=\dfrac{\sin^2 t}{\cos^4 t}$ d) $1-\tan^4 t=2\sec^2 t-\sec^4 t$

  e) $(\cot t+\csc t)^2=\dfrac{1+\cos t}{1-\cos t}$ f) $\tan^2 t-\sin^2 t=\dfrac{\sin^4 t}{\cos^2 t}$

  g) $\dfrac{\cos^2 t}{1+\sin t}=1-\sin t$ h) $\dfrac{1-\tan^2 t}{1+\tan^2 t}=1-2\sin^2 t$

  i) $\cot t-\tan t=2\cot 2t$ j) $\cos^2 t(1-\tan^2 t)=\cos 2t$

k) $\dfrac{\tan s + \tan t}{\tan s - \tan t} = \dfrac{\sin (s+t)}{\sin (s-t)}$

l) $\dfrac{\tan (s-t) + \tan t}{1 - \tan (s-t)\tan t} = \tan s$

m) $\tan\left(t + \dfrac{\pi}{3}\right) = \dfrac{\tan t + \sqrt{3}}{1 - \sqrt{3}\tan t}$

n) $\cot\left(t + \dfrac{\pi}{4}\right) + \tan\left(t - \dfrac{\pi}{4}\right) = 0$

o) $\dfrac{1 + \cos 2t}{\sin 2t} = \cot t$

p) $\tan t = \dfrac{\sin 2t}{1 + \cos 2t}$

q) $2 \csc 2t = \csc t \sec t$

r) $\dfrac{\sin 4t}{\sin 2t} = 2\cos 2t$

s) $\dfrac{\sin 5t}{\sin t} - \dfrac{\cos 5t}{\cos t} = 4\cos 2t$

t) $\dfrac{\cos 2t}{1 + \sin 2t} = \dfrac{1 - \tan t}{1 + \tan t}$

u) $4\sin^2 t \cos^2 t + \cos^2 2t = 1$

v) $\sec t \csc t - 2\cos t \csc t = \tan t - \cot t$

w) $\tan t + \cot t = 2\csc 2t$

x) $\dfrac{\sin 3t}{\sin t} - \dfrac{\cos 3t}{\cos t} = 2$

16 a) Use $\cot (t+s) = 1/\tan (t+s)$ to derive a formula for $\cot (t+s)$ in terms of $\cot s$ and $\cot t$.

b) Use the formula derived in part a) to prove that $f(x) = \cot x$ is a period function of fundamental period π.

17 Find the exact values of the following expressions.
a) $\cos (\sin^{-1}(-\tfrac{3}{5}))$
b) $\sin (2 \cos^{-1}(-\tfrac{24}{25}))$
c) $\tan (2 \tan^{-1}\tfrac{4}{3})$

18 Prove each of the following identities.
a) $-\sin^2 \theta(1 - \csc^2 \theta) = \cos^2 \theta$
b) $\cot^2 \theta \sec^2 \theta = 1 + \cot^2 \theta$
c) $\sec^2 \theta \cot^2 \theta - \cos^2 \theta \csc^2 \theta = 1$
d) $\cos^4 \theta - \sin^4 \theta = \cos 2\theta$
e) $4\sin^2 \theta \cos^2 \theta = 1 - \cos^2 2\theta$
f) $2\cos \theta - \sin 2\theta \csc \theta = 0$
g) $\cos 2\theta + 2\sin^2 \theta = 1$

5 Trigonometric Equations

So far we have dealt only with identities, that is, equations that are true for all real numbers or angles in which the functions are defined. Now we consider *conditional equations*, that is, equations that are true only for some values of the variable. In particular, we will use the properties of the trigonometric functions, together with the trigonometric identities to develop techniques to solve conditional equations involving trigonometric or circular expressions. As before, the set of all possible solutions is called the *solution set* of the equation.

For example, the equation $\sin t = 1$ is true for $t = \pi/2$ since $\sin(\pi/2) = 1$. But it is also true that $\sin t = \sin(t + 2\pi n)$, where n is any integer, because the sine function is a periodic function of period 2π. Therefore, $\sin t = 1$ is true for $t = \pi/2, 5\pi/2, 9\pi/2, \ldots$. The solution set of this equation can be expressed as the set $\{t \mid t = [(4n+1)/2]\pi, \, n \in I\}$.

EXAMPLES

Solve each of the following trigonometric equations.

1 $\quad \sin t = \frac{1}{2}$, for $0 \leq t < 2\pi$.

SOLUTION.   (See Figure 1.) We know that $\sin(\pi/6) = \frac{1}{2}$ because $P(\pi/6) = (\sqrt{3}/2, \frac{1}{2})$. Here $\pi/6$ is the reference angle, and, since the sine is positive also in quadrant II, we have $\sin(5\pi/6) = \frac{1}{2}$. Thus the solution set of the equation $\sin t = \frac{1}{2}$, for $0 \leq t < 2\pi$ is $\{\pi/6, 5\pi/6\}$.

Figure 1

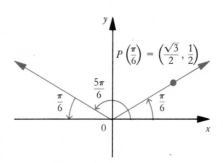

2 $\quad \cos\theta = \sqrt{3}/2$, for $0 \leq \theta < 360°$.

Figure 2

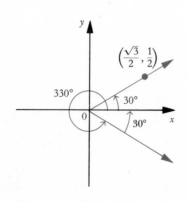

SOLUTION.   (See Figure 2.) One number in the solution set is 30°, since $\cos 30° = \sqrt{3}/2$. Also, since the cosine is positive in quadrant IV, we have $\cos 330° = \sqrt{3}/2$. Thus the solution set is $\{30°, 330°\}$.

3   $\tan t = \sqrt{3}$, for $0 \leq t < 2\pi$.

SOLUTION.   (See Figure 3.) One number in the solution set is $\pi/3$ since $\tan(\pi/3) = \sqrt{3}$. Since the tangent is also positive in quadrant III, we have $\tan(4\pi/3) = \sqrt{3}$. Thus the solution set is $\{\pi/3, 4\pi/3\}$.

Figure 3

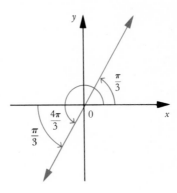

4   $\sin t = -0.8$, for $0 \leq t < 2\pi$ (use $\pi = 3.142$).

SOLUTION.   (See Figure 4.) We can find the reference angle t_R by using linear interpolation to solve $\sin t_R = 0.8$.

Figure 4

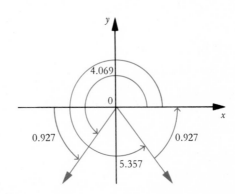

$$0.01 \begin{bmatrix} \sin 0.92 = 0.7956 \\ d \begin{bmatrix} \sin t & = 0.8000 \\ \sin 0.93 = 0.8016 \end{bmatrix} 0.0016 \end{bmatrix} 0.0060$$

$$\frac{d}{0.01} = \frac{0.0016}{0.0060}$$

so that

$d = 0.003$ (approximately)

Hence

$t_R = 0.927$

Since the sine is negative in quadrants III and IV, we have $\sin 4.069 = -0.8$ and $\sin 5.357 = -0.8$. Thus the solution set is $\{4.069, 5.357\}$.

5   $\sin 3\theta = 1$.

SOLUTION.   Since $\sin 3\theta = 1$,

$$3\theta = 90°, 450°, 810°, \ldots$$

or

$$3\theta = -270°, -630°, -990°, \ldots$$

so that

$$\theta = 30°, 150°, 270°, \ldots$$

or

$$\theta = -90°, -210°, -330°, \ldots$$

Hence the solution set is

$$\{\theta \mid \theta = 30° + 120°k \text{ or } \theta = -90° - 120°k, \ k \in I_p\}$$

6   $4\cos^2 t - 3 = 0$, for $0 \leq t < 2\pi$.

SOLUTION.   (See Figure 5.) $4\cos^2 t = 3$ is equivalent to $\cos^2 t = \frac{3}{4}$, so that $\cos t = \sqrt{3}/2$ or $\cos t = -\sqrt{3}/2$. The solution set of $\cos t = \sqrt{3}/2$ is $\{\pi/6, 11\pi/6\}$, since the cosine is positive in quadrants I and IV. The solution set of $\cos t = -\sqrt{3}/2$ is $\{5\pi/6, 7\pi/6\}$ since the cosine is negative

Figure 5

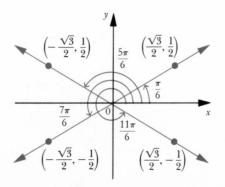

in quadrants II and III. Thus the solution of both $\cos t = \sqrt{3}/2$ and $\cos t = -\sqrt{3}/2$ are the solutions of $\cos^2 t = \frac{3}{4}$. Hence the solution set of the equation $\cos^2 t = \frac{3}{4}$, for $0 \leq t < 2\pi$, is $\{\pi/6, 5\pi/6, 7\pi/6, 11\pi/6\}$.

7 $\sec^2 \theta + \sec \theta - 2 = 0$, for $0° \leq \theta < 360°$.

SOLUTION. This equation can be factored as $(\sec \theta + 2)(\sec \theta - 1) = 0$. Thus $\sec^2 \theta + \sec \theta - 2 = 0$ if and only if $\sec \theta + 2 = 0$ or $\sec \theta - 1 = 0$, that is, if $\sec \theta = -2$ or $\sec \theta = 1$. The solution set of $\sec \theta = -2$, $0° \leq \theta < 360°$, is $\{120°, 240°\}$ (Figure 6). The solution set of $\sec \theta = 1$, $0° \leq \theta \leq 360°$ is $\{0°\}$ (Figure 6). Therefore, the solution set of $\sec^2 \theta + \sec \theta - 2 = 0$, for $0° \leq \theta < 360°$, is $\{0°, 120°, 240°\}$.

Figure 6

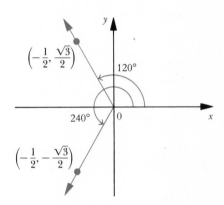

8 $\sin 2x - 2 \sin x = 0$, for $0 \leq x < 2\pi$.

SOLUTION. First we use the identity for $\sin 2x$ to rewrite the equation as $2 \sin x \cos x - 2 \sin x = 0$, so that $2 \sin x = 0$ or $\cos x - 1 = 0$. Thus $\sin 2x - 2 \sin x = 0$ if and only if $\sin x = 0$ or $\cos x = 1$. Hence the solution set of $\sin 2x - 2 \sin x = 0$ is the union of the solutions of $\sin x = 0$ and $\cos x = 1$. The solution set of $\sin x = 0$, for $0 \leq x < 2\pi$, is $\{0, \pi\}$ and the solution set of $\cos x = 1$, for $0 \leq x < 2\pi$, is $\{0\}$. Thus the solution set of $\sin 2x - 2 \sin x = 0$, for $0 \leq x < 2\pi$, is $\{0, \pi\}$.

9 $2 \sin^2 t - 11 \sin t - 6 = 0$.

SOLUTION. This equation can be factored as $(2 \sin t + 1)(\sin t - 6) = 0$. Thus $2 \sin^2 t - 11 \sin t - 6 = 0$ if and only if $2 \sin t + 1 = 0$ or $\sin t - 6 = 0$, that is, if $\sin t = -\frac{1}{2}$ or $\sin t = 6$. $\sin t = 6$ has no solution, since 6 is not in the range of the sine function. The solution set of

$$\sin t = -\frac{1}{2} \text{ is } \left\{ t \,\middle|\, t = -\frac{\pi}{6} + 2\pi k \text{ or } t = \frac{7\pi}{6} + 2\pi k, \, k \in I \right\}$$

Hence the solution set of $2 \sin^2 t - 11 \sin t - 6 = 0$ is

$$\left\{ t \mid t = -\frac{\pi}{6} + 2\pi k \text{ or } t = \frac{7\pi}{6} + 2\pi k,\ k \in I \right\}$$

PROBLEM SET 4

1 Solve each of the following trigonometric equations for $0 \le t < 2\pi$ (use $\pi = 3.14$).

a) $\sin t = \dfrac{1}{\sqrt{2}}$

b) $\cos t = \dfrac{1}{2}$

c) $\tan 3t = -\sqrt{3}$

d) $\cot t = -1$

e) $\sec 4t = 2$

f) $\csc t = -\dfrac{2}{\sqrt{3}}$

g) $\cos t = 0$

h) $\cos t = 2$

i) $\cos t = 0.3248$

j) $\sin t = -0.8866$

k) $\sin t = 0.70$

l) $\cos t = 0.55$

2 Solve for x and y, if $0 \le x \le \pi/2$ and $0 \le y \le \pi/2$, when $\tan(x-y) = 1$ and $\sin(x+y) = 1$.

3 Solve each of the following equations for θ to the nearest 10 minutes if $0 \le \theta < 360°$.

a) $\sin \theta = -\dfrac{\sqrt{3}}{2}$

b) $\cos \theta = -0.8880$

c) $3 \tan \theta - \sqrt{3} = 0$

d) $8\sqrt{3} \csc \theta - 16 = 0$

e) $4 \sin^2 \theta - 1 = 0$

f) $\csc^2 \theta - 2 \cot \theta = 0$

g) $\tan \theta - 3 \cot \theta = 0$

h) $\tan^2 \theta + 3 \sec \theta - 3 = 0$

i) $\sec^2 \theta - \sec \theta - 2 = 0$

j) $\tan^2 \theta - 2 \tan \theta + 1 = 0$

4 Use the identity $\cos s - \cos t = -2 \sin[(s+t)/2] \sin[(s-t)/2]$ to solve the equation $\cos 2x - \cos 3x = 0$ for $0 \le x < 2\pi$.

5 Solve each of the following equations for x if $0 \le x < 2\pi$ (use $\pi = 3.14$).

a) $2 \sec x + 4 = 0$

b) $4 \csc x - 8 = 0$

c) $3 \cot^2 x - 1 = 0$

d) $2 \sin^2 x - \sin x - 1 = 0$

e) $\sin^2 x - 2 \sin x + 1 = 0$

f) $4 \cos^2 x - 5 \sin x \cot x - 6 = 0$

g) $2 - \sin x = 2 \cos^2 x$

h) $2 \cos x - \sin x = 1$

6 Solve for θ if $\sin \theta = \cos(45° - 2\theta)$.

7 Solve each of the following equations for t.

a) $\sin t - \cos t = 0$

b) $2 \cos t + \sqrt{3} = 0$

c) $4 \cos^2 t - 1 = 0$

d) $\sec^2 t + 2 \tan t = 0$

e) $\sin 2t - \sqrt{3} \cos t = 0$

f) $\cos 2t + \sin^2 t = 1$

6 Triangle Trigonometry

We conclude this chapter with a study of applications of the trigonometric functions in determining unknown parts of triangles.

Two standard formulas of trigonometry, the *law of sines* and *law of cosines*, will be derived. Before deriving the formulas, it is convenient to extablish some standard notation for representing the angles and corresponding sides of triangles. If a triangle is determined by points A, B, and C ($\triangle ABC$), the angle at vertex A is denoted by α; the angle at vertex B is denoted by β; the angle at vertex C is denoted by γ; the side opposite angle α is a; the side opposite angle β is b; and the side opposite angle γ is c (Figure 1).

Figure 1

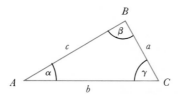

6.1 Right Triangles

Suppose we are given a right triangle, $\triangle ABC$, with $\gamma = 90°$ (Figure 2). If we consider $\triangle ABC$ to be on a coordinate system with α in standard position, then it is not difficult to describe the values of the six trigonometric functions in terms of the sides of the right triangle:

Figure 2

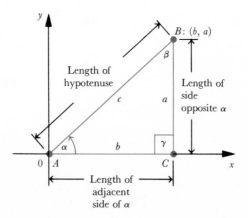

$$\sin \alpha = \frac{a}{c} = \frac{\text{length of side opposite } \alpha}{\text{length of hypotenuse}}$$

$$\cos \alpha = \frac{b}{c} = \frac{\text{length of adjacent side of } \alpha}{\text{length of hypotenuse}}$$

$$\tan \alpha = \frac{\sin \alpha}{\cos \alpha} = \frac{a}{b} = \frac{\text{length of side opposite } \alpha}{\text{length of adjacent side of } \alpha}$$

$$\csc \alpha = \frac{1}{\sin \alpha} = \frac{c}{a} = \frac{\text{length of hypotenuse}}{\text{length of side opposite } \alpha}$$

$$\sec \alpha = \frac{1}{\cos \alpha} = \frac{c}{b} = \frac{\text{length of hypotenuse}}{\text{length of adjacent side of } \alpha}$$

$$\cot \alpha = \frac{\cos \alpha}{\sin \alpha} = \frac{b}{a} = \frac{\text{length of adjacent side of } \alpha}{\text{length of side opposite } \alpha}$$

Now if any of α, β, a, b, or c are unknown, we may use these formulas and the trigonometric tables to find the unknown value. Notice that, in general, the results of such calculations are usually approximate values if the tables are used.

EXAMPLES

1 Determine the unknown parts of the right triangle in Figure 3.

SOLUTION. $\tan \alpha = \frac{1}{2}$, so that α is approximately $26° 34'$ and β is approximately $63° 26'$. Finally, we can use the Pythagorean theorem to determine c as $c = \sqrt{4 + 1} = \sqrt{5}$.

Figure 3

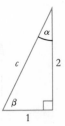

2 Determine the measure of the angle α of the right triangle, $\triangle ABC$, if $a = 4$, $b = 3$, and $c = 5$ as illustrated in Figure 4.

SOLUTION.   From Figure 4 we see that $\sin \alpha = \frac{4}{5} = 0.8$, so that by using Table IV with linear interpolation, we have $\alpha = 53° 8'$.

Figure 4

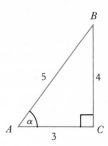

3 A man is standing 150 feet from the foot of a flagpole, which is at his eye level, and the angle of elevation (acute angle formed by the line of sight and a horizontal line passing through the position of the sighting) to the top of the flagpole is 45°. Find the height of the pole.

SOLUTION. Let y be the height of the flagpole (Figure 5); then $\tan 45° = y/150 = 1$, so that $y = 150$ feet. Therefore, the height of the flagpole is 150 feet.

Figure 5

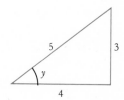

4 Use a right triangle to evaluate $\sin(\tan^{-1}\frac{3}{4})$.

SOLUTION. Let $y = \tan^{-1}\frac{3}{4}$; then using the definition of the inverse tangent (Chapter 5, Section 7.3) we have $\tan y = \frac{3}{4}$ (Figure 6). By the Pythagorean theorem, $c = \sqrt{3^2 + 4^2} = 5$. Hence $\sin(\tan^{-1}\frac{3}{4}) = \sin y = \frac{3}{5}$.

Figure 6

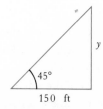

150 ft

5 From a mountain top 10,000 feet above a horizontal plane two towns were observed with angles of depression 60° and 45°, respectively, as given in Figure 7. How far apart are the two towns?

SOLUTION.   In Figure 7, A represents the position of the observer, C the position of the first town, and D the position of the second town. The angles of depression are $\angle DAE$ and $\angle CAE$. Using the fact that alternate interior angles formed by the transversal of two parallel lines are equal, we have

$$\angle BCA = \angle CAE = 60° \qquad \text{and} \qquad \angle BDA = \angle DAE = 45°$$

Figure 7

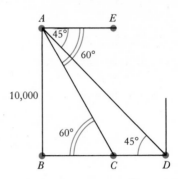

Hence $\overline{BC} = 10{,}000 \cot 60° = \left(10{,}000/\sqrt{3}\right)$ feet, and $\overline{BD} = 10{,}000 \cot 45° = 10{,}000$ feet, so the distance between C and D is

$$\overline{BD} - \overline{BC} = 10{,}000 - \frac{10{,}000}{\sqrt{3}} = 4.23 \times 10^3 \text{ feet (approximately)}$$

We have seen above how it is possible to use the trigonometric functions to relate the sides and angles of right triangles. Now, let us use the trigonometric functions to derive two formulas that relate the sides and angles of triangles that are not necessarily right triangles.

THEOREM 1   LAW OF SINES

In any $\triangle ABC$,

$$\frac{\sin \alpha}{a} = \frac{\sin \beta}{b} = \frac{\sin \gamma}{c}$$

PROOF.   (See Figure 8.) We can use the fact that the area of a triangle is equal to the product of one-half the base and altitude, together with the fact that the sine of an acute angle of a right triangle is equal to the ratio of the length of the side opposite the angle to the length of the hypotenuse of the right triangle. First, draw altitudes h_1 and h_2.

The area of $\triangle ABC = \frac{1}{2}ch_2 = \frac{1}{2}ah_1$, but

$$\tfrac{1}{2}ch_2 = \tfrac{1}{2}c\,(a \sin \beta) \qquad \text{since} \qquad \sin \beta = \frac{h_2}{a}$$

Figure 8

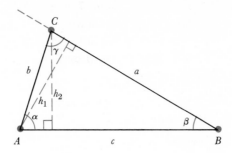

and

$$\tfrac{1}{2}ch_2 = \tfrac{1}{2}c\,(b\sin\alpha) \qquad \text{since} \qquad \sin\alpha = \frac{h_2}{b}$$

Also,

$$\tfrac{1}{2}ah_1 = \tfrac{1}{2}a\,(b\sin\gamma) \qquad \text{since} \qquad \sin\gamma = \frac{h_1}{b}$$

Hence

$$\tfrac{1}{2}ca\sin\beta = \tfrac{1}{2}cb\sin\alpha = \tfrac{1}{2}ab\sin\gamma$$

so that

$$ca\sin\beta = cb\sin\alpha = ab\sin\gamma$$

After dividing the latter equation by abc, we get

$$\frac{\sin\beta}{b} = \frac{\sin\alpha}{a} = \frac{\sin\gamma}{c}$$

If any two angles and a side of a triangle are known (denoted by *ASA*), the law of sines can be used to determine either of the two remaining sides. Let us consider a few examples of this situation.

EXAMPLES

1 In $\triangle ABC$, $a = 12$, $\alpha = 45°$, and $\beta = 105°$. Find c.
SOLUTION. (See Figure 9.) Since $\alpha + \beta + \gamma = 180°$, $\gamma = 30°$. Using the law of sines, we have

$$\frac{\sin 45°}{12} = \frac{\sin 30°}{c}$$

so that

$$c = \frac{12\sin 30°}{\sin 45°} = 6\sqrt{2}$$

Figure 9

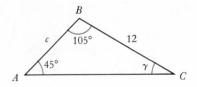

2   If in $\triangle ABC$, $a = 20$, $\gamma = 51°$, and $\beta = 42°$, find b.

SOLUTION.   (See Figure 10.) Since $\alpha + \beta + \gamma = 180°$, $\alpha = 87°$. Using the law of sines, we have

$$\frac{\sin 87°}{20} = \frac{\sin 42°}{b}$$

so that

$$b = \frac{20 \sin 42°}{\sin 87°} = \frac{20(0.6691)}{0.9986} = 13.4$$

Figure 10

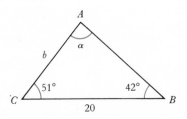

3   A surveyor wishes to find the distance between the city of Detroit D and the city of Windsor W (the two cities are on the opposite banks of the Detroit river). From D a line $\overline{DC} = 550$ feet is laid off and the angles $\angle CDW = 125°40'$ and $\angle DCW = 48°50'$ are measured. Find the distance between the two cities, that is, the length of $\overline{DW}$.

SOLUTION.   Let the length $\overline{DW}$ be c feet. In $\triangle WDC$ (Figure 11), the measure of $\angle DWC$ is

$$180° - (125°40' + 48°50') = 5°30'$$

Using the law of sines, we have

$$\frac{\sin 48°50'}{c} = \frac{\sin 5°30'}{550}$$

so that

$$c = \frac{550(\sin 48°50')}{\sin 5°30'} = \frac{550(0.7528)}{0.0958} = 4321.92 \text{ feet}$$

Therefore, the distance is 4321.92 feet.

Figure 11

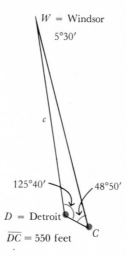

W = Windsor
5°30′
c
125°40′
48°50′
D = Detroit
C
$\overline{DC} = 550$ feet

Let us examine the various cases that may occur if we are given two sides and an angle opposite one of them (denoted by *SSA*). It may be that there is no possible triangle with these parts (Figure 12a); there may be two different triangles with these parts (Figure 12b); or there may be only one triangle with these parts (Figure 12c). Because there are three different

Figure 12

a
b
h
α
C
b
a
h
a
α
A
B_1
B
b
a
b
α

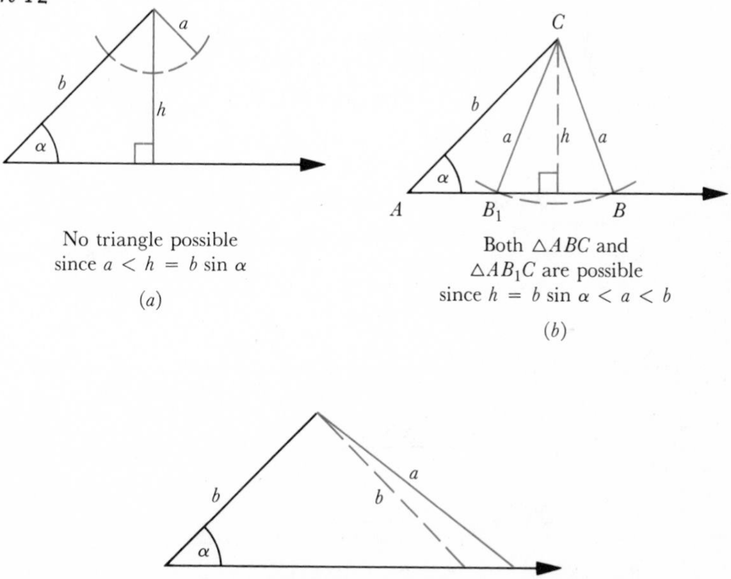

No triangle possible since $a < h = b \sin \alpha$

(*a*)

Both $\triangle ABC$ and $\triangle AB_1C$ are possible since $h = b \sin \alpha < a < b$

(*b*)

Only one triangle is possible since $a > b$

(*c*)

possibilities here, we sometimes refer to this situation as the *ambiguous case*
In the next example we will see how the law of sines can be used to solve
any situation that may occur in the ambiguous case.

EXAMPLES

1   In each of the following problems, find the indicated unknown parts of all
possible triangles.
a)   If $a = 20$, $b = 15$, and $\alpha = 30°$, find β to the nearest degree.
b)   If $a = 5$, $b = 20$, and $\alpha = 30°$, find β to the nearest degree.
c)   If $a = 30$, $b = 50$, and $\alpha = 30°$, find β and γ to the nearest minute.

SOLUTION

a)   We shall attempt to find β, keeping in mind that there may be zero,
one, or two triangles. If we assume that at least one triangle exists,
then, by the law of sines, we have

$$\frac{\sin \beta}{15} = \frac{\sin 30°}{20}$$

so that

$$\sin \beta = \tfrac{3}{8}$$

Since β must be less than $180°$ (it is an angle of a triangle) and since
the sine is positive for acute (first quadrant) and obtuse (second
quadrant) angles, β is approximately $22°$ or $158°$. (Why?) Since it is
impossible to have a $158°$ angle in a triangle that is already known to
have a $30°$ angle, there is only one triangle possible, and $\beta = 22°$
(Figure 13).

Figure 13

b)   Here again we assume that at least one such triangle exists, so that

$$\frac{\sin \beta}{20} = \frac{\sin 30°}{5}$$

Hence

$$\sin \beta = 2$$

But, since $|\sin \theta| \leq 1$, no such β exists; therefore, no triangle exists
with the given parts (Figure 14).

Figure 14

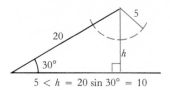

$5 < h = 20 \sin 30° = 10$

c)   If we assume that at least one such triangle exists, we have

$$\frac{\sin \beta}{50} = \frac{\sin 30°}{30}$$

so that

$$\sin \beta = \tfrac{5}{6}$$

Hence β is approximately $56°26'$ or $123°34'$. Since either value of β satisfies $\alpha + \beta < 180°$, we have two possible triangles. One of the two possible triangles has angles $30°$, $56°26'$, and $93°34'$, whereas the other triangle has angles $30°$, $123°34'$, and $26°26'$ (Figure 15).

Figure 15

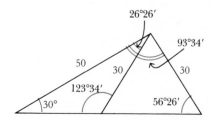

2   A tower is 125 feet high on a mountain on the bank of a river. It is observed that the angle of depression from the top of the tower to a point on the opposite shore is $28°40'$ and the angle of depression from the base of the tower to the same point on the shore is $18°20'$. How wide is the river and how high is the mountain to the nearest foot?

SOLUTION.   In Figure 16, $\overline{BC}$ represents the height of the tower, $\overline{BD}$ represents the height of the mountain, and A is the point on the opposite shore. $\overline{BC} = 125$ feet. In $\triangle ABC$,

$$\gamma = 90° - 28°40' = 61°20'$$

$$\beta = 90° + 18°20' = 108°20'$$

$$\alpha = 180° - (\beta + \gamma) = 10°20'$$

Figure 16

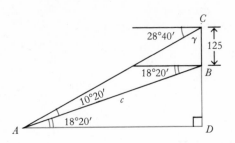

By the law of sines,

$$c = \frac{125 \sin 61°20'}{\sin 10°20'} = \frac{125(0.8774)}{0.1794} = 611 \text{ feet}$$

so that

$$\overline{BD} = c \sin 18°20' = 611(0.3145) = 192 \text{ feet}$$

and

$$\overline{AD} = c \cos 18°20' = 611(0.9492) = 580 \text{ feet}$$

Therefore, the river is 580 feet wide and the mountain is 192 feet high.

Now let us derive another formula which relates the sides and the angles of triangles, the *law of cosines*.

THEOREM 2 LAW OF COSINES

In any $\triangle ABC$,

i $c^2 = a^2 + b^2 - 2ab \cos \gamma$
ii $b^2 = a^2 + c^2 - 2ac \cos \beta$
iii $a^2 = b^2 + c^2 - 2bc \cos \alpha$

That is, the square of the length of any side of a triangle is equal to the sum of the squares of the lengths of the other two sides minus twice their product times the cosine of the angle included between these other two sides.

PROOF.   First, consider $\triangle ABC$ on a Cartesian coordinate system with γ in standard position (Figure 17). A has coordinates $(b \cos \gamma, b \sin \gamma)$ (why?) and B has coordinates $(a, 0)$ (why?), so, by the distance formula, we have

$$\begin{aligned}
c^2 &= (b \cos \gamma - a)^2 + (b \sin \gamma - 0)^2 \\
&= b^2 \cos^2 \gamma - 2ab \cos \gamma + a^2 + b^2 \sin^2 \gamma \\
&= b^2 \cos^2 \gamma + b^2 \sin^2 \gamma + a^2 - 2ab \cos \gamma \\
&= a^2 + b^2 (\cos^2 \gamma + \sin^2 \gamma) - 2ab \cos \gamma \\
&= a^2 + b^2 - 2ab \cos \gamma
\end{aligned}$$

Figure 17

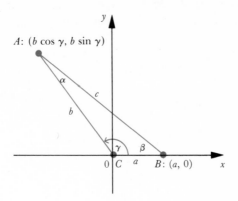

In view of the fact that the location of the coordinate axes in the plane is merely a matter of convenience, the other two formulas are obtained in a similar manner, since any letters may be used to denote the unknown parts of the triangle.

If two sides and an included angle of a triangle are known (*SAS*), the law of cosines can be used to determine the third side (see Example 1 below), or, if the three sides of a triangle are known (*SSS*), any of the angles can be found by using the law of cosines (see Example 2 below).

EXAMPLES

1 In $\triangle ABC$, $a = 8$, $b = 6$, and $\gamma = 60°$. Find c.

SOLUTION. (See Figure 18.) Using the law of cosines,

$$c^2 = a^2 + b^2 - 2ab \cos \gamma$$
$$= 64 + 36 - 2(8)(6)(\tfrac{1}{2})$$
$$= 100 - 48 = 52$$

so that

$$c = \sqrt{52} = 2\sqrt{13}$$

Figure 18

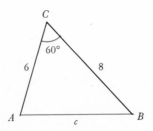

2   In $\triangle ABC$, $a = 5$, $b = 10$, and $c = 8$. Find α to the nearest minute.
    SOLUTION.   (See Figure 19.) Using the law of cosines,

$$a^2 = b^2 + c^2 - 2bc \cos \alpha$$
$$25 = 100 + 64 - 2(10)(8) \cos \alpha$$

so that

$$\cos \alpha = \tfrac{139}{160} = 0.8688$$

Using Table IV and linear interpolation, we find that α is approximately
$29°41'$.

Figure 19

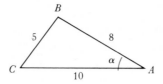

3   A motorboat is moving west and is propelled 30 feet per second. Find the
    strength of the tide that alters the course of the boat 20° south of west and
    reduces the speed to 25 feet per second.

    SOLUTION.   (See Figure 20.) Let $\overline{AB}$ represent the strength of the tide in
    feet per second. Using the law of cosines, we have

$$\overline{AB}^2 = \overline{OA}^2 + \overline{OB}^2 - 2\overline{OA} \cdot \overline{OB} \cdot \cos 20°$$
$$= (30)^2 + (25)^2 - 2(30)(25)(0.9397)$$
$$= 900 + 625 - 1500(0.9397)$$
$$= 1525 - 1409.6$$
$$= 115.4$$

Therefore, $\overline{AB} = \sqrt{115.4} = 10.73$ feet per second.

Figure 20

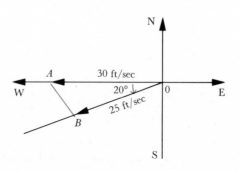

4   The straight-line distance between two cities A and B cannot be measured directly because of a swamp. A surveyor at city C is able to measure $\overline{CA} = 5.73$ miles, $\overline{CB} = 8.19$ miles, and $\angle ACB = 113°$. What is the distance from city A to city B?

SOLUTION.   Using the law of cosines (Figure 21), we have

$$\begin{aligned}
\overline{AB}^2 &= \overline{BC}^2 + \overline{AC}^2 - 2\overline{BC}\cdot\overline{AC}\cos\gamma \\
&= (8.19)^2 + (5.73)^2 - 2(8.19)(5.73)\cos 113° \\
&= 67.0761 + 32.8329 - (93.8574)(-0.3907) \\
&= 136.58
\end{aligned}$$

Therefore, $\overline{AB} = 11.69$ miles (approximately).

Figure 21

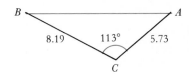

PROBLEM SET 5

1   Assume that $\triangle ABC$ is a right triangle, with the parts as labeled in Figure 22. Find the unknown parts, given the following parts. Find angle measures to the nearest 10 minutes.

a) $a = 5, b = 3$
b) $a = 10, \alpha = 30°$
c) $a = 17, \beta = 43°$
d) $a = 7, c = 12$
e) $b = 8, \beta = 15°$
f) $b = 3a$
g) $a = b$

Figure 22

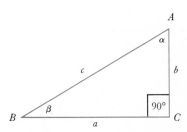

2   Use the right triangle with the parts as labeled in Figure 22. Find the unknown parts, given the following parts.

a)   If $a = 5$ and $\sin \alpha = \frac{4}{5}$, find b and c.
b)   If $c = 50$ and $\tan \alpha = 2$, find a, b, $\sin \alpha$, and $\cos \alpha$.
c)   If $c = 10$ and $\sin \alpha = \frac{4}{7}$, find a, b, and $\cos \alpha$.

d) If $b = 10$ and $\cos \alpha = \frac{3}{5}$, find a, c, $\sin \alpha$, and $\tan \alpha$.

e) If $a = 6$ and $b = 8$, find c, $\sin \alpha$, and $\cos \alpha$.

3 Use the law of sines to determine whether there is one triangle, two triangles, or no triangle in each of the following cases. Find the unknown parts of all possible triangles. Find angle measures to the nearest 10 minutes.

a) $b = 17.5$, $c = 15.2$, and $\gamma = 45°$

b) $a = 10$, $b = 50$, and $\alpha = 22°$

c) $c = 18.9$, $a = 10.4$, and $\gamma = 65°$

d) $a = 182.5$, $b = 82.5$, and $\alpha = 72°$

e) $b = 17.41$, $c = 19.32$, and $\gamma = 45°$

f) $a = 3$, $b = 5$, and $\beta = 45°$

4 Use the law of cosines to prove that if $\alpha = 90°$ in $\triangle ABC$, then $a^2 = b^2 + c^2$.

5 Evaluate each of the following expressions.

a) $\sin(\cos^{-1} \frac{3}{5})$

b) $\cos[\sin^{-1}(-\frac{3}{5})]$

c) $\sec[\tan^{-1}(-\frac{3}{4})]$

d) $\sin(\tan^{-1} \frac{3}{7})$

e) $\sec\left(\tan^{-1} \frac{\sqrt{5}}{2}\right)$

f) $\cos(\tan^{-1} \frac{9}{2})$

6 Prove the following identities for $\triangle ABC$.

a) If $\beta = 2\alpha$, then $\cos \alpha = b/2a$.

b) $b^2 - c^2 = a(b \cos \gamma - c \cos \beta)$

7 Use the law of cosines to find the indicated unknown part of each of the following triangles. Find angle measures to the nearest 10 minutes.

a) $\alpha = 60°$, $b = 20$, $c = 6$; find a.

b) $a = 3$, $b = 4$, $c = 6$; find α.

c) $a = 4$, $b = 7$, $c = 5$; find α, β, and γ.

d) $a = 5$, $b = 8$, $c = 10$; find β.

e) $c = 39$, $b = 98$, and $\alpha = 17°$; find a.

f) $a = 144$, $b = 180$, $c = 108$; find β.

8 a) Prove Hero's formula: The area A of a triangle with sides a, b, and c and semiperimeter $s = \frac{1}{2}(a+b+c)$, is

$$A = \sqrt{s(s-a)(s-b)(s-c)}$$

(*Hint:* Use the law of sines.)

b) Use part a to find the area of $\triangle ABC$ with

  i $a = 810$, $b = 990$, and $c = 360$

  ii $a = 492$, $b = 369$, and $c = 246$

  iii Vertices $A:(2,3)$, $B:(5,7)$, and $C:(6,1)$

9 Find the indicated unknown part of each of the following triangles.
   a) If $b = 4$, $c = 6$, $\beta = 30°$; find $\sin \gamma$.
   b) $\alpha = 56°$, $b = 7$, $c = 12$; find a.
   c) $a = 20$, $b = 4$, $c = 18$; find α to the nearest 10 minutes.
   d) If $\alpha = 60°$, $\beta = 75°$, $a = 10\sqrt{3}$; find b.
   e) $a = 12$, $b = \sqrt{109}$, $c = 5$; find α, γ, and β to the nearest 10 minutes.

10 A ladder is placed against a building with its foot 10 feet away from the building, on level ground. If it just reaches a window 20 feet above the ground, what angle will the ladder make with the ground?

11 A man standing 10 feet from a wall in an art gallery observes that the angle of elevation (acute angle between the line of sight and the horizontal line through the position of sighting) to the top of a picture in the art gallery is 30° and the angle of elevation to the bottom of the picture is 15°. What is the height of the picture?

12 From the top of an observation post the angle of depression (acute angle determined by the plane of the observer and the line of sight) of a ship is 30°. If the distance between the observer and the ship is 450 feet, find the height of the observation post.

13 From a tower 108 feet high, a man observed that the angles of depression of a street light and its base are 30° and 60°, respectively. How high is the light above the ground?

14 While flying at a height of 1000 feet, a pilot observes that the angle of depression of an airport is 15°. At that moment what is the distance between the plane and the sirport?

15 An observer on the ground views a kite from the east with an angle of elevation of 59°20′. Another person observes that its angle of elevation from the west is 34°15′. If the observers are 200 feet apart, how high is the kite?

16 An airplane heads directly east with an airspeed of 170 miles per hour, while a wind is blowing 30° south of east. Find the speed of the wind after 2 hours if the plane has traveled 600 miles.

17 A balloon is rising vertically at the rate of 15 feet per second; at the same time it is blown horizontally by a wind of 20 feet per second. Find the angle (to the nearest 10 minutes) its path makes with the horizon after 3 seconds.

18 Two railroads meet at an angle of 35°20′. If two trains leave from the railway station (where the two railroads meet) at the same time and if the speed of the first is 30 miles per hour, find the speed of the second train if they are 50 miles apart after $2\frac{1}{2}$ hours from their departure.

19 Assume that the diagonal of a parallelogram is 80 inches long and at one end makes angles measuring 35° and 27° with the sides of the parallelogram. Find the lengths of the sides of the parallelogram.

20 A tower 150 feet high is situated at the top of a hill. At a point 650 feet down the hill the angle between the surface of the hill and the line of sight to the top of the tower is 12° 30′. Find the inclination of the hill to a horizontal plane.

REVIEW PROBLEM SET

1 a) Change the following radian measures to degree measures. Find the reference number for each part.

i $5\pi/3$	ii $-3\pi/4$	iii $5\pi/16$
iv $11\pi/6$	v $13\pi/6$	vi $-13\pi/4$
vii $8\pi/3$	viii $-18\pi/7$	ix $17\pi/5$
x $27\pi/6$	xi $33\pi/4$	xii $-41\pi/4$

  b) Change the following degree measures to radian measures. Find the reference angle for each part.

i 290°	ii $-70°$	iii 135°
iv 460°	v 190°	vi $-105°$
vii 820°	viii $-30°$	ix 12°

2 Use the results from Problem 1 above together with Table III and Table IV (if necessary) to find the value of the sine of each part of Problem 1.

3 Find the length of the circular arc and the area of the circular sector that is generated by a central angle θ in a circle of radius r if
  a) $r = 7$ and $\theta = 75°$ b) $r = 2$ and $\theta = 160°$
  c) $r = 10$ and $\theta = 340°$ d) $r = 4$ and $\theta = -30°$
  e) $r = 6$ and $\theta = 36°$

4 a) A circle of radius 4 inches contains a central angle θ that intercepts an arc 10 inches long. What size is θ in radians? In degrees? Find the area of the sector whose central angle is θ.
  b) A central angle is 310° and has an intercepted arc of 35π inches. Find the radius of the circle and the area of the sector.

5 Find the values of the six trigonometric functions for each of the following angles without using a table.
  a) $-330°$ b) 270° c) 180°
  d) 150° e) 210° f) 135°
  g) $-120°$ h) $-225°$ i) $-45°$

6 Given an angle θ whose terminal side does not lie either along the x axis or the y axis, prove that the functions $f(\theta) = \tan\theta$ and $g(\theta) = \cot\theta$ are defined and $f(\theta)g(\theta) = 1$.

7 Find the values of all six trigonometric functions of the angle θ if the terminal side of θ in its standard position contains the following points.

    a) $(3,4)$ b) $(5,12)$ c) $(7,25)$

    d) $(6,8)$ e) $(8,17)$ f) $(-3,4)$

    g) $(-4,-3)$ h) $(5,6)$

8 Let $f(\theta) = \cot\theta + 4$.

    a) Find the domain of f.   b) Find the range of f.

    c) Is f an even or an odd function?

9 Let $\csc\theta = -\frac{7}{4}$, $0° < \theta < 360°$. Find

    a) $\sin\theta$ b) $\cos\theta$ c) $\tan\theta$

    d) $\cot\theta$ e) $\sec\theta$

10 Use the trigonometric identities to evaluate each of the following expressions.

    a) $\sin 330° \cos 180° - \cos 330° \sin 180°$

    b) $\cos 270° \cos 150° + \sin 270° \sin 150°$

    c) $\sin 120° \cos 30° + \cos 120° \sin 30°$

    d) $\cos 60° \cos 30° - \sin 60° \sin 30°$

11 Prove each of the following equations by using the identities.

    a) $\cos(90° - \theta) \tan(90° - \theta) = \cos\theta$

    b) $\tan\theta + \cot\theta = \csc(90° - \theta) \csc\theta$

    c) $(\sin 62°/\sec 62°)(\cot 28°/\cos 28°) = \sin 62°$

    d) $\cot 67° \cot 23° \cos 67° \tan 67° = \cos 23°$

    e) $\cos 13° \tan 13° \tan 77° \csc 77° = 1$

    f) $\sin^2\theta \csc(90° - \theta) - \cot^2(90° - \theta) \cos\theta = 0$

12 Use appropriate trigonometric identities to determine each of the following values.

    a) $\sin 15°$ b) $\cos 15°$

    c) $\tan 75°$ d) $\sec 75°$

13 Suppose that $\sin t = \frac{4}{5}$ and $\cos s = \frac{4}{5}$, where $0 < s < t < \pi/2$. Evaluate each of the following trigonometric functions.

    a) $\sin(t+s)$ b) $\cos(t+s)$

    c) $\sin(t-s)$ d) $\cos(t-s)$

    e) $\tan(t+s)$ f) $\cot(t-s)$

    g) $\sec(t+s)$ h) $\csc(t-s)$

14 Prove or disprove whether the first expression is equal to the second expression in each of the following problems.

    a) $(1+\tan t)^2/(1+\tan^2 t)$ and $\sin 2t + 1$

    b) $\cot 4t + \tan 2t$ and $\csc 4t$

    c) $\tan 4t - \tan 2t$ and $\tan 2t \sec 4t$

d) $\tan 2t - \tan t$ and $\tan t \sec 2t$

e) $\tan (t/2 + \pi/4)$ and $(1 + \sin t)/\cos t$

f) $\sin 5t/\sin t - \cos 5t/\cos t$ and $\cos 2t$

g) $\sin t + \cos t \tan s$ and $\sec s \sin (t + s)$

h) $(4 \tan t - 4 \tan^3 t)/(1 - 6 \tan^2 t + \tan^4 t)$ and $\tan 4t$

i) $\tan (t + \pi/3) + \tan (t - \pi/3)$ and $\tan t/[1 - (3 \tan^2 t)]$

j) $\cos 2t \sin 3t + \cos 3t \sin 2t$ and $\sin 5t$

k) $\sin 5t \cos t - \cos 5t \sin t$ and $2 \sin 2t(\cos^2 t - \sin^2 t)$

l) $\sin 4t + \sin 2t$ and $\sin 2t(2 \cos 2t + 1)$

15   Find the measure of θ in each of the following equations if $0° \leqq \theta < 360°$.

a) $\sin \theta = -1$ b) $\cos \theta = \sqrt{3}/2$

c) $\tan \theta = \sqrt{3}$ d) $\cos \theta = -\sqrt{2}/2$

e) $\csc \theta = -2$ f) $\cot \theta = \sqrt{3}/3$

g) $\sin \theta = \cos 2\theta$

16   Evaluate each of the following expressions.

a) $\cos (\cos^{-1} \frac{7}{24})$ b) $\sin [\tan^{-1}(-\frac{3}{4})]$ c) $\tan (2 \tan^{-1} \frac{3}{5})$

17   Solve the following trigonometric equations for $0 \leqq t < 2\pi$.

a) $\sin t = \sqrt{2}/2$ b) $4 \sin^2 t - 3 = 0$

c) $2 \cos^2 t - \cos t - 1 = 0$ d) $2 \sin^2 t + \sqrt{2} \sin t = 0$

e) $3 \tan^2 t - \sqrt{3} \tan t = 0$

18   Solve for the unknown parts of $\triangle ABC$ if $\gamma = 90°$.

a) $a = 5$ and $b = 12$ b) $c = 15$ and $\alpha = 37°$

c) $b = 25$ and $\beta = 65°$ d) $a = 5$ and $c = 14$

e) $a = 17$ and $\beta = 51°$ f) $\sin \alpha = \frac{4}{5}$ and $c = 25$

g) $\tan \alpha = \frac{3}{4}$ and $a = 12$ h) $a = 6$ and $b = 8$

19   Find the unknown parts of each of the following triangles. Find angle measures to the nearest 10 minutes.

a) $a = 5$, $b = 7$, and $\gamma = 30°$

b) $\alpha = 120°$, $c = 8$, and $b = 3$

c) $c = 10$, $\alpha = 45°$, and $\beta = 75°$

d) $a = 162$, $b = 215$, and $\beta = 110°$

e) $b = 4$, $c = 6$, and $\beta = 30°$

f) $a = 13.6$, $b = 7.82$, and $\alpha = 60°$

g) $a = 4.8$, $c = 4.3$, and $\alpha = 115°$

h) $b = 66.2$, $c = 42.3$, and $\alpha = 30°$

i) $a = 10$, $\beta = 42°$, and $\gamma = 51°$

j) $a = 4$, $c = 10$, and $\beta = 150°$

20   Given $\triangle ABC$, assume that $c = b \cos \alpha + a \cos \beta$. Then use the law of sines to express b and a in terms of c to obtain the formula

$$\sin (\alpha + \beta) = \sin \alpha \cos \beta + \sin \beta \cos \alpha$$

21 Two men 600 feet apart observe a balloon in the sky between them. The respective angles of elevation of the balloon are 75° and 48°. Find the height of the balloon above the ground.

22 Two points A and B are 50 feet apart on one side of a river. A point C across the river is located so that $\angle CAB$ is 70° and angle $\angle ABC$ is 80°. How wide is the river?

23 A guy line from the top of a pole is 40 feet long and forms a 50° angle with the ground. How long is the pole if it is tilted 15° out of line away from the guy line?

24 A diagonal of a parallelogram is 16 inches long and forms angles of 43° and 15° with each of the two sides. How long are the two sides of the parallelogram?

Vectors in the Plane

1 Introduction

Vectors can be used to explain and integrate many of the basic notions of algebra, geometry, and trigonometry. Here we shall investigate vectors in the plane from both a geometric and algebraic viewpoint. Also, we shall survey the applications of vectors to solve problems in geometry. By defining rotations as functions of vectors, we shall see how it is possible to use vectors to prove the trigonometric identities for $\cos(s+t)$ and $\sin(s+t)$.

2 Geometric Approach to Vectors in the Plane

Suppose that we are given the task of categorizing air trips within the United States according to the following scheme. Each category is determined by both the distance traveled and the direction in which the distance is traveled, and two trips are considered equivalent if they are in the same category. One such category would be a 100-mile trip north; another category would be a 435-mile trip southwest; a third category would be a 535-mile trip northeast (Figure 1).

Figure 1

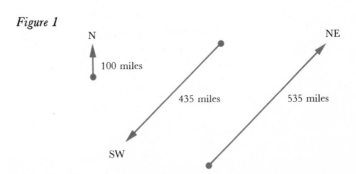

Notice that a 100-mile trip north from Detroit, *according to this scheme*, is considered to be equivalent to a 100-mile trip north from Cleveland (Figure 2). Here the equivalence of trips does not depend on where the trip

Figure 2

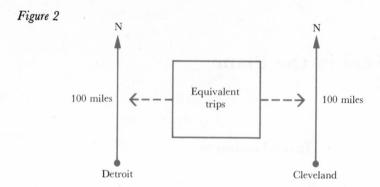

originates but only on the direction and length of the trip. For example, a 100-mile trip south is not considered to be equivalent to a 100-mile trip north because the directions are different (Figure 3).

Figure 3

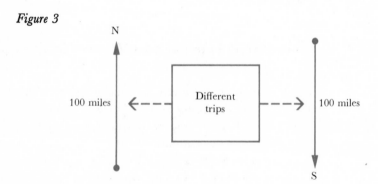

A 100-mile trip north is not considered to be equivalent to a 200-mile trip north according to this scheme because the lengths are different (Figure 4). The trips must agree both in length and direction in order to be considered equivalent.

Figure 4

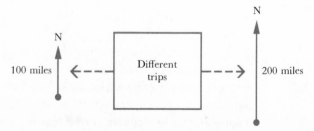

This illustration is a specific example of a situation that can be modeled by *vectors* in the plane. A *vector* in the plane is a line segment with a direction usually denoted by an arrowhead at the end of the segment (Figure 5). The length of the line segment is called the *length or magnitude* of the vector, and the direction of the line segment is called the *direction* of the vector. The end point of the vector containing the arrowhead is called the *terminal point* of the vector, and the other end point is called the *initial point* of the vector (Figure 5).

Figure 5

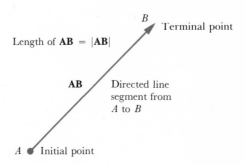

Length of **AB** = |**AB**|

AB

B Terminal point

Directed line
segment from
A to *B*

A Initial point

A *zero vector*, denoted by **0**, is a vector whose initial and terminal points are the same. In the example above, a 100-mile trip north is represented by a vector of length 100 pointing north (Figure 6). If a vector is determined by a directed line segment with the initial point *A* and the terminal point *B*, then it is denoted by **AB**; its magnitude is denoted by |**AB**| (Figure 5). Notice that |**0**| = 0. (Why?)

Figure 6

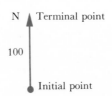

N ▲ Terminal point

100

Initial point

The notation **AB** suggests that the vector is determined by line segment $\overline{AB}$ *from point A to point B* (Figure 7*a*). **BA**, on the other hand, represents a vector with the same end points as vector **AB** but is in the opposite direction (Figure 7*b*).

As in the example, two vectors are considered to be equal if the vectors agree both in magnitude and in direction; for instance, **AB** and **BA** (Figure 7) are not equal (even though their lengths are the same) because they are opposite in direction; that is,

$$|\mathbf{AB}| = |\mathbf{BA}| \qquad \text{but} \qquad \mathbf{AB} \neq \mathbf{BA}$$

Figure 7

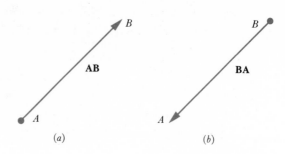

(a)   (b)

Finally, lowercase boldface letters will be used to denote vectors. Hence we speak of "vector u" and write it as **u**.

2.1 Addition of Vectors

Suppose that we are given two vectors **u** and **v** (Figure 8). First, we "shift" or "translate" **v**, so that the initial point of **v** coincides with the terminal

Figure 8

point of **u** (Figure 9). The vector having the same initial point as **u** and the same terminal point as **v** is defined to be the sum **u**+**v** (Figure 9).

Figure 9

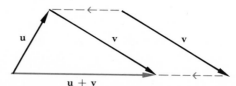

EXAMPLE

Use the geometry of vector addition to show that vector addition is commutative; that is, show that **u**+**v** = **v**+**u**, where **u** and **v** are given in Figure 10.

Figure 10

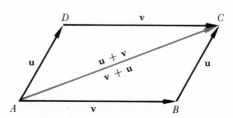

SOLUTION. (Figure 10.) Here **u** and **v** determine parallelogram $ABCD$, with **u**+**v** and **v**+**u** coinciding with the same diagonal vector **AC**. Hence **u**+**v** = **v**+**u**.

2.2 Scalar Multiplication and Subtraction of Vectors

Suppose that we are given a real number a and a vector **u**. The real number a in this context is called a *scalar*; a**u** is defined to be a vector with magnitude $|a||\mathbf{u}|$, with the same direction as **u** if $a > 0$ and with opposite direction as **u** if $a < 0$ (Figure 11). In other words, a**u** is a vector collinear

Figure 11

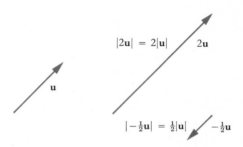

with vector **u** and $|a\mathbf{u}| = |a||\mathbf{u}|$. If $a = 0$, $a\mathbf{u} = \mathbf{0}$, the zero vector. If $0 < |a| < 1$ and $\mathbf{u} \neq \mathbf{0}$, $|a\mathbf{u}| < |\mathbf{u}|$; if $|a| > 1$, $|a\mathbf{u}| > |\mathbf{u}|$; if $|a| = 1$, $|a\mathbf{u}| = |\mathbf{u}|$.

For example, **AB** = − **BA**. (Why?) 5**u** is a vector in the direction of **u**, with magnitude five times the magnitude of **u**; − $\frac{1}{3}$**u** is opposite in direction to **u** and has a length $\frac{1}{3}$ that of **u**.

Now we can use vector addition and scalar multiplication to define *vector subtraction* as **u** − **v** = **u** + (−**v**) (Figure 12).

Figure 12

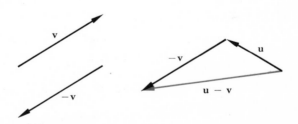

EXAMPLES

1 Compare **u** + **v** to **u** − **v** geometrically.

SOLUTION. **u** + **v** and **u** − **v** are opposite diagonals of the parallelogram determined by **u** and **v** (Figure 13).

Figure 13

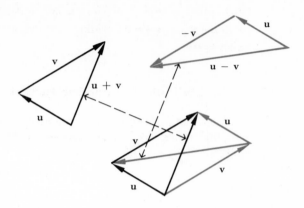

2   Show geometrically that $2(\mathbf{u}+\mathbf{v}) = 2\mathbf{u}+2\mathbf{v}$.

SOLUTION.   First sketch $\mathbf{u}+\mathbf{v}$ (Figure 14a). Then construct $\triangle EKL$ similar to the original triangle, $\triangle CDF$, such that $\mathbf{EK} = 2\mathbf{u}$ and $\mathbf{KL} = 2\mathbf{v}$, so that $\mathbf{EL} = 2\mathbf{u}+2\mathbf{v}$ by vector addition (Figure 14b); and $\mathbf{EL} = 2\mathbf{CF} = 2(\mathbf{u}+\mathbf{v})$ by the similarity of the two triangles.

Figure 14

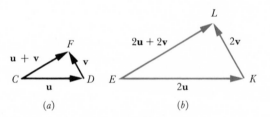

(a)   (b)

3   Let $ABCD$ be a parallelogram. Use vectors to prove that the diagonals $\overline{AC}$ and $\overline{BD}$ bisect each other (Figure 15).

Figure 15

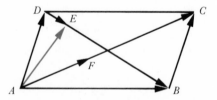

SOLUTION.   Assume that E is the midpoint of $\overline{BD}$ and F is the midpoint of $\overline{AC}$. We wish to show that E and F coincide, that is, we wish to show that $\mathbf{AF} = \mathbf{AE}$.

$$\mathbf{AF} = \tfrac{1}{2}(\mathbf{AC}) \qquad \text{and} \qquad \mathbf{AC} = \mathbf{AB} + \mathbf{BC}$$

so that

$$\mathbf{AF} = \tfrac{1}{2}(\mathbf{AB} + \mathbf{BC})$$

Also,

(1) $\mathbf{AE} = \mathbf{AD} + \mathbf{DE}$

but

$$\mathbf{DE} = \tfrac{1}{2}(\mathbf{DB}) \qquad \text{where} \qquad \mathbf{DB} = \mathbf{AB} - \mathbf{AD}$$

Hence

$$\mathbf{DE} = \tfrac{1}{2}(\mathbf{AB} - \mathbf{AD})$$

and, by substitution into Equation (1), we have

$$\mathbf{AE} = \mathbf{AD} + \tfrac{1}{2}(\mathbf{AB} - \mathbf{AD}) = \tfrac{1}{2}(\mathbf{AB} + \mathbf{AD}) = \tfrac{1}{2}(\mathbf{AB} + \mathbf{BC}) = \tfrac{1}{2}\mathbf{AC} = \mathbf{AF}$$

Hence F and E coincide.

3   Analytic Representation of Vectors in the Plane

Suppose that vector $\mathbf{u}$ is positioned in a plane with a Cartesian coordinate system so that the initial point of $\mathbf{u}$ is the origin and the terminal point is point (a, b) (Figure 1). $\mathbf{u}$ is called a *radius vector* or *position vector*, and we identify such a vector $\mathbf{u}$ as $\mathbf{u} = \langle a, b \rangle$. a and b are called the *x component and y component of* $\mathbf{u}$, respectively.

Figure 1

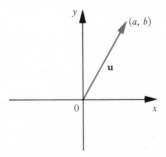

For example, $\mathbf{0} = \langle 0, 0 \rangle$; $\mathbf{u} = \langle -4, 5 \rangle$ has x component -4 and y component 5 (Figure 2); all vectors with x component 0 have terminal points on the y axis. (Why?) For example, $\mathbf{v} = \langle 0, -3 \rangle$ has a terminal point on the y axis.

We will see that the components of position vectors completely characterize these types of vectors in the sense that the determination of the magnitude of a position vector, the addition of position vectors, the scalar

Figure 2

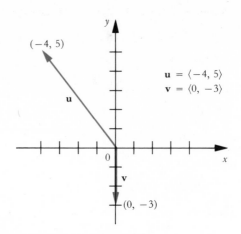

$$\mathbf{u} = \langle -4, 5 \rangle$$
$$\mathbf{v} = \langle 0, -3 \rangle$$

multiplication of position vectors, and the subtraction of position vectors can be performed by using the components.

3.1 Vector Equality in Terms of Components

We have already seen that $\mathbf{u} = \mathbf{v}$ whenever $|\mathbf{u}| = |\mathbf{v}|$ and $\mathbf{u}$ and $\mathbf{v}$ have the same direction. If $\mathbf{u}_1$ and $\mathbf{u}_2$ are position vectors, they have the same initial point $(0, 0)$, and the question of equality is reduced to a comparison of the components; that is, if $\mathbf{u}_1 = \langle a_1, b_1 \rangle$ and $\mathbf{u}_2 = \langle a_2, b_2 \rangle$, then $\mathbf{u}_1 = \mathbf{u}_2$ whenever $a_1 = a_2$ and $b_1 = b_2$.

3.2 Magnitude in Terms of Components

The magnitude or length of a position vector $\mathbf{u}$, $|\mathbf{u}|$, can be determined by using the components and the distance formula, for if $\mathbf{u} = \langle a, b \rangle$, then

$$|\mathbf{u}| = \sqrt{a^2 + b^2} \qquad \text{(Figure 3)}$$

Figure 3

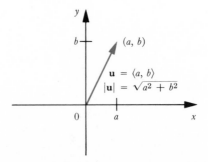

$$\mathbf{u} = \langle a, b \rangle$$
$$|\mathbf{u}| = \sqrt{a^2 + b^2}$$

For example, if $\mathbf{u} = \langle -4, 3 \rangle$, then $|\mathbf{u}| = \sqrt{(-4)^2 + 3^2} = \sqrt{25} = 5$, and if $\mathbf{v} = \langle 5, -12 \rangle$, then

$$|\mathbf{v}| = \sqrt{5^2 + (-12)^2} = \sqrt{169} = 13$$

3.3   Vector Algebra in Terms of Components

Addition and scalar multiplication of position vectors can be accomplished by using the components. Assume that $\mathbf{u}_1 = \langle a_1, b_1 \rangle$, $\mathbf{u}_2 = \langle a_2, b_2 \rangle$, and c is a scalar; then $\mathbf{u}_1 + \mathbf{u}_2$ and $c\mathbf{u}_1$ can be represented geometrically as in Figures 4a and b. Hence

$$\mathbf{u}_1 + \mathbf{u}_2 = \langle a_1, b_1 \rangle + \langle a_2, b_2 \rangle = \langle a_1 + a_2, b_1 + b_2 \rangle$$

and

$$c\mathbf{u}_1 = c \langle a_1, b_1 \rangle = \langle ca_1, cb_1 \rangle$$

Figure 4a

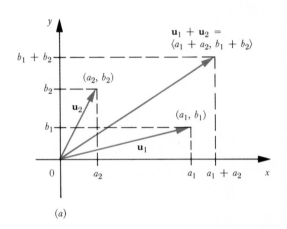

(a)

Figure 4b

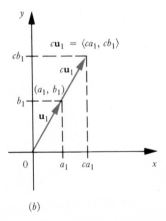

(b)

For example, if $\mathbf{u} = \langle -3, 4 \rangle$ and $\mathbf{v} = \langle 5, 7 \rangle$, then

$$\mathbf{u} + \mathbf{v} = \langle -3+5, 4+7 \rangle = \langle 2, 11 \rangle \quad \text{and} \quad 2\mathbf{u} = 2\langle -3, 4 \rangle = \langle -6, 8 \rangle$$

As before, $\mathbf{u} - \mathbf{v} = \mathbf{u} + (-\mathbf{v})$, so that if $\mathbf{u} = \langle a_1, b_1 \rangle$ and $\mathbf{v} = \langle a_2, b_2 \rangle$,

$$
\begin{aligned}
\mathbf{u} - \mathbf{v} &= \mathbf{u} + (-\mathbf{v}) \\
&= \langle a_1, b_1 \rangle + (-\langle a_2, b_2 \rangle) \\
&= \langle a_1, b_1 \rangle + \langle -a_2, -b_2 \rangle \\
\mathbf{u} - \mathbf{v} &= \langle a_1 - a_2, b_1 - b_2 \rangle \qquad \text{(Figure 5)}
\end{aligned}
$$

Figure 5

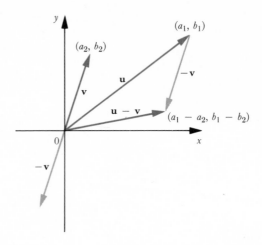

Thus, if $\mathbf{u} = \langle 3, 4 \rangle$ and $\mathbf{v} = \langle -1, 5 \rangle$, then $\mathbf{u} - \mathbf{v} = \langle 3 - (-1), 4 - 5 \rangle = \langle 4, -1 \rangle$.

EXAMPLES

1 Characterize all position vectors $\mathbf{u}$ such that $|\mathbf{u}| = 1$.

SOLUTION.   Assume that $\mathbf{u} = \langle x, y \rangle$; then $|\mathbf{u}| = \sqrt{x^2 + y^2} = 1$; that is, $x^2 + y^2 = 1$. Hence all such vectors have their terminal points on the unit circle (Figure 6).

Figure 6

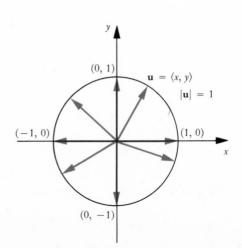

2 Let $\mathbf{u}_1 = \langle 3, 4 \rangle$ and $\mathbf{u}_2 = \langle 5, 6 \rangle$. Find

a) $\mathbf{u}_1 + \mathbf{u}_2$ b) $\mathbf{u}_1 - \mathbf{u}_2$

c) $5\mathbf{u}_1$ d) $|\mathbf{u}_2|$

e) $4\mathbf{u}_1 - 3\mathbf{u}_2$ f) $\left| (1/|\mathbf{u}_1|) \mathbf{u}_1 \right|$

SOLUTION

a) $\mathbf{u}_1 + \mathbf{u}_2 = \langle 3, 4 \rangle + \langle 5, 6 \rangle = \langle 8, 10 \rangle$

b) $\mathbf{u}_1 - \mathbf{u}_2 = \langle 3, 4 \rangle - \langle 5, 6 \rangle = \langle -2, -2 \rangle$

c) $5\mathbf{u}_1 = 5(3, 4) = \langle 15, 20 \rangle$

d) $|\mathbf{u}_2| = \sqrt{25 + 36} = \sqrt{61}$

e) $4\mathbf{u}_1 - 3\mathbf{u}_2 = 4\langle 3, 4 \rangle - 3\langle 5, 6 \rangle$
$$= \langle 12, 16 \rangle + \langle -15, -18 \rangle$$
$$= \langle -3, -2 \rangle$$

f) $\left| (1/|\mathbf{u}_1|) \mathbf{u}_1 \right| = \left| (1/|\langle 3,4 \rangle|) \langle 3, 4 \rangle \right| = \left| \tfrac{1}{5} \langle 3, 4 \rangle \right| = \left| \langle \tfrac{3}{5}, \tfrac{4}{5} \rangle \right| = 1$

3.4 Basis Vectors

A vector of length 1 is called a *unit vector*; hence if $\mathbf{u} = \langle a, b \rangle$ is a unit vector, $|\mathbf{u}| = \sqrt{a^2 + b^2} = 1$ (see Example 1 above). For example, $\mathbf{u}_1 = \langle 1, 0 \rangle$, $\mathbf{u}_2 = \langle 0, 1 \rangle$, $\mathbf{u}_3 = \langle -\tfrac{1}{2}, \sqrt{3}/2 \rangle$, $\mathbf{u}_4 = \langle -1/\sqrt{2}, -1/\sqrt{2} \rangle$, and $\mathbf{u}_5 = \langle \tfrac{3}{5}, -\tfrac{4}{5} \rangle$ are unit vectors (Figure 7).

Figure 7

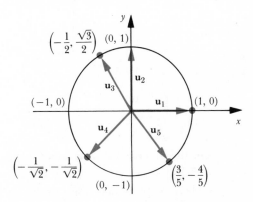

If $\mathbf{u}$ is not a unit vector and if $\mathbf{u} \neq \mathbf{0}$, we can form a unit vector in the same direction as $\mathbf{u}$ by multiplying $\mathbf{u}$ by $1/|\mathbf{u}|$ (see Example 2f above).

THEOREM 1

If $\mathbf{u} \neq \mathbf{0}$, $(1/|\mathbf{u}|)\mathbf{u}$ is a unit vector in the same direction as $\mathbf{u}$.

PROOF. Suppose that $\mathbf{u} = \langle a, b \rangle$. Then $|\mathbf{u}| = \sqrt{a^2 + b^2}$, so that

$$\frac{1}{|\mathbf{u}|}\mathbf{u} = \frac{1}{\sqrt{a^2 + b^2}} \langle a, b \rangle = \left\langle \frac{a}{\sqrt{a^2 + b^2}}, \frac{b}{\sqrt{a^2 + b^2}} \right\rangle$$

Hence

$$\left|\frac{1}{|\mathbf{u}|}\mathbf{u}\right| = \sqrt{\frac{a^2}{a^2+b^2}+\frac{b^2}{a^2+b^2}} = \sqrt{\frac{a^2+b^2}{a^2+b^2}} = 1$$

Since $1/|\mathbf{u}| > 0$, $(1/|\mathbf{u}|)\mathbf{u}$ has the same direction as $\mathbf{u}$. $(1/|\mathbf{u}|)\mathbf{u}$ is often called the *normalized* $\mathbf{u}$ vector.

Next, we define the vectors $\mathbf{i} = \langle 1,0 \rangle$ and $\mathbf{j} = \langle 0,1 \rangle$. $\mathbf{i}$ and $\mathbf{j}$ are unit vectors (why?) and are in the direction of the positive x axis and y axis, respectively (Figure 8). $\mathbf{i}$ and $\mathbf{j}$ form a *basis* for the system of vectors in the plane in the sense that each vector in the plane can be written in terms of $\mathbf{i}$ and $\mathbf{j}$ as follows.

If $\mathbf{u}$ is a radius vector, then $\mathbf{u}$ is of the form $\mathbf{u} = \langle x,y \rangle$, so that

$$\begin{aligned}\mathbf{u} = \langle x,y \rangle &= \langle x,0 \rangle + \langle 0,y \rangle \qquad \text{(Figure 8)}\\ &= x\langle 1,0 \rangle + y\langle 0,1 \rangle \\ &= x\mathbf{i} + y\mathbf{j}\end{aligned}$$

Figure 8

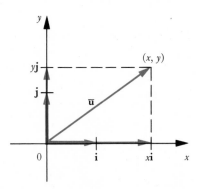

$x\mathbf{i}$ and $y\mathbf{j}$ are called the vector *projections* of $\mathbf{u}$ along the x axis and the y axis, respectively.

On the other hand, if $\mathbf{u}$ is not a radius vector, then $\mathbf{u}$ is of the form $\mathbf{u} = \mathbf{P_1P_2}$, where $P_1 = (x_1,y_1)$ and $P_2 = (x_2,y_2)$ are any two points in the plane representing, respectively, the initial and terminal points of $\mathbf{u}$. But $\mathbf{u} = \mathbf{u_2} - \mathbf{u_1}$, where $\mathbf{u_1}$ and $\mathbf{u_2}$ are the radius vectors determined by P_1 and P_2 (Figure 9).

Since we have $\mathbf{u_1} = \langle x_1,y_1 \rangle$ and $\mathbf{u_2} = \langle x_2,y_2 \rangle$,

$$\mathbf{u} = \langle x_2,y_2 \rangle - \langle x_1,y_1 \rangle = \langle x_2-x_1, y_2-y_1 \rangle$$

and

$$\mathbf{u} = (x_2-x_1)\mathbf{i} + (y_2+y_1)\mathbf{j} = u_x\mathbf{i} + u_y\mathbf{j}$$

where (x_1,y_1) is the initial point of $\mathbf{u}$ and (x_2,y_2) is the terminal point of $\mathbf{u}$ and $u_x = x_2-x_1$ and $u_y = y_2-y_1$. u_x and u_y represent the x and y components of $\mathbf{u}$, respectively. Notice that $|\mathbf{u}| = \sqrt{(x_1-x_2)^2+(y_1-y_2)^2}$.

Figure 9

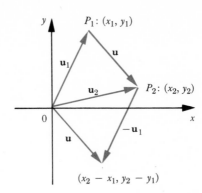

EXAMPLES

1 Let **u** be the vector with initial point $(-1, 3)$ and terminal point $(2, 7)$.
a) Write **u** in the form $u_x\mathbf{i} + u_y\mathbf{j}$.
b) Find $|\mathbf{u}|$.
c) Find $\mathbf{u}_N$, the normalized **u** vector.

SOLUTION. (See Figure 10.)

a) $\mathbf{u} = (x_2 - x_1)\mathbf{i} + (y_2 - y_1)\mathbf{j} = (2-(-1))\mathbf{i} + (7-3)\mathbf{j} = 3\mathbf{i} + 4\mathbf{j}$
b) $|\mathbf{u}| = \sqrt{u_x^2 + u_y^2} = \sqrt{9+16} = \sqrt{25} = 5$
c) $\mathbf{u}_N = (1/|\mathbf{u}|)\mathbf{u} = \frac{1}{5}(3\mathbf{i}+4\mathbf{j}) = \frac{3}{5}\mathbf{i} + \frac{4}{5}\mathbf{j}$.

Figure 10

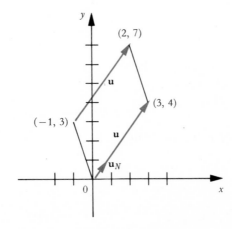

2 Find a point $\frac{1}{3}$ of the way from $(1, 4)$ to $(6, 6)$.
SOLUTION. (See Figure 11.) Let $P = (x, y)$ be $\frac{1}{3}$ of the way from A to B;
then $\mathbf{AP} = (x-1)\mathbf{i} + (y-4)\mathbf{j} = \frac{1}{3}(\mathbf{AB}) = \frac{1}{3}(5\mathbf{i}+2\mathbf{j})$, from which it follows
that $x-1 = \frac{5}{3}$ and $y-4 = \frac{2}{3}$, or $x = \frac{8}{3}$ and $y = \frac{14}{3}$. Therefore, the coordinates of P are given by $(\frac{8}{3}, \frac{14}{3})$.

Figure 11

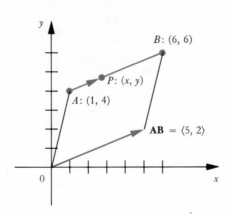

ALTERNATE SOLUTION.   (See Figure 12.) Let $\mathbf{v} = \langle x, y \rangle$, where $P = (x, y)$ is $\frac{1}{3}$ of the way from A to B. $\mathbf{v} = \mathbf{0A} + \mathbf{AP}$, but $\mathbf{AP} = \frac{1}{3}\mathbf{AB}$. Therefore,

$$\mathbf{v} = \langle 1, 4 \rangle + \frac{1}{3}\langle 5, 2 \rangle$$
$$= \langle 1, 4 \rangle + \langle \tfrac{5}{3}, \tfrac{2}{3} \rangle$$
$$= \langle \tfrac{8}{3}, \tfrac{14}{3} \rangle$$

so that $x = \frac{8}{3}$ and $y = \frac{14}{3}$ are the coordinates of P.

Figure 12

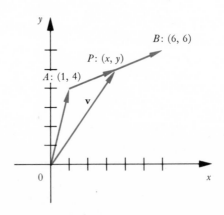

PROBLEM SET 1

1   Let $\mathbf{u}$ be a vector in the positive direction of the x axis and $\mathbf{v}$ be a vector that makes an angle of 30° with the vector $\mathbf{u}$ by a counterclockwise rotation from the positive x axis; assume that $|\mathbf{u}| = 3$ and $|\mathbf{v}| = 5$. Make a sketch of each of the following vectors.

a) $2\mathbf{u} + \mathbf{v}$ b) $2\mathbf{u} + \frac{1}{2}\mathbf{v}$

c) $\frac{1}{2}\mathbf{u} - 2\mathbf{v}$ d) $-\mathbf{u} - 3\mathbf{v}$

2 **a)** Use the geometry of vector addition to show that vector addition is associative; that is, show that $(\mathbf{u}+\mathbf{v})+\mathbf{w} = \mathbf{u}+(\mathbf{v}+\mathbf{w})$.

b) If $|\mathbf{u}| = |\mathbf{v}|$, does it follow that $\mathbf{u} = \mathbf{v}$? Explain.

3 Let $PQRS$ be a parallelogram as shown in Figure 13.

    a) Find $\mathbf{w}$ in terms of

        i $\mathbf{u}_1$ and $\mathbf{v}_1$

        ii $\mathbf{u}_2$ and $\mathbf{v}_1$

        iii $\mathbf{u}_2$ and $\mathbf{v}_2$

    b) Sketch $\mathbf{v}_2+\mathbf{w}+\mathbf{u}_2$.

    c) Sketch $\mathbf{u}_1+\mathbf{v}_1+\mathbf{u}_2+\mathbf{v}_2$.

Figure 13

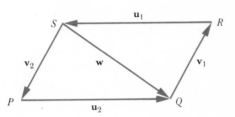

4 Use vectors to prove that the line segment joining the midpoints of two sides of a triangle in the plane is parallel to the third side and has a length equal to one-half the length of the third side.

5 Use $\mathbf{u}_1 = \langle 3,4\rangle$, $\mathbf{u}_2 = \langle -2,4\rangle$, and $\mathbf{u}_3 = \langle 7,8\rangle$ to determine each of the following.

  a) $2\mathbf{u}_1+3\mathbf{u}_2$ **b)** $3\mathbf{u}_2-\mathbf{u}_1$

  c) $|\mathbf{u}_1|+|\mathbf{u}_2|$ **d)** $|\mathbf{u}_1+\mathbf{u}_2|$

  e) $\mathbf{u}_1+(\mathbf{u}_2+\mathbf{u}_3)$ **f)** $(\mathbf{u}_1+\mathbf{u}_2)+\mathbf{u}_1$

  g) $2\mathbf{u}_1-3\mathbf{u}_2$ **h)** $\mathbf{u}_2-\mathbf{u}_1-\mathbf{u}_3$

  i) $3|\mathbf{u}_1-\mathbf{u}_2|$ **j)** $3|\mathbf{u}_1|-3|\mathbf{u}_2|$

6 Write each of the following vectors in the form of $r(\cos\theta\mathbf{i}+\sin\theta\mathbf{j})$, where $r = |\mathbf{u}|$ and θ is the radian measure of the angle giving the direction of the vector.

  a) $\mathbf{u} = -4\mathbf{i}$ **b)** $\mathbf{u} = -\mathbf{i}-\mathbf{j}$ **c)** $\mathbf{u} = -\mathbf{i}+3\mathbf{j}$

7 Let $\mathbf{u} = 3\mathbf{i}+4\mathbf{j}$ and $\mathbf{v} = 5\mathbf{i}-2\mathbf{j}$. Find a unit vector having the same direction as $\mathbf{u}+\mathbf{v}$.

8 Let $\mathbf{u} = -2\mathbf{i}+\mathbf{j}$, $\mathbf{v} = 2\mathbf{i}-3\mathbf{j}$, and $\mathbf{w} = 5\mathbf{i}+2\mathbf{j}$. Find scalar numbers a and b such that $\mathbf{w} = a\mathbf{u}+b\mathbf{v}$.

9 Write $\mathbf{u}$ in the form $\mathbf{u} = u_x\mathbf{i}+u_y\mathbf{j}$, find $|\mathbf{u}|$, and normalize $\mathbf{u}$ if $\mathbf{u}$ is a vector whose initial point is the first point and whose terminal point is the second point given in each of the following cases.

a) $(8,6), (3,4)$ b) $(7,6), (2,3)$
c) $(-2,6), (3,-5)$ d) $(-3,2), (1,-3)$
e) $(3,7), (-3,1)$ f) $(1,-3), (5,-1)$

10 a) Sketch V where $V = \{\mathbf{u} \mid \mathbf{u} = t\langle -1,2\rangle, t \in R\}$.
   b) Sketch V where $V = \{\mathbf{u} \mid \mathbf{u} = (1-t)\langle 0,2\rangle + t\langle 3,4\rangle, t \in [0,1]\}$.
   c) Sketch the graph determined by the terminal points of vectors $\mathbf{u}$ if $\mathbf{u} = (5\mathbf{i}-\mathbf{j})t+(4\mathbf{i}+2\mathbf{j})$ and t is any real number.

11 Let $P_1 = (-6,-10)$ and $P_2 = (8,8)$ be two points in the plane.
   a) Use vectors to find the point that is $\frac{2}{3}$ of the way from P_1 to P_2.
   b) Use vectors to find the point that is $\sqrt{130}$ units from P_1 along the directed line from P_1 to P_2.

4 Inner Product

In this section we shall examine a multiplicative operation between two vectors that results in a scalar. This multiplication, called *inner product* or *dot product*, will enable us to study the notion of orthogonality (perpendicularity) from an algebraic viewpoint.

If $\mathbf{u}$ and $\mathbf{v}$ are two vectors in the plane, we define the *inner product* or *dot product* of $\mathbf{u}$ and $\mathbf{v}$, denoted by $\mathbf{u} \cdot \mathbf{v}$, as $\mathbf{u} \cdot \mathbf{v} = |\mathbf{u}||\mathbf{v}| \cos \theta$, $0° \leq \theta \leq 180°$, where θ is the angle between $\mathbf{u}$ and $\mathbf{v}$. The product $|\mathbf{u}| \cos \theta$ is called the *scalar projection of* $\mathbf{u}$ *onto* $\mathbf{v}$. A geometric interpretation is given in Figure 1.

Figure 1

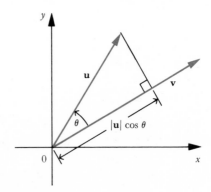

Notice that if $\mathbf{v} \neq \mathbf{0}$, it follows from

$$\mathbf{u} \cdot \mathbf{v} = |\mathbf{u}||\mathbf{v}| \cos \theta$$

that

$$|\mathbf{u}| \cos \theta = \frac{\mathbf{u} \cdot \mathbf{v}}{|\mathbf{v}|}$$

EXAMPLES

1 Find $\mathbf{u} \cdot \mathbf{v}$ if $|\mathbf{u}| = 3$ and $|\mathbf{v}| = 4$ and
   a) $\theta = 0°$ b) $\theta = 30°$ c) $\theta = 45°$
   d) $\theta = 90°$ e) $\theta = 180°$

SOLUTION

a) $\mathbf{u} \cdot \mathbf{v} = |\mathbf{u}| \, |\mathbf{v}| \cos\theta$
$$= 3 \cdot 4 \cos 0° = 12(1) = 12$$

b) $\mathbf{u} \cdot \mathbf{v} = 3 \cdot 4 \cos 30° = 12\left(\dfrac{\sqrt{3}}{2}\right) = 6\sqrt{3}$

c) $\mathbf{u} \cdot \mathbf{v} = 3 \cdot 4 \cos 45° = 12\left(\dfrac{\sqrt{2}}{2}\right) = 6\sqrt{2}$

d) $\mathbf{u} \cdot \mathbf{v} = 3 \cdot 4 \cos 90° = 12(0) = 0$
e) $\mathbf{u} \cdot \mathbf{v} = 3 \cdot 4 \cos 180° = 12(-1) = -12$

2 Find the scalar projection of $\mathbf{u}$ onto $\mathbf{v}$ for each part in Example 1.

SOLUTION

a) $\dfrac{\mathbf{u} \cdot \mathbf{v}}{|\mathbf{v}|} = \dfrac{12}{4} = 3$

b) $\dfrac{\mathbf{u} \cdot \mathbf{v}}{|\mathbf{v}|} = \dfrac{6\sqrt{3}}{4} = \dfrac{3\sqrt{3}}{2}$

c) $\dfrac{\mathbf{u} \cdot \mathbf{v}}{|\mathbf{v}|} = \dfrac{6\sqrt{2}}{4} = \dfrac{3\sqrt{2}}{2}$

d) $\dfrac{\mathbf{u} \cdot \mathbf{v}}{|\mathbf{v}|} = \dfrac{0}{4} = 0$

e) $\dfrac{\mathbf{u} \cdot \mathbf{v}}{|\mathbf{v}|} = \dfrac{-12}{4} = -3$

3 Suppose that $\mathbf{u} \neq 0$, $\mathbf{v} \neq 0$, and $\mathbf{u}$ is parallel to $\mathbf{v}$. Find $\mathbf{u} \cdot \mathbf{v}$.

SOLUTION. Since $\mathbf{u}$ is parallel to $\mathbf{v}$, the angle between $\mathbf{u}$ and $\mathbf{v}$ is $0°$ or $180°$. Hence

$$\mathbf{u} \cdot \mathbf{v} = |\mathbf{u}| \, |\mathbf{v}| \cos\theta$$

so that $\mathbf{u} \cdot \mathbf{v} = |\mathbf{u}| \, |\mathbf{v}|$ if $\theta = 0°$ and $\mathbf{u} \cdot \mathbf{v} = -|\mathbf{u}| \, |\mathbf{v}|$ if $\theta = 180°$.
   Now the dot product can be used to characterize orthogonality or perpendicularity.

THEOREM 1

Two nonzero vectors **u** and **v** are orthogonal (perpendicular) if and only if $\mathbf{u} \cdot \mathbf{v} = 0$.

PROOF.   Assume that **u** is orthogonal to **v**; then

$$\mathbf{u} \cdot \mathbf{v} = |\mathbf{u}|\,|\mathbf{v}| \cos 90° = |\mathbf{u}|\,|\mathbf{v}| \cdot 0 = 0$$

Conversely, assume that $\mathbf{u} \cdot \mathbf{v} = 0$; then $0 = |\mathbf{u}|\,|\mathbf{v}| \cos \theta$, where θ is the angle between **u** and **v**. But **u** and **v** are nonzero, so $\cos \theta = 0$. This implies that $\theta = 90°$; that is, **u** is orthogonal to **v**.

It is easy to find $\mathbf{u} \cdot \mathbf{v}$ if **u**, **v**, and θ are known; however, if **u** and **v** are known, the ease of the computation of $\mathbf{u} \cdot \mathbf{v}$ depends on how difficult it is to find θ. It is our purpose here to derive a *second* method of computing $\mathbf{u} \cdot \mathbf{v}$, a method that depends only on the components of **u** and **v**.

THEOREM 2

If $\mathbf{u} = u_x \mathbf{i} + u_y \mathbf{j}$ and $\mathbf{v} = v_x \mathbf{i} + v_y \mathbf{j}$, then $\mathbf{u} \cdot \mathbf{v} = u_x v_x + u_y v_y$.

PROOF.   Suppose that θ is the angle between **u** and **v** and the distance between the terminal points of **u** and **v** is L. (Figure 2.)

Figure 2

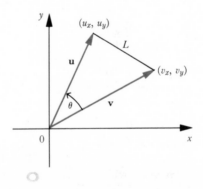

By the law of cosines,

$$L^2 = |\mathbf{u}|^2 + |\mathbf{v}|^2 - 2\,|\mathbf{u}|\,|\mathbf{v}| \cos \theta$$

so that

$$L^2 = |\mathbf{u}|^2 + |\mathbf{v}|^2 - 2\,\mathbf{u} \cdot \mathbf{v}$$

where

$$|\mathbf{u}|^2 = u_x^2 + u_y^2 \qquad \text{and} \qquad |\mathbf{v}|^2 = v_x^2 + v_y^2$$

Then

(1) $L^2 = u_x^2 + u_y^2 + v_x^2 + v_y^2 - 2\mathbf{u} \cdot \mathbf{v}$

But, by the distance formula,

(2) $L^2 = (u_x - v_x)^2 + (u_y - v_y)^2$

so that, after equating the right hand sides of equations (1) and (2), we have

$$(u_x - v_x)^2 + (u_y - v_y)^2 = u_x^2 + u_y^2 + v_x^2 + v_y^2 - 2\mathbf{u} \cdot \mathbf{v}$$

After simplifying this expression, we get

$$-2u_x v_x - 2u_y v_y = -2\mathbf{u} \cdot \mathbf{v}$$

that is,

$$\mathbf{u} \cdot \mathbf{v} = u_x v_x + u_y v_y$$

EXAMPLES

1 Compute each of the following values.

a) $\mathbf{i} \cdot \mathbf{j}$ b) $\mathbf{i} \cdot \mathbf{i}$ c) $\mathbf{j} \cdot \mathbf{j}$

SOLUTION

a) $\mathbf{i} \cdot \mathbf{j} = (1\mathbf{i} + 0\mathbf{j}) \cdot (0\mathbf{i} + 1\mathbf{j}) = 1 \cdot 0 + 0 \cdot 1 = 0$;
    thus $\mathbf{i}$ and $\mathbf{j}$ are orthogonal.
b) $\mathbf{i} \cdot \mathbf{i} = (1\mathbf{i} + 0\mathbf{j}) \cdot (1\mathbf{i} + 0\mathbf{j}) = 1 \cdot 1 + 0 \cdot 0 = 1$
c) $\mathbf{j} \cdot \mathbf{j} = (0\mathbf{i} + 1\mathbf{j}) \cdot (0\mathbf{i} + 1\mathbf{j}) = 0 \cdot 0 + 1 \cdot 1 = 1$

2 Show that $u_x = \mathbf{u} \cdot \mathbf{i}$ and $u_y = \mathbf{u} \cdot \mathbf{j}$, if $\mathbf{u} = u_x \mathbf{j} + u_y \mathbf{j}$.

SOLUTION

$$\mathbf{u} \cdot \mathbf{i} = (u_x \mathbf{i} + u_y \mathbf{j}) \cdot (1\mathbf{i} + 0\mathbf{j}) = u_x \cdot 1 + 0 = u_x$$

and

$$\mathbf{u} \cdot \mathbf{j} = (u_x \mathbf{i} + u_y \mathbf{j}) \cdot (0\mathbf{i} + 1\mathbf{j}) = 0 + u_y \cdot 1 = u_y$$

3 Determine the following dot products.

a) $\langle 2, 3 \rangle \cdot \langle 4, 5 \rangle$ b) $\langle \sqrt{2}, \sqrt{8} \rangle \cdot \langle \sqrt{8}, \sqrt{2} \rangle$
c) $\langle 3, -2 \rangle \cdot \langle 1, 3 \rangle$ d) $\langle -2, -2 \rangle \cdot \langle 3, 2 \rangle$

SOLUTION

a) $\langle 2, 3 \rangle \cdot \langle 4, 5 \rangle = 2 \cdot 4 + 3 \cdot 5 = 8 + 15 = 23$
b) $\langle \sqrt{2}, \sqrt{8} \rangle \cdot \langle \sqrt{8}, \sqrt{2} \rangle = \sqrt{2} \cdot \sqrt{8} + \sqrt{8} \cdot \sqrt{2} = 4 + 4 = 8$
c) $\langle 3, -2 \rangle \cdot \langle 1, 3 \rangle = 3 \cdot 1 + (-2) \cdot 3 = 3 - 6 = -3$
d) $\langle -2, -2 \rangle \cdot \langle 3, 2 \rangle = (-2) \cdot 3 + (-2) \cdot 2 = -6 - 4 = -10$

4 Let $\mathbf{u} = \langle 2, -4 \rangle$ and $\mathbf{v} = \langle 6, 3 \rangle$. Show that $\mathbf{u}$ and $\mathbf{v}$ are orthogonal.

PROOF.   (See Figure 3.)

$$\mathbf{u} \cdot \mathbf{v} = \langle 2, -4 \rangle \cdot \langle 6, 3 \rangle = 12 - 12 = 0$$

Figure 3

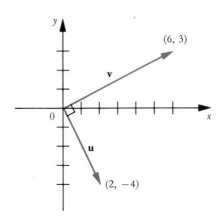

5   Find to the nearest degree the angle θ between $\mathbf{u} = 3\mathbf{i} + 5\mathbf{j}$ and $\mathbf{v} = 2\mathbf{i} - 7\mathbf{j}$.

SOLUTION.   Since $\mathbf{u} \cdot \mathbf{v} = 3 \cdot 2 - 7 \cdot 5 = -29$, $|\mathbf{u}| = \sqrt{34}$, and $|\mathbf{v}| = \sqrt{53}$, it follows from $\mathbf{u} \cdot \mathbf{v} = |\mathbf{u}| |\mathbf{v}| \cos \theta$ that

$$\cos \theta = \frac{-29}{\sqrt{34} \sqrt{53}} = -0.6831 \qquad \text{(approximately)}$$

so that $\theta = 133°$.

6   Prove that $|\mathbf{u}| = \sqrt{\mathbf{u} \cdot \mathbf{u}}$.

PROOF.   Assume that $\mathbf{u} = \langle u_x, u_y \rangle$; then

$$\mathbf{u} \cdot \mathbf{u} = u_x^2 + u_y^2 = |\mathbf{u}|^2$$

so that

$$|\mathbf{u}| = \sqrt{\mathbf{u} \cdot \mathbf{u}}$$

PROBLEM SET 2

1   Find $\mathbf{u} \cdot \mathbf{v}$ if $|\mathbf{u}| = 4$, $|\mathbf{v}| = 5$ and θ is the angle between $\mathbf{u}$ and $\mathbf{v}$.
    a) $\theta = 60°$ b) $\theta = 135°$
    c) $\theta = 150°$ d) $\theta = 90°$

2   Find the scalar projections of $\mathbf{u}$ onto $\mathbf{v}$ in Problem 1.

3   Find $\mathbf{u} \cdot \mathbf{v}$ and the angle θ (to the nearest degree) between $\mathbf{u}$ and $\mathbf{v}$ if
    a) $\mathbf{u} = \langle 1, 1 \rangle$ and $\mathbf{v} = \langle 1, -1 \rangle$ b) $\mathbf{u} = \langle 2, 3 \rangle$ and $\mathbf{v} = \langle 4, -5 \rangle$

c) $\mathbf{u} = \langle 7,0 \rangle$ and $\mathbf{v} = \langle 0,8 \rangle$ d) $\mathbf{u} = \langle 1,2 \rangle$ and $\mathbf{v} = \langle 2,-3 \rangle$

e) $\mathbf{u} = \langle 4,1 \rangle$ and $\mathbf{v} = \langle 1,-2 \rangle$

4 Find the scalar projections of $\mathbf{u}$ onto $\mathbf{v}$ in Problem 3.

5 In each of the following, find x so that $\mathbf{u}$ and $\mathbf{v}$ are orthogonal if $\mathbf{u} = 3\mathbf{i} + 4\mathbf{j}$ and

a) $\mathbf{v} = x\mathbf{i} + 4\mathbf{j}$ b) $\mathbf{v} = x\mathbf{i} - 4\mathbf{j}$ c) $\mathbf{v} = 4\mathbf{i} + x\mathbf{j}$

6 Use vectors to show that $\triangle ABC$ is a right triangle and identify the right angle if $A = (-2,1)$, $B = (4,-4)$, and $C = (-1,-10)$.

7 Suppose that $\mathbf{u}$ and $\mathbf{v}$ are orthogonal vectors in the plane. Express the following values in terms of $|\mathbf{u}|$ and $|\mathbf{v}|$.

a) $|\mathbf{u} + \mathbf{v}|$ b) $|\mathbf{u} - \mathbf{v}|$ c) $|3\mathbf{u} - 4\mathbf{v}|$

d) $|-3\mathbf{u} + 4\mathbf{v}|$

8 Let $\mathbf{u}$ and $\mathbf{v}$ be two vectors in the plane. Show that

a) $|\mathbf{u} + \mathbf{v}|^2 = |\mathbf{u}|^2 + 2\mathbf{u} \cdot \mathbf{v} + |\mathbf{v}|^2$

b) $|\mathbf{u} - \mathbf{v}|^2 = |\mathbf{u}|^2 - 2\mathbf{u} \cdot \mathbf{v} + |\mathbf{v}|^2$

c) $|\mathbf{u} + \mathbf{v}|^2 + |\mathbf{u} - \mathbf{v}|^2 = 2|\mathbf{u}|^2 + 2|\mathbf{v}|^2$

9 Let $\mathbf{u}$ and $\mathbf{v}$ be unit position vectors and let θ be the angle between them. Show that $\frac{1}{2}|\mathbf{v} - \mathbf{u}| = \sin(\theta/2)$.

10 Let $\mathbf{u} = -1\mathbf{i} + 3\mathbf{j}$, $\mathbf{v} = 3\mathbf{i} + 5\mathbf{j}$, and $\mathbf{w} = -2\mathbf{i} - 3\mathbf{j}$. Use $\mathbf{u}$, $\mathbf{v}$, and $\mathbf{w}$ to verify each of the following equations.

a) $\mathbf{v} \cdot \mathbf{v} = |\mathbf{v}|^2$

b) $\mathbf{w} \cdot \mathbf{w} = |\mathbf{w}|^2$

c) $\mathbf{u} \cdot (\mathbf{v} + \mathbf{w}) = \mathbf{u} \cdot \mathbf{v} + \mathbf{u} \cdot \mathbf{w}$

d) $(2\mathbf{u}) \cdot (3\mathbf{v}) = 6(\mathbf{u} \cdot \mathbf{v})$

e) $\mathbf{u} \cdot (4\mathbf{v} + 2\mathbf{w}) = 2\mathbf{u} \cdot 2\mathbf{v} + 2\mathbf{u} \cdot \mathbf{w}$

5 Applications

In this section let us consider a few examples of the application of vectors to solving problems in geometry and trigonometry.

5.1 Rotations

A rotation about the origin can be considered as a function that maps position vectors into position vectors by "turning" the given vectors through the *angle of rotation*. For example, a 90° rotation f maps position vectors into other position vectors by a "90° counterclockwise turn" about the origin (Figure 1). In the figure $f(\mathbf{u}) = \mathbf{v}$, where $\mathbf{u}$ and $\mathbf{v}$ determine a 90° angle, and $|\mathbf{u}| = |\mathbf{v}|$.

Figure 1

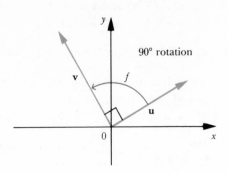

If f is a 90° rotation, then

$$f(\mathbf{i}) = \mathbf{j} \qquad f(-\mathbf{i}) = f(-\mathbf{j}) \qquad f(-\mathbf{j}) = \mathbf{i} \qquad \text{(Figure 2)}$$

Figure 2

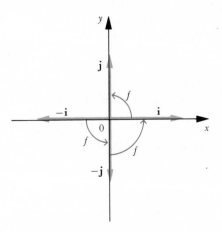

Suppose that f is a rotation through angle θ such that $f(\mathbf{u}) = \mathbf{v}$ (Figure 3).

Figure 3

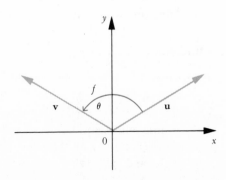

The rotation *f* has the following two properties:

1 The rotation of *a***u** by *f* is equivalent to the rotation of **u** by *f* with the result multiplied by *a*; that is,

$$f(a\mathbf{u}) = af(\mathbf{u}) \qquad \text{(Figure 4)}$$

Figure 4

(a) Multiply **u** by *a* to get *a***u**, and then rotate the result through θ to get $f(a\mathbf{u})$.

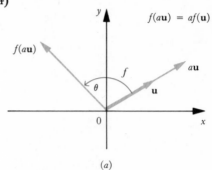

(*a*)

(b) Rotate **u** through θ to get $f(\mathbf{u})$, and then multiply the result by *a* to get $af(\mathbf{u})$.

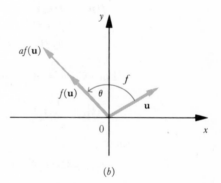

(*b*)

2 Another important property of rotations is that the rotation of the sum of two vectors is equivalent to the sum of the rotations; that is,

$$f(\mathbf{u}+\mathbf{v}) = f(\mathbf{u}) + f(\mathbf{v}) \qquad \text{(Figure 5)}$$

Figure 5

(a) Add **u** and **v** to get **u**+**v**, then rotate the result through θ to get $f(\mathbf{u}+\mathbf{v})$.

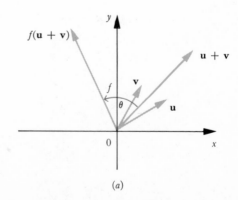

(*a*)

$$f(\mathbf{u} + \mathbf{v}) = f(\mathbf{u}) + f(\mathbf{v})$$

(b) Rotate **v** and **u** through θ to get $f(\mathbf{u})$ and $f(\mathbf{v})$, and then add the results to get $f(\mathbf{u})+f(\mathbf{v})$.

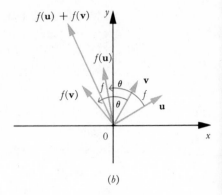

(b)

If any function f satisfies the above two properties, then f is called a *linear transformation*; hence a rotation function is a linear transformation.

The two properties above tell us that if we know what a rotation f does to **i** and **j**, then we can find the rotation on any vector **u**; for, if $\mathbf{u} = u_x\mathbf{i}+u_y\mathbf{j}$, then

$$
\begin{aligned}
f(\mathbf{u}) &= f(u_x\mathbf{i}+u_y\mathbf{j}) \\
&= f(u_x\mathbf{i}) + f(u_y\mathbf{j}) \quad \text{(Property 2)} \\
&= u_x f(\mathbf{i}) + u_y f(\mathbf{j}) \quad \text{(Property 1)}
\end{aligned}
$$

For example, if f is a rotation such that

$$f(\mathbf{i}) = \frac{1}{\sqrt{2}}\mathbf{i} + \frac{1}{\sqrt{2}}\mathbf{j} \quad \text{and} \quad f(\mathbf{j}) = \frac{-1\mathbf{i}}{\sqrt{2}} + \frac{1\mathbf{j}}{\sqrt{2}}$$

then

$$
\begin{aligned}
f(3\mathbf{i}+4\mathbf{j}) &= 3f(\mathbf{i}) + 4f(\mathbf{j}) \\
&= 3\left(\frac{1\mathbf{i}}{\sqrt{2}} + \frac{1\mathbf{j}}{\sqrt{2}}\right) + 4\left(\frac{-1\mathbf{i}}{\sqrt{2}} + \frac{1\mathbf{j}}{\sqrt{2}}\right) \\
&= -\frac{1\mathbf{i}}{\sqrt{2}} + \frac{7\mathbf{j}}{\sqrt{2}} \quad \text{(Figure 6)}
\end{aligned}
$$

Figure 6

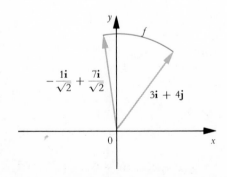

Notice that if g is a rotation function through an angle α and f is a rotation function through an angle β, then we may consider the composite mappings $(f \circ g)$ or $(g \circ f)$ as a rotation function through the angle $\alpha + \beta$. Also, if $f(\mathbf{i}) = x\mathbf{i} + y\mathbf{j}$, then, by using Properties 1 and 2, we have

$$(g \circ f)(\mathbf{i}) = g[f(\mathbf{i})] = g(x\mathbf{i} + y\mathbf{j}) = xg(\mathbf{i}) + yg(\mathbf{j})$$

EXAMPLE

Use rotations of vectors to prove the identities

$$\cos(t+s) = \cos t \cos s - \sin t \sin s$$

and

$$\sin(t+s) = \sin t \cos s + \cos t \sin s$$

where s and t are real numbers.

PROOF. Let f be a rotation that maps $\mathbf{i}$ through an angle of t radians to the vector $x\mathbf{i} + y\mathbf{j}$ (Figure 7). By the definition of circular functions (see Chapter 5, Section 3), $x = \cos t$ and $y = \sin t$, so that

$$f(\mathbf{i}) = x\mathbf{i} + y\mathbf{j} = (\cos t)\mathbf{i} + (\sin t)\mathbf{j}$$

Figure 7

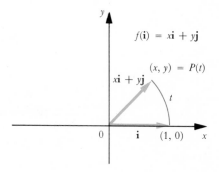

By symmetry (Figure 8),

$$f(\mathbf{j}) = -y\mathbf{i} + x\mathbf{j} = (-\sin t)\mathbf{i} + (\cos t)\mathbf{j}$$

Similarly, if g is a rotation through an angle of s radians, then

$$g(\mathbf{i}) = (\cos s)\mathbf{i} + (\sin s)\mathbf{j} \qquad \text{and} \qquad g(\mathbf{j}) = (-\sin s)\mathbf{i} + (\cos s)\mathbf{j}$$

Figure 8

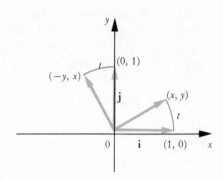

Hence

$$(g \circ f)(\mathbf{i}) = g[f(\mathbf{i})]$$
$$= g(x\mathbf{i}+y\mathbf{j})$$
$$= xg(\mathbf{i}) + yg(\mathbf{j})$$
$$= \cos t[(\cos s)\mathbf{i}+(\sin s)\mathbf{j}] + \sin t[(-\sin s)\mathbf{i}+(\cos s)\mathbf{j}]$$
$$= (\cos t \cos s - \sin t \sin s)\mathbf{i} + (\cos t \sin s + \sin t \cos s)\mathbf{j}$$

Furthermore, $(g \circ f)$ can be regarded as a single rotation through an angle of $t+s$ radians, so that

$$(g \circ f)(\mathbf{i}) = \cos(t+s)\mathbf{i} + \sin(t+s)\mathbf{j} \qquad \text{(Figure 9)}$$

Figure 9

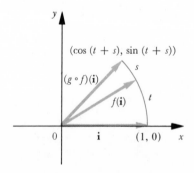

By comparing the two representations for $(g \circ f)(\mathbf{i})$, we can equate the components to get

$$\cos(t+s) = \cos t \cos s - \sin t \sin s$$

and

$$\sin(t+s) = \cos t \sin s + \sin t \cos s$$

5.2 Vector Equation of a Line

It is possible to use vectors to write the equation of a line. Assume that $P_1 = (x_1, y_1)$ and $P_2 = (x_2, y_2)$ are two points on a line. Let $\mathbf{b}$ be the position vector $\mathbf{OP_1}$ and let $\mathbf{m}$ be the vector $\mathbf{P_1 P_2}$, so that

$$\mathbf{b} = \langle x_1, y_1 \rangle \quad \text{and} \quad \mathbf{m} = \langle x_2 - x_1, y_2 - y_1 \rangle$$

Now, if t is any real number, then $\mathbf{m}t$ denotes all vectors parallel to $\mathbf{m}$, so that $\mathbf{u} = \mathbf{m}t + \mathbf{b}$ denotes all vectors with terminal points on the given line. The equation $\mathbf{u} = \mathbf{m}t + \mathbf{b}$, where t is any real number, is called a *vector equation* of the line (Figure 10).

Figure 10

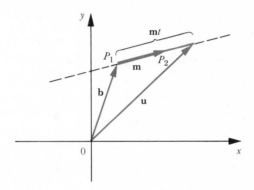

Since $\mathbf{m} = \langle x_2 - x_1, y_2 - y_1 \rangle$ and $\mathbf{b} = \langle x_1, y_1 \rangle$, we can also write the vector equation as

$$\begin{aligned} \mathbf{u} &= \langle x_2 - x_1, y_2 - y_1 \rangle t + \langle x_1, x_2 \rangle \\ &= \langle t(x_2 - x_1), t(y_2 - y_1) \rangle + \langle x_1, x_2 \rangle \\ &= \langle t(x_2 - x_1) + x_1, t(y_2 - y_1) + x_2 \rangle \end{aligned}$$

EXAMPLES

1   Find a vector equation of the line containing points $(3, 4)$ and $(-1, 2)$.

SOLUTION. The form of the equation is $\mathbf{u} = \mathbf{m}t + \mathbf{b}$, where $\mathbf{b}$ is a position vector with terminal point on the line and $\mathbf{m}$ is a position vector parallel to the line. Thus we can let $\mathbf{b} = \langle -1, 2 \rangle$ and $\mathbf{m} = \langle 3 - (-1), 4 - 2 \rangle = \langle 4, 2 \rangle$ (Figure 11). Hence

$$\begin{aligned} \mathbf{u} &= \langle 4, 2 \rangle t + \langle -1, 2 \rangle \\ &= \langle 4t, 2t \rangle + \langle -1, 2 \rangle \\ &= \langle 4t - 1, 2t + 2 \rangle \end{aligned}$$

where t is any real number.

Figure 11

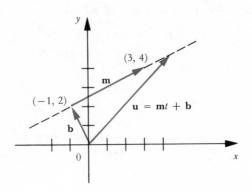

2   Find a vector equation of the line $y = 3x + 2$.

SOLUTION.   Here we can use the two points on the line $(0, 2)$ and $(1, 5)$ to define **b** as $\mathbf{b} = \langle 0, 2 \rangle$ and **m** as $\mathbf{m} = \langle 1 - 0, 5 - 2 \rangle = \langle 1, 3 \rangle$ (Figure 12).

Figure 12

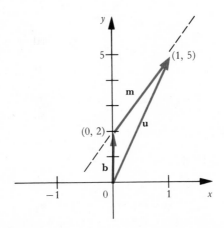

Hence the vector equation of the line is given by

$$\begin{aligned}\mathbf{u} &= \langle 1, 3 \rangle t + \langle 0, 2 \rangle \\ &= \langle t, 3t + 2 \rangle \end{aligned}$$

where t is any real number.

We conclude this section with a few more examples displaying the use of vectors in solving problems in geometry and trigonometry.

EXAMPLES

1   Prove that the diagonals of a rhombus $ADCB$ (Figure 13) (parallelogram with equal sides) are perpendicular to each other.

Figure 13

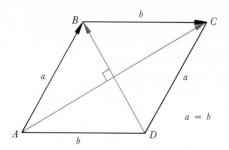

PROOF. Let $|\mathbf{AB}| = a$ and $|\mathbf{BC}| = b$. We see from Figure 13 that $\mathbf{AC} = \mathbf{AB} + \mathbf{BC}$ and $\mathbf{DB} = \mathbf{AB} - \mathbf{BC}$. Thus

$$
\begin{aligned}
(\mathbf{AC} \cdot \mathbf{DB}) &= (\mathbf{AB} + \mathbf{BC}) \cdot (\mathbf{AB} - \mathbf{BC}) \\
&= (\mathbf{AB}) \cdot (\mathbf{AB}) + (\mathbf{BC}) \cdot (\mathbf{AB}) - (\mathbf{AB}) \cdot (\mathbf{BC}) - (\mathbf{BC}) \cdot (\mathbf{BC}) \\
&= |\mathbf{AB}|^2 - |\mathbf{BC}|^2 \\
&= a^2 - b^2
\end{aligned}
$$

Since $a = b$, $\mathbf{AC} \cdot \mathbf{DB} = 0$, so $\mathbf{AC}$ is orthogonal (perpendicular) to $\mathbf{DB}$.

2   Prove that the three altitudes of any triangle intersect.

PROOF. Suppose that altitudes $\overline{AE}$ and $\overline{BD}$ of $\triangle ABC$ intersect at O, as shown in Figure 14. We must show that the line segment $\overline{CF}$ formed by extending $\overline{CO}$ through O to F is orthogonal (perpendicular) to $\overline{AB}$.

Figure 14

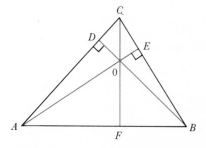

We know that $\mathbf{CB} \cdot \mathbf{OA} = 0$ and $\mathbf{AC} \cdot \mathbf{OB} = 0$; but $\mathbf{CB} = \mathbf{OB} - \mathbf{OC}$ and $\mathbf{AC} = \mathbf{OC} - \mathbf{OA}$, so that $(\mathbf{OB} - \mathbf{OC}) \cdot \mathbf{OA} = 0$ and $(\mathbf{OC} - \mathbf{OA}) \cdot \mathbf{OB} = 0$. Hence

$$(\mathbf{OB} \cdot \mathbf{OA}) - (\mathbf{OC} \cdot \mathbf{OA}) + (\mathbf{OC} \cdot \mathbf{OB}) - (\mathbf{OA} \cdot \mathbf{OB}) = 0$$

That is,

$$-(\mathbf{OC \cdot OA}) + (\mathbf{OC \cdot OB}) = 0$$

so that

$$(\mathbf{OB - OA}) \cdot \mathbf{OC} = 0$$

However, since $\mathbf{OB - OA} = \mathbf{AB}$, this means that $\mathbf{AB \cdot OC} = 0$, so that $\overleftrightarrow{CF}$ is perpendicular to $\overline{AB}$.

3   Use the inner product to derive the law of cosines.

PROOF.   Let $|\mathbf{AB}| = c$, $|\mathbf{AC}| = b$, and let θ be the angle between $\mathbf{AC}$ and $\mathbf{AB}$ in $\triangle ABC$ as shown in Figure 15.

Figure 15

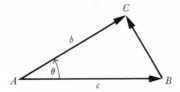

$\mathbf{BC = AC - AB}$. Hence

$$\begin{aligned}
|\mathbf{BC}|^2 = \mathbf{BC \cdot BC} &= (\mathbf{AC - AB}) \cdot (\mathbf{AC - AB}) \\
&= |\mathbf{AC}|^2 - [(\mathbf{AC}) \cdot (\mathbf{AB})] - [(\mathbf{AB}) \cdot (\mathbf{AC})] + |\mathbf{AB}|^2 \\
&= b^2 - 2(\mathbf{AC}) \cdot (\mathbf{AB}) + c^2 \\
&= b^2 + c^2 - 2bc \cos \theta
\end{aligned}$$

since $\mathbf{AC \cdot AB} = |\mathbf{AC}||\mathbf{AB}| \cos \theta = bc \cos \theta$.

4   Let $\mathbf{u}$ and $\mathbf{v}$ be *unit* radius vectors that make angles θ and ϕ with the x axis, respectively. Use the inner product to prove that $\cos(\theta - \phi) = \cos \theta \cos \phi + \sin \theta \sin \phi$.

SOLUTION.   First, we can use trigonometry to write

$$\mathbf{u} = \langle u_x, u_y \rangle = \langle \cos \theta, \sin \theta \rangle$$

and

$$\mathbf{v} = \langle v_x, v_y \rangle = \langle \cos \phi, \sin \phi \rangle \qquad \text{(Figure 16)}$$

Figure 16

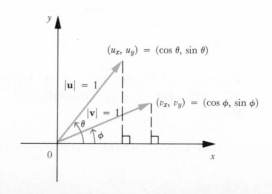

so that

$$\mathbf{u} \cdot \mathbf{v} = \cos \theta \cos \phi + \sin \theta \sin \phi$$

Also,

$$\mathbf{u} \cdot \mathbf{v} = |\mathbf{u}| \, |\mathbf{v}| \cos (\theta - \phi)$$
$$= 1 \cdot 1 \, \cos (\theta - \phi)$$

since $\mathbf{u}$ and $\mathbf{v}$ are unit vectors. Hence

$$\cos (\theta - \phi) = \cos \theta \cos \phi + \sin \phi \sin \theta$$

PROBLEM SET 3

1 Let f be a rotation through $30°$.
   a) Use trigonometry to find $f(\mathbf{i})$.
   b) Use symmetry to find $f(\mathbf{j})$ from part a.
   c) Use Properties 1 and 2 on page 405 to find $f(4\mathbf{i})$, $f(-7\mathbf{i}+5\mathbf{j})$, $f(-8\mathbf{j})$, and $f(\mathbf{i}-\mathbf{j})$.
   d) Does f^{-1} exist? Explain.

2 Assume that f is a rotation through a positive angle of measure less than $360°$ such that $f(\mathbf{i}) = -\mathbf{i}$.
   a) What is the angle of rotation of f?
   b) What is $f(\mathbf{j})$?
   c) If g is a $45°$ rotation, describe the rotations $g \circ f$ and $f \circ g$.

3 Assume that

$$f(\mathbf{i}) = \left(\frac{1}{\sqrt{2}}\right)\mathbf{i} + \left(\frac{1}{\sqrt{2}}\right)\mathbf{j}$$

Explain why

$$f(\mathbf{j}) = \left(\frac{-1}{\sqrt{2}}\right)\mathbf{i} + \left(\frac{1}{\sqrt{2}}\right)\mathbf{j}$$

Use $\mathbf{u} = \mathbf{i}+2\mathbf{j}$ and $\mathbf{v} = -3\mathbf{i}+\mathbf{j}$ to verify each of the following equations.
   a) $f(\mathbf{u}+\mathbf{v}) = f(\mathbf{u})+f(\mathbf{v})$
   b) $f(8\mathbf{u}) = 8f(\mathbf{u})$
   c) $f(5\mathbf{u} - 7\mathbf{v}) = 5f(\mathbf{u}) - 7f(\mathbf{v})$

4 Use vectors to derive the formula for the distance between points (x_1, y_1) and (x_2, y_2).

5 If $\mathbf{u} = 3\mathbf{i} - 6\mathbf{j}$ and $\mathbf{v} = 4\mathbf{i}+2\mathbf{j}$ are the sides of a right triangle, prove that the midpoint of the hypotenuse is equidistant from the vertices.

6 Use vectors to prove that the median to the base of an isosceles triangle is perpendicular to the base.

7 Find a vector equation and graph each of the following lines.
   a) A line containing points $(1, 4)$ and $(4, 3)$
   b) The line $y = 3$
   c) The line $x = 2$
   d) The line $y = -3x - 1$
   e) The line containing point $(1, 2)$ and parallel to the line $y = -\frac{1}{2}x$

8 Use vectors to prove that if the midpoints of the consecutive sides of any quadrilateral are joined, the resulting quadrilateral is a parallelogram.

REVIEW PROBLEM SET

1 Determine the components of $\mathbf{u}$ in each of the following cases if θ is the angle that $\mathbf{u}$ makes with the positive x axis.
   a) $|\mathbf{u}| = 5$ and $\theta = 30°$
   b) $|\mathbf{u}| = 6$ and $\theta = 45°$
   c) $|\mathbf{u}| = 8$ and $\theta = 150°$

2 Let O, A, and B be collinear points. Find n such that $\mathbf{OB} = n\mathbf{OA}$ if
   a) B is the midpoint of $\mathbf{OA}$
   b) A is the midpoint of $\mathbf{OB}$
   c) O is the midpoint of $\mathbf{AB}$
   d) B is $\frac{3}{4}$ of the way from O to A

3 Let $\mathbf{u} = \langle 3, 4 \rangle$ and $\mathbf{v} = \langle 4, 3 \rangle$. Find each of the following vectors and represent them graphically.
   a) $3\mathbf{u}$ b) $\mathbf{v} - \mathbf{u}$ c) $-2\mathbf{u} + \mathbf{v}$
   d) $3\mathbf{u} + 2\mathbf{v}$ e) $\mathbf{u} - 2\mathbf{v}$

4 Let $\mathbf{u} = \langle 1, 1 \rangle$, $\mathbf{v} = \langle 1, 1 \rangle$, and $\mathbf{w} = \langle 1, -1 \rangle$. Determine s, t, and r such that $\langle 4, 6 \rangle = s\mathbf{u} + t\mathbf{v} + r\mathbf{w}$.

5 Let $\mathbf{u} = \langle -3, 4 \rangle$, $\mathbf{v} = \langle 4, 3 \rangle$, and $\mathbf{w} = \langle 1, -3 \rangle$; find each of the following expressions.
   a) $|\mathbf{u}| - |\mathbf{v}|$ b) $|3\mathbf{u}| + |5\mathbf{v}|$
   c) $|\mathbf{u}|^2 + |\mathbf{v}|^2$ d) $|\mathbf{u} + \mathbf{v}|^2$
   e) $|\mathbf{u} - \mathbf{v} - \mathbf{w}|^2 - |\mathbf{u}|^2$ f) $|2\mathbf{u} + 3\mathbf{v} + 4\mathbf{w}|^2$
   g) $|\mathbf{u}|^2 + |\mathbf{v}|^2 + |\mathbf{w}|^2$ h) $2|\mathbf{u}|^2 + 3|\mathbf{u}||\mathbf{v}| + 6|\mathbf{v}|^2$

6 Assume that $\mathbf{u} \neq 0$, $\mathbf{v} \neq 0$, and $c \neq 0$. Show that $\mathbf{u}$ is collinear with $\mathbf{v}$ if and only if $\mathbf{u} = c\mathbf{v}$.

7 Write $\mathbf{u}$ in the form $\mathbf{u} = u_x\mathbf{i} + u_y\mathbf{j}$ and find $|\mathbf{u}|$, if $\mathbf{u}$ is a vector whose initial point is the first point and whose terminal point is the second point.
   a) $(-6, 8)$, $(4, 3)$ b) $(-7, 6)$, $(3, -1)$
   c) $(6, -1)$, $(5, 3)$ d) $(7, 1)$, $(-3, 5)$
   e) $(1, -3)$, $(5, -1)$

8   Let $\mathbf{u} = \langle 2,3 \rangle$, $\mathbf{v} = \langle -1,4 \rangle$, $\mathbf{w} = \langle 4,5 \rangle$. Determine $\mathbf{z}$ in each of the following parts.

a) $\mathbf{u} + \mathbf{v} = \mathbf{w} + \mathbf{z}$ b) $2(\mathbf{u} - \mathbf{v}) = 3(\mathbf{w} - \mathbf{z})$
c) $3(\mathbf{z} + \mathbf{v}) = 5(\mathbf{z} - \mathbf{w})$ d) $\mathbf{z} + 3(\mathbf{z} + \mathbf{u}) + 4(\mathbf{z} + \mathbf{v}) = \mathbf{0}$

9   Use $\mathbf{u}$, $\mathbf{v}$, and $\mathbf{w}$ of Problem 8 to find

a) $\mathbf{v} \cdot \mathbf{u}$ b) $3\mathbf{u} \cdot 2\mathbf{v}$
c) $4\mathbf{u} \cdot (\mathbf{v} + \mathbf{w})$ d) $2\mathbf{v} \cdot (3\mathbf{u} + 2\mathbf{w})$
e) $(\mathbf{u} - \mathbf{v}) \cdot 6(\mathbf{u} + \mathbf{v})$ f) $(4\mathbf{v} + 3\mathbf{w}) \cdot (4\mathbf{v} - 3\mathbf{w})$
g) $(\mathbf{u} + \mathbf{v} - \mathbf{w}) \cdot (\mathbf{u} + \mathbf{v} + \mathbf{w})$

10   Let $\mathbf{u}$ and $\mathbf{v}$ be vectors in the plane with $|\mathbf{u}| = a$ and $|\mathbf{v}| = b$. Show that

a) $(a\mathbf{v} + b\mathbf{u}) \cdot (a\mathbf{v} - b\mathbf{u}) = 0$
b) $(\mathbf{u} + \mathbf{v}) \cdot (\mathbf{u} - \mathbf{v}) = a^2 - b^2$

11   Find the angle θ (to the nearest degree) between $\mathbf{u}$ and $\mathbf{v}$ if $|\mathbf{u}| = 2$, $|\mathbf{v}| = 5$, and $\mathbf{u} \cdot \mathbf{v}$ is

a) 0   b) 1   c) 4   d) -2

12   Find k so that $\mathbf{u} = 3\mathbf{i} - 4\mathbf{j}$ is orthogonal to $\mathbf{v}$ in each of the following.

a) $\mathbf{v} = k\mathbf{i} + 3\mathbf{j}$ b) $\mathbf{v} = 5k\mathbf{i} - 2\mathbf{j}$ c) $\mathbf{v} = 4\mathbf{i} - 3k\mathbf{j}$

13   Find a unit vector orthogonal to each of the following vectors.

a) $\langle -3,4 \rangle$ b) $4\mathbf{i} - 3\mathbf{j}$ c) $2\mathbf{i} - 7\mathbf{j}$

14   Use vectors to prove the Pythagorean theorem for a right triangle.

15   Find the three angles of a triangle whose vertices are $(2, -1)$, $(1, -3)$, and $(3, -4)$.

16   Suppose that f and g are rotations of the plane about the origin. Show that $f \circ g = g \circ f$.

17   Write $f(\mathbf{i})$ in the form $x\mathbf{i} + y\mathbf{j}$ if f corresponds to a rotation through each of the following angles.

a) $2\pi/3$ b) π
c) $-45°$ d) $100°$

18   Use vectors to prove that the median to the base of an isosceles triangle bisects the vertex angle.

19   Let $\mathbf{u}$ be the position vector $\mathbf{OP}$, where $P = (\frac{1}{2}, \sqrt{3}/2)$.

a) Write $\mathbf{u}$ in the form $x\mathbf{i} + y\mathbf{j}$.
b) If $\mathbf{u} = f(\mathbf{i})$, where f is a rotation, find the angle of rotation θ.
c) Determine $f(\mathbf{j})$.
d) Use the results of parts b and c to determine $f(2\mathbf{i} - 5\mathbf{j})$.

20   Show that the line that joins one vertex of a parallelogram to the midpoint of an opposite side divides the diagonal in the ratio 2:1.

21   Find a vector equation of each of the following lines.

a) The line containing the points $(-3, 5)$ and $(2, -6)$
b) The line $y = 6x - 1$

CHAPTER 8

Complex Numbers and Theory of Equations

1 Introduction

In Chapter 3 we discussed the solution of a quadratic equation, an equation that has an equivalent form $ax^2 + bx + c = 0$, where $a \neq 0$ and a, b, and c are real numbers. The solution of such an equation is real if the discriminant, $b^2 - 4ac$, is nonnegative. If the discriminant is negative, it is impossible to solve the quadratic equation *in the real number system*. For example, $\{x \mid x^2 + 1 = 0\} = \emptyset$ if we restrict ourselves to R. Our purpose here is to extend the real number system to a "new" system—the *complex number system*—which contains numbers that will satisfy equations such as $x^2 + 1 = 0$.

2 Complex Numbers

An ordered pair of real numbers (a, b), which shall be denoted as $a + bi$, is called a *complex number*. The symbol i is a constant denoting $\sqrt{-1}$. i is not a real number. The algebra of complex numbers is defined in terms of the algebra of real numbers as follows.

Assume that $z_1 = a_1 + b_1 i$ and $z_2 = a_2 + b_2 i$ are complex numbers. Then

Equality. $z_1 = z_2$ if and only if $a_1 = a_2$ and $b_1 = b_2$.

Addition. $z_1 + z_2 = (a_1 + a_2) + (b_1 + b_2)i$.

Multiplication. $z_1 \cdot z_2 = (a_1 a_2 - b_1 b_2) + (a_1 b_2 + a_2 b_1)i$.

For $z = a + bi$, a is called the *real part* of z and b is called the *imaginary part* of z. It would perhaps be better to identify a as the "non-i part" and b as the "i part" of the complex number, but the choice of words "real part" and "imaginary part" is accepted today for historical reasons. Hence two complex numbers are equal if and only if the real and imaginary parts are equal; the real part of the sum of two complex numbers is the sum of the real parts and the imaginary part of the sum of two complex numbers is the sum of the imaginary parts.

The set of complex numbers will be denoted by C, so in set notation we have $C = \{a + bi \mid a, b \in R\}$.

If the imaginary part of a complex number is 0, we will consider the number to be real; hence $a + 0i$ will be considered to be the real number a; in this sense, R can be considered to be a proper subset of C.

EXAMPLES

1 Find the sum and product of each of the following pairs of complex numbers and identify the real and imaginary parts of each result.

    a) i, i b) $5 + 6i, 9 + 3i$ c) $4 - 2i, -3 + i$

SOLUTION

    a) $i + i = 2i$, so the real part of $2i$ is 0 and the imaginary part is 2.

$$i \cdot i = i^2 = (0 + 1i)(0 + 1i)$$
$$= (0 \cdot 0 - 1 \cdot 1) + (0 \cdot 1 + 1 \cdot 0)i$$
$$= -1$$

The real part of -1 is -1 and the imaginary part is 0.

    b) $(5 + 6i) + (9 + 3i) = 14 + 9i$; 14 is the real part and 9 is the imaginary part of $14 + 9i$.

$$(5 + 6i)(9 + 3i) = (45 - 18) + (54 + 15)i$$
$$= 27 + 69i$$

27 is the real part and 69 is the imaginary part of $27 + 69i$.

    c) $(4 - 2i) + (-3 + i) = 1 - i$; here 1 is the real part and -1 is the imaginary part of $1 - i$.

$$(4 - 2i)(-3 + i) = (-12 + 2) + (6 + 4)i$$
$$= -10 + 10i$$

The real part of $-10 + 10i$ is -10 and the imaginary part is 10.

2 Write each of the following complex numbers in the form of $a + bi$. Identify the real and imaginary parts of each result.

    a) $(1 + 2i)^2$ b) $i^4 + 3i^3$

SOLUTION

    a) $(1 + 2i)^2 = (1 + 2i)(1 + 2i) = (1 - 4) + (2 + 2)i = -3 + 4i$. The real part of $-3 + 4i$ is -3 and the imaginary part is 4.

    b) $i^4 + 3i^3 = i^2 \cdot i^2 + 3i^2 i = (-1)(-1) + 3(-1)i = 1 - 3i$. The real part of $1 - 3i$ is 1 and the imaginary part is -3.

We have seen in Example 1a one of the properties that does not hold in R but holds in C: If $x \in R$, $x^2 \geq 0$, whereas it is possible to have $z \in C$ such that $z^2 < 0$ $(i^2 = -1)$.

If the order properties of R did hold in C, then, by trichotomy,

$$i = 0 \quad \text{or} \quad i < 0 \quad \text{or} \quad i > 0$$

But, if $i = 0$, then $i \cdot i = 0$ implies that $-1 = 0$; hence $i \neq 0$. On the other hand, if $i > 0$, then $i^2 > 0$ implies that $-1 > 0$ since $i^2 = -1$; hence $i \not> 0$. Finally, if $i < 0$, then $i^2 > 0$, so $-1 > 0$ since $i^2 = -1$; hence $i \not< 0$. Consequently, trichotomy does *not* hold in C; that is, the order relation that exists in R does not exist in C.

The domain of functions can be extended to the set of complex numbers. For example, consider the function f whose domain is the set of complex numbers and whose rule of formation is $f(z) = 3z + 1$; then $f(i) = 3i + 1$ and $f(1 - i) = 3(1 - i) + 1 = 4 - 3i$.

2.1 Properties of Addition and Multiplication

Since we defined the operations of addition and multiplication on C in terms of the corresponding operations on R, it is not surprising that the properties of addition and multiplication on C are the same as the properties of addition and multiplication on R. (These properties are listed in Appendix B.)

Assume that $z_1, z_2, z_3 \in C$; then the following properties hold.

1 CLOSURE OF ADDITION AND MULTIPLICATION

   i $z_1 + z_2 \in C$

  ii $z_1 \cdot z_2 \in C$

2 COMMUTATIVITY OF ADDITION AND MULTIPLICATION

   i $z_1 + z_2 = z_2 + z_1$

  ii $z_1 z_2 = z_2 z_1$

3 ASSOCIATIVITY OF ADDITION AND MULTIPLICATION

   i $z_1 + (z_2 + z_3) = (z_1 + z_2) + z_3$

  ii $z_1 \cdot (z_2 \cdot z_3) = (z_1 \cdot z_2) \cdot z_3$

4 DISTRIBUTIVE PROPERTIES

   i $z_1 \cdot (z_2 + z_3) = (z_1 \cdot z_2) + (z_1 \cdot z_3)$

  ii $(z_1 + z_2) \cdot z_3 = (z_1 \cdot z_3) + (z_2 \cdot z_3)$

5 IDENTITY

   i   There exists $0 \in C$ such that $z + 0 = 0 + z = z$ for every $z \in C$.

  ii   There exists $1 \in C$ such that $z \cdot 1 = 1 \cdot z = z$ for every $z \in C$.

6   INVERSE

i   If $z \in C$, then there exists $-z \in C$ such that $z + (-z) = (-z) + z = 0$.

ii If $z \in C, z \neq 0$, then there exists $z^{-1} \in C$ such that $z \cdot z^{-1} = z^{-1} \cdot z = 1$.

PROOF OF PROPERTY 6, i.   Suppose that $z = a + bi$. Then, by letting $-z = -a - bi$, we have $(a + bi) + (-a - bi) = 0 + 0i = 0$.

For the proofs of the other properties, see Problem 6.

2.2   Subtraction of Complex Numbers

If $z_1, z_2 \in C$, then $z_1 - z_2$, that is, the *difference* of z_1 and z_2, is defined as $z_1 - z_2 = z_1 + (-z_2)$. For example, if $z_1 = 7 + 2i$ and $z_2 = 1 - 5i$, then

$$z_1 - z_2 = (7 + 2i) - (1 - 5i) = (7 + 2i) + (-1 + 5i)$$
$$= (7 - 1) + (2 + 5)i = 6 + 7i$$

EXAMPLES

Find $z_1 - z_2$ for each of the following pairs of complex numbers.

1   $z_1 = 7 + 4i$ and $z_2 = 3 + 5i$

SOLUTION

$$z_1 - z_2 = (7 + 4i) - (3 + 5i) = (7 + 4i) + (-3 - 5i)$$
$$= 4 - i$$

2   $z_1 = 4 - 5i$ and $z_2 = -5 + 7i$

SOLUTION

$$z_1 - z_2 = (4 - 5i) - (-5 + 7i) = (4 - 5i) + (5 - 7i)$$
$$= 9 - 12i$$

2.3   Division of Complex Numbers

The *conjugate* of a complex number $z = a + bi$, written $\bar{z}$ (read "z bar"), is defined as $\bar{z} = a - bi$. For example, the conjugate of $3 + 3i$ is $3 - 3i$, the conjugate of $-4 = -4 + 0i$ is $-4 - 0i = -4$, and the conjugate of $5i = 0 + 5i$ is $0 - 5i = -5i$.

If $z_1 = a_1 + b_1 i, z_1 \neq 0$, and $z_2 = a_2 + b_2 i$, then the *quotient* $z_2 \div z_1$ is given by

$$\frac{z_2}{z_1} = \frac{z_2 \cdot \bar{z}_1}{z_1 \cdot \bar{z}_1}$$

or, equivalently, by

$$\frac{a_2 + b_2 i}{a_1 + b_1 i} = \frac{(a_2 + b_2 i)(a_1 - b_1 i)}{(a_1 + b_1 i)(a_1 - b_1 i)}$$
$$= \frac{(a_2 a_1 + b_2 b_1) + (a_1 b_2 - b_1 a_2)i}{a_1^2 + b_1^2}$$

Notice that the result is a complex number of the form $a + bi$, with

$$a = \frac{a_2 a_1 + b_2 b_1}{a_1^2 + b_1^2} \qquad \text{and} \qquad b = \frac{a_1 b_2 - b_1 a_2}{a_1^2 + b_1^2}$$

and that the denominator is a real number.

For example, if $z_1 = 2 + 3i$, then

$$\frac{1}{z} = \frac{1}{2 + 3i} = \frac{1}{2 + 3i} \cdot \frac{2 - 3i}{2 - 3i} = \frac{2 - 3i}{4 + 9} = \frac{2}{13} - \frac{3}{13} i$$

and if $z_1 = -2 + 3i$ and $z_2 = 1 - i$, then

$$\frac{z_1}{z_2} = \frac{-2 + 3i}{1 - i} \cdot \frac{1 + i}{1 + i} = \frac{-5 + i}{1 + 1} = \frac{-5}{2} + \frac{1}{2} i$$

EXAMPLES

1 Show that if z is a complex number, then that
a) $z\bar{z} \in R$.
b) $z + \bar{z}$ is a real number.

SOLUTION. Let $z = a + bi$; then $\bar{z} = a - bi$, so that
a) $z\bar{z} = (a + bi)(a - bi)$
$= (a^2 + b^2) + (ab - ab)i$
$= (a^2 + b^2) + (0 \cdot i)$
$= a^2 + b^2 \in R$ since a and b are real numbers.

Since $a^2 \geqq 0$ and $b^2 \geqq 0$ for any $a, b \in R$, we have $a^2 + b^2 \geqq 0$.

b) $z + \bar{z} = (a + a) + (b - b)i$
$= 2a \in R$

2 Let $z = 3 + 5i$. Find $\bar{z}, \bar{z} + z$, and $(z - \bar{z})/i$.

SOLUTION. Since $z = 3 + 5i$, $\bar{z} = 3 - 5i$, so that

$$\bar{z} + z = (3 - 5i) + (3 + 5i) = 6$$

and

$$\frac{z - \bar{z}}{i} = \frac{(3 + 5i) - (3 - 5i)}{i} = \frac{10i}{i} = 10$$

3 Write each of the following quotients in the form $a + bi$.
a) $1/(3 - 2i)$ b) $(1 + i)/(1 - i)$ c) $(6 + 9i)/(1 - 2i)$

SOLUTION

a) $\dfrac{1}{3 - 2i} = \dfrac{1}{3 - 2i} \cdot \dfrac{3 + 2i}{3 + 2i} = \dfrac{3 + 2i}{9 + 4} = \dfrac{3}{13} + \dfrac{2}{13} i$

b) $\dfrac{1+i}{1-i} = \dfrac{1+i}{1-i} \cdot \dfrac{1+i}{1+i} = \dfrac{(1+i)^2}{2} = \dfrac{2i}{2} = i$

c) $\dfrac{6+9i}{1-2i} = \dfrac{6+9i}{1-2i} \cdot \dfrac{1+2i}{1+2i} = \dfrac{-12+21i}{5} = -\dfrac{12}{5} + \dfrac{21}{5}i$

4   If $z \neq 0$, show that $1/z$ is the multiplicative inverse of z; that is, $z \cdot z^{-1} = 1$, where $z^{-1} = 1/z$.

SOLUTION.   If $z = a+bi \neq 0$, then

$$\frac{1}{z} = \frac{1}{a+bi} \cdot \frac{a-bi}{a-bi} = \frac{a-bi}{a^2+b^2}$$

so that

$$z \cdot \frac{1}{z} = (a+bi) \cdot \frac{a-bi}{a^2+b^2} = \frac{a^2+b^2}{a^2+b^2} = 1$$

PROBLEM SET 1

1   Perform the indicated operations and write the answer in the form $a+bi$.

a) $(2+3i)+(4+5i)$ b) $(-1+2i)+(3+5i)$

c) $(2+i)-(4+3i)$ d) $(-4+2i)-(3+2i)$

e) $(6+i)(5-3i)$ f) $(3-4i)(7-3i)$

g) $(-7+2i)(-7-2i)$ h) $(-2+3i)(-2-3i)$

i) $i^{27}+i^5-i^9$ j) $i^{18}-3i^7$

k) $i^{14}-3i^5$ l) $i^{102}+5i^{51}$

m) $(3-i)^2$ n) $(1+3i)^2$

o) $(2-3i)^4$ p) $(1-\sqrt{2i})^3$

q) $\dfrac{6}{7i}$ r) $-\dfrac{3}{5i^3}$

s) $4i^{-13}$ t) $\dfrac{5+8i}{3i}$

u) $\dfrac{7-3i}{5i}$ v) $\dfrac{3-i}{2+5i}$

w) $\dfrac{3+5i}{4-3i}$ x) $\dfrac{7+4i}{3+5i}$

y) $\dfrac{2i}{(1+i)^4}$ z) $\dfrac{2i^4}{(6-i)^2}$

2   Find the real numbers x and y that satisfy each of the following equations.

a) $x-3+2yi = 8i$

b) $3x-y+ix-2iy = 6-3i$

c) $3x+2yi = 6+11i$

3 Find $\bar{z}$, the real part of z, the imaginary part of z, and $1/z$ for each of the following numbers.

a) $z = 2 + \sqrt{3}\,i$ b) $z = 1 - \frac{1}{2}i$

c) $z = (3 - 2i)^2$ d) $z = 2i$

4 Prove each of the following statements.

a) If $\bar{z} = z$, then z is a real number.

b) $z + \bar{z} = 0$ if and only if the real part of z is 0.

c) $\overline{z_1 + z_2} = \bar{z}_1 + \bar{z}_2$

d) $\overline{z_1 z_2} = \bar{z}_1 \bar{z}_2$

e) $\overline{z_1/z_2} = \bar{z}_1/\bar{z}_2, \; z_2 \neq 0$

f) $\bar{\bar{z}} = z$

5 Let z_1 and z_2 be complex numbers. Show that

$$\mathrm{Re}\left(\frac{z_1}{z_1 + z_2}\right) + \mathrm{Re}\left(\frac{z_2}{z_1 + z_2}\right) = 1$$

where $\mathrm{Re}\,z$ represents the real part of z.

6 Assume the properties of addition and multiplication on R that are listed in Appendix B.

a) Prove that C is closed under addition and multiplication.

b) Prove the commutativity of addition and multiplication on C.

c) Prove the associativity of addition and multiplication on C.

d) Prove the distributive properties on C.

e) Prove the identity properties of addition and multiplication on C.

7 Let f be a function whose domain is the set of complex numbers and whose rule of formation is given by $f(z) = z^2 + 5z + i$. Find

a) $f(2 - i)$ b) $f(2 + i)$

c) $f(1 - i)$ d) $f(1 + i)$

8 We are given the set $T = \{1, -1, i, -i\}$.

a) Construct a *multiplication* table as shown and fill in the blanks.

$\cdot$	1	-1	i	$-i$
1				
-1				
i				
$-i$				

b) Is the set T closed under multiplication?

c) Is multiplication on T commutative?

d) Is multiplication on T associative?

e) What is the identity element?

f) What are the reciprocals of each of the elements of T? (Two elements are reciprocals of each other if their product equals 1.)

3 Geometric Representation of Complex Numbers

Each ordered pair of real numbers (a, b) can be associated with the complex number $z = a + bi$, and each complex number $z = a + bi$ can be associated with the ordered pair of real numbers (a, b). Because of this one-to-one correspondence between the set of complex numbers and the set of ordered pairs of real numbers, we use the points in the plane associated with the ordered pairs of real numbers to represent the complex numbers. For example, the ordered pairs $(2, -3)$, $(5, 2)$, and (e, π) are used to represent complex numbers $z_1 = 2 - 3i$, $z_2 = 5 + 2i$, and $z_3 = e + \pi i$, respectively, as points in the plane (Figure 1). The plane on which the complex numbers are represented is called the *complex plane*; the horizontal axis (x axis) is called the *real axis*, and the vertical axis (y axis) is called the *imaginary axis* (Figure 1).

Figure 1

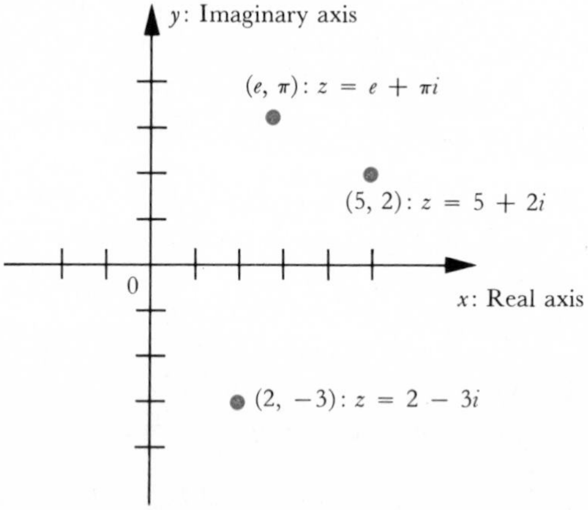

Thus, complex numbers of the form $z = bi$ are represented by points of the form $(0, b)$, that is, by points on the imaginary axis, whereas complex numbers of the form $z = a$ are represented by points of the form $(a, 0)$, that is, by points on the real axis.

3.1 The Modulus of a Complex Number

If $z = a + bi$, then the *absolute value* or *length* or *modulus* of z, written $|z|$, is defined by

$$|z| = \sqrt{a^2 + b^2}$$

The modulus of $z = a + bi$ is the distance between the origin and the point (a, b) (Figure 2). Notice that $|z| = \sqrt{z \cdot \bar{z}}$ (see Problem 4).

Figure 2

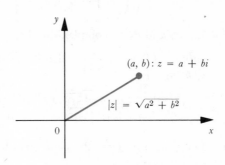

EXAMPLES

1   Let $z = 1 + \sqrt{3}\, i$. Find $|z|$ and show that $z\bar{z} = |z|^2$.

SOLUTION

$$|z| = \sqrt{1^2 + (\sqrt{3})^2} = \sqrt{1+3} = \sqrt{4} = 2$$

Also

$$z\bar{z} = \left(1 + \sqrt{3}\, i\right)\left(1 - \sqrt{3}\, i\right) = 1^2 + (\sqrt{3})^2 = 4 = |z|^2 \qquad \text{(Figure 3)}$$

Figure 3

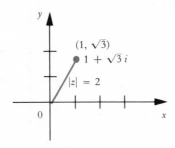

2   Let $z_1 = 4 + 3i$ and $z_2 = \sqrt{3} - i$. Find

a) $|z_1|$ b) $|z_2|$ c) $|z_1 z_2|$ d) $\left|\dfrac{z_1}{z_2}\right|$

SOLUTION

a) $|z_1| = \sqrt{16+9} = \sqrt{25} = 5$

b) $|z_2| = \sqrt{(\sqrt{3})^2 + (-1)^2} = \sqrt{4} = 2$

c) $|z_1 z_2| = \left|(4+3i)(\sqrt{3}-i)\right|$

$= \left|(4\sqrt{3}+3) + (3\sqrt{3}-4)i\right|$

$= \sqrt{(4\sqrt{3}+3)^2 + (3\sqrt{3}-4)^2} = 10$

d) $\left|\dfrac{z_1}{z_2}\right| = \left|\dfrac{4+3i}{\sqrt{3}-i}\right| = \left|\dfrac{(4+3i)(\sqrt{3}+i)}{3+1}\right| = \left|\dfrac{(4\sqrt{3}-3)+(3\sqrt{3}+4)i}{4}\right|$

$= \sqrt{\left(\dfrac{4\sqrt{3}-3}{4}\right)^2 + \left(\dfrac{3\sqrt{3}+4}{4}\right)^2} = \dfrac{10}{4} = \dfrac{5}{2}$

3   Let z_1 and z_2 be complex numbers; then $|z_1 z_2| = |z_1||z_2|$.

PROOF.   (This is a generalization of the result in Example 2 above, parts
a, b, and c.)

$$|z_1 z_2| = \sqrt{(z_1 z_2)(\overline{z_1}\,\overline{z_2})} = \sqrt{(z_1 \overline{z_1})(z_2 \overline{z_2})} \qquad \text{(Problem 4)}$$
$$= \sqrt{|z_1|^2 |z_2|^2} = |z_1||z_2|$$

4   Let $z = x+iy$. Describe geometrically the set of all complex numbers z
such that $|z-1| = 1$.

SOLUTION

$$|z-1| = |(x+iy)-1| = |(x-1)+iy|$$
$$= \sqrt{(x-1)^2 + y^2} = 1$$

so that

$$(x-1)^2 + y^2 = 1$$

In other words, the distance between any point (x, y) and the point $(1, 0)$
is always 1. Thus the points (x, y) lie on a circle with center $(1, 0)$ and
radius 1 (Figure 4).

Figure 4

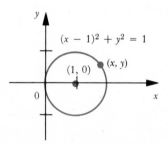

4   Polar Coordinates

We have seen that points in the plane can be referenced or associated with
pairs of real numbers by using the Cartesian coordinate system. Before
we proceed further with the discussion of complex numbers we need
another way of associating pairs of numbers with points in the plane based
upon a "grid" composed of concentric circles and rays emanating from
the common center of the circles (Figure 1).

Figure 1

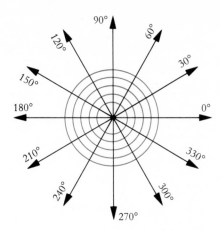

Such a system is called a *polar coordinate system*. The frame of reference for this coordinate system consists of a fixed point O, called the *pole*, and a fixed ray called the *polar axis*. The position of a point P is uniquely determined by r and θ, where θ is any angle in standard position having the ray $\overrightarrow{OA}$ as its initial side and a ray on line OP as its terminal side, and r is the directed distance along the terminal side of θ between P and the pole. The pair (r, θ) is called the *polar coordinates* of P (**Figure 2**).

Figure 2

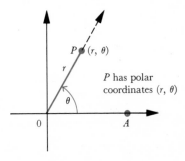

P has polar coordinates (r, θ)

If the angle θ is measured in degrees, then (r, θ) clearly indicates that the ordered pair represents polar coordinates. On the other hand, if θ is given in radians, then the ordered pair of real numbers (r, θ) is indistinguishable from the notation used in Cartesian coordinates. For example, clearly $(2, 30°)$ represents polar coordinates, whereas $(2, 3)$ could be rectangular (Figure 3a) or 3 could be the radian measure of an angle (Figure 3b). In specifying polar coordinates, we use the phrase "plot the polar point (r, θ)" as short for "plot the point whose polar coordinates are (r, θ)." If the context does not make clear that the coordinates are polar, we will assume that they are rectangular coordinates.

Figure 3

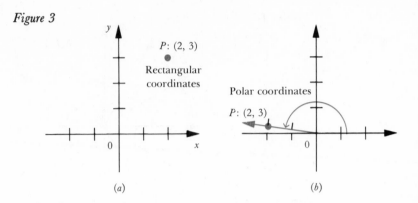

(a)   (b)

It is important to observe that a polar coordinate system does not establish a one-to-one correspondence between points in a plane and ordered pairs (r, θ). In fact, each point can be represented by infinitely many ordered pairs of numbers. For example, $(2, 30°)$, $(2, 390°)$, and $(2, -330°)$ each represent the same point (Figure 4).

Figure 4

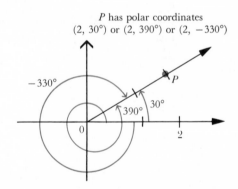

P has polar coordinates
$(2, 30°)$ or $(2, 390°)$ or $(2, -330°)$

Also, r need not be positive. If $r < 0$, the point (r, θ) is determined by plotting $(|r|, \theta + 180°)$ or $(|r|, \theta + \pi)$, depending on whether θ is measured in degrees or radians. For example, $(-2, 30°)$ is the same as $(2, 210°)$ (Figure 5).

Figure 5

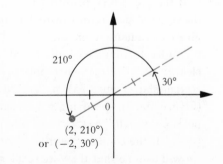

$(2, 210°)$
or $(-2, 30°)$

EXAMPLES

1 Locate the points which have the following polar coordinates.

a) $(3, 70°)$ b) $(0, 0)$ c) $(7, 7\pi/5)$

d) $(3, 100°)$ e) (π, π) f) $(3, 5)$

g) $(5, 0)$

SOLUTION. The points are shown in Figure 6.

Figure 6

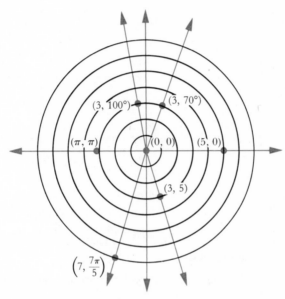

2 Locate polar point $(4, 45°)$, and then give five other polar representations of the same point.

SOLUTION. $(4, -315°)$, $(4, 405°)$, $(-4, -135°)$, $(-4, 225°)$, and $(4, \pi/4)$ are other polar representations of the same point (Figure 7).

Figure 7

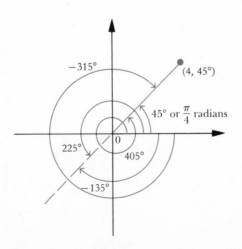

4.1 Conversion of Coordinates

If (r, θ) is a polar representation of a point P, the trigonometric functions can be used to find the rectangular coordinates (x, y) of the same point (Figure 8).

Figure 8

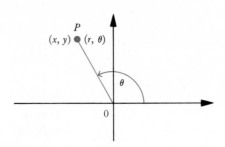

We know from trigonometry that $\sin \theta = y/r$ and $\cos \theta = x/r$; hence

$$y = r \sin \theta \qquad \text{and} \qquad x = r \cos \theta$$

These formulas are often referred to as the *transformation* or *conversion* formulas; they enable us to convert from *polar to rectangular coordinates.*

EXAMPLE

Convert the given polar coordinates to rectangular coordinates.

a) $(3, 60°)$ b) $(-2, 180°)$ c) $(4, -150°)$ d) (π, π)

SOLUTION.   (See Figure 9.)

a) $x = r \cos \theta = 3 \cos 60° = \frac{3}{2}$ and $y = r \sin \theta = 3 \sin 60° = 3\sqrt{3}/2$. Hence the rectangular coordinates are $\left(\frac{3}{2}, 3\sqrt{3}/2\right)$.

Figure 9

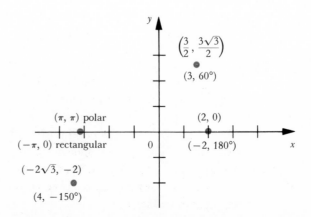

b) $x = r \cos \theta = -2 \cos 180° = 2$ and $y = r \sin \theta = -2 \sin 180° = 0$. Hence the rectangular coordinates are $(2, 0)$.

c) $x = r \cos \theta = 4 \cos(-150°) = 4(-\sqrt{3}/2) = -2\sqrt{3}$ and $y = r \sin \theta = 4 \sin(-150°) = 4(-\frac{1}{2}) = -2$. Hence the rectangular coordinates are $(-2\sqrt{3}, -2)$.

d) $x = r \cos \theta = \pi \cos \pi = -\pi$ and $y = r \sin \theta = \pi \sin \pi = 0$. Hence the rectangular coordinates are $(-\pi, 0)$.

Now, assume that the rectangular coordinates of a point P are given by (x, y). Then the tangent function can be used to find polar coordinates (r, θ) of the same point. More precisely, $r = \sqrt{x^2 + y^2}$ and $\tan \theta = y/x$ can be used to transform the *rectangular coordinates to polar coordinates* (Figure 10). Notice that a point can be represented by a unique set of polar coordinates for which $r > 0$ and $0 \le \theta < 2\pi$. For example, the origin is represented as $(0, 0)$, even though $(0, \theta)$ would do for any θ.

Figure 10

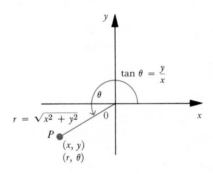

$$\tan \theta = \frac{y}{x}$$

$$r = \sqrt{x^2 + y^2}$$

$$P$$
$$(x, y)$$
$$(r, \theta)$$

EXAMPLE

Convert the given rectangular coordinates to polar coordinates.

a) $(-1, 1)$ b) $(3, -3/\sqrt{3})$ c) $(-3, 4)$

SOLUTION. (See Figure 11.)

a) $r = \sqrt{x^2 + y^2} = \sqrt{(-1)^2 + 1} = \sqrt{2}$, and, since the point is in quadrant II, $\tan \theta = -1$ implies that $\theta = 135°$; hence one pair of polar coordinates is $(\sqrt{2}, 135°)$. Another pair is $(\sqrt{2}, -225°)$.

b) $r = \sqrt{3^2 + (-3/\sqrt{3})^2} = \sqrt{9 + 3} = \sqrt{12} = 2\sqrt{3}$ and $\tan \theta = -\sqrt{3}/3$, so that $\theta = -30°$, since the point is in quadrant IV. Hence one possible pair of polar coordinates is given by $(2\sqrt{3}, -30°)$.

c) $r = \sqrt{(-3)^2 + 4^2} = 5$ and $\tan \theta = -\frac{4}{3} = -1.3333$. From Table IV we find that θ is approximately equal to $127°$, so $(5, 127°)$ is one pair of polar coordinates.

Figure 11

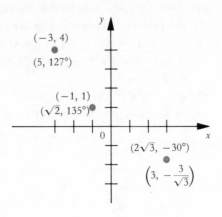

4.2   Polar Form of Complex Numbers

A complex number $z = x + iy$ can be written in the form

$$z = r \cos \theta + ir \sin \theta = r(\cos \theta + i \sin \theta)$$

where $x = r \cos \theta$, $y = r \sin \theta$, and r is the modulus of z (Figure 12). $r(\cos \theta + i \sin \theta)$ is called the *polar form* or *trigonometric form* of the complex number z. The number θ in this representation is called an *argument* of the complex number z. ("Argument" means the polar angle associated with z and has nothing to do with its meaning in English.) Notice that θ is not unique, since $r(\cos \theta + i \sin \theta) = r(\cos \theta_1 + i \sin \theta_1)$ holds whenever $\theta - \theta_1$ is an integral multiple of 2π. Hence *two complex numbers are equal if and only if their moduli are equal and their arguments differ by a multiple of 2π.* Thus we can write the complex number z in the form

$$z = x + iy = r[\cos(\theta + 2\pi k) + i \sin(\theta + 2\pi k)], \qquad k \in I.$$

Figure 12

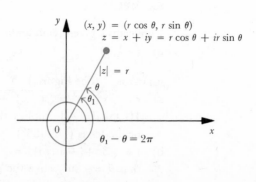

For example, the complex number $z = 1 + i$ can be represented in the polar form as $z = \sqrt{2}[\cos(\pi/4) + i \sin(\pi/4)]$, or as $z = \sqrt{2}[\cos(9\pi/4) + i \sin(9\pi/4)]$, since $r = \sqrt{1^2 + 1^2} = \sqrt{2}$ and since both $\pi/4$

and $9\pi/4$ have the same terminal sides and both satisfy $\tan \theta = 1$. Also, $z = 1+i$ can be represented in the polar form by

$$z = \sqrt{2}\left[\cos\left(\frac{-7\pi}{4}\right) + i \sin\left(\frac{-7\pi}{4}\right)\right]$$

$$= \sqrt{2}\left(\cos\frac{7\pi}{4} - i \sin\frac{7\pi}{4}\right) \quad \text{(Figure 13)}$$

Figure 13

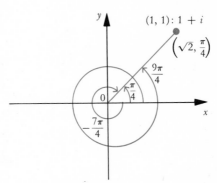

Notice that $(\sqrt{2}, \pi/4)$, $(\sqrt{2}, 9\pi/4)$, and $(\sqrt{2}, -7\pi/4)$ are possible polar coordinates of $(1, 1)$.

EXAMPLES

1 Change each of the following complex numbers from polar form to rectangular form.
   a) $z = 2[\cos(\pi/3) + i \sin(\pi/3)]$
   b) $z = 4[\cos(-\pi/6) + i \sin(-\pi/6)]$
   c) $z = 8[\cos(\pi/2) + i \sin(\pi/2)]$

SOLUTION

a) $z = x + iy = 2 \cos\frac{\pi}{3} + i2 \sin\frac{\pi}{3} = 2 \cdot \frac{1}{2} + i \cdot 2\frac{\sqrt{3}}{2} = 1 + i\sqrt{3}$

b) $z = x + iy = 4 \cos\left(-\frac{\pi}{6}\right) + i4 \sin\left(-\frac{\pi}{6}\right) = 4\frac{\sqrt{3}}{2} + i \cdot 4\left(-\frac{1}{2}\right)$

$= 2\sqrt{3} - 2i$

c) $z = x + iy = 8 \cos\frac{\pi}{2} + 8i \sin\frac{\pi}{2} = 8 \cdot 0 + i8 \cdot 1 = 8i$

2 Change each of the following complex numbers from rectangular form to polar form.
   a) $z = 2$ b) $z = 2i$
   c) $z = -\sqrt{3} - i$

SOLUTION.   (See Figure 14.)

a) $2+0i = r(\cos\theta + i\sin\theta)$, where $r = \sqrt{2^2+0^2} = 2$ and an argument is $\theta = 0$, so that $z = 2(\cos 0 + i\sin 0)$.

b) $z = 2i = 0 + 2i = r(\cos\theta + i\sin\theta)$, where $r = \sqrt{0+2^2} = 2$. One measure of θ is $\pi/2$. Hence $z = 2[\cos(\pi/2) + i\sin(\pi/2)]$.

c) $z = r(\cos\theta + i\sin\theta)$, where $r = \sqrt{(-\sqrt{3})^2 + (-1)^2} = 2$, and θ satisfies $\tan\theta = 1/\sqrt{3}$ with θ in quadrant III, so that one value of θ is $\theta = 7\pi/6$. Hence $z = 2[\cos(7\pi/6) + i\sin(7\pi/6)]$.

Figure 14

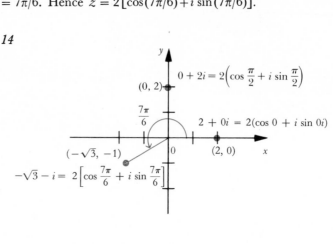

4.3   Multiplication and Division of Complex Numbers in Polar Form

Consider the complex numbers $z_1 = 1+i$ and $z_2 = 1+\sqrt{3}i$. The polar representations of z_1 and z_2 can be given by

$$z_1 = \sqrt{2}\left(\cos\frac{\pi}{4} + i\sin\frac{\pi}{4}\right) \quad \text{and} \quad z_2 = 2\left(\cos\frac{\pi}{3} + i\sin\frac{\pi}{3}\right)$$

The modulus of z_1 is $r_1 = \sqrt{2}$ and an argument is $\pi/4$; and the modulus of z_2 is $r_2 = 2$ and an argument is $\pi/3$. By multiplication, we find that

$$z_1 z_2 = 2\sqrt{2}\left[\left(\cos\frac{\pi}{4}\cos\frac{\pi}{3} - \sin\frac{\pi}{4}\sin\frac{\pi}{3}\right) + i\left(\sin\frac{\pi}{4}\cos\frac{\pi}{3} + \sin\frac{\pi}{3}\cos\frac{\pi}{4}\right)\right]$$

This product can be simplified, by using the trigonometric identities, as

$$2\sqrt{2}\left[\cos\left(\frac{\pi}{4}+\frac{\pi}{3}\right) + i\sin\left(\frac{\pi}{4}+\frac{\pi}{3}\right)\right] = 2\sqrt{2}\left(\cos\frac{7\pi}{12} + i\sin\frac{7\pi}{12}\right)$$

(Figure 15)

Notice in this example that the modulus of the product $z_1 z_2$ satisfies

Figure 15

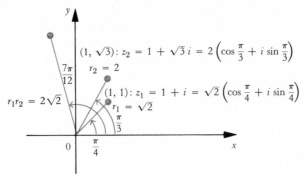

the equation $|z_1 z_2| = |z_1||z_2| = r_1 r_2 = 2\sqrt{2}$ and an argument of $z_1 z_2$ is $\pi/4 + \pi/3 = 7\pi/12$. The following theorem generalizes the results of this example.

THEOREM 1

Suppose that z_1 and z_2 are complex numbers in polar form such that

$$z_1 = r_1 (\cos\theta_1 + i\sin\theta_1)$$

and

$$z_2 = r_2 (\cos\theta_2 + i\sin\theta_2)$$

Then

$$z_1 z_2 = r_1 r_2 [\cos(\theta_1 + \theta_2) + i\sin(\theta_1 + \theta_2)]$$

(This theorem states that the product of two complex numbers is the complex number whose modulus is the product of the moduli of the two complex numbers, and with an argument that is the sum of the arguments of the two complex numbers.)

PROOF

$$z_1 z_2 = r_1 (\cos\theta_1 + i\sin\theta_1) r_2 (\cos\theta_2 + i\sin\theta_2)$$
$$= r_1 r_2 [(\cos\theta_1 \cos\theta_2 - \sin\theta_1 \sin\theta_2) + i(\cos\theta_1 \sin\theta_2 + \cos\theta_2 \sin\theta_1)]$$

Using the trigonometric identities for $\cos(\theta_1 + \theta_2)$ and $\sin(\theta_1 + \theta_2)$, we can write this latter result as

$$z_1 z_2 = r_1 r_2 [\cos(\theta_1 + \theta_2) + i\sin(\theta_1 + \theta_2)]$$

EXAMPLES

1 Let $z_1 = 7(\cos 25° + i\sin 25°)$ and $z_2 = 3(\cos 35° + i\sin 35°)$. Find
a) $z_1 z_2$ in rectangular form
b) z_1^2 in polar form

SOLUTION

a) The modulus of $z_1 z_2$ is $r_1 r_2 = 7(3) = 21$ and an argument of $z_1 z_2$ is $\theta_1 + \theta_2 = 25° + 35° = 60°$. Hence

$$z_1 z_2 = 21 (\cos 60° + i \sin 60°)$$

$$= 21 \left(\frac{1}{2} + i \frac{\sqrt{3}}{2} \right)$$

$$= \frac{21}{2} + i \frac{21\sqrt{3}}{2}$$

b) The modulus of z_1^2 is $r_1^2 = 49$, and an argument is $2\theta_1 = 2(25°) = 50°$. Hence $z_1^2 = 49 (\cos 50° + i \sin 50°)$.

2   Convert $z_1 = 1 + i$ and $z_2 = 2 - 2\sqrt{3} i$ to polar form, and then compute $z_1 z_2$ in polar form and in rectangular form.

SOLUTION.   (See Figure 16.) The modulus of z_1 is $r_1 = \sqrt{1+1} = \sqrt{2}$ and the modulus of z_2 is $r_2 = \sqrt{4+12} = 4$. An argument of z_1 is $\theta_1 = \pi/4$, and an argument of z_2 is $\theta_2 = -\pi/3$. Hence

$$z_1 z_2 = 4\sqrt{2} \left[\cos \left[\left(\frac{\pi}{4} \right) + \left(-\frac{\pi}{3} \right) \right] + i \sin \left[\left(\frac{\pi}{4} \right) + \left(-\frac{\pi}{3} \right) \right] \right]$$

$$= 4\sqrt{2} \left[\cos \left(\frac{-\pi}{12} \right) + i \sin \left(\frac{-\pi}{12} \right) \right]$$

so that the rectangular form is

$$z_1 z_2 = 4\sqrt{2} [0.9659 - i(0.2588)]$$

Figure 16

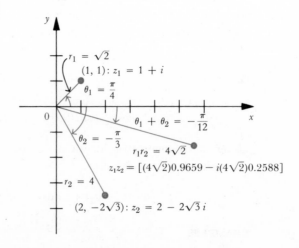

To find the quotient z_1/z_2 of two complex numbers $z_1 = 6 + 6\sqrt{3} i$ and $z_2 = 2\sqrt{2} + 2\sqrt{2} i$, we can first locate z_1 and z_2 as shown in Figure 17.

Two polar representations of z_1 and z_2 are $z_1 = 12[\cos(\pi/3) + i \sin(\pi/3)]$ and $z_2 = 4[\cos(\pi/4) + i\sin(\pi/4)]$. We are looking for a complex number $z_3 = r_3(\cos\theta_3 + i \sin\theta_3)$ such that $z_1 = z_2 z_3$. Hence, if we use Theorem 1, we see that r_3 and θ_3 should be chosen so that

$$r_1(\cos\theta_1 + i \sin\theta_1) = r_2 r_3[\cos(\theta_2 + \theta_3) + i \sin(\theta_2 + \theta_3)]$$

That is,

$$12\left(\cos\frac{\pi}{3} + i \sin\frac{\pi}{3}\right) = 4r_3\left[\cos\left(\frac{\pi}{4} + \theta_3\right) + i \sin\left(\frac{\pi}{4} + \theta_3\right)\right]$$

But two complex numbers are equal if and only if their moduli are equal and their arguments differ by a multiple of 2π. Thus

$$12 = 4r_3 \quad \text{and} \quad \frac{\pi}{3} = \frac{\pi}{4} + \theta_3 + 2\pi k \quad k \in I$$

That is,

$$r_3 = 3 \quad \text{and} \quad \theta_3 = \frac{\pi}{3} - \frac{\pi}{4} - 2\pi k \quad k \in I$$

Since $\cos(\pi/3 - \pi/4 - 2\pi k) = \cos(\pi/3 - \pi/4)$ and $\sin(\pi/3 - \pi/4 - 2\pi k) = \sin(\pi/3 - \pi/4)$, the polar form of the quotient z_1/z_2 can be written as

$$\frac{z_1}{z_2} = \frac{12}{4}\left[\cos\left(\frac{\pi}{3} - \frac{\pi}{4}\right) + i \sin\left(\frac{\pi}{3} - \frac{\pi}{4}\right)\right]$$

$$= 3\left(\cos\frac{\pi}{12} + i \sin\frac{\pi}{12}\right) \qquad \text{(Figure 17)}$$

Figure 17

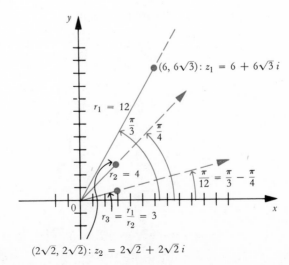

$(2\sqrt{2}, 2\sqrt{2})$: $z_2 = 2\sqrt{2} + 2\sqrt{2}\,i$

In general, we have the following result.

THEOREM 2

Let $z_1 = r_1(\cos \theta_1 + i \sin \theta_1)$ and $z_2 = r_2(\cos \theta_2 + i \sin \theta_2)$; then

$$\frac{z_1}{z_2} = \frac{r_1}{r_2}[\cos(\theta_1 - \theta_2) + i \sin(\theta_1 - \theta_2)] \qquad z_2 \neq 0$$

The proof of this theorem is left as an exercise for the student (see Problem 6).

EXAMPLES

1   Let $z_1 = 4(\cos 80° + i \sin 80°)$ and $z_2 = 2(\cos 50° + i \sin 50°)$. Find z_1/z_2 and express the result in rectangular form.

SOLUTION.   Since $z_2 \neq 0$, we use Theorem 2 to get

$$\frac{z_1}{z_2} = \frac{4}{2}[\cos(80° - 50°) + i \sin(80° - 50°)]$$

$$= 2(\cos 30° + i \sin 30°)$$

$$= 2\left(\frac{\sqrt{3}}{2} + i\frac{1}{2}\right) = \sqrt{3} + i$$

2   Express $z_1 = -1 + i$ and $z_2 = -4i$ in polar form, and then compute z_1/z_2 in polar form. Write your answer in rectangular form also.

SOLUTION.   (See Figure 18.) The modulus of z_1 is $r_1 = \sqrt{2}$ and an argument is $3\pi/4$. (Why?) The modulus of z_2 is $r_2 = 4$, and an argument is $3\pi/2$. Using Theorem 2, we have

$$\frac{z_1}{z_2} = \frac{\sqrt{2}}{4}\left[\cos\left(\frac{3\pi}{4} - \frac{3\pi}{2}\right) + i \sin\left(\frac{3\pi}{4} - \frac{3\pi}{2}\right)\right]$$

$$= \frac{\sqrt{2}}{4}\left[\cos\left(-\frac{3\pi}{4}\right) + i \sin\left(-\frac{3\pi}{4}\right)\right]$$

$$= \frac{\sqrt{2}}{4}\left(\cos\frac{3\pi}{4} - i \sin\frac{3\pi}{4}\right)$$

$$= \frac{\sqrt{2}}{4}\left(-\frac{1}{\sqrt{2}} - i\frac{1}{\sqrt{2}}\right)$$

$$= -\frac{1}{4} - \frac{1}{4}i$$

Figure 18

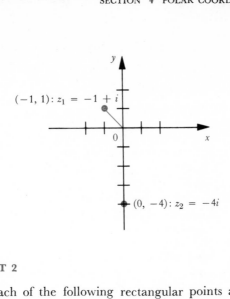

PROBLEM SET 2

1 a) Locate each of the following rectangular points and then convert to polar coordinates.

  i $(-1, \sqrt{3})$ ii $(-3, 0)$
  iii $(4, 3)$ iv $(-6, 6\sqrt{3})$
  v $(5, 5)$ vi $(0, -2)$
  vii $(-3, 3\sqrt{3})$ viii $(2\sqrt{3}, -2)$
  ix $(-5, -5)$ x $(-2, 5)$

  b) Locate the following polar points and convert the representation to rectangular form.

  i $(6, 30°)$ ii $(10, \pi/3)$
  iii $(7, 120°)$ iv $(4, -\pi/6)$
  v $(4, 90°)$ vi $(8, 45°)$
  vii $(2\pi, -\pi/6)$ viii $(5, \pi/2)$
  ix $(4, 210°)$ x $(4, 330°)$

2 Let $z = x + iy$. Describe the following sets geometrically.

  a) $\{z \,|\, |z| = 1\}$ b) $\{z \,|\, |z| = 2\}$
  c) $\{z \,|\, |z - i| = 1\}$ d) $\{z \,|\, |z - 1| = |z - 2|\}$

3 For each of the following complex numbers, find the modulus and an argument. Express these numbers in polar form, and then represent them graphically.

  a) $-1 - i$ b) 7

  c) $-2i$ d) $\dfrac{\sqrt{3}}{2} + \dfrac{1}{2}i$

  e) $(-\sqrt{3} - i)$ f) $3 + 4i$
  g) $-\tfrac{1}{2} + (\sqrt{3}/2)i$ h) $(1 - i)^2$
  i) $(1 - \sqrt{3}i)^3$

4 Prove that $|z| = \sqrt{z\bar{z}}$.

5   Represent each of the following complex numbers graphically, and then express the number in the rectangular form $a+bi$.

a) $z = 2(\cos 10° + i \sin 10°)$

b) $z = 3[\cos(-75°) + i \sin(-75°)]$

c) $z = 4(\cos 0° + i \sin 0°)$

d) $z = 2[\cos(\pi/4) + i \sin(\pi/4)]$

e) $z = 10[\cos(3\pi/4) + i \sin(3\pi/4)]$

f) $z = 2[\cos(\pi/2) + i \sin(\pi/2)]$

g) $z = 7[\cos(-3\pi/2) + i \sin(-3\pi/2)]$

6   Prove Theorem 2 by generalizing the example preceding the statement of Theorem 2.

7   Find $z_1 z_2$ and z_1/z_2 for each of the following numbers, and express the answer in both polar and rectangular form.

a) $z_1 = 5(\cos 170° + i \sin 170°)$ and $z_2 = (\cos 55° + i \sin 55°)$

b) $z_1 = 2(\cos 50° + i \sin 50°)$ and $z_2 = 3(\cos 40° + i \sin 40°)$

c) $z_1 = 4[\cos(3\pi/4) + i \sin(3\pi/4)]$ and $z_2 = 2(\cos \pi + i \sin \pi)$

d) $z_1 = 5(\cos 30° + i \sin 30°)$ and $z_2 = 6(\cos 240° + i \sin 240°)$

8   Let $z_1 = \cos 30° + i \sin 30°$ and $z_2 = \cos 60° + i \sin 60°$. Find $z_1 z_2$ and z_2/z_1 and represent them graphically.

9   Convert $z_1 = -1 - i$ and $z_2 = -4 + 4\sqrt{3}\,i$ to polar form, and then compute each of the following values in polar form. Convert the answers to rectangular form.

a) $z_1 z_2$

b) z_1/z_2

c) $(z_1 z_2)^2$

5   Powers and Roots of Complex Numbers

If $z = r(\cos\theta + i \sin\theta)$, then

$$z^2 = r \cdot r[\cos(\theta+\theta) + i \sin(\theta+\theta)] = r^2(\cos 2\theta + i \sin 2\theta).$$

But, $z^3 = z^2 \cdot z$, so that

$$z^3 = r^2 \cdot r[\cos(2\theta+\theta) + i \sin(2\theta+\theta)]$$
$$= r^3(\cos 3\theta + i \sin 3\theta).$$

If we repeat the process once more, we get

$$z^4 = z^3 \cdot z = r^4(\cos 4\theta + i \sin 4\theta).$$

This scheme for repeated multiplication of complex numbers in polar form is generalized in the following theorem.

THEOREM 1 DEMOIVRE'S THEOREM

Let $z = r(\cos\theta + i\sin\theta)$. Then $z^n = r^n(\cos n\theta + i\sin n\theta)$ for n a positive integer.

PROOF.   We use mathematical induction to prove this theorem. Let S_n be the statement

$$[r(\cos\theta + i\sin\theta)]^n = r^n(\cos n\theta + i\sin n\theta)$$

i   The statement S_1 is true, since

$$r(\cos\theta + i\sin\theta) = r(\cos\theta + i\sin\theta)$$

ii   We must show that if S_n is true, then the statement S_{n+1} is also true. If S_n is true, that is, if $[r(\cos\theta + i\sin\theta)]^n = r^n(\cos n\theta + i\sin n\theta)$, then, after multiplying both sides by $r(\cos\theta + i\sin\theta)$, we get

$$\begin{aligned}
[r(\cos\theta + i\sin\theta)]^{n+1} &= [r(\cos\theta + i\sin\theta)]^n[r(\cos\theta + i\sin\theta)] \\
&= [r^n(\cos n\theta + i\sin n\theta)][r(\cos\theta + i\sin\theta)] \\
&= r^{n+1}[\cos(n+1)\theta + i\sin(n+1)\theta]
\end{aligned}$$

which shows that S_{n+1} is true, so that S_n is true for any positive integer n; that is,

$$[r(\cos\theta + i\sin\theta)]^n = r^n(\cos n\theta + i\sin n\theta)$$

EXAMPLES

1   Use DeMoivre's theorem to determine each of the following values.
a) $[3(\cos 60° + i\sin 60°)]^4$ in rectangular form
b) $(1+i)^{20}$ in rectangular form

SOLUTION

a)   Using DeMoivre's theorem, we have

$$[3(\cos 60° + i\sin 60°)]^4 = 3^4(\cos 240° + i\sin 240°)$$

$$= 81\left(-\frac{1}{2} - \frac{i\sqrt{3}}{2}\right) = -\frac{81}{2} - \frac{81\sqrt{3}}{2}i$$

b)   The complex number $1+i$ can be expressed in the polar form as

$$\sqrt{2}\left(\cos\frac{\pi}{4} + i\sin\frac{\pi}{4}\right)$$

where the modulus is $\sqrt{2}$ and the argument is $\pi/4$. Using DeMoivre's theorem, we have

$$(1+i)^{20} = \left[\sqrt{2}\left(\cos\frac{\pi}{4} + i\sin\frac{\pi}{4}\right)\right]^{20}$$

$$= 2^{10}(\cos 5\pi + i\sin 5\pi)$$

$$= 1024(-1 + 0i) = -1024$$

2   Use DeMoivre's theorem to express $\cos 2\theta$ and $\sin 2\theta$ in terms of $\sin\theta$ and $\cos\theta$.

SOLUTION.   By DeMoivre's theorem for $n = 2$, we have

$$\cos 2\theta + i\sin 2\theta = (\cos\theta + i\sin\theta)^2$$
$$= \cos^2\theta + i2\sin\theta\cos\theta - \sin^2\theta$$
$$= (\cos^2\theta - \sin^2\theta) + i2\sin\theta\cos\theta$$

Since the two complex numbers are equal, the real parts are equal; that is,

$$\cos 2\theta = \cos^2\theta - \sin^2\theta$$

and the imaginary parts are also equal; that is,

$$\sin 2\theta = 2\sin\theta\cos\theta$$

5.1   Roots

DeMoivre's theorem is useful in finding the nth roots of a complex number. If $w = R(\cos\phi + i\sin\phi)$ and $z = r(\cos\theta + i\sin\theta)$ is any solution of $z^n = w$, where n is a positive integer, then, by DeMoivre's theorem, it follows that

$$[r(\cos\theta + i\sin\theta)]^n = r^n(\cos n\theta + i\sin n\theta)$$
$$= R(\cos\phi + i\sin\phi)$$

so that

$$r^n = R$$

or, equivalently,

$$r = \sqrt[n]{R} \quad \text{(notice that } R \geq 0\text{)}$$

and

$$n\theta = \phi + 2k\pi \quad (n\theta = \phi + 360°k \text{ if degrees are used})$$

or, equivalently,

$$\theta = \frac{\phi}{n} + \frac{2k\pi}{n} \quad \left(\theta = \frac{\phi}{n} + \frac{360°k}{n} \text{ if degrees are used}\right)$$

where $k = 0, \pm 1, \pm 2, \ldots$ gives n different angles equally spaced around the unit circle.

We will see in the examples below that those values of k from 0 to $n-1$ give us all *distinct* n roots that exist.

Also for each of the angles the complex number z with modulus r and argument θ satisfies $z^n = R(\cos \phi + i \sin \phi)$ and they are given by

$$z_k = \sqrt[n]{R}\left\{\cos\left(\frac{\phi}{n} + \frac{2\pi k}{n}\right) + i \sin\left(\frac{\phi}{n} + \frac{2\pi k}{n}\right)\right\} \qquad k = 0, 1, 2, \ldots, n-1$$

EXAMPLES

1 Determine the square roots of $-i$; that is, solve $z^2 = -i$ and represent the two roots geometrically.

SOLUTION. First, we determine a polar representation, $R(\cos \phi + i \sin \phi)$, of $-i$. Here $R = \sqrt{0^2 + (-1)^2} = 1$ and $\phi = 270°$ (Figure 1). Thus we obtain the roots of $z = r(\cos \theta + i \sin \theta)$ by using the formula $r = \sqrt{1} = 1$ and $\theta = (270°/2) + (360°k/2) = 135° + 180°k$, $k = 0, 1$, so the roots are

$$z_0 = 1(\cos 135° + i \sin 135°) = -\frac{\sqrt{2}}{2} + i\frac{\sqrt{2}}{2} \qquad \text{for} \quad k = 0$$

$$z_1 = 1(\cos 315° + i \sin 315°) = \frac{\sqrt{2}}{2} - i\frac{\sqrt{2}}{2} \qquad \text{for} \quad k = 1$$

The square roots of $-i$ are equally spaced on the circumference of a circle of radius 1 and differ by an angle of π radians or 180° (Figure 1).

Figure 1

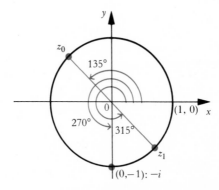

2 Find the fourth roots of $1 + i$ in polar form and represent them geometrically.

SOLUTION. As our first step, we determine the polar representation, $R(\cos \phi + i \sin \phi)$, of $1 + i$. Here $R = \sqrt{2}$ and $\phi = \pi/4$. Finally, we obtain the roots $z = r(\cos\theta + i\sin\theta)$ by using the formulas for r and θ.

$$r = \sqrt[4]{\sqrt{2}} \quad \text{and} \quad \theta = \frac{\pi}{4\cdot 4} + \frac{2\pi k}{4} = \frac{\pi}{16} + \frac{\pi k}{2}$$

$$k = 0, 1, 2, 3, \ldots$$

so that the roots are

$$z_0 = \sqrt[8]{2}\left(\cos\frac{\pi}{16} + i\sin\frac{\pi}{16}\right) \qquad \text{for } k = 0$$

$$z_1 = \sqrt[8]{2}\left(\cos\frac{9\pi}{16} + i\sin\frac{9\pi}{16}\right) \qquad \text{for } k = 1$$

$$z_2 = \sqrt[8]{2}\left(\cos\frac{17\pi}{16} + i\sin\frac{17\pi}{16}\right) \qquad \text{for } k = 2$$

$$z_3 = \sqrt[8]{2}\left(\cos\frac{25\pi}{16} + i\sin\frac{25\pi}{16}\right) \qquad \text{for } k = 3$$

The fourth roots of $1 + i$ are equally spaced on the circumference of a circle of radius $\sqrt[8]{2}$ and differ by angles of $\pi/2$ radians (Figure 2). Notice that if we were to set $k = 4$, we would get

$$z_4 = \sqrt[8]{2}\left(\cos\frac{33\pi}{16} + i\sin\frac{33\pi}{16}\right)$$

Figure 2

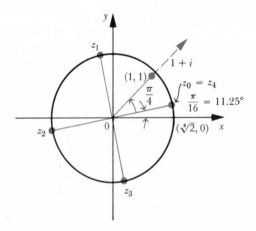

which is the same number as z_0. In general, if $k = 4, 5, 6, \ldots$, we would get a repetition of the roots that we have already found.

3   The solutions to $z^n = 1$ are called the *n*th *roots of unity*. Find the fifth roots of unity in polar form. Represent the roots geometrically.

SOLUTION. The number 1 can be written in polar form as $1 = 1(\cos 0 + i \sin 0)$. We obtain the five roots $z = r(\cos \theta + i \sin \theta)$ by using the formulas

$$r = \sqrt[5]{1} \quad \text{and} \quad \theta = \frac{0}{5} + \frac{2\pi k}{5} \quad k = 0, 1, 2, 3, 4$$

so the roots are

$$z_0 = 1(\cos 0 + i \sin 0) \qquad \text{for } k = 0$$

$$z_1 = 1\left(\cos \frac{2\pi}{5} + i \sin \frac{2\pi}{5}\right) \qquad \text{for } k = 1$$

$$z_2 = 1\left(\cos \frac{4\pi}{5} + i \sin \frac{4\pi}{5}\right) \qquad \text{for } k = 2$$

$$z_3 = 1\left(\cos \frac{6\pi}{5} + i \sin \frac{6\pi}{5}\right) \qquad \text{for } k = 3$$

$$z_4 = 1\left(\cos \frac{8\pi}{5} + i \sin \frac{8\pi}{5}\right) \qquad \text{for } k = 4$$

Observe that all the fifth roots of unity are on a unit circle and they are equally spaced at angles of $2\pi/5$ radians (Figure 3). Again, $k = 5, 6, \ldots$ would give us a repetition of the roots we have already found.

Figure 3

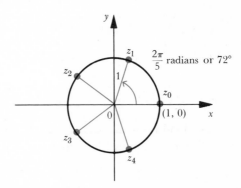

PROBLEM SET 3

1 Use DeMoivre's theorem to compute each of the following powers. Express the answer in the rectangular form $a + bi$.

a) $(\cos 30° + i \sin 30°)^7$

b) $(\cos 15° + i \sin 15°)^8$

c) $\left[2\left(\cos \frac{\pi}{6} + i \sin \frac{\pi}{6}\right)\right]^{10}$

d) $\left[3\left(\cos \frac{\pi}{18} + i \sin \frac{\pi}{18}\right)\right]^6$

e) $\left[2\left(\cos \frac{5\pi}{4} + i \sin \frac{5\pi}{4}\right)\right]^8$

f) $[4(\cos 36° + i \sin 36°)]^5$

2   Let $z_1 = [2(\cos(\pi/8) + i \sin(\pi/8))]^4$ and $z_2 = [4(\cos(\pi/12) + i \sin(\pi/12))]^6$, find $z_1 z_2$.

3   Express each of the following complex numbers in polar form, then use DeMoivre's theorem to calculate the indicated powers. Express the result in the rectangular form $a + bi$.

a) $(5 + 5i)^6$ b) $(1 + i\sqrt{3})^5$

c) $(\sqrt{3} - i)^4$ d) $\left(-\dfrac{1}{2} - i\dfrac{\sqrt{3}}{2}\right)^8$

e) $(\sqrt{3} + i)^{30}$ f) $(1 + i)^{50}$

g) $\left(\dfrac{1}{\sqrt{2}} + i\dfrac{1}{\sqrt{2}}\right)^{100}\left(\dfrac{1}{2} + i\dfrac{\sqrt{3}}{2}\right)^{30}$ h) $\dfrac{(\sqrt{3} + i)^3}{(1 - \sqrt{3}i)^3}$

4   Use DeMoivre's theorem to derive formulas for $\cos 3\theta$ and $\sin 3\theta$. [*Hint*: $\cos 3\theta + i \sin 3\theta = (\cos \theta + i \sin \theta)^3$.]

5   Find all the roots in the following equations and represent them geometrically.

a) $z^2 = i$ b) $z^2 = 3 - 3i$

c) $z^3 = 8$ d) $z^3 = i$

e) $z^4 = -16$ f) $z^4 = -8 - 8\sqrt{3}i$

6   a)   Use DeMoivre's theorem, together with Theorem 2 on page 438, to show that if $z = r(\cos \theta + i \sin \theta)$, then

$$z^{-n} = r^{-n}[\cos(-n\theta) + i \sin(-n\theta)]$$

b)   Use the result of part a to find

i $\left(\dfrac{\sqrt{3}}{2} + i\dfrac{1}{2}\right)^{-5}$ ii $(-2 + 2i)^{-3}$

7   a)   Find the fifth roots of -32.
    b)   Graph the five fifth roots of -32.

6   Complex Zeros of Polynomial Functions

We have seen (Chapter 3) that a polynomial function with real coefficients does not always have real number zeros. In particular, a quadratic polynomial function $f(x) = ax^2 + bx + c$, $a, b, c \in R$, $a \neq 0$, has real zeros if and only if the discriminant, $b^2 - 4ac$, is nonnegative. These real roots can be found by the quadratic formula

$$x = \frac{-b \pm \sqrt{b^2 - 4ac}}{2a}$$

If the zeros of the quadratic polynomial function are not real numbers, that is, if $b^2 - 4ac < 0$, the zeros are complex numbers and the quadratic formula can still be used. For example, the zeros of $f(x) = 2x^2 + x + 1$ can be found by using the quadratic formula to get

$$x = \frac{-1 \pm \sqrt{1-8}}{4} = \frac{-1 \pm \sqrt{-7}}{4} = -\frac{1 \pm \sqrt{7}\,i}{4}$$

The polynomial function $f(x) = x^3 - 6x^2 + 13x - 10$ can be factored as $f(x) = (x-2)(x^2 - 4x + 5)$, so that after using the quadratic formula, we find that the zeros of f are 2, $2 - i$, and $2 + i$, and

$$f(x) = (x-2)[x - (2-i)][x - (2+i)]$$

In general, all polynomial functions are factorable as the product of linear factors in the complex domain, and the factorization is based on the zeros of the polynomial functions.

Notice, for example, that

$$2x^2 + x + 1 = 2\left(x - \frac{-1 + \sqrt{7}\,i}{4}\right)\left(x - \frac{-1 - \sqrt{7}\,i}{4}\right)$$

This result follows as a corollary of the *fundamental theorem of algebra*, whose proof depends upon methods generally considered beyond the scope of this text. We shall state the theorem without proof.

THEOREM 1   FUNDAMENTAL THEOREM OF ALGEBRA

If $f(x)$ is a polynomial of degree $n \geq 1$ with complex coefficients, then there is a complex number r such that $f(r) = 0$.

Assuming this fundamental theorem, we can prove the following theorem on factoring polynomials.

THEOREM 2   THE FACTORIZATION THEOREM

If $f(x) = a_n x^n + a_{n-1} x^{n-1} + \cdots + a_1 x + a_0$ and $a_n \neq 0$, n a positive integer, then

$$f(x) = a_n(x - r_1)(x - r_2) \cdots (x - r_n)$$

where the numbers r_j are complex numbers.

PROOF.   By the fundamental theorem of algebra, $f(x) = 0$ has a root r_1, so by the factor theorem (see Chapter 3),

$$f(x) = (x - r_1) Q_1(x)$$

$Q_1(x)$ is a polynomial of degree $n-1$, so it has a zero r_2 if $n-1 \geq 1$, and, as above,

$$Q_1(x) = (x-r_2) Q_2(x)$$

so that

$$f(x) = (x-r_1)(x-r_2) Q_2(x)$$

where $Q_2(x)$ has degree $n-2$. Continuing the process we get

$$f(x) = (x-r_1)(x-r_2) \cdots (x-r_n) Q_n(x)$$

where $Q_n(x)$ has degree 0; that is, $Q_n(x)$ is a constant. Multiplying out this expression for $f(x)$, it is seen that the coefficient of x^n is Q_n; hence $Q_n = a_n$ and the theorem is proved.

THEOREM 3

If $f(x)$ is a polynomial of degree n, $n \neq 0$, then $f(x) = 0$ has n roots. (Not all n roots are necessarily different.)

PROOF.   By the factorization theorem,

$$f(x) = a_n(x-r_1)(x-r_2) \cdots (x-r_n)$$

Clearly, the numbers $r_1, r_2, ..., r_n$ are roots of $f(x) = 0$. Moreover, if $f(r) = 0$ for $r \neq r_i$, $i = 1, ..., n$, then

$$f(x) = a_n(x-r_1)(x-r_2) \cdots (x-r_n)(x-r)$$

so the degree of $f(x)$ is $n+1$ (why?), which contradicts our assumption that $f(x)$ is a polynomial of degree n.

Notice that the roots need not be distinct. For example, $x^2 - 4x + 4 = 0$ has two roots, both of which are equal to 2 and we say that $f(x) = x^2 - 4x + 4$ has $x = 2$ as a double root. In general, if

$$f(x) = (x-r)^s Q(x) \qquad \text{and} \qquad Q(r) \neq 0$$

we say that r is a zero of *multiplicity* s. For example, $f(x) = (x-1)^2 (x-2)$ has $x = 1$ as a root of multiplicity 2 and $x = 2$ as a root of multiplicity 1.

THEOREM 4   CONJUGATE ROOT THEOREM

If $f(z)$ is a polynomial of degree n, $n \neq 0$, has *real* coefficients, and $f(z_0) = 0$, where $z_0 = a + bi$, then $f(\bar{z}_0) = 0$.

PROOF.   Let $f(z) = a_n z^n + a_{n-1} z^{n-1} + \cdots + a_1 z + a_0$ be a polynomial with real coefficients. Since $z_0 = a + bi$ is a root of $f(z)$, then

$$f(z_0) = a_n z_0^n + a_{n-1} z_0^{n-1} + \cdots + a_1 z_0 + a_0 = 0$$

so that

$$\overline{f(z_0)} = \overline{a_n z_0^n + a_{n-1} z_0^{n-1} + \cdots + a_1 z_0 + a_0} = \overline{0} = 0$$

However,

$$\overline{a_n z_0^n + a_{n-1} z_0^{n-1} + \cdots + a_1 z_0 + a_0} = \overline{a_n z_0^n} + \overline{a_{n-1} z_0^{n-1}} + \cdots + \overline{a_1 z_0} + \overline{a_0}$$

because the conjugate of the sum of complex numbers is the same as the sum of the conjugates of the complex numbers (see Problem Set 1, Problem 4c). Also,

$$\overline{a_n z_0^n} = \overline{a_n}\, \overline{z_0^n}, \qquad \overline{a_{n-1} z_0^{n-1}} = \overline{a_{n-1}}\, \overline{z_0^{n-1}}, \ldots, \overline{a_1 z_0} = \overline{a_1}\, \overline{z_0}$$

and

$$\overline{a_n}\, \overline{z_0^n} = \overline{a_n}\, \overline{z_0}^n, \qquad \overline{a_{n-1}}\, \overline{z_0^{n-1}} = \overline{a_{n-1}}\, \overline{z_0}^{n-1}, \ldots, \overline{a_1}\, \overline{z_0} = \overline{a_1}\, \overline{z_0}$$

(See Problem Set 1, Problem 4d.) Since the conjugate of a real number is the real number itself, we have $\overline{a_0} = a_0, \overline{a_1} = a_1, \ldots, \overline{a_n} = a_n$. Hence

$$f(\overline{z_0}) = a_n \overline{z_0}^n + a_{n-1} \overline{z_0}^{n-1} + \cdots + a_1 \overline{z_0} + a_0 = \overline{f(z_0)} = 0$$

That is, $\overline{z_0}$ is also a root of $f(z)$.

For example, the polynomial function $f(x) = x^2 - 4x + 5$ has two zeros, one of which is the complex number $2 + i$. By the conjugate root theorem, $2 - i$ is also a zero of $f(x) = x^2 - 4x + 5$, as shown by the following multiplication:

$$\begin{aligned}
[x - (2+i)][x - (2-i)] &= x^2 - (2+i)x - (2-i)x + (2+i)(2-i) \\
&= x^2 - (2+i+2-i)x + 2^2 + 1^2 \\
&= x^2 - 4x + 5
\end{aligned}$$

EXAMPLES

1 Form a polynomial $f(x)$ that has the following numbers as roots: $-\frac{1}{2}$, $1 + i$, and 1 as a double zero.

SOLUTION. Since $1 + i$ is a root of $f(x) = 0$, it follows from the conjugate root theorem that $1 - i$ is also a root; therefore,

$$f(x) = (x + \tfrac{1}{2})(x - 1)^2 [x - (1+i)][x - (1-i)]$$

has the given roots. Simplifying the equation, we get

$$f(x) = x^5 - \tfrac{7}{2}x^4 + 5x^3 - \tfrac{5}{2}x^2 - x + 1$$

2   Determine the multiplicity of the zeros of the polynomial

$$f(x) = x^4 - 4x^3 + 5x^2 - 4x + 4$$

SOLUTION.   Using synthetic division, the polynomial can be factored as $x^4 - 4x^3 + 5x^2 - 4x + 4 = (x-2)^2(x+i)(x-i)$, so that 2 is a double zero and i and $-i$ are each zeros of multiplicity 1.

PROBLEM SET 4

1   Determine whether the given numbers are zeros of the given polynomial functions. If they are, find their multiplicities.
   a) $f(x) = x^4 - x^3 - 18x^2 + 52x - 40, \ x = 2$
   b) $f(x) = 4x^6 + 4x^5 + 9x^4 + 8x^3 + 6x^2 + 4x + 1, \ x = i$
   c) $f(x) = 9x^4 - 12x^3 + 13x^2 - 12x + 4, \ x = \frac{2}{3}$

2   Show that the equation

$$\frac{1}{x-3} + \frac{1}{x-2} - \frac{x-2}{x-3} = 0$$

has no solution. Why does this not contradict the fundamental theorem of algebra?

3   a)   Given that $-1 + i$ is a zero of $f(x) = x^4 + 2x^3 - 4x - 4$, find all other zeros of f.
   b)   Given that i is a double zero of $f(x) = 2x^6 + x^5 + 2x^3 - 6x^2 + x - 4$, find all other zeros of f.

4   Use the roots to write each of the following polynomial functions in factored form
   a) $f(x) = 2x^2 - x - 2$
   b) $f(x) = x^2 - 3x - 3$

5   Find polynomials having the following numbers as their roots.
   a)   2, 3, i b)   2, 2, 1+i, 1-i
   c)   1-3i, 1-3i, 1+3i, 1+3i d) i, i, 0, 1, 2i
   e)   2, 2, $\frac{1}{2}(-1 + i\sqrt{3})$

REVIEW PROBLEM SET

1   Perform the indicated operations and write the answer in the form $a + bi$.
   a) $(3 - 2i) + (7 - 3i)$
   b) $(3 - \sqrt{7}i)(3 + \sqrt{7}i)$
   c) $(5 + 12i) + (-5 - 3i) + (1 + 2i)$
   d) $(4 - i) - (7 - 3i) + (2 + i)$
   e) $(5 + 3i)(3 - 5i)$

f) $(3+7i)/(2-3i)$

g) $(4-\sqrt{3}i)/(2+\sqrt{3}i)$

h) $(3-5i)/4i$

2 Solve each of the following for x and y, where $x, y \in R$.

a) $5x+15i = 15-yi$

b) $(2x+3)+(y-3)i = 0$

c) $-3+17i = x+3yi$

d) $3x-2i+7+3yi = 0$

e) $x+iy = (3-2i)/(2-3i)$

f) $3x+5yi = (1+3i)/(2+i)$

3 a) Locate each of the following polar points, and then find the rectangular coordinates that correspond to each of them.

i $(5, \pi/4)$ ii $(2, -90°)$ iii $(\sqrt{2}, -135°)$

iv $(3, \pi)$ v $(3, 270°)$ vi $(2, 4\pi/3)$

b) Locate each of the following rectangular points and then find polar coordinates that correspond to each of them.

i $(-3, 0)$ ii $(-2, 2\sqrt{3})$ iii $(-5\sqrt{3}, -5)$

iv $(10, 10)$ v $(0, -14)$ vi $(-15, 0)$

4 a) Suppose that $z = x+yi$. Under what conditions does $z^2 = \bar{z}^2$?

b) If $z = x+iy$ and $\bar{z} = -z$, show that the real part of z is zero.

5 Find the modulus and an argument of each of the following complex numbers.

a) $z = 5+5i$

b) $z = \sqrt{3}+i$

c) $z = 6\sqrt{3}+6i$

d) $z = 8i$

e) $z = 2+2i$

f) $z = -1+\sqrt{3}i$

g) $z = 3+4i$

h) $z = (2+2\sqrt{3}i)^{10}$

6 Which of the following statements is true? If the statement is false, give a counterexample.

a) $z+\bar{z} = 0$ if and only if $\text{Re}\,z = 0$, where $\text{Re}\,z$ denotes the real part of z.

b) $z+1/z$ is real if and only if $\text{Im}\,z = 0$ or $|z| = 1$, where $\text{Im}\,z$ denotes the imaginary part of z.

c) If $\text{Im}\,z \neq 0$, then $z/(1+z^2)$ is real if and only if $|z| = 1$.

7 Find $z_1 z_2$ and z_1/z_2 for each of the following pairs of complex numbers. Write the results in polar form and in rectangular form.

a) $z_1 = 2(\cos \pi + i \sin \pi)$ and $z_2 = 3[\cos(\pi/2)+i\sin(\pi/2)]$

b) $z_1 = 6(\cos 230° + i \sin 230°)$ and $z_2 = 3(\cos 75° + i \sin 75°)$

c) $z_1 = 6(\cos 110° + i \sin 110°)$ and $z_2 = 2(\cos 212° + i \sin 212°)$

d) $z_1 = 14(\cos 305° + i \sin 305°)$ and $z_2 = 7(\cos 65° + i \sin 65°)$

8 Let $z = a+bi$. Show that each of the following equations are true.

a) $|(a+bi)/(a-bi)|^2 = 1$

b) $|(a+bi)^3/(a-bi)^2| = |z|$

9   Express each of the following powers in polar form and rectangular form.

a) $(\cos 60° + i \sin 60°)^5$ b) $(1+i)^{40}$

c) $[\sqrt{2}/2 + i(\sqrt{2}/2)]^{100}$ d) $(\cos 0° + i \sin 0°)^{150}$

e) $(\sqrt{3}+i)^{30}$

10   Use DeMoivre's theorem to find expressions for $\cos 5\theta$ and $\sin 5\theta$. [*Hint:* $\cos 5\theta + i \sin 5\theta = (\cos \theta + i \sin \theta)^5$.]

11   Let $z = 10(\cos 17° + i \sin 17°)^{10}/(1+i)^2$. Find $\operatorname{Re} z$, $\operatorname{Im} z$, and $|z|$.

12   Let $z = \cos(2\pi/5) + i \sin(2\pi/5)$. Show that

$$\left| \frac{(z^2 - z^3)}{(z^4 - z^5)} \right| = 1$$

13   Find all the roots of the following equations.

a) $z^3 = -64$ b) $z^4 = -8i$

c) $z^4 = 1+i$ d) $z^4 = 8-8i$

14   Show that the *sum* of the complex cube roots of 1 is 0.

15   Form an equation of a polynomial that has the given roots:

a) $-2, -2, -3, 3$

b) $2, 2-i$

c) $2, 3, -3, -4$

d) $1, 1, 3, -2, 5$

e) $i, 1+i, 1-i$

CHAPTER 9

Analytic Geometry

1 Introduction

In this chapter we shall continue to use the Cartesian coordinate system (see Chapter 1, Section 5) to relate the geometry of certain graphs to the algebraic representation of the graphs. We will investigate the *circle*, the

Figure 1

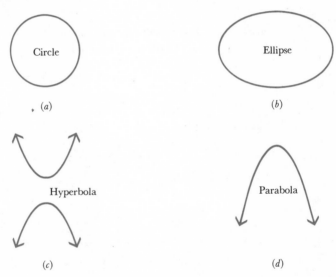

Circle

(a)

Ellipse

(b)

Hyperbola

(c)

Parabola

(d)

ellipse, the *hyperbola*, and the *parabola* (Figures 1a, b, c, and d). [For convenience, we will use the notation $P = (a, b)$ to identify points in the plane rather than using the notation $P:(a, b)$.]

2 Circle

Geometrically, a *circle* in a plane can be defined as the set of all points that are at a fixed distance r, called the length of the *radius*, from a fixed point called the *center* C. The length of a diameter of a circle is equal to $2r$. In Figure 1, P is a point of the circle, C is the center, and r is the length of the radius. $\overline{AB} = 2r$ is the length of a diameter. The distance formula can be used to write an equation of a relation that has as its graph a circle.

Figure 1

THEOREM 1   CIRCLE EQUATION

Let (h, k) be the center of a circle whose radius is r; then the equation of the circle is

$$(x-h)^2 + (y-k)^2 = r^2$$

PROOF.   Let $P = (x,y)$ represent *any* point on the circle whose center C is (h, k). Then, by the distance formula,

$$r = [(x-h)^2 + (y-k)^2]^{1/2} \qquad \text{(Figure 2)}$$

Figure 2

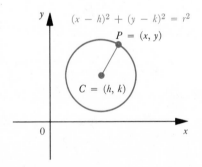

After squaring both sides of this equation, we get

(1) $r^2 = (x-h)^2 + (y-k)^2$

Thus any point on the circle has coordinates that satisfy equation (1).

Conversely, if $P = (x, y)$ is a point that satisfies equation (1), we have

$$\sqrt{(x-h)^2 + (y-k)^2} = \sqrt{r^2} = r$$

so that P is r units from the point (h, k) and is a point on the circle.

Notice that the equation of the circle gives us an algebraic characterization that depends only on the center and the radius of the circle; hence an equation of the form $(x-h)^2 + (y-k)^2 = r^2$ is a relation whose graph is a circle with center (h, k) and radius r. Using set notation, this means that a circle with center (h, k) of radius r can be considered as the relation $\{(x, y) \mid (x-h)^2 + (y-k)^2 = r^2\}$.

EXAMPLES

1 Find the equation of a circle if $P_1 = (3, 7)$ and $P_2 = (-3, -1)$ are the end points of a diameter.

SOLUTION.   (See Figure 3.) The center of the circle, the midpoint of $\overline{P_1 P_2}$, can be determined by the formula $[(x_1 + x_2)/2, (y_1 + y_2)/2]$ (see Chapter 1, Problem Set 5, Problem 13). Hence we have the center $C = (0, 3)$. The radius is

$$r = \tfrac{1}{2}\overline{P_1 P_2} = \tfrac{1}{2}\sqrt{(3+3)^2 + (7+1)^2} = \tfrac{1}{2}\sqrt{36+64} = 5$$

Therefore, the equation of the circle is given by $(x-0)^2 + (y-3)^2 = 5^2$ or $x^2 + (y-3)^2 = 25$.

Figure 3

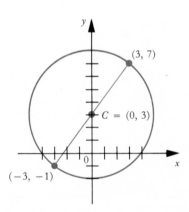

2 Find the center and the radius of the circle whose equation is $x^2 + y^2 - 4x + 6y - 12 = 0$.

SOLUTION.   (See Figure 4.) First, we will rewrite the equation in the standard form of Theorem 1. To do this, we "complete the square" as follows.

$$(x^2 - 4x +) + (y^2 + 6y +) = 12$$
$$(x^2 - 4x + 4) + (y^2 + 6y + 9) = 12 + 4 + 9$$
$$(x-2)^2 + (y+3)^2 = 25$$

so that the graph is a circle with center at $(2, -3)$ and radius 5.

Figure 4

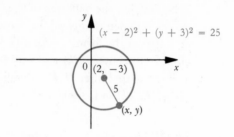

3   Find the equation of the circle that contains the three points $P_1 = (5, 4)$, $P_2 = (3, 2)$, and $P_3 = (-3, 0)$.

SOLUTION.   From plane geometry, we know that three noncollinear points determine a unique circle (Figure 5). The equation of the circle is $(x-h)^2 + (y-k)^2 = r^2$. We must determine h, k, and r.

Figure 5

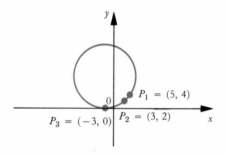

Since P_1 is on the circle, its coordinates satisfy the equation $(5-h)^2 + (4-k)^2 = r^2$. Similarly, P_2 and P_3 are on the circle, so that there are three equations for the three unknowns, and we proceed to solve the system of equations

$$\begin{cases} 25 - 10h + h^2 + 16 - 8k + k^2 = r^2 \\ 9 - 6h + h^2 + 4 - 4k + k^2 = r^2 \\ 9 + 6h + h^2 \qquad\qquad + k^2 = r^2 \end{cases}$$

simultaneously to get $h = -3$, $k = 10$, and $r = 10$. (Why?) Hence the equation of the circle is

$$(x+3)^2 + (y-10)^2 = 100$$

PROBLEM SET 1

1   Find the center and radius of each of the following circles. Also, sketch the circle in each case.
a) $(x-3)^2 + (y-1)^2 = 4$
b) $(x+5)^2 + (y-3)^2 = 9$

c) $(x-1)^2+(y+2)^2 = 16$
d) $x^2+y^2+4x-6y = 5$
e) $x^2+y^2-3x+4y+4 = 0$
f) $x^2+y^2-6x+8y-25 = 0$
g) $x^2+y^2-2x+3y+3 = 0$

2 a) Prove that any circle with a center at the origin has an equation of the form $x^2+y^2 = r^2$.

  b) We say that a relation R is symmetric with respect to the x axis if $(x, y) \in R$ and $(x, -y) \in R$ for all x in the domain of R. Discuss the symmetry of the circle $x^2+y^2 = r^2$ with respect to the x axis, with respect to the y axis, and with respect to the origin.

3 Find the equation for each of the following circles.

  a) The circle with radius 3 and center $(2, 1)$
  b) The circle with center $(-3, -2)$ and radius 3
  c) The circle that contains the points $(2, -6)$, $(6, 4)$, and $(-3, 1)$
  d) The circle whose center lies on the line $x-4y = 1$ and passes through the points $(3, 7)$ and $(5, 5)$

4 Use the graph of a circle to show that a circle is a relation which is not a function.

5 a) Find the equations of the circles of radius $\sqrt{10}$ that are tangent to $3x+y = 6$ at $(3, -3)$.

  b) Find the equation of a circle of radius 2 that contains point $(3, 4)$ and is tangent to $x^2+y^2 = 25$.

  c) Find the equations of the circles that contain point $(3, -2)$ with center on $2x-y+2 = 0$ and with radius 5.

6 For each of the relations of Problem 1 indicate the domain and range.

7 Compare $y = \sqrt{9-x^2}$, $y = -\sqrt{9-x^2}$, and $x^2+y^2 = 9$. Graph each of the three relations on different coordinate axes. Indicate the domain and range range of each. Which, if any, of the three is a function? Are any of the three one-to-one? If any of the three is one-to-one, find the inverse.

8 Show that if the matrix

$$C = \begin{bmatrix} x-h & y-k \\ -(y-k) & x-h \end{bmatrix}$$

then $\det(C) = r^2$ is the general equation of a circle. Find the equation of each of the following circles in determinant form.

  a) Radius of 1, center $(0, 0)$
  b) Radius of 1, center (a, b)
  c) Radius of 1, with center on the line $y = x$
  d) Radius of 1, with center on $y = x^2$

3 Translations

The circles $x^2 + y^2 = 9$ and $(x-3)^2 + (y+4)^2 = 9$ have the same radius 3, but different centers. $x^2 + y^2 = 9$ has center at $(0,0)$, whereas $(x-3)^2 + (y+4)^2 = 9$ has center at $(3, -4)$. In this section we shall investigate a method of changing the coordinate system, called a *translation*. With this method, equations such as $(x-3)^2 + (y+4)^2 = 9$ take on the simpler form $\overline{x}^2 + \overline{y}^2 = 9$ in the new coordinate system.

Suppose that P is a point in the xy plane (Figure 1a). The coordinates of P depend on how the coordinate axes are placed in the plane. If the axes are positioned so that the origin coincides with P, then $P = (0,0)$, as shown in Figure 1b.

Figure 1

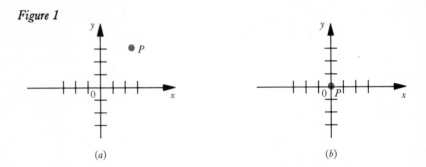

(a)   (b)

If another pair of axes are formed by "shifting" or "translating" the y axis 5 units to the left to get a $\overline{y}$ axis and the x axis 2 units up to get an $\overline{x}$ axis, the coordinates of P in this new coordinate system, the $\overline{xy}$ system, are given by $(5, -2)$ (why?) (Figure 2). Also, $(5, 0)$ in the xy system is $(10, -2)$ in the $\overline{xy}$ system, and $(0, 0)$ in the $\overline{xy}$ system is $(-5, 2)$ in the xy system (why?).

Figure 2

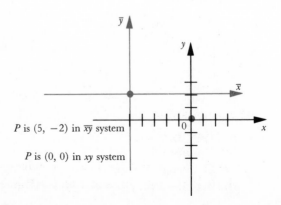

P is $(5, -2)$ in $\overline{xy}$ system

P is $(0, 0)$ in xy system

It is not difficult to see that a point in the plane has different coordinates depending on how the coordinate system has been established. In the example above, a "new" coordinate system was formed from a given coordinate system by "translating" the x axis 2 units up to get the $\bar{x}$ axis and by "translating" the y axis 5 units to the left to get the $\bar{y}$ axis. A translation of axes results in new coordinate axes that are parallel to the given axes.

Suppose that the xy coordinate axes have been translated to form the $\overline{xy}$ coordinate system, so that the origin $\bar{O}$ of the $\overline{xy}$ system has coordinates (h, k) in the xy system, as shown in Figure 3. $\bar{O} = (0,0)$ in the $\overline{xy}$ system; $O = (h, k)$ in the xy system.

Figure 3

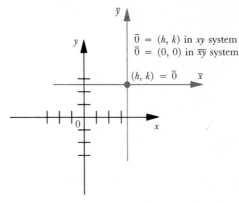

$\bar{O} = (h, k)$ in xy system
$\bar{O} = (0, 0)$ in $\overline{xy}$ system

$(h, k) = \bar{O}$

Assume that P is a point in the plane such that

$$P = (x, y) \qquad \text{in the } xy \text{ system}$$

whereas

$$P = (\bar{x}, \bar{y}) \qquad \text{in the } \overline{xy} \text{ system}$$

From our knowledge of vectors, we know that if $\mathbf{i}$ and $\mathbf{j}$ are the basis vectors in the xy system,

$$\mathbf{OP} = x\mathbf{i} + y\mathbf{j}$$
$$\mathbf{O\bar{O}} = h\mathbf{i} + k\mathbf{j}$$

and

$$\mathbf{\overline{OP}} = \bar{x}\mathbf{i} + \bar{y}\mathbf{j} \qquad \text{(why?)}$$

but, equivalently,

$$\mathbf{O\bar{O}} + \mathbf{\overline{OP}} = \mathbf{OP} \qquad \text{(why?)}$$

so that

$$(h\mathbf{i} + k\mathbf{j}) + (\bar{x}\mathbf{i} + \bar{y}\mathbf{j}) = x\mathbf{i} + y\mathbf{j} \qquad \text{(Figure 4)}$$

Figure 4

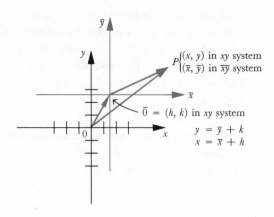

Hence

$$(h+\bar{x})\mathbf{i} + (k+\bar{y})\mathbf{j} = x\mathbf{i} + y\mathbf{j}$$

That is,

$$(h+\bar{x}, k+\bar{y}) = (x, y)$$

from which we conclude that

$$x = \bar{x} + h \qquad \text{and} \qquad y = \bar{y} + k$$

or, equivalently,

$$\bar{x} = x - h \qquad \text{and} \qquad \bar{y} = y - k$$

Hence we have proved the following theorem.

THEOREM 1 TRANSLATION EQUATIONS

Suppose that the coordinates axes xy are translated to coordinate axes $\overline{xy}$, so that the origin $\overline{O}$ of the $\overline{xy}$ system has coordinates (h, k) in the xy system. If P has coordinates (x, y) in the xy system and coordinates $(\bar{x}, \bar{y})$ in the $\overline{xy}$ system, then

$$x = \bar{x} + h \qquad \text{and} \qquad y = \bar{y} + k$$

or, equivalently,

$$\bar{x} = x - h \qquad \text{and} \qquad \bar{y} = y - k$$

EXAMPLES

1 Translate the xy axes to form the $\overline{xy}$ axes, so that the origin $\overline{O}$ in the $\overline{xy}$ system corresponds to $(4, -3)$ in the xy system. If $P_1 = (2, 1)$, $P_2 = (0, 1)$, $P_3 = (-2, 3)$, and $P_4 = (-3, 5)$ are given in the xy system, find their representations in the $\overline{xy}$ system.

SOLUTION. (See Figure 5.) From Theorem 1, we have $\bar{x} = x - h$ and $\bar{y} = y - k$. In this situation $h = 4$ and $k = -3$; hence, for $(2, 1)$,

$$\bar{x} = 2 - 4 = -2 \quad \text{and} \quad \bar{y} = 1 - (-3) = 4$$

Figure 5

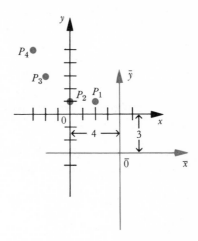

so that $(2, 1)$ in the xy system is represented by $(-2, 4)$ in the $\bar{x}\bar{y}$ system.

For $(0, 1)$, $\bar{x} = 0 - 4 = -4$ and $\bar{y} = 1 - (-3) = 4$ so that $(0, 1)$ in the xy system is represented by $(-4, 4)$ in the $\bar{x}\bar{y}$ system.

For $(-2, 3)$, $\bar{x} = -2 - 4 = -6$ and $\bar{y} = 3 - (-3) = 6$, so that $(-2, 3)$ in the xy system is represented by $(-6, 6)$ in the $\bar{x}\bar{y}$ system.

For $(-3, 5)$, $\bar{x} = -3 - 4 = -7$ and $\bar{y} = 5 - (-3) = 8$, so that $(-3, 5)$ in the xy system is represented by $(-7, 8)$ in the $\bar{x}\bar{y}$ system.

2   Transform the equation $x^2 + y^2 + 6x - 8y - 11 = 0$ to the form $\bar{x}^2 + \bar{y}^2 = r^2$, using Theorem 1 with $h = -3$ and $k = 4$.

SOLUTION. (See Figure 6.) We have $h = -3$, and $k = 4$. Thus $x = \bar{x} - 3$ and $y = \bar{y} + 4$. Upon substituting these values of x and y into the given equation, we obtain

$$(\bar{x} - 3)^2 + (\bar{y} + 4)^2 + 6(\bar{x} - 3) - 8(\bar{y} + 4) - 11 = 0$$

so that

$$\bar{x}^2 - 6\bar{x} + 9 + \bar{y}^2 + 8\bar{y} + 16 + 6\bar{x} - 18 - 8\bar{y} - 32 - 11 = 0$$

After simplifying this equation, we obtain $\bar{x}^2 + \bar{y}^2 = 36$.

Figure 6

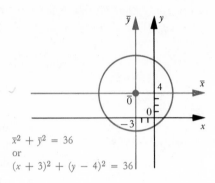

$\bar{x}^2 + \bar{y}^2 = 36$
or
$(x + 3)^2 + (y - 4)^2 = 36$

3   Find a translation of axes that reduces the equation $x^2 + y^2 - 4x + 2y - 4 = 0$ to the form $\bar{x}^2 + \bar{y}^2 = r^2$.

SOLUTION.   (See Figure 7.) By completing the square for $x^2 + y^2 - 4x + 2y - 4 = 0$, we get the equation of the circle,

$$(x - 2)^2 + (y + 1)^2 = 9$$

so if we let $\bar{x} = x - 2$ and $\bar{y} = y + 1$, we get the equation of the circle in the simple form

$$\bar{x}^2 + \bar{y}^2 = 9$$

Figure 7

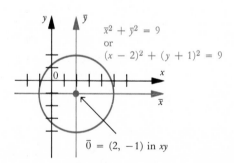

$\bar{x}^2 + \bar{y}^2 = 9$
or
$(x - 2)^2 + (y + 1)^2 = 9$

$\bar{0} = (2, -1)$ in xy

PROBLEM SET 2

1   Indicate the xy coordinates of the point whose $\bar{x}\bar{y}$ coordinates are
   a) $(3, -4)$ b) $(2, 2)$ c) $(5, 7)$
   d) $(3, -5)$ e) $(0, 0)$ f) $(1, -3)$
     i   If $\bar{x} = x + 1$ and $\bar{y} = \overset{*}{y} - 3$
     ii   If the translation is the one that takes $(5, 6)$ in the xy system to $(-1, 3)$ in the $\bar{x}\bar{y}$ system

2   If $\bar{x} = x - 1$ and $\bar{y} = y + 2$, show that

$$\{(x, y) \mid y = 2x + 1\} = \{(\bar{x}, \bar{y}) \mid \bar{y} = 2\bar{x} + 5\}$$

3 Indicate the $\overline{xy}$ coordinates of the point whose xy coordinates are

a) $(2, -7)$ b) $(-5, 3)$ c) $(-3, 4)$

d) $(\frac{7}{4}, -\frac{13}{4})$ e) $(-7, -5)$ f) $(3, -7)$

    i If $\overline{x} = x + 1$ and $\overline{y} = y - 3$

    ii If the translation is the one that takes $(3, 4)$ in the xy system to $(0, 0)$ in the $\overline{xy}$ system

4 We say that a quantity is *invariant* under a process if the quantity remains the same after the process has been carried out. Prove that each of the following is invariant under a translation.

a) The distance between $P_1 = (x_1, y_1)$ and $P_2 = (x_2, y_2)$

b) The area of a triangle whose base is b and whose height is h

c) The slope of a line joining $P_1 = (x_1, y_1)$ and $P_2 = (x_2, y_2)$

5 Find the translation of axes that will reduce each of the following equations to the form $\overline{x}^2 + \overline{y}^2 = r^2$.

a) $2x^2 + 2y^2 + 16x - 7y = 0$

b) $3x^2 + 3y^2 + 7x - 5y + 3 = 0$

c) $x^2 + y^2 - 8x - 10y + 40 = 0$

6 Suppose that the coordinate axes xy are translated, so that the origin of the $\overline{xy}$ coordinate system is the point whose coordinates are the indicated point. Write each of the following equations in terms of $\overline{x}$ and $\overline{y}$.

a) $3x - 4y + 13 = 0, (-2, 3)$

b) $-2x + 3y + 5 = 0, (3, 4)$

c) $x^2 - 2x + y^2 - 4y - 16 = 0, (1, -1)$

d) $3x^2 + 3y^2 - 9x - 7y - 36 = 0, (1, 0)$

4 Ellipse

With the tool of translations available, let us study the *ellipse*. The planets follow an elliptical orbit about the sun, and satellites follow a very nearly elliptical orbit about the earth.

Geometrically, an *ellipse* in a plane can be defined as the set of all points each of which has the property that the sum of its distances from two fixed points, called *foci*, is a constant. In Figure 1, F_1 and F_2 are the foci, P_1, P_2, and P_3 are points of the conic, and

$$d_1 + c_1 = d_2 + c_2 = d_3 + c_3 = k$$

where k is the constant of the ellipse. The midpoint of line segment $\overline{F_1 F_2}$ is called the *center of the ellipse*.

Figure 1

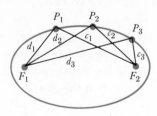

The ellipse is symmetric with respect to each of two perpendicular lines that intersect at its center (Figure 2).

Figure 2

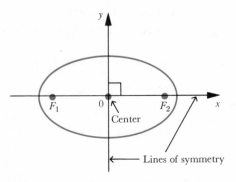

The four points of intersection of the lines of symmetry and the ellipse are called *vertices* of the ellipse; the longer line segment determined by the vertices is called the *major axis*, whereas the shorter line segment determined by the vertices is called the *minor axis*. In Figure 3, V_1, V_2, V_3, and V_4 are the vertices, $\overline{V_1 V_2}$ is the major axis, $\overline{V_3 V_4}$ is the minor axis, and O is the center.

Figure 3

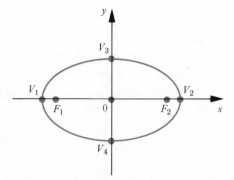

In order to express the geometric description of an ellipse in analytic terms, let us choose the coordinate system so that the coordinates of the foci are $(-c, 0)$ and $(c, 0)$, where $c > 0$. Also, we will assume that the constant,

k, which is equal to the sum of the distances between a point on the ellipse and the foci is $2a$ (Figure 4). (This constant is written in the form $2a$ so that the equation of the ellipse will have a simpler form.) Notice that the major axis lies on the x axis; the minor axis lies on the y axis; the center of the ellipse is $(0,0)$.

It is not difficult to show that the major axis, that is, the line segment $\overline{V_1 V_2}$, has length $2a$. $(\overline{F_1 O} + \overline{OV_2}) + (\overline{OV_2} - \overline{OF_2}) = 2a$, since V_2 is a point on the ellipse; therefore, $c + \overline{OV_2} + \overline{OV_2} - c = \overline{2OV_2} = 2a$, so that $\overline{OV_2} = a$. By symmetry, $\overline{OV_1} = a$; hence $\overline{V_1 V_2} = 2a$.

Figure 4

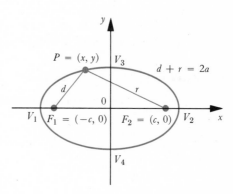

Next we shall show the relationship between the length of the major and the length of the minor axis. Let the distance $\overline{V_3 V_4} = 2b$ (Figure 5). Since V_3 is a point on the ellipse, $\overline{V_3 F_1} + \overline{V_3 F_2} = 2a$. By symmetry, $\overline{V_3 F_1} = \overline{V_3 F_2}$, so $\overline{V_3 F_2} = a$. From the right triangle, $\triangle OF_2 V_3$, we get $a^2 = b^2 + c^2$, from which we conclude that $b < a$, and the major axis is longer than the minor axis.

Figure 5

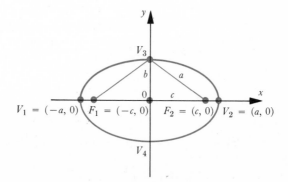

The important distances associated with the ellipse are displayed in Figure 6.

Figure 6

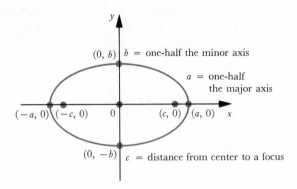

$(0, b)$ b = one-half the minor axis

a = one-half
the major axis

$(-a, 0)$ $(-c, 0)$ 0 $(c, 0)$ $(a, 0)$ x

$(0, -b)$ c = distance from center to a focus

Now the equation of the ellipse will be derived in the following theorem.

THEOREM 1 ELLIPSE EQUATION

An equation for the ellipse with foci at $F_2 = (c, 0)$ and $F_1 = (-c, 0)$ is

$$\frac{x^2}{a^2} + \frac{y^2}{b^2} = 1$$

where $b^2 = a^2 - c^2$, $a > b$, $2a$ is the length of the major axis, and $2b$ is the length of the minor axis.

PROOF.   Assume that $P = (x, y)$ is any point on the ellipse (Figure 7).

Figure 7

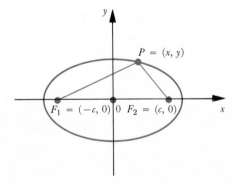

$P = (x, y)$

$F_1 = (-c, 0)$ 0 $F_2 = (c, 0)$ x

Then $\overline{PF}_1 + \overline{PF}_2 = 2a$, so that by the distance formula

$$\sqrt{(x+c)^2 + y^2} + \sqrt{(x-c)^2 + y^2} = 2a$$

That is,

$$\sqrt{(x+c)^2 + y^2} = 2a - \sqrt{(x-c)^2 + y^2}$$

Squaring both sides of the latter equation we get

$$x^2 + 2xc + c^2 + y^2 = 4a^2 - 4a\sqrt{(x-c)^2 + y^2} + x^2 - 2cx + c^2 + y^2$$

so that

$$4cx - 4a^2 = -4a\sqrt{(x-c)^2 + y^2}$$

That is,

$$cx - a^2 = -a\sqrt{(x-c)^2 + y^2}$$

Squaring both sides of the equation again, we get

$$c^2x^2 - 2a^2cx + a^4 = a^2(x^2 - 2cx + c^2 + y^2)$$

so that

$$a^4 - a^2c^2 = (a^2 - c^2)x^2 + a^2y^2$$

and

$$(a^2 - c^2)x^2 + a^2y^2 = a^2(a^2 - c^2)$$

Dividing both sides of the equation by $a^2(a^2 - c^2)$, we get

$$\frac{x^2}{a^2} + \frac{y^2}{a^2 - c^2} = 1$$

Since $a^2 - c^2 = b^2$, we have $x^2/a^2 + y^2/b^2 = 1$. The graph of the ellipse is shown in Figure 8.

Figure 8

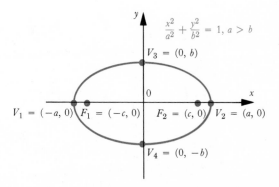

$$\frac{x^2}{a^2} + \frac{y^2}{b^2} = 1, a > b$$

$V_3 = (0, b)$

$V_1 = (-a, 0)$ $F_1 = (-c, 0)$ $F_2 = (c, 0)$ $V_2 = (a, 0)$

$V_4 = (0, -b)$

EXAMPLES

1 Given the equation of the ellipse $x^2 + 9y^2 = 9$, find the vertices and the foci, and sketch the graph.

SOLUTION.   (See Figure 9.) After dividing both sides of the equation by 9, we get the equation

$$\frac{x^2}{9} + \frac{y^2}{1} = 1$$

This equation is of the form $x^2/a^2 + y^2/b^2 = 1$, with $a^2 = 9$ and $b^2 = 1$. Thus the graph is an ellipse whose vertices are $(3,0)$, $(-3,0)$, $(0,1)$, and $(0,-1)$. Since $c^2 = a^2 - b^2 = 9 - 1 = 8$, we have $c = 2\sqrt{2}$, so that the coordinates of the foci are $(2\sqrt{2},0)$ and $(-2\sqrt{2},0)$.

Figure 9

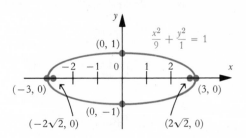

2   Find the equation of the ellipse with foci $(2,0)$ and $(-2,0)$ and vertices $(3,0)$ and $(-3,0)$. Also sketch the graph of the ellipse.

SOLUTION.   (See Figure 10.) The ellipse has its center at the origin and its foci on the x axis; therefore, its equation is of the form $x^2/a^2 + y^2/b^2 = 1$, with $b^2 = c^2 - a^2 = 9 - 4 = 5$, so that $b = \sqrt{5}$; thus the ellipse has an equation of the form

$$\frac{x^2}{9} + \frac{y^2}{5} = 1$$

Figure 10

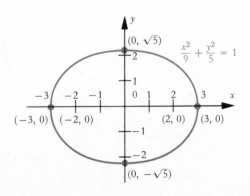

Since we denoted a to be half of the major axis and b to be half of the minor axis, the equation of the ellipse takes the form $x^2/a^2 + y^2/b^2 = 1$. Its major axis is along the x axis if $a > b$; and its foci are on the x axis.

If the foci of an ellipse centered at the origin are on the y axis and $2a$ is the sum of the distances to the foci from each point on the ellipse, then, by symmetry, its equation will become $x^2/b^2 + y^2/a^2 = 1$, and, since $a > b$, we still have $c^2 = a^2 - b^2$. The graph is shown in Figure 11.

Figure 11

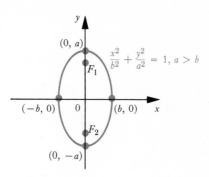

EXAMPLE

Given the equation of the ellipse $4x^2 + y^2 = 4$, find the vertices and the foci, and sketch the graph.

SOLUTION.   (See Figure 12.) After dividing both sides of the equation by 4, we get the equation

$$\frac{x^2}{1} + \frac{y^2}{4} = 1$$

Figure 12

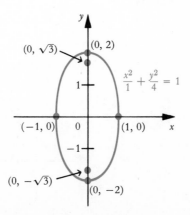

This equation is of the form $x^2/b^2 + y^2/a^2 = 1$, with $b^2 = 1$ and $a^2 = 4$. Thus the graph is an ellipse whose vertices are $(0, 2)$, $(0, -2)$, $(1, 0)$, and $(-1, 0)$. Since $c^2 = a^2 - b^2 = 4 - 1 = 3$, $c = \sqrt{3}$, and the coordinates of the foci are $(0, -\sqrt{3})$ and $(0, \sqrt{3})$.

When we use the translation equations $\overline{x} = x - h$ and $\overline{y} = y - k$, then the equations of the ellipses with centers at (h, k) in the xy coordinate system are as follows:

If the major axis is horizontal (Figure 13):

$$\frac{(x-h)^2}{a^2} + \frac{(y-k)^2}{b^2} = 1 \qquad a > b$$

Figure 13

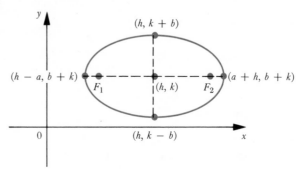

If the major axis is vertical (Figure 14):

$$\frac{(x-h)^2}{b^2} + \frac{(y-k)^2}{a^2} = 1 \qquad a > b$$

Figure 14

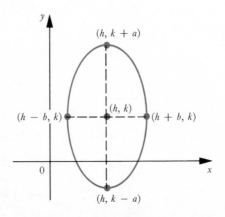

EXAMPLES

1 Given the equation of the ellipse $4x^2 + 9y^2 + 16x - 18y - 11 = 0$, find the coordinates of the center, the coordinates of the vertices, and the coordinates of the foci. Also sketch the graph.

SOLUTION. (See Figure 15.) First, complete the square to get

$$4(x^2 + 4x + 4) + 9(y^2 - 2y + 1) = 11 + 16 + 9$$

so that

$$4(x + 2)^2 + 9(y - 1)^2 = 36$$

or, equivalently,

$$\frac{(x+2)^2}{9} + \frac{(y-1)^2}{4} = 1$$

If we let $\bar{x} = x + 2$ and $\bar{y} = y - 1$, then $\bar{x}^2/9 + \bar{y}^2/4 = 1$. The center is $(-2, 1)$. $a^2 = 9$ and $b^2 = 4$, so that $a = 3$ and $b = 2$ and the coordinates of the vertices are $(1, 1)$, $(-5, 1)$, $(-2, -1)$, and $(-2, 3)$, since the major axis is parallel to the x axis. $c^2 = a^2 - b^2 = 9 - 4 = 5$; therefore, $c = \sqrt{5}$, so that the coordinates of the foci are $(-2 + \sqrt{5}, 1)$ and $(-2 - \sqrt{5}, 1)$.

Figure 15

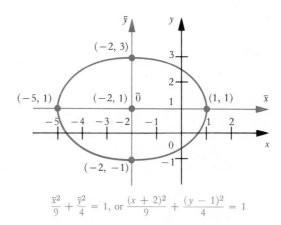

$$\frac{\bar{x}^2}{9} + \frac{\bar{y}^2}{4} = 1, \text{ or } \frac{(x+2)^2}{9} + \frac{(y-1)^2}{4} = 1$$

2 Given the equation of the ellipse $9x^2 + 4y^2 - 18x + 16y - 11 = 0$, find the coordinates of the center, the coordinates of the vertices, and the coordinates of the foci. Also sketch the graph.

SOLUTION. (See Figure 16.) First, complete the square to get

$$9(x^2 - 2x + 1) + 4(y^2 + 4y + 4) = 11 + 9 + 16$$

Hence

$$9(x-1)^2 + 4(y+2)^2 = 36$$

That is,

$$\frac{(x-1)^2}{4} + \frac{(y+2)^2}{9} = 1$$

If we let $\bar{x} = x-1$ and $\bar{y} = y+2$, we get $\bar{x}^2/4 + \bar{y}^2/9 = 1$; the center is $(1, -2)$ in the xy system.

$a^2 = 9$ and $b^2 = 4$; therefore, $a = 3$ and $b = 2$, and the coordinates of the vertices are $(1,1)$, $(1, -5)$, $(-1, -2)$, and $(3, -2)$, since the major axis is parallel to the y axis. $c^2 = a^2 - b^2 = 9 - 4 = 5$; therefore, $c = \sqrt{5}$ and the coordinates of the foci are $\left(1, -2+\sqrt{5}\right)$ and $\left(1, -2-\sqrt{5}\right)$.

Figure 16

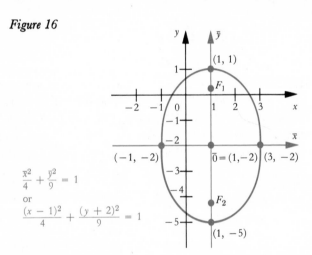

3   Write the equation of the ellipse whose vertices are $(-2, -3)$, $(-2, 5)$, $(-7, 1)$, and $(3, 1)$ and sketch its graph.

SOLUTION.   First, locate the given vertices as shown in Figure 17. These four points are the ends of the major axis and the minor axis; the major axis is parallel to the x axis and is 10 units long (why?), and the minor axis is parallel to the y axis and is 8 units long (why?). Thus $a = 5$ and $b = 4$. In this case the center is at $(-2, 1)$ (why?), so that $h = -2$ and $k = 1$; hence the equation in the $\overline{xy}$ system is

$$\frac{\bar{x}^2}{25} + \frac{\bar{y}^2}{16} = 1 \qquad \text{where } \bar{x} = x + 2 \text{ and } \bar{y} = y - 1$$

or

$$\frac{(x+2)^2}{25} + \frac{(y-1)^2}{16} = 1 \qquad \text{in the } xy \text{ system}$$

Figure 17

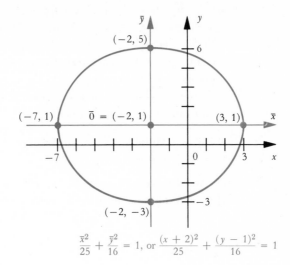

$$\frac{\bar{x}^2}{25} + \frac{\bar{y}^2}{16} = 1, \text{ or } \frac{(x+2)^2}{25} + \frac{(y-1)^2}{16} = 1$$

PROBLEM SET 3

1 For each of the following ellipses find the coordinates of the vertices and the coordinates of the foci, and sketch the graph.

a) $x^2/16 + y^2/9 = 1$ b) $y^2/25 + x^2/16 = 1$

c) $y^2/16 + x^2/4 = 1$ d) $4x^2 + 9y^2 = 36$

e) $4x^2 + 16y^2 = 64$ f) $25x^2 + 9y^2 = 1$

2 Use the graph of an ellipse to show that the equation of an ellipse is a relation that is not a function.

3 For each of the following ellipses find the coordinates of the center, the coordinates of the vertices, the coordinates of the foci, and sketch the graph.

a) $3(x-1)^2 + 4(y+2)^2 = 192$

b) $25(x-3)^2 + 4(y-1)^2 = 100$

c) $16(x+2)^2 + 25(y-1)^2 = 400$

d) $x^2 + 4y^2 - 2x - 16y + 11 = 0$

e) $9x^2 + 4y^2 + 18x - 16y - 11 = 0$

f) $4x^2 + 9y^2 - 24x + 36y + 36 = 0$

4 For each of the relations of Problem 1, indicate the domain and range.

5 Find the equations of the ellipse whose vertices are:

a) $(1, -2)$, $(5, -2)$, $(3, -7)$, and $(3, 3)$

b) $(0, -1)$, $(12, -1)$, $(6, -4)$, and $(6, 2)$

c) $(1, 1)$, $(5, 1)$, $(3, 6)$, and $(3, -4)$

6 Discuss the symmetry of an ellipse of the form $x^2/a^2 + y^2/b^2 = 1$.

7 Find the equation of the ellipse for each of the following cases.

a) Vertices $(-5,0)$ and $(5,0)$ and containing the point $(4, \frac{12}{5})$

b) Foci $(1,4)$ and $(3,4)$ and with major axis of length 4.

c) Center at $(0,0)$, axes parallel to coordinate axes, and containing the points $(3\sqrt{3}/2, 1)$ and $(2, 2\sqrt{5}/3)$

d) Center at $(-3,1)$, major axis parallel to the y axis and 10 units long, and minor axis 2 units long.

8 Consider the equation of the ellipse $x^2/a^2 + y^2/b^2 = 1$. Describe the graph of the ellipse if $a = b$.

9 Compare the graphs of $y = \sqrt{1-(x^2/4)}$, $y = -\sqrt{1-(x^2/4)}$, and $y^2 = 1-(x^2/4)$. Which of the three equations defines a function?

5 Hyperbola

The *hyperbola* will be developed in a manner similar to that used to develop the ellipse. **Geometrically,** a hyperbola in a plane can be defined as the set of points each of which has the property that the absolute value of the difference of the distances from each point on the hyperbola to two distinct fixed points, called *foci*, is equal to a constant.

The line determined by the foci is a line of symmetry; the midpoint of the line segment determined by the foci is the *center* of the hyperbola; the two points of intersection of the hyperbola with the line of symmetry are called the *vertices* of the hyperbola; the line segment determined by the vertices is called the *transverse axis*. In Figure 1 F_1 and F_2 are the foci, V_1 and V_2 are the vertices, C is the center, $V_1 V_2$ is the transverse axis, and

$$|d_2 - c_2| = |d_1 - c_1| = k$$

where k is the constant of the hyperbola.

Figure 1

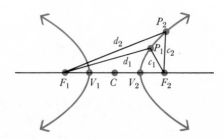

We will determine the equation of the hyperbola by choosing the coordinate system so that the foci are $(c,0)$ and $(-c,0)$, with $c > 0$, and we will assume the constant difference k to be $2a$. Thus, if $P = (x, y)$ is a point on the hyperbola, we have $|\overline{PF}_1 - \overline{PF}_2| = 2a$, as shown in Figure 2.

Figure 2

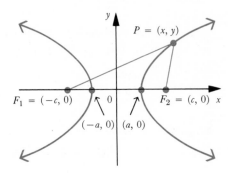

By using the distance formula, we get

$$\left| \sqrt{(x+c)^2 + y^2} - \sqrt{(x-c)^2 + y^2} \right| = 2a$$

Hence

$$\sqrt{(x+c)^2 + y^2} - \sqrt{(x-c)^2 + y^2} = \pm 2a$$

so that

$$\sqrt{(x+c)^2 + y^2} = \pm 2a + \sqrt{(x-c)^2 + y^2}$$

After squaring both sides of the equation we get

$$x^2 + 2cx + c^2 + y^2 = 4a^2 \pm 4a\sqrt{(x-c)^2 + y^2} + x^2 - 2cx + c^2 + y^2$$

or, equivalently,

$$4cx - 4a^2 = \pm 4a\sqrt{(x-c)^2 + y^2}$$

so that

$$cx - a^2 = \pm a\sqrt{(x-c)^2 + y^2}$$

Again, square both sides of the equation to get

$$c^2x^2 - 2a^2cx + a^4 = a^2(x^2 - 2cx + c^2 + y^2)$$

That is

$$(c^2 - a^2)x^2 - a^2y^2 = a^2c^2 - a^4 = a^2(c^2 - a^2)$$

so that, after dividing both sides by $a^2(c^2 - a^2)$, we get

$$\frac{x^2}{a^2} - \frac{y^2}{c^2 - a^2} = 1$$

If we let $b^2 = c^2 - a^2$, the equation of the hyperbola becomes

$$\frac{x^2}{a^2} - \frac{y^2}{b^2} = 1$$

Notice that by letting $y = 0$, we get $(a, 0)$ and $(-a, 0)$, the vertices of the hyperbola. Also notice that $a^2 + b^2 = c^2$, because we defined $b^2 = c^2 - a^2$.

We have just proved the following theorem.

THEOREM 1 HYPERBOLA EQUATION

The equation of the hyperbola with foci $(c, 0)$ and $(-c, 0)$ and constant distance $2a$ is

$$\frac{x^2}{a^2} - \frac{y^2}{b^2} = 1 \qquad \text{where } b^2 = c^2 - a^2$$

5.1 Properties of the Hyperbola

The graph of the hyperbola $x^2/a^2 - y^2/b^2 = 1$ is symmetric with respect to the x axis and the y axis. (Why?)

The x intercepts of the hyperbola $x^2/a^2 - y^2/b^2 = 1$ are found by letting $y = 0$ and solving for x to get $(-a, 0)$ and $(a, 0)$. If $x = 0$, then we have $y^2 = -b^2$. But such a y is not a real number (why?); hence we conclude that the hyperbola does not intersect the y axis.

If we write $x^2/a^2 - y^2/b^2 = 1$ as $y^2 = (b^2/a^2)(x^2 - a^2)$, then

$$y = \frac{\pm bx}{a}\sqrt{1 - \frac{a^2}{x^2}}$$

Figure 3

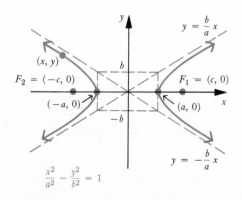

$$\frac{x^2}{a^2} - \frac{y^2}{b^2} = 1$$

Now, $1 - a^2/x^2$ approaches 1 as $|x|$ gets very large, so that the larger x is in absolute value, the closer the graph of the hyperbola is to the lines whose equations are given by $y = \pm(b/a)x$. These lines are called the *asymptotes of the hyperbola*. The asymptotes are not part of the hyperbola; they are easy to draw if we construct the rectangle of dimensions $2a$ and $2b$, as in Figure 3.

EXAMPLES

1 Given the equation of the hyperbola $x^2/16 - y^2/9 = 1$, find the coordinates of the foci, the coordinates of the vertices, and the equations of the asymptotes. Also sketch the graph.

SOLUTION. From Theorem 1, since $a = 4$ and $b = 3$, $c = \sqrt{4^2 + 3^2} = 5$, so that the coordinates of the foci are $(5, 0)$ and $(-5, 0)$ and the coordinates of the vertices are $(4, 0)$ and $(-4, 0)$. The equations of the asymptotes are given by $y = \pm(b/a)x = \pm\frac{3}{4}x$ (Figure 4).

Figure 4

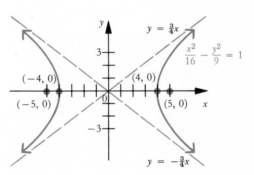

2 Given the hyperbola $25x^2 - 16y^2 = 400$, find the coordinates of the vertices, the coordinates of the foci, and the equations of the asymptotes. Also sketch the graph.

SOLUTION. Divide both sides of the equation by 400 to get

$$\frac{x^2}{16} - \frac{y^2}{25} = 1$$

The vertices are $(4, 0)$ and $(-4, 0)$. Since $c^2 = a^2 + b^2 = 16 + 25 = 41$, $c = \sqrt{41}$, so that the foci are $(-\sqrt{41}, 0)$ and $(\sqrt{41}, 0)$. The equations of the asymptotes are given by $y = \pm(b/a)x = \pm\frac{5}{4}x$ (Figure 5).

The equations of the hyperbola with center $(0, 0)$ are as follows:
If its transverse axis is on the x axis:

$$\frac{x^2}{a^2} - \frac{y^2}{b^2} = 1$$

Figure 5

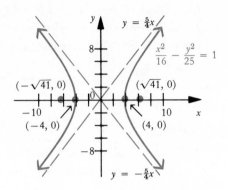

If its transverse axis is on the y axis (by symmetry):

$$\frac{y^2}{a^2} - \frac{x^2}{b^2} = 1$$

As with the ellipse, if the center of the hyperbola is translated so that the point (h, k) is the center, the forms of the equation are as follows:
If its transverse axis is parallel to the x axis (**Figure 6**):

$$\frac{(x-h)^2}{a^2} - \frac{(y-k)^2}{b^2} = 1$$

Figure 6

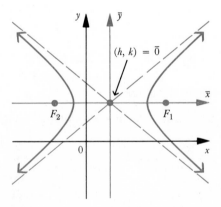

If its transverse axis is parallel to the y axis (**Figure 7**):

$$\frac{(y-k)^2}{a^2} - \frac{(x-h)^2}{b^2} = 1$$

Figure 7

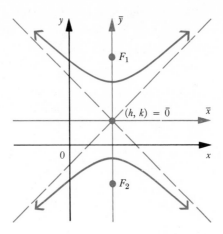

EXAMPLES

1   Given the equation of the hyperbola $4x^2 - y^2 - 8x + 2y + 7 = 0$, find the coordinates of the center, the coordinates of the foci, the coordinates of the vertices, and the equations of the asymptotes. Also sketch the graph.

SOLUTION.   (See Figure 8.) First, complete the square to get

$$4(x^2 - 2x + 1) - (y^2 - 2y + 1) = -7 + 4 - 1$$

That is,

$$4(x - 1)^2 - (y - 1)^2 = -4$$

so that

$$\frac{(y-1)^2}{4} - \frac{(x-1)^2}{1} = 1$$

Figure 8

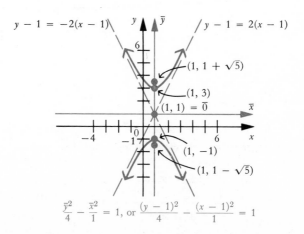

Hence the center $(h, k) = (1, 1)$. The coordinates of the foci are $(1, 1 + \sqrt{5})$ and $(1, 1 - \sqrt{5})$, since $c^2 = 4 + 1 = 5$. The coordinates of the vertices are $(1, -1)$ and $(1, 3)$, since $a = 2$ and the transverse axis is on the line $x = 1$. $y - k = \pm(a/b)(x - h)$ gives the asymptotes $y - 1 = \pm 2(x - 1)$.

2   Find the equation of the hyperbola whose foci are the points $(-\frac{9}{2}, 3)$ and $(\frac{1}{2}, 3)$ and whose vertices are the points $(0, 3)$ and $(-4, 3)$; also, sketch the graph.

SOLUTION.   (See Figure 9.) First, locate these points on the coordinate axes; the transverse axis is parallel to the x axis. The center is $[(-\frac{9}{2} + \frac{1}{2})/2, (3 + 3)/2] = (-2, 3)$; $a = 2$ and $c = \frac{5}{2}$. Since $b^2 = c^2 - a^2 = \frac{25}{4} - 4 = \frac{9}{4}$, $b = \frac{3}{2}$. Hence the equation in the $\overline{xy}$ system is

$$\frac{\overline{x}^2}{4} - \frac{\overline{y}^2}{\frac{9}{4}} = 1 \qquad \text{with } x = \overline{x} + 2 \text{ and } \overline{y} = y - 3$$

or, equivalently, in the xy system the equation is

$$\frac{(x+2)^2}{4} - \frac{(y-3)^2}{\frac{9}{4}} = 1$$

Figure 9

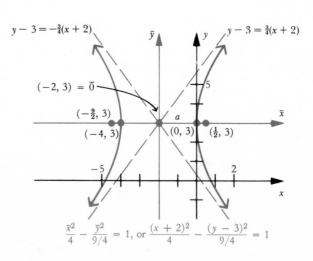

$$\frac{\overline{x}^2}{4} - \frac{\overline{y}^2}{9/4} = 1, \text{ or } \frac{(x+2)^2}{4} - \frac{(y-3)^2}{9/4} = 1$$

PROBLEM SET 4

1   For each of the following hyperbolas find the coordinates of the vertices, the coordinates of the foci, and the equations of the asymptotes, and sketch the graph.

a) $x^2/49 - y^2/36 = 1$ b) $16y^2 - 4x^2 = 48$

c) $y^2 - x^2 = 9$ d) $36x^2 - 9y^2 = 1$

e) $x^2 - 9y^2 = 9$ f) $9x^2 - y^2 = 9$

2 Find the equation of the hyperbola with vertices at $(-16, 0)$ and $(16, 0)$ and asymptotes $y = \pm\frac{5}{4}x$.

3 For each of the following hyperbolas find the coordinates of the center, the coordinates of the vertices, the coordinates of the foci, and the equations of the asymptotes, and sketch the graph.
a) $(x-5)^2/25 - (y-4)^2/4 = 1$
b) $(x+2)^2/16 - (y-3)^2/9 = 1$
c) $(y-2)^2/16 - (x+3)^2/25 = 1$
d) $4x^2 - y^2 + 8x - 2y + 6 = 0$
e) $4x^2 - 3y^2 - 32x + 6y + 73 = 0$
f) $4x^2 - 9y^2 - 32x + 36y + 27 = 0$

4 a) Discuss the symmetry of the graph of a hyperbola.
b) Use the graph to show that the equation of the hyperbola is not a function.

5 Find the equation of the hyperbola with vertices at $(-1, 4)$ and $(-1, 6)$ and foci at $(-1, 3)$ and $(-1, 7)$.

6 The line segment with end points on a hyperbola that contains a focus and is perpendicular to the transverse axis is called a *focal chord*, or *latus rectum*, of the hyperbola. Show that the length of the focal chord of the hyperbola $x^2/a^2 - y^2/b^2 = 1$ is $2b^2/a$.

7 Find the equation of the hyperbola in each of the following cases.
a) Center at $(2, 3)$, a focus at $(2, 5)$, and a focal chord of length 6
b) Center at $(-2, 1)$, a focus at $(-2, 6)$, and a focal chord of length $\frac{32}{3}$

8 Find the domain and range of each of the relations of Problem 1.

9 Write the general equation for a hyperbola in determinant form if the transverse axis is parallel to the x axis.

6 Parabola

Geometrically, a *parabola* in a plane can be defined as the set of points, each of which is equidistant from a given point called the *focus* and from a given line called the *directrix*. In Figure 1, P_1, P_2, and P_3 are on the parabola; hence

$$d_1 = c_1 \qquad d_2 = c_2 \qquad d_3 = c_3$$

It is not difficult to derive the equation of a parabola from this geometric description.

Figure 1

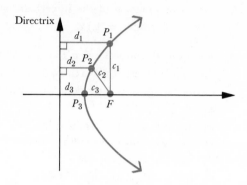

THEOREM 1 PARABOLA EQUATION

Let $P = (x,y)$ be any point on a parabola with focus $(c, 0)$ and directrix $x = -c$; then its equation is $y^2 = 4cx$, where $c > 0$ **(Figure 2)**.

PROOF.   (See Figure 2.) Since $P = (x, y)$ is equidistant from the focus and directrix, we can write $\overline{PF} = \overline{PD}$, so that

$$\sqrt{(x-c)^2 + y^2} = |x + c|$$

After squaring both sides of the equation, we get

$$(x-c)^2 + y^2 = (x+c)^2$$

or, equivalently,

$$x^2 - 2cx + c^2 + y^2 = x^2 + 2cx + c^2$$

That is,

$$y^2 = 4cx$$

Figure 2

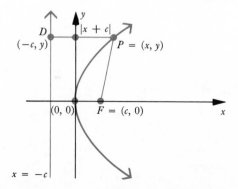

The line passing through the focus of the parabola perpendicular to the directrix is called the *axis of symmetry*; the point on the axis of symmetry midway between the directrix and focus is called the *center*, or *vertex*, of the parabola; the line segment with end points on the parabola perpendicular to the axis at the focus is called the *focal chord*, or *latus rectum*, of the parabola. **In Figure 3, V is the vertex, F is the focus, and segment $\overline{AB}$ is the focal chord.**

Figure 3

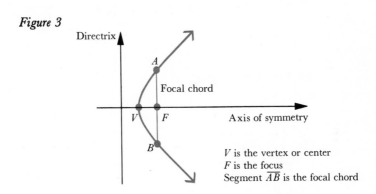

V is the vertex or center
F is the focus
Segment $\overline{AB}$ is the focal chord

Notice that the graph of $y^2 = 4cx$ has the x axis as its axis of symmetry. Also, no point of the graph lies to the left of the y axis, for, if $x < 0$, we have $y^2 < 0$ (why?) and y cannot be real. The directrix is parallel to the y axis, and the focus is a point to the right of the vertex. This case, which might be described by saying that the parabola "opens to the right," is one of four possible cases that could have been developed here. These cases are as follows (assume that $c > 0$ in all cases):

CASE 1. $y^2 = 4cx$. The focus is on the x axis, the directrix is parallel to the y axis, and the vertex is $(0, 0)$. The parabola opens to the right **(Figure 4).**

Figure 4

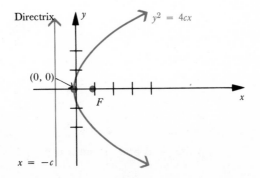

CASE 2. $y^2 = -4cx$. The focus is on the x axis, the directrix is parallel to the y axis, and the vertex is $(0, 0)$. The parabola opens to the left **(Figure 5)**.

Figure 5

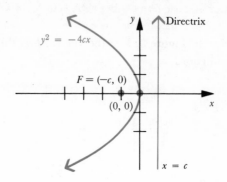

CASE 3. $x^2 = 4cy$. The focus is on the y axis, the directrix is parallel to the x axis, and the vertex is $(0, 0)$. The parabola opens upward **(Figure 6)**.

Figure 6

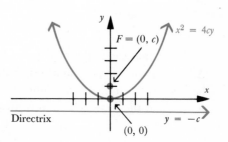

CASE 4. $x^2 = -4cy$. The focus is on the y axis, the directrix is parallel to the x axis, and the vertex is $(0, 0)$. The parabola opens downward **(Figure 7)**.

Figure 7

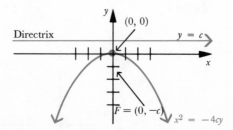

EXAMPLES

1 Find the focus and directrix of $x^2 = 12y$ and sketch the graph.

SOLUTION. This is an example of case 3. The graph is symmetric with respect to the y axis because $(-x, y)$ lies on the graph whenever (x, y) lies on the graph. Since x^2 in $x^2 = 12y$ is always nonnegative, $y \geq 0$, so the parabola opens upward. Since $4c = 12$, $c = 3$; therefore, the focus is $(0, 3)$, and the equation of the directrix is $y = -3$ (Figure 8).

Figure 8

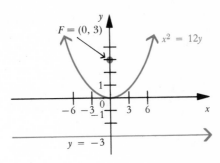

2 A parabola with its vertex at the origin has its focus at $(6, 0)$. Find its equation and sketch the graph.

SOLUTION. Since the focus is $(6, 0)$, the directrix is the line $x = -6$. The axis of symmetry in this case is the x axis; hence its equation is $y^2 = 4cx$. Since $c = 6$, the equation is $y^2 = 24x$. This is an example of case 1. The graph is shown in Figure 9.

Figure 9

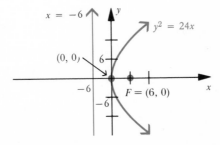

If the vertex of a parabola is translated to the point (h, k), then the equation of the parabola can take any one of the following four forms (assume that $c > 0$ in all cases):

1 $(y - k)^2 = 4c(x - h)$. The vertex of the parabola is the point (h, k) and the directrix is the line $x = h - c$ (Figure 10).

Figure 10

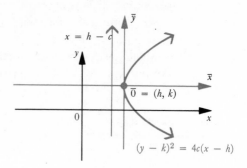

$$(y - k)^2 = 4c(x - h)$$

2 $(y-k)^2 = -4c(x-h)$. The vertex of the parabola is the point (h,k), and the directrix is the line $x = h+c$ (**Figure 11**).

Figure 11

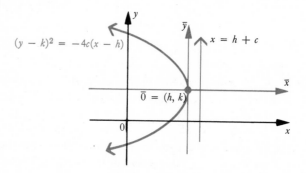

3 $(x-h)^2 = 4c(y-k)$. The vertex of the parabola is the point (h,k), and the directrix is the line $y = k-c$ (**Figure 12**).

Figure 12

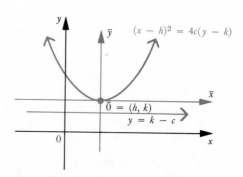

4 $(x-h)^2 = -4c(y-k)$. The vertex of the parabola is the point (h,k), and the directrix is the line $y = k+c$ (**Figure 13**).

Figure 13

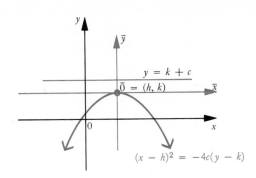

EXAMPLES

1 For the parabola $y^2+2y-8x-3 = 0$, find the vertex, the focus, the equation of the directrix, and the length of the focal chord. Also sketch the graph.

SOLUTION. (See Figure 14.) First, complete the square in $y^2+2y-8x-3 = 0$ to get $y^2+2y+1 = 8x+3+1$. Hence $(y+1)^2 = 8x+4 = 4(2x+1)$; that is, $(y+1)^2 = 8(x+\frac{1}{2})$ or, equivalently, $\bar{y}^2 = 8\bar{x}$, where $\bar{y} = y+1$ and $\bar{x} = x+\frac{1}{2}$; therefore, the vertex is $(-\frac{1}{2},-1)$. Since $4c = 8$, it follows that $c = 2$; hence, we have the focus at $(-\frac{1}{2}+2,-1) = (\frac{3}{2},-1)$. The equation of the directrix is $x = -\frac{1}{2}-2 = -\frac{5}{2}$. The length of the focal chord is 8.

Figure 14

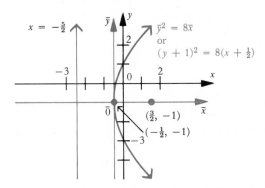

2 Consider the equation of the parabola $x^2+2x+4y-7 = 0$. Find the vertex, the focus, and the equation of the directrix. Also sketch the graph.

SOLUTION. (See Figure 15.) After completing the square, we have

$$x^2 + 2x + 1 = -4y + 7 + 1 = -4y + 8$$

That is,

$$(x+1)^2 = -4(y-2)$$

Thus the vertex is $(-1, 2)$, and the parabola opens downward because of the negative sign. Since $4c = 4$, $c = 1$ and the focus is $(-1, 1)$; the equation of the directrix is $y = 2 + 1 = 3$.

Figure 15

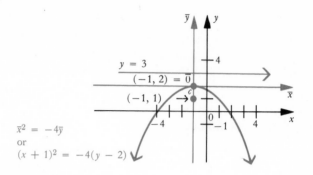

$$\bar{x}^2 = -4\bar{y}$$
or
$$(x + 1)^2 = -4(y - 2)$$

PROBLEM SET 5

1 For each of the following parabolas find the vertex, the focus, the equation of the directrix, and the length of the focal chord, and sketch the graph.
   a) $y^2 = 8x$ b) $x^2 + 2y = 0$
   c) $x^2 - 4y = 0$ d) $y^2 + 4x = 0$
   e) $y^2 + 5x = 0$ f) $3x^2 - 2y = 0$

2 If the vertex of a parabola is translated to the point (h, k), then the equation of the parabola can take on the following forms, where $c > 0$.
   a) $(y - k)^2 = 4c(x - h)$ b) $(y - k)^2 = -4c(x - h)$
   c) $(x - h)^2 = 4c(y - k)$ d) $(x - h)^2 = -4c(y - k)$
   Sketch the graph of a specific parabola for each of the above cases.

3 For each of the following parabolas find the vertex, the focus, the equation of the directrix, and the length of the focal chord, and sketch the graph.
   a) $(x - 2)^2 = -6(y - 1)$ b) $(y + 3)^2 = 4(x + 1)$
   c) $(x + 3)^2 = y + 2$ d) $y^2 + 2x + 2y + 7 = 0$
   e) $x^2 - 2x - 4y + 5 = 0$ f) $x^2 + 4x + 5y - 11 = 0$

4 Find the equation of the parabola whose axis is parallel to the y axis, with vertex $(1, 3)$ and containing the point $(5, 7)$.

5 Find the equation of the parabola in each of the following cases.
   a) Vertex at $(1, 2)$ and focus at $(3, 2)$
   b) Focus at $(0, 4)$ and directrix $x = 2$
   c) Vertex at $(3, 4)$ and directrix $x = 1$
   d) Vertex at $(2, -5)$ and directrix $y = 3$

6 Find the equation of the parabola with focus $(-3, -5)$ and directrix $x = 5$.

7   Discuss the symmetry of the graph of a parabola.

8   Use the graphs of parabolas to determine the cases when parabolas are functions. If a parabola is a function, is it one-to-one? Explain.

9   Find the equation of the line through the points on the parabola $y^2 = 3x$ whose ordinates are 2 and 3.

10   Find the domain and range of each of the relations in Problem 1.

11   For each of the four cases in which the vertex is $(0, 0)$, write the equation of the parabola in determinant form.

7   Conics

We have developed *analytically* the equations of the circle, the ellipse, the hyperbola, and the parabola, and we have also discussed their properties. In our approach, we derived the equations using the geometric properties. In this section we shall give a general approach that simultaneously applies to the ellipse, the hyperbola, and the parabola. These three graphs are called *conics* because they are determined by the intersections of planes with cones of two nappes (Figures 1a, b, and c).

Figure 1

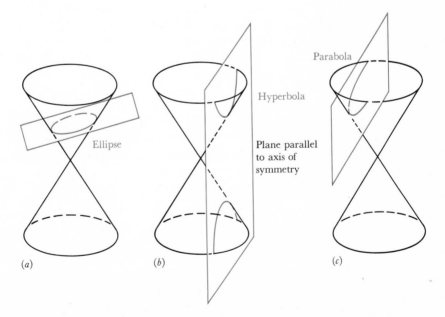

(a)   Ellipse

(b)   Hyperbola   Plane parallel to axis of symmetry

(c)   Parabola

In general, a conic is determined by a given point F, called the *focus*, a given line d, associated with F but not containing F, called the *directrix*, and a positive constant number e, called the *eccentricity*.

DEFINITION CONIC

Assume that e is a positive constant (*eccentricity*), F is a fixed point (*focus*), and $\overline{PD}$ is the distance between a point P and a given line d (*directrix*) (Figure 2). Then a *conic* is a set of points C where

$$C = \left\{ P \mid P = (x, y) \quad \text{where } \frac{\overline{FP}}{\overline{PD}} = e \right\}$$

Figure 2

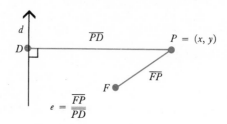

EXAMPLES

1 Assume that the eccentricity $e = 1$, the focus is $(-2, 0)$, and the directrix is $x = 3$. Find the equation of the conic and sketch the conic.

SOLUTION. (See Figure 3.) If $P = (x, y)$ is a point of the conic, then $r_1/r_2 = 1$ (why?); that is, $r_1 = r_2$ or, equivalently, $r_1^2 = r_2^2$. By the distance formula, $(x+2)^2 + y^2 = (x-3)^2$; that is, the conic is the parabola $y^2 = -10(x - \frac{1}{2})$.

Figure 3

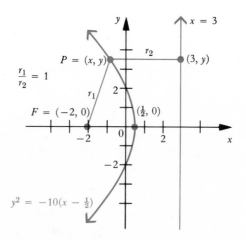

2 Find the equation of the conic that has eccentricity $\frac{2}{3}$ and has the point $(4, 0)$ as a focus, and has as its directrix the y axis.

SOLUTION. (See Figure 4.) By the definition of a conic we have $e = r_1/r_2 = \frac{2}{3}$ or, equivalently, $3r_1 = 2r_2$. But, $r_1^2 = (x-4)^2 + y^2$ and $r_2^2 = x^2$ so that $9r_1^2 = 4r_2^2$ implies that

$$9(x^2 - 8x + 16 + y^2) = 4x^2$$

or

$$5x^2 - 72x + 9y^2 + 144 = 0$$

Figure 4

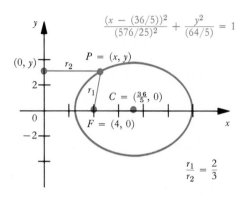

$$\frac{(x-(36/5))^2}{(576/25)^2} + \frac{y^2}{(64/5)} = 1$$

$$\frac{r_1}{r_2} = \frac{2}{3}$$

However, upon completing the square of the left side of the equation, we have

$$5\left[x^2 - \tfrac{72}{5}x + \left(\tfrac{36}{5}\right)^2\right] + 9y^2 = -144 + \frac{(36)^2}{5}$$

That is,

$$5\left(x - \tfrac{36}{5}\right)^2 + 9y^2 = \tfrac{576}{5}$$

Hence the conic is the ellipse

$$\frac{\left(x - \tfrac{36}{5}\right)^2}{\left(\tfrac{576}{25}\right)} + \frac{y^2}{\left(\tfrac{64}{5}\right)} = 1$$

3   Find the equation of the conic with focus $(2,0)$, directrix $x = 0$, and eccentricity 2.

SOLUTION. (See Figure 5.) Since $e = r_1/r_2 = 2$, $r_1 = 2r_2$; that is, $r_1^2 = 4r_2^2$, so that $(x-2)^2 + y^2 = 4x^2$ (why?) or, equivalently, $3x^2 + 4x - y^2 = 4$. After completing the square, we get $3\left(x + \tfrac{2}{3}\right)^2 - y^2 = \tfrac{16}{3}$ or, equivalently, the hyperbola

$$\frac{\left(x + \tfrac{2}{3}\right)^2}{\left(\tfrac{16}{9}\right)} - \frac{y^2}{\left(\tfrac{16}{3}\right)} = 1$$

Figure 5

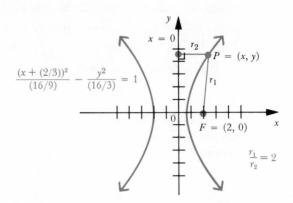

$$\frac{(x + (2/3))^2}{(16/9)} - \frac{y^2}{(16/3)} = 1$$

From the geometric definition of a *parabola* (see Section 6), it can be seen that the eccentricity $e = 1$ (see Example 1 above). Also, it can be shown that if $e < 1$, the conic is an ellipse (see Example 2), and if $e > 1$, the conic is a hyperbola **(see Example 3).**

Finally, we state without proof the following theorem.

THEOREM 1

Consider ellipses and hyperbolas of the form

$$\frac{x^2}{a^2} \pm \frac{y^2}{b^2} = 1$$

where $c^2 = a^2 - b^2$ for the ellipse and $c^2 = a^2 + b^2$ for the hyperbola. The eccentricity $e = c/a$ and the equations of the directrices are given by $x = -a^2/c$ and $x = a^2/c$.

EXAMPLES

1   Find the eccentricity and the directrices of the hyperbola whose equation is $x^2 - 2y^2 = 8$.

SOLUTION.   (See Figure 6.) $x^2 - 2y^2 = 8$ in standard form is $x^2/8 - y^2/4 = 1$. This is an equation of a hyperbola where $a = 2\sqrt{2}$ and $b = 2$; hence $c = \sqrt{a^2 + b^2} = \sqrt{8+4} = 2\sqrt{3}$. Therefore, by Theorem 1, the eccentricity is

$$e = \frac{c}{a} = \frac{2\sqrt{3}}{2\sqrt{2}} = \frac{\sqrt{6}}{2}$$

The equations of the directrices are

$$x = -\frac{a^2}{c} = -\frac{8}{2\sqrt{2}} = -2\sqrt{2} \quad \text{and} \quad x = \frac{a^2}{c} = \frac{8}{2\sqrt{2}} = 2\sqrt{2}$$

Figure 6

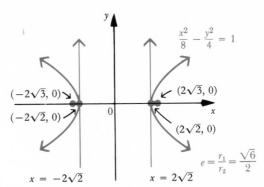

$(-2\sqrt{3}, 0)$

$(-2\sqrt{2}, 0)$

$(2\sqrt{3}, 0)$

$(2\sqrt{2}, 0)$

$\dfrac{x^2}{8} - \dfrac{y^2}{4} = 1$

$e = \dfrac{r_1}{r_2} = \dfrac{\sqrt{6}}{2}$

$x = -2\sqrt{2}$ $x = 2\sqrt{2}$

2   Find the eccentricity of the ellipse whose equation is $4x^2 + 9y^2 = 36$.

SOLUTION.   (See Figure 7.) $4x^2 + 9y^2 = 36$ in standard form is $x^2/9 + y^2/4 = 1$. This is an equation of an ellipse where $a = 3$ and $b = 2$; hence

$$c = \sqrt{a^2 - b^2} = \sqrt{9 - 4} = \sqrt{5}$$

Therefore, by Theorem 1, the eccentricity is

$$e = \frac{c}{a} = \frac{\sqrt{5}}{3}$$

and the equations of the directrices are

$$x = -\frac{a^2}{c} = -\frac{9}{\sqrt{5}} = -\frac{9\sqrt{5}}{5} \quad \text{and} \quad x = \frac{a^2}{c} = \frac{9}{\sqrt{5}} = \frac{9\sqrt{5}}{5}$$

Figure 7

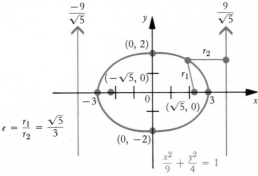

$\dfrac{-9}{\sqrt{5}}$

$\dfrac{9}{\sqrt{5}}$

$(0, 2)$

r_2

$(-\sqrt{5}, 0)$

r_1

-3

$(\sqrt{5}, 0)$ 3

$e = \dfrac{r_1}{r_2} = \dfrac{\sqrt{5}}{3}$

$(0, -2)$

$\dfrac{x^2}{9} + \dfrac{y^2}{4} = 1$

3   Find the equation of the parabola that has focus $(1,3)$ and directrix $y = -2$, and graph the parabola.

SOLUTION.   (See Figure 8.) Since the conic is a parabola, the eccentricity is 1, so that $r_1^2 = r_2^2$ (why?); that is,

$$(x-1)^2 + (y-3)^2 = (y+2)^2$$

Hence

$$(x-1)^2 = 4(\tfrac{5}{2})(y - \tfrac{1}{2})$$

Figure 8

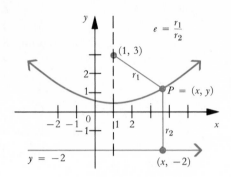

PROBLEM SET 6

1 Use the definition of conics to determine the equation of each of the following conics. Also sketch the graph.

a) Focus at $(3,0)$, directrix $y = \frac{5}{3}$, and eccentricity $e = 3/\sqrt{5}$

b) Focus at $(5,0)$, directrix $x = \frac{16}{5}$, and eccentricity $e = \frac{5}{4}$

c) Focus at $(3,0)$, directrix $x = \frac{25}{3}$, and eccentricity $e = \frac{3}{5}$

d) Focus at $(1,2)$, directrix $y = -2$, and eccentricity $e = \frac{1}{2}$

e) Focus at $(2,0)$, directrix $x = 4$, and eccentricity $e = 1$

2 Discuss the way in which the shape of an ellipse changes as the eccentricity varies from near 0 to near 1.

3 Find the eccentricity and the equations of the directrices for each of the following conics.

a) $25x^2 + 16y^2 = 400$ b) $x^2 + 2y^2 = 1$

c) $x^2 + 3y^2 = 4$ d) $y^2 - 8y + 3x + 5 = 0$

e) $9x^2 - 16y^2 + 144 = 0$ f) $4x^2 - 4y^2 + 1 = 0$

g) $3x^2 - 5x + y^2 + 22y = 1$ h) $x^2 + 16x - y + 7 = 0$

4 We know that the eccentricity $e = c/a < 1$ for the ellipse $x^2/a^2 + y^2/b^2 = 1$, where $c^2 = a^2 - b^2$. Now, as e becomes closer and closer to zero in value, $c/a = e$ approaches zero; hence c^2 approaches zero or a approaches b in value (why?). What is the shape of an ellipse in which a and b are close in value? What if $a = b$? Does this give an indication of how to define a conic with eccentricity $e = 0$? Use sketches and examples to answer these questions.

5 Given the equation $25x^2 - 4y^2 + 50x - 12y + 116 = 0$, find the center, the foci, the vertices, the equations of the asymptotes, and the eccentricity. Also sketch the graph.

REVIEW PROBLEM SET

1 Find the equation of the circle in each of the following cases and sketch the graph.

a)   Center at $(4, -3)$ and radius 5
b)   Containing the points $(4, 5)$, $(3, -2)$, and $(1, -4)$
c)   Center at $(1, 2)$ and containing the point $(3, 1)$

2 a) If $\overline{x} = x - 2$ and $\overline{y} = y + 4$, name the xy coordinates of the point whose $\overline{xy}$ coordinates are:

   i $(-5, 2)$ ii $(5, -3)$ iii $(-7, -1)$
   iv $(4, 6)$ v $(-7, 0)$

   b) If $\overline{x} = x + 3$, and $\overline{y} = y - 1$, name the $\overline{xy}$ coordinates of the point whose xy coordinates are:

   i $(6, 1)$ ii $(-2, 9)$ iii $(-1, -4)$
   iv $(8, 0)$ v $(0, -9)$

3 Find the equation of the parabola for each of the following cases.
   a)   Focus $(-4, 0)$ and directrix $y = 3$
   b)   Focus $(0, 3)$ and directrix $x = \frac{5}{2}$
   c)   Focus $(1, -\frac{5}{2})$ and directrix $y = -4$
   d)   Focus $(3, 0)$, containing point $(2, 2\sqrt{2})$ with the axis of symmetry the x axis and opens to the right

4 The end points of the base of a triangle are $(-2, 0)$ and $(2, 0)$ and the sum of the lengths of the other two sides is 6. Find the equation of the set of all points that are possible vertices.

5 Find the equation of the ellipse satisfying the given conditions and sketch the graph in each case.
   a)   Vertices $(-5, 0)$ and $(5, 0)$ and foci $(-3, 0)$ and $(3, 0)$
   b)   Foci $(-1, 0)$ and $(1, 0)$ and length of minor axis $2\sqrt{2}$
   c)   Foci $(0, -6)$ and $(0, 6)$ and eccentricity $e = \frac{1}{2}$
   d)   Vertices $(-5, 0)$ and $(5, 0)$ and eccentricity $e = \frac{3}{5}$

6 Find the points on the parabola $y^2 + 20x = 0$ for which the sum of two focal radii is equal in length to the latus rectum. (*Note:* The focal radii are the line segments with one end point on the parabola and the other end point the focus.)

7 Find the equation of each of the following ellipses.
   a)   Focus $(2, 0)$, eccentricity $e = \frac{2}{3}$, and directrix $2x - 9 = 0$
   b)   Focus $(0, -4)$, eccentricity $e = \frac{2}{3}$, and directrix $y + 9 = 0$
   c)   Vertex $(10, 0)$, center $(0, 0)$, and eccentricity $e = \frac{3}{5}$
   d)   Center $(0, 0)$, vertex $(0, 3)$, and eccentricity $e = \frac{2}{5}$

8 Find the equation of the line containing the points on the parabola $y^2 = 8x$ whose y coordinates are 2 and 8, respectively.

9 Find the equation of the hyperbola in each of the following cases and sketch the graph.
a) Vertices $(-2,0)$ and $(2,0)$ and foci $(-3,0)$ and $(3,0)$
b) Foci $(-10,0)$ and $(10,0)$ and eccentricity $e = \frac{5}{4}$
c) Vertices $(-15,0)$ and $(15,0)$ and asymptotes $5y = \pm 4x$
d) Contains the point $(5,9)$ and asymptotes $y = \pm x$

10 Consider the equation $(b^2 - k)x^2 + (a^2 - k)y^2 = 1$, where $a \geq b$.
a) For what choices of k is the graph an ellipse?
b) For what choices of k is the graph a hyperbola?
c) For what choices of k is the graph a circle?

11 Show that the hyperbolas whose equations are $x^2/4 - y^2/9 = 1$ and $y^2/9 - x^2/4 = 1$ have the same set of asymptotes, and sketch them on the same coordinate system.

12 A square with sides parallel to the coordinate axes is inscribed in the ellipse $9x^2 + 16y^2 = 100$. Find the coordinates of its vertices and its area.

13 Show that if a point P is equidistant from the y axis and the point $(6,0)$, then its coordinates (x, y) must satisfy the equation $y^2 = 12(x-3)$. Is the converse true?

14 Let d_1 be the distance from the point $P = (x, y)$ to the point $A = (-10,0)$ and d_2 be the distance from the point $P = (x, y)$ to the point $B = (10,0)$. Show that if $d_1 - d_2 = 12$, then the coordinates of P must satisfy the equation $x^2/36 - y^2/64 = 1$. Is it true that every point P whose coordinates satisfy this equation has $d_1 - d_2 = 12$?

15 Let d_1 be the distance from the point $P = (x, y)$ to the point $A = (1,0)$, and d_2 be the distance from the point $P = (x, y)$ to the point $B = (9,0)$. Derive the equation that the point must satisfy if $d_1 + d_2 = 4$.

16 Write the equation of an ellipse whose center is at the origin, whose axes of symmetry are on the coordinate axes, and whose major axis is three times the minor axis and contains the point $(3, 1)$.

17 The Earth moves in an elliptical orbit with eccentricity 0.017, major axis 185.8 million miles, and the sun at one focus. How close does the Earth come to the sun when it is positioned on the major axis?

APPENDIX

APPENDIX A

Tables

TABLE I   COMMON LOGARITHMS

n	0	1	2	3	4	5	6	7	8	9
10	0000	0043	0086	0128	0170	0212	0253	0294	0334	0374
11	0414	0453	0492	0531	0569	0607	0645	0682	0719	0755
12	0792	0828	0864	0899	0934	0969	1004	1038	1072	1106
13	1139	1173	1206	1239	1271	1303	1335	1367	1399	1430
14	1461	1492	1523	1553	1584	1614	1644	1673	1703	1732
15	1761	1790	1818	1847	1875	1903	1931	1959	1987	2014
16	2041	2068	2095	2122	2148	2175	2201	2227	2253	2279
17	2304	2330	2355	2380	2405	2430	2455	2480	2504	2529
18	2553	2577	2601	2625	2648	2672	2695	2718	2742	2765
19	2788	2810	2833	2856	2878	2900	2923	2945	2967	2989
20	3010	3032	3054	3075	3096	3118	3139	3160	3181	3201
21	3222	3243	3263	3284	3304	3324	3345	3365	3385	3404
22	3424	3444	3464	3483	3502	3522	3541	3560	3579	3598
23	3617	3636	3655	3674	3692	3711	3729	3747	3766	3784
24	3802	3820	3838	3856	3874	3892	3909	3927	3945	3962
25	3979	3997	4014	4031	4048	4065	4082	4099	4116	4133
26	4150	4166	4183	4200	4216	4232	4249	4265	4281	4298
27	4314	4330	4346	4362	4378	4393	4409	4425	4440	4456
28	4472	4487	4502	4518	4533	4548	4564	4579	4594	4609
29	4624	4639	4654	4669	4683	4698	4713	4728	4742	4757
30	4771	4786	4800	4814	4829	4843	4857	4871	4886	4900
31	4914	4928	4942	4955	4969	4983	4997	5011	5024	5038
32	5051	5065	5079	5092	5105	5119	5132	5145	5159	5172
33	5185	5198	5211	5224	5237	5250	5263	5276	5289	5302
34	5315	5328	5340	5353	5366	5378	5391	5403	5416	5428
35	5441	5453	5465	5478	5490	5502	5514	5527	5539	5551
36	5563	5575	5587	5599	5611	5623	5635	5647	5658	5670
37	5682	5694	5705	5717	5729	5740	5752	5763	5775	5786
38	5798	5809	5821	5832	5843	5855	5866	5877	5888	5899
39	5911	5922	5933	5944	5955	5966	5977	5988	5999	6010
40	6021	6031	6042	6053	6064	6075	6085	6096	6107	6117
41	6128	6138	6149	6160	6170	6180	6191	6201	6212	6222
42	6232	6243	6253	6263	6274	6284	6294	6304	6314	6325
43	6335	6345	6355	6365	6375	6385	6395	6405	6415	6425
44	6435	6444	6454	6464	6474	6484	6493	6503	6513	6522
45	6532	6542	6551	6561	6571	6580	6590	6599	6609	6618
46	6628	6637	6646	6656	6665	6675	6684	6693	6702	6712
47	6721	6730	6739	6749	6758	6767	6776	6785	6794	6803
48	6812	6821	6830	6839	6848	6857	6866	6875	6884	6893
49	6902	6911	6920	6928	6937	6946	6955	6964	6972	6981

TABLE I **COMMON LOGARITHMS** — Cont.

n	0	1	2	3	4	5	6	7	8	9
50	6990	6998	7007	7016	7024	7033	7042	7050	7059	7067
51	7076	7084	7093	7101	7110	7118	7126	7135	7143	7152
52	7160	7168	7177	7185	7193	7202	7210	7218	7226	7235
53	7243	7251	7259	7267	7275	7284	7292	7300	7308	7316
54	7324	7332	7340	7348	7356	7364	7372	7380	7388	7396
55	7404	7412	7419	7427	7435	7443	7451	7459	7466	7474
56	7482	7490	7497	7505	7513	7520	7528	7536	7543	7551
57	7559	7566	7574	7582	7589	7597	7604	7612	7619	7627
58	7634	7642	7649	7657	7664	7672	7679	7686	7694	7701
59	7709	7716	7723	7731	7738	7745	7752	7760	7767	7774
60	7782	7789	7796	7803	7810	7818	7825	7832	7839	7846
61	7853	7860	7868	7875	7882	7889	7896	7903	7910	7917
62	7924	7931	7938	7945	7952	7959	7966	7973	7980	7987
63	7993	8000	8007	8014	8021	8028	8035	8041	8048	8055
64	8062	8069	8075	8082	8089	8096	8102	8109	8116	8122
65	8129	8136	8142	8149	8156	8162	8169	8176	8182	8189
66	8195	8202	8209	8215	8222	8228	8235	8241	8248	8254
67	8261	8267	8274	8280	8287	8293	8299	8306	8312	8319
68	8325	8331	8338	8344	8351	8357	8363	8370	8376	8382
69	8388	8395	8401	8407	8414	8420	8426	8432	8439	8445
70	8451	8457	8463	8470	8476	8482	8488	8494	8500	8506
71	8513	8519	8525	8531	8537	8543	8549	8555	8561	8567
72	8573	8579	8585	8591	8597	8603	8609	8615	8621	8627
73	8633	8639	8645	8651	8657	8663	8669	8675	8681	8686
74	8692	8698	8704	8710	8716	8722	8727	8733	8739	8745
75	8751	8756	8762	8768	8774	8779	8785	8791	8797	8802
76	8808	8814	8820	8825	8831	8837	8842	8848	8854	8859
77	8865	8871	8876	8882	8887	8893	8899	8904	8910	8915
78	8921	8927	8932	8938	8943	8949	8954	8960	8965	8971
79	8976	8982	8987	8993	8998	9004	9009	9015	9020	9025
80	9031	9036	9042	9047	9053	9058	9063	9069	9074	9079
81	9085	9090	9096	9101	9106	9112	9117	9122	9128	9133
82	9138	9143	9149	9154	9159	9165	9170	9175	9180	9186
83	9191	9196	9201	9206	9212	9217	9222	9227	9232	9238
84	9243	9248	9253	9258	9263	9269	9274	9279	9284	9289
85	9294	9299	9304	9309	9315	9320	9325	9330	9335	9340
86	9345	9350	9355	9360	9365	9370	9375	9380	9385	9390
87	9395	9400	9405	9410	9415	9420	9425	9430	9435	9440
88	9445	9450	9455	9460	9465	9469	9474	9479	9484	9489
89	9494	9499	9504	9509	9513	9518	9523	9528	9533	9538
90	9542	9547	9552	9557	9562	9566	9571	9576	9581	9586
91	9590	9595	9600	9605	9609	9614	9619	9624	9628	9633
92	9638	9643	9647	9652	9657	9661	9666	9671	9675	9680
93	9685	9689	9694	9699	9703	9708	9713	9717	9722	9727
94	9731	9736	9741	9745	9750	9754	9759	9763	9768	9773
95	9777	9782	9786	9791	9795	9800	9805	9809	9814	9818
96	9823	9827	9832	9836	9841	9845	9850	9854	9859	9863
97	9868	9872	9877	9881	9886	9890	9894	9899	9903	9908
98	9912	9917	9921	9926	9930	9934	9939	9943	9948	9952
99	9956	9961	9965	9969	9974	9978	9983	9987	9991	9996

TABLE II NATURAL LOGARITHMS

t	0.00	0.01	0.02	0.03	0.04	0.05	0.06	0.07	0.08	0.09
1.0	0.0000	0.0100	0.0198	0.0296	0.0392	0.0488	0.0583	0.0677	0.0770	0.0862
1.1	0.0953	0.1044	0.1133	0.1222	0.1310	0.1398	0.1484	0.1570	0.1655	0.1740
1.2	0.1823	0.1906	0.1989	0.2070	0.2151	0.2231	0.2311	0.2390	0.2469	0.2546
1.3	0.2624	0.2700	0.2776	0.2852	0.2927	0.3001	0.3075	0.3148	0.3221	0.3293
1.4	0.3365	0.3436	0.3507	0.3577	0.3646	0.3716	0.3784	0.3853	0.3920	0.3988
1.5	0.4055	0.4121	0.4187	0.4253	0.4318	0.4383	0.4447	0.4511	0.4574	0.4637
1.6	0.4700	0.4762	0.4824	0.4886	0.4947	0.5008	0.5068	0.5128	0.5188	0.5247
1.7	0.5306	0.5365	0.5423	0.5481	0.5539	0.5596	0.5653	0.5710	0.5766	0.5822
1.8	0.5878	0.5933	0.5988	0.6043	0.6098	0.6152	0.6206	0.6259	0.6313	0.6366
1.9	0.6419	0.6471	0.6523	0.6575	0.6627	0.6678	0.6729	0.6780	0.6831	0.6881
2.0	0.6931	0.6981	0.7031	0.7080	0.7130	0.7178	0.7227	0.7275	0.7324	0.7372
2.1	0.7419	0.7467	0.7514	0.7561	0.7608	0.7655	0.7701	0.7747	0.7793	0.7839
2.2	0.7885	0.7930	0.7975	0.8020	0.8065	0.8109	0.8154	0.8198	0.8242	0.8286
2.3	0.8329	0.8372	0.8416	0.8459	0.8502	0.8544	0.8587	0.8629	0.8671	0.8713
2.4	0.8755	0.8796	0.8838	0.8879	0.8920	0.8961	0.9002	0.9042	0.9083	0.9123
2.5	0.9163	0.9203	0.9243	0.9282	0.9322	0.9361	0.9400	0.9439	0.9478	0.9517
2.6	0.9555	0.9594	0.9632	0.9670	0.9708	0.9746	0.9783	0.9821	0.9858	0.9895
2.7	0.9933	0.9969	1.0006	1.0043	1.0080	1.0116	1.0152	0.0188	1.0225	1.0260
2.8	1.0296	1.0332	1.0367	1.0403	1.0438	1.0473	1.0508	1.0543	1.0578	1.0613
2.9	1.0647	1.0682	1.0716	1.0750	1.0784	1.0818	1.0852	1.0886	1.0919	1.0953
3.0	1.0986	1.1019	1.1053	1.1086	1.1119	1.1151	1.1184	1.1217	1.1249	1.1282
3.1	1.1314	1.1346	1.1378	1.1410	1.1442	1.1474	1.1506	1.1537	1.1569	1.1600
3.2	1.1632	1.1663	1.1694	1.1725	1.1756	1.1787	1.1817	1.1848	1.1878	1.1909
3.3	1.1939	1.1970	1.2000	1.2030	1.2060	1.2090	1.2119	1.2149	1.2179	1.2208
3.4	1.2238	1.2267	1.2296	1.2326	1.2355	1.2384	1.2413	1.2442	1.2470	1.2499
3.5	1.2528	1.2556	1.2585	1.2613	1.2641	1.2669	1.2698	1.2726	1.2754	1.2782
3.6	1.2809	1.2837	1.2865	1.2892	1.2920	1.2947	1.2975	1.3002	1.3029	1.3056
3.7	1.3083	1.3110	1.3137	1.3164	1.3191	1.3218	1.3244	1.3271	1.3297	1.3324
3.8	1.3350	1.3376	1.3403	1.3429	1.3455	1.3481	1.3507	1.3533	1.3558	1.3584
3.9	1.3610	1.3635	1.3661	1.3686	1.3712	1.3737	1.3762	1.3788	1.3813	1.3838
4.0	1.3863	1.3888	1.3913	1.3938	1.3962	1.3987	1.4012	1.4036	1.4061	1.4085
4.1	1.4110	1.4134	1.4159	1.4183	1.4207	1.4231	1.4255	1.4279	1.4303	1.4327
4.2	1.4351	1.4375	1.4398	1.4422	1.4446	1.4469	1.4493	1.4516	1.4540	1.4563
4.3	1.4586	1.4609	1.4633	1.4656	1.4679	1.4702	1.4725	1.4748	1.4770	1.4793
4.4	1.4816	1.4839	1.4861	1.4884	1.4907	1.4929	1.4952	1.4974	1.4996	1.5019
4.5	1.5041	1.5063	1.5085	1.5107	1.5129	1.5151	1.5173	1.5195	1.5217	1.5239
4.6	1.5261	1.5282	1.5304	1.5326	1.5347	1.5369	1.5390	1.5412	1.5433	1.5454
4.7	1.5476	1.5497	1.5518	1.5539	1.5560	1.5581	1.5602	1.5623	1.5644	1.5665
4.8	1.5686	1.5707	1.5728	1.5748	1.5769	1.5790	1.5810	1.5831	1.5851	1.5872
4.9	1.5892	1.5913	1.5933	1.5953	1.5974	1.5994	1.6014	1.6034	1.6054	1.6074
5.0	1.6094	1.6114	1.6134	1.6154	1.6174	1.6194	1.6214	1.6233	1.6253	1.6273
5.1	1.6292	1.6312	1.6332	1.6351	1.6371	1.6390	1.6409	1.6429	1.6448	1.6467
5.2	1.6487	1.6506	1.6525	1.6544	1.6563	1.6582	1.6601	1.6620	1.6639	1.6658
5.3	1.6677	1.6696	1.6715	1.6734	1.6752	1.6771	1.6790	1.6808	1.6827	1.6845
5.4	1.6864	1.6882	1.6901	1.6919	1.6938	1.6956	1.6974	1.6993	1.7011	1.7029
5.5	1.7047	1.7066	1.7084	1.7102	1.7120	1.7138	1.7156	1.7174	1.7192	1.7210
5.6	1.7228	1.7246	1.7263	1.7281	1.7299	1.7317	1.7334	1.7352	1.7370	1.7387
5.7	1.7405	1.7422	1.7440	1.7457	1.7475	1.7492	1.7509	1.7527	1.7544	1.7561
5.8	1.7579	1.7596	1.7613	1.7630	1.7647	1.7664	1.7682	1.7699	1.7716	1.7733
5.9	1.7750	1.7766	1.7783	1.7800	1.7817	1.7834	1.7851	1.7867	1.7884	1.7901

TABLE II   **NATURAL LOGARITHMS** — Cont.

t	0.00	0.01	0.02	0.03	0.04	0.05	0.06	0.07	0.08	0.09
6.0	1.7918	1.7934	1.7951	1.7967	1.7984	1.8001	1.8017	1.8034	1.8050	1.8066
6.1	1.8083	1.8099	1.8116	1.8132	1.8148	1.8165	1.8181	1.8197	1.8213	1.8229
6.2	1.8245	1.8262	1.8278	1.8294	1.8310	1.8326	1.8342	1.8358	1.8374	1.8390
6.3	1.8406	1.8421	1.8437	1.8453	1.8469	1.8485	1.8500	1.8516	1.8532	1.8547
6.4	1.8563	1.8579	1.8594	1.8610	1.8625	1.8641	1.8656	1.8672	1.8687	1.8703
6.5	1.8718	1.8733	1.8749	1.8764	1.8779	1.8795	1.8810	1.8825	1.8840	1.8856
6.6	1.8871	1.8886	1.8901	1.8916	1.8931	1.8946	1.8961	1.8976	1.8991	1.9006
6.7	1.9021	1.9036	1.9051	1.9066	1.9081	1.9095	1.9110	1.9125	1.9140	1.9155
6.8	1.9169	1.9184	1.9199	1.9213	1.9228	1.9242	1.9257	1.9272	1.9286	1.9301
6.9	1.9315	1.9330	1.9344	1.9359	1.9373	1.9387	1.9402	1.9416	1.9430	1.9445
7.0	1.9459	1.9473	1.9488	1.9502	1.9516	1.9530	1.9544	1.9559	1.9573	1.9587
7.1	1.9601	1.9615	1.9629	1.9643	1.9657	1.9671	1.9685	1.9699	1.9713	1.9727
7.2	1.9741	1.9755	1.9769	1.9782	1.9796	1.9810	1.9824	1.9838	1.9851	1.9865
7.3	1.9879	1.9892	1.9906	1.9920	1.9933	1.9947	1.9961	1.9974	1.9988	2.0001
7.4	2.0015	2.0028	2.0042	2.0055	2.0069	2.0082	2.0096	2.0109	2.0122	2.0136
7.5	2.0149	2.0162	2.0176	2.0189	2.0202	2.0215	2.0229	2.0242	2.0255	2.0268
7.6	2.0282	2.0295	2.0308	2.0321	2.0334	2.0347	2.0360	2.0373	2.0386	2.0399
7.7	2.0412	2.0425	2.0438	2.0451	2.0464	2.0477	2.0490	2.0503	2.0516	2.0528
7.8	2.0541	2.0554	2.0567	2.0580	2.0592	2.0605	2.0618	2.0631	2.0643	2.0665
7.9	2.0669	2.0681	2.0694	2.0707	2.0719	2.0732	2.0744	2.0757	2.0769	2.0782
8.0	2.0794	2.0807	2.0819	2.0832	2.0844	2.0857	2.0869	2.0882	2.0894	2.0906
8.1	2.0919	2.0931	2.0943	2.0956	2.0968	2.0980	2.0992	2.1005	2.1017	2.1029
8.2	2.1041	2.1054	2.1066	2.1078	2.1090	2.1102	2.1114	2.1126	2.1138	2.1150
8.3	2.1163	2.1175	2.1187	2.1199	2.1211	2.1223	2.1235	2.1247	2.1258	2.1270
8.4	2.1282	2.1294	2.1306	2.1318	2.1330	2.1342	2.1353	2.1365	2.1377	2.1389
8.5	2.1401	2.1412	2.1424	2.1436	2.1448	2.1459	2.1471	2.1483	2.1494	2.1506
8.6	2.1518	2.1529	2.1541	2.1552	2.1564	2.1576	2.1587	2.1599	2.1610	2.1622
8.7	2.1633	2.1645	2.1656	2.1668	2.1679	2.1691	2.1702	2.1713	2.1725	2.1736
8.8	2.1748	2.1759	2.1770	2.1782	2.1793	2.1804	2.1815	2.1827	2.1838	2.1849
8.9	2.1861	2.1872	2.1883	2.1894	2.1905	2.1917	2.1928	2.1939	2.1950	2.1961
9.0	2.1972	2.1983	2.1994	2.2006	2.2017	2.2028	2.2039	2.2050	2.2061	2.2072
9.1	2.2083	2.2094	2.2105	2.2116	2.2127	2.2138	2.2148	2.2159	2.2170	2.2181
9.2	2.2192	2.2203	2.2214	2.2225	2.2235	2.2246	2.2257	2.2268	2.2279	2.2289
9.3	2.2300	2.2311	2.2322	2.2332	2.2343	2.2354	2.2364	2.2375	2.2386	2.2396
9.4	2.2407	2.2418	2.2428	2.2439	2.2450	2.2460	2.2471	2.2481	2.2492	2.2502
9.5	2.2513	2.2523	2.2534	2.2544	2.2555	2.2565	2.2576	2.2586	2.2597	2.2607
9.6	2.2618	2.2628	2.2638	2.2649	2.2659	2.2670	2.2680	2.2690	2.2701	2.2711
9.7	2.2721	2.2732	2.2742	2.2752	2.2762	2.2773	2.2783	2.2793	2.2803	2.2814
9.8	2.2824	2.2834	2.2844	2.2854	2.2865	2.2875	2.2885	2.2895	2.2905	2.2915
9.9	2.2925	2.2935	2.2946	2.2956	2.2966	2.2976	2.2986	2.2996	2.3006	2.3016

TABLE III   VALUES OF CIRCULAR FUNCTIONS

t	$\sin t$	$\cos t$	$\tan t$	$\cot t$	$\sec t$	$\csc t$
.00	.0000	1.0000	.0000	—	1.000	—
.01	.0100	1.0000	.0100	99.997	1.000	100.00
.02	.0200	.9998	.0200	49.993	1.000	50.00
.03	.0300	.9996	.0300	33.323	1.000	33.34
.04	.0400	.9992	.0400	24.987	1.001	25.01
.05	.0500	.9988	.0500	19.983	1.001	20.01
.06	.0600	.9982	.0601	16.647	1.002	16.68
.07	.0699	.9976	.0701	14.262	1.002	14.30
.08	.0799	.9968	.0802	12.473	1.003	12.51
.09	.0899	.9960	.0902	11.081	1.004	11.13
.10	.0998	.9950	.1003	9.967	1.005	10.02
.11	.1098	.9940	.1104	9.054	1.006	9.109
.12	.1197	.9928	.1206	8.293	1.007	8.353
.13	.1296	.9916	.1307	7.649	1.009	7.714
.14	.1395	.9902	.1409	7.096	1.010	7.166
.15	.1494	.9888	.1511	6.617	1.011	6.692
.16	.1593	.9872	.1614	6.197	1.013	6.277
.17	.1692	.9856	.1717	5.826	1.015	5.911
.18	.1790	.9838	.1820	5.495	1.016	5.586
.19	.1889	.9820	.1923	5.200	1.018	5.295
.20	.1987	.9801	.2027	4.933	1.020	5.033
.21	.2085	.9780	.2131	4.692	1.022	4.797
.22	.2182	.9759	.2236	4.472	1.025	4.582
.23	.2280	.9737	.2341	4.271	1.027	4.386
.24	.2377	.9713	.2447	4.086	1.030	4.207
.25	.2474	.9689	.2553	3.916	1.032	4.042
.26	.2571	.9664	.2660	3.759	1.035	3.890
.27	.2667	.9638	.2768	3.613	1.038	3.749
.28	.2764	.9611	.2876	3.478	1.041	3.619
.29	.2860	.9582	.2984	3.351	1.044	3.497
.30	.2955	.9553	.3093	3.233	1.047	3.384
.31	.3051	.9523	.3203	3.122	1.050	3.278
.32	.3146	.9492	.3314	3.018	1.053	3.179
.33	.3240	.9460	.3425	2.920	1.057	3.086
.34	.3335	.9428	.3537	2.827	1.061	2.999
.35	.3429	.9394	.3650	2.740	1.065	2.916
.36	.3523	.9359	.3764	2.657	1.068	2.839
.37	.3616	.9323	.3879	2.578	1.073	2.765
.38	.3709	.9287	.3994	2.504	1.077	2.696
.39	.3802	.9249	.4111	2.433	1.081	2.630
.40	.3894	.9211	.4228	2.365	1.086	2.568
.41	.3986	.9171	.4346	2.301	1.090	2.509
.42	.4078	.9131	.4466	2.239	1.095	2.452
.43	.4169	.9090	.4586	2.180	1.100	2.399
.44	.4259	.9048	.4708	2.124	1.105	2.348
.45	.4350	.9004	.4831	2.070	1.111	2.299
.46	.4439	.8961	.4954	2.018	1.116	2.253
.47	.4529	.8916	.5080	1.969	1.122	2.208
.48	.4618	.8870	.5206	1.921	1.127	2.166
.49	.4706	.8823	.5334	1.875	1.133	2.125

TABLE III   VALUES OF CIRCULAR FUNCTIONS — Cont.

t	$\sin t$	$\cos t$	$\tan t$	$\cot t$	$\sec t$	$\csc t$
.50	.4794	.8776	.5463	1.830	1.139	2.086
.51	.4882	.8727	.5594	1.788	1.146	2.048
.52	.4969	.8678	.5726	1.747	1.152	2.013
$\dfrac{\pi}{6}$	.5000	.8660	.5774	1.732	1.155	2.000
.53	.5055	.8628	.5859	1.707	1.159	1.978
.54	.5141	.8577	.5994	1.668	1.166	1.945
.55	.5227	.8525	.6131	1.631	1.173	1.913
.56	.5312	.8473	.6269	1.595	1.180	1.883
.57	.5396	.8419	.6410	1.560	1.188	1.853
.58	.5480	.8365	.6552	1.526	1.196	1.825
.59	.5564	.8309	.6696	1.494	1.203	1.797
.60	.5646	.8253	.6841	1.462	1.212	1.771
.61	.5729	.8196	.6989	1.431	1.220	1.746
.62	.5810	.8139	.7139	1.401	1.229	1.721
.63	.5891	.8080	.7291	1.372	1.238	1.697
.64	.5972	.8021	.7445	1.343	1.247	1.674
.65	.6052	.7961	.7602	1.315	1.256	1.652
.66	.6131	.7900	.7761	1.288	1.266	1.631
.67	.6210	.7838	.7923	1.262	1.276	1.610
.68	.6288	.7776	.8087	1.237	1.286	1.590
.69	.6365	.7712	.8253	1.212	1.297	1.571
.70	.6442	.7648	.8423	1.187	1.307	1.552
.71	.6518	.7584	.8595	1.163	1.319	1.534
.72	.6594	.7518	.8771	1.140	1.330	1.517
.73	.6669	.7452	.8949	1.117	1.342	1.500
.74	.6743	.7385	.9131	1.095	1.354	1.483
.75	.6816	.7317	.9316	1.073	1.367	1.467
.76	.6889	.7248	.9505	1.052	1.380	1.452
.77	.6961	.7179	.9697	1.031	1.393	1.437
.78	.7033	.7109	.9893	1.011	1.407	1.422
$\dfrac{\pi}{4}$	.7071	.7071	1.000	1.000	1.414	1.414
.79	.7104	.7038	1.009	.9908	1.421	1.408
.80	.7174	.6967	1.030	.9712	1.435	1.394
.81	.7243	.6895	1.050	.9520	1.450	1.381
.82	.7311	.6822	1.072	.9331	1.466	1.368
.83	.7379	.6749	1.093	.9146	1.482	1.355
.84	.7446	.6675	1.116	.8964	1.498	1.343
.85	.7513	.6600	1.138	.8785	1.515	1.331
.86	.7578	.6524	1.162	.8609	1.533	1.320
.87	.7643	.6448	1.185	.8437	1.551	1.308
.88	.7707	.6372	1.210	.8267	1.569	1.297
.89	.7771	.6294	1.235	.8100	1.589	1.287
.90	.7833	.6216	1.260	.7936	1.609	1.277
.91	.7895	.6137	1.286	.7774	1.629	1.267
.92	.7956	.6058	1.313	.7615	1.651	1.257
.93	.8016	.5978	1.341	.7458	1.673	1.247
.94	.8076	.5898	1.369	.7303	1.696	1.238

TABLE III   VALUES OF CIRCULAR FUNCTIONS — Cont.

t	$\sin t$	$\cos t$	$\tan t$	$\cot t$	$\sec t$	$\csc t$
.95	.8134	.5817	1.398	.7151	1.719	1.229
.96	.8192	.5735	1.428	.7001	1.744	1.221
.97	.8249	.5653	1.459	.6853	1.769	1.212
.98	.8305	.5570	1.491	.6707	1.795	1.204
.99	.8360	.5487	1.524	.6563	1.823	1.196
1.00	.8415	.5403	1.557	.6421	1.851	1.188
1.01	.8468	.5319	1.592	.6281	1.880	1.181
1.02	.8521	.5234	1.628	.6142	1.911	1.174
1.03	.8573	.5148	1.665	.6005	1.942	1.166
1.04	.8624	.5062	1.704	.5870	1.975	1.160
$\dfrac{\pi}{3}$	.8660	.5000	1.732	.5774	2.000	1.155
1.05	.8674	.4976	1.743	.5736	2.010	1.153
1.06	.8724	.4889	1.784	.5604	2.046	1.146
1.07	.8772	.4801	1.827	.5473	2.083	1.140
1.08	.8820	.4713	1.871	.5344	2.122	1.134
1.09	.8866	.4625	1.917	.5216	2.162	1.128
1.10	.8912	.4536	1.965	.5090	2.205	1.122
1.11	.8957	.4447	2.014	.4964	2.249	1.116
1.12	.9001	.4357	2.066	.4840	2.295	1.111
1.13	.9044	.4267	2.120	.4718	2.344	1.106
1.14	.9086	.4176	2.176	.4596	2.395	1.101
1.15	.9128	.4085	2.234	.4475	2.448	1.096
1.16	.9168	.3993	2.296	.4356	2.504	1.091
1.17	.9208	.3902	2.360	.4237	2.563	1.086
1.18	.9246	.3809	2.427	.4120	2.625	1.082
1.19	.9284	.3717	2.498	.4003	2.691	1.077
1.20	.9320	.3624	2.572	.3888	2.760	1.073
1.21	.9356	.3530	2.650	.3773	2.833	1.069
1.22	.9391	.3436	2.733	.3659	2.910	1.065
1.23	.9425	.3342	2.820	.3546	2.992	1.061
1.24	.9458	.3248	2.912	.3434	3.079	1.057
1.25	.9490	.3153	3.010	.3323	3.171	1.054
1.26	.9521	.3058	3.113	.3212	3.270	1.050
1.27	.9551	.2963	3.224	.3102	3.375	1.047
1.28	.9580	.2867	3.341	.2993	3.488	1.044
1.29	.9608	.2771	3.467	.2884	3.609	1.041
1.30	.9636	.2675	3.602	.2776	3.738	1.038
1.31	.9662	.2579	3.747	.2669	3.878	1.035
1.32	.9687	.2482	3.903	.2562	4.029	1.032
1.33	.9711	.2385	4.072	.2456	4.193	1.030
1.34	.9735	.2288	4.256	.2350	4.372	1.027
1.35	.9757	.2190	4.455	.2245	4.566	1.025
1.36	.9779	.2092	4.673	.2140	4.779	1.023
1.37	.9799	.1994	4.913	.2035	5.014	1.021
1.38	.9819	.1896	5.177	.1931	5.273	1.018
1.39	.9837	.1798	5.471	.1828	5.561	1.017
1.40	.9854	.1700	5.798	.1725	5.883	1.015
1.41	.9871	.1601	6.165	.1622	6.246	1.013
1.42	.9887	.1502	6.581	.1519	6.657	1.011
1.43	.9901	.1403	7.055	.1417	7.126	1.010
1.44	.9915	.1304	7.602	.1315	7.667	1.009

TABLE III   VALUES OF CIRCULAR FUNCTIONS — Cont.

t	$\sin t$	$\cos t$	$\tan t$	$\cot t$	$\sec t$	$\csc t$
1.45	.9927	.1205	8.238	.1214	8.299	1.007
1.46	.9939	.1106	8.989	.1113	9.044	1.006
1.47	.9949	.1006	9.887	.1011	9.938	1.005
1.48	.9959	.0907	10.983	.0910	11.029	1.004
1.49	.9967	.0807	12.350	.0810	12.390	1.003
1.50	.9975	.0707	14.101	.0709	14.137	1.003
1.51	.9982	.0608	16.428	.0609	16.458	1.002
1.52	.9987	.0508	19.670	.0508	19.695	1.001
1.53	.9992	.0408	24.498	.0408	24.519	1.001
1.54	.9995	.0308	32.461	.0308	32.476	1.000
1.55	.9998	.0208	48.078	.0208	48.089	1.000
1.56	.9999	.0108	92.620	.0108	92.626	1.000
1.57	1.0000	.0008	1255.8	.0008	1255.8	1.000
$\frac{\pi}{2}$	1.0000	.0000	—	.0000	—	1.000

TABLE IV   VALUES OF TRIGONOMETRIC FUNCTIONS

Degrees	Radians	Sin	Csc	Tan	Cot	Sec	Cos		
0° 0′	.0000	.0000	——	.0000	——	1.000	1.0000	1.5708	90° 0′
10′	029	029	343.8	029	343.8	000	000	679	50′
20′	058	058	171.9	058	171.9	000	000	650	40′
30′	.0087	.0087	114.6	.0087	114.6	1.000	1.0000	1.5621	30′
40′	116	116	85.95	116	85.94	000	0.9999	592	20′
50′	145	145	68.76	145	68.75	000	999	563	10′
1° 0′	.0175	.0175	57.30	.0175	57.29	1.000	.9998	1.5533	89° 0′
10′	204	204	49.11	204	49.10	000	998	504	50′
20′	233	233	42.98	233	42.96	000	997	475	40′
30′	.0262	.0262	38.20	.0262	38.19	1.000	.9997	1.5446	30′
40′	291	291	34.38	291	34.37	000	996	417	20′
50′	320	320	31.26	320	31.24	001	995	388	10′
2° 0′	.0349	.0349	28.65	.0349	28.64	1.001	.9994	1.5359	88° 0′
10′	378	378	26.45	378	26.43	001	993	330	50′
20′	407	407	24.56	407	24.54	001	992	301	40′
30′	.0436	.0436	22.93	.0437	22.90	1.001	.9990	1.5272	30′
40′	465	465	21.49	466	21.47	001	989	243	20′
50′	495	494	20.23	495	20.21	001	988	213	10′
3° 0′	.0524	.0523	19.11	.0524	19.08	1.001	.9986	1.5184	87° 0′
10′	553	552	18.10	553	18.07	002	985	155	50′
20′	582	581	17.20	582	17.17	002	983	126	40′
30′	.0611	.0610	16.38	.0612	16.35	1.002	.9981	1.5097	30′
40′	640	640	15.64	641	15.60	002	980	068	20′
50′	669	669	14.96	670	14.92	002	978	039	10′
4° 0′	.0698	.0698	14.34	.0699	14.30	1.002	.9976	1.5010	86° 0′
10′	727	727	13.76	729	13.73	003	974	1.4981	50′
20′	756	756	13.23	758	13.20	003	971	952	40′
30′	.0785	.0785	12.75	.0787	12.71	1.003	.9969	1.4923	30′
40′	814	814	12.29	816	12.25	003	967	893	20′
50′	844	843	11.87	846	11.83	004	964	864	10′
5° 0′	.0873	.0872	11.47	.0875	11.43	1.004	.9962	1.4835	85° 0′
10′	902	901	11.10	904	11.06	004	959	806	50′
20′	931	929	10.76	934	10.71	004	957	777	40′
30′	.0960	.0958	10.43	.0963	10.39	1.005	.9954	1.4748	30′
40′	989	.0987	10.13	.0992	10.08	005	951	719	20′
50′	.1018	.1016	9.839	.1022	9.788	005	948	690	10′
6° 0′	.1047	.1045	9.567	.1051	9.514	1.006	.9945	1.4661	84° 0′
10′	076	074	9.309	080	9.255	006	942	632	50′
20′	105	103	9.065	110	9.010	006	939	603	40′
30′	.1134	.1132	8.834	.1139	8.777	1.006	.9936	1.4573	30′
40′	164	161	8.614	169	8.556	007	932	544	20′
50′	193	190	8.405	198	8.345	007	929	515	10′
7° 0′	.1222	.1219	8.206	.1228	8.144	1.008	.9925	1.4486	83° 0′
10′	251	248	8.016	257	7.953	008	922	457	50′
20′	280	276	7.834	287	7.770	008	918	428	40′
30′	.1309	.1305	7.661	.1317	7.596	1.009	.9914	1.4399	30′
40′	338	334	7.496	346	7.429	009	911	370	20′
50′	367	363	7.337	376	7.269	009	907	341	10′
8° 0′	.1396	.1392	7.185	.1405	7.115	1.010	.9903	1.4312	82° 0′
		Cos	Sec	Cot	Tan	.Csc	Sin	Radians	Degrees

TABLE IV VALUES OF TRIGONOMETRIC FUNCTIONS — Cont.

Degrees	Radians	Sin	Csc	Tan	Cot	Sec	Cos		
8° 0'	.1396	.1392	7.185	.1405	7.115	1.010	.9903	1.4312	82° 0'
10'	425	421	7.040	435	6.968	010	899	283	50'
20'	454	449	6.900	465	6.827	011	894	254	40'
30'	.1484	.1478	6.765	.1495	6.691	1.011	.8980	1.4224	30'
40'	513	507	6.636	524	6.561	012	886	195	20'
50'	542	536	6.512	554	6.435	012	881	166	10'
9° 0'	.1571	.1564	6.392	.1584	6.314	1.012	.9877	1.4137	81° 0'
10'	600	593	277	614	197	013	872	108	50'
20'	629	622	166	644	6.084	013	868	079	40'
30'	.1658	.1650	6.059	.1673	5.976	014	.9863	1.4050	30'
40'	687	679	5.955	703	871	014	858	1.4021	20'
50'	716	708	855	733	769	015	853	1.3992	10'
10° 0'	.1745	.1736	5.759	.1763	5.671	1.015	.9848	1.3963	80° 0'
10'	774	765	665	793	576	016	843	934	50'
20'	804	794	575	823	485	016	838	904	40'
30'	.1833	.1822	5.487	.1853	5.396	1.017	.9833	1.3875	30'
40'	862	851	403	883	309	018	827	846	20'
50'	891	880	320	914	226	018	822	817	10'
11° 0'	.1920	.1908	5.241	.1944	5.145	1.019	.9816	1.3788	79° 0'
10'	949	937	164	.1974	5.066	019	811	759	50'
20'	978	965	089	.2004	4.989	020	805	730	40'
30'	.2007	.1994	5.016	.2035	4.915	1.020	.9799	1.3701	30'
40'	036	.2022	4.945	065	843	021	793	672	20'
50'	065	051	876	095	773	022	787	643	10'
12° 0'	.2094	.2079	4.810	.2126	4.705	1.022	.9781	1.3614	78° 0'
10'	123	108	745	156	638	023	775	584	50'
20'	153	136	682	186	574	024	769	555	40'
30'	.2182	.2164	4.620	.2217	4.511	1.024	.9763	1.3526	30'
40'	211	193	560	247	449	025	757	497	20'
50'	240	221	502	278	390	026	750	468	10'
13° 0'	.2269	.2250	4.445	.2309	4.331	1.026	.9744	1.3439	77° 0'
10'	298	278	390	339	275	027	737	410	50'
20'	327	306	336	370	219	028	730	381	40'
30'	.2356	.2334	4.284	.2401	4.165	1.028	.9724	1.3352	30'
40'	385	363	232	432	113	029	717	323	20'
50'	414	391	182	462	061	030	710	294	10'
14° 0'	.2443	.2419	4.134	.2493	4.011	1.031	.9703	1.3265	76° 0'
10'	473	447	086	524	3.962	031	696	235	50'
20'	502	476	4.039	555	914	032	698	206	40'
30'	.2531	.2504	3.994	.2586	3.867	1.033	.9681	1.3177	30'
40'	560	532	950	617	821	034	674	148	20'
50'	589	560	906	648	776	034	667	119	10'
15° 0'	.2618	.2588	3.864	.2679	3.732	1.035	.9659	1.3090	75° 0'
10'	647	616	822	711	689	036	652	061	50'
20'	676	644	782	742	647	037	644	032	40'
30'	.2705	.2672	3.742	.2773	3.606	1.038	.9636	1.3003	30'
40'	734	700	703	805	566	039	628	1.2974	20'
50'	763	728	665	836	526	039	621	945	10'
16° 0'	.2793	.2756	3.628	.2867	3.487	1.040	.9613	1.2915	74° 0'
	Cos	Sec	Cot	Tan	Csc	Sin	Radians	Degrees	

TABLE IV   **VALUES OF TRIGONOMETRIC FUNCTIONS** — Cont.

Degrees	Radians	Sin	Csc	Tan	Cot	Sec	Cos		
16° 0′	.2793	.2756	3.628	.2867	3.487	1.040	.9613	1.2915	74° 0′
10′	822	784	592	899	450	041	605	886	50′
20′	851	812	556	931	412	042	596	857	40′
30′	.2880	.2840	3.521	.2962	3.376	1.043	.9588	1.2828	30′
40′	909	868	487	.2994	340	044	580	799	20′
50′	938	896	453	3026	305	045	572	770	10′
17° 0′	.2967	.2924	3.420	.3057	3.271	1.046	.9563	1.2741	73° 0′
10′	996	952	388	089	237	047	555	712	50′
20′	.3025	.2979	357	121	204	048	546	683	40′
30′	.3054	.3007	3.326	.3153	3.172	1.048	.9537	1.2654	30′
40′	083	035	295	185	140	049	528	625	20′
50′	113	062	265	217	108	050	520	595	10′
18° 0′	.3142	.3090	3.236	.3249	3.078	1.051	.9511	1.2566	72° 0′
10′	171	118	207	281	047	052	502	537	50′
20′	200	145	179	314	3.018	053	492	508	40′
30′	.3229	.3173	3.152	.3346	2.989	1.054	.9483	1.2479	30′
40′	258	201	124	378	960	056	474	450	20′
50′	287	228	098	411	932	057	465	421	10′
19° 0′	.3316	.3256	3.072	.3443	2.904	1.058	.9455	1.2392	71° 0′
10′	345	283	046	476	877	059	446	363	50′
20′	374	311	3.021	508	850	060	436	334	40′
30′	.3403	.3338	2.996	.3541	2.824	1.061	.9426	1.2305	30′
40′	432	365	971	574	798	062	417	275	20′
50′	462	393	947	607	773	063	407	246	10′
20° 0′	.3491	.3420	2.924	.3640	2.747	1.064	.9397	1.2217	70° 0′
10′	520	448	901	673	723	065	387	188	50′
20′	549	475	878	706	699	066	377	159	40′
30′	.3578	.3502	2.855	.3739	2.675	1.068	.9367	1.2130	30′
40′	607	529	833	772	651	069	356	101	20′
50′	636	557	812	805	628	070	346	072	10′
21° 0′	.3665	.3584	2.790	.3839	2.605	1.071	.9336	1.2043	69° 0′
10′	694	611	769	872	583	072	325	1.2014	50′
20′	723	638	749	906	560	074	315	1.1985	40′
30′	.3752	.3665	2.729	.3939	2.539	1.075	.9304	1.1956	30′
40′	782	692	709	.3973	517	076	293	926	20′
50′	811	719	689	.4006	496	077	283	897	10′
22° 0′	.3840	.3746	2.669	.4040	2.475	1.079	.9272	1.1868	68° 0′
10′	869	.773	650	074	455	080	261	839	50′
20′	898	800	632	108	434	081	250	810	40′
30′	.3927	.3827	2.613	.4142	2.414	1.082	.9239	1.1781	30′
40′	956	854	595	176	394	084	228	752	20′
50′	985	881	577	210	375	085	216	723	10′
23° 0′	.4014	.3907	2.559	.4245	2.356	1.086	.9205	1.1694	67° 0′
10′	043	934	542	279	337	088	194	665	50′
20′	072	961	525	314	318	089	182	636	40′
30′	.4102	.3987	2.508	.4348	2.300	1.090	.9171	1.1606	30′
40′	131	.4014	491	383	282	092	159	577	20′
50′	160	041	475	417	264	093	147	548	10′
24° 0′	.4189	.4067	2.459	.4452	2.246	1.095	.9135	1.1519	66° 0′
		Cos	Sec	Cot	Tan	Csc	Sin	Radians	Degrees

TABLE IV   VALUES OF TRIGONOMETRIC FUNCTIONS — Cont.

Degrees	Radians	Sin	Csc	Tan	Cot	Sec	Cos		
24° 0′	.4189	.4067	2.459	.4452	2.246	1.095	.9135	1.1519	66° 0′
10′	218	094	443	487	229	096	124	490	50′
20′	247	120	427	522	211	097	112	461	40′
30′	.4276	.4147	2.411	.4557	2.194	1.099	.9100	1.1432	30′
40′	305	173	396	592	177	100	088	403	20′
50′	334	200	381	628	161	102	075	374	10′
25° 0′	.4363	.4226	2.366	.4663	2.145	1.103	.9063	1.1345	65° 0′
10′	392	253	352	699	128	105	051	316	50′
20′	422	279	337	734	112	106	038	286	40′
30′	.4451	.4305	2.323	.4770	2.097	1.108	.9026	1.1257	30′
40′	480	331	309	806	081	109	013	228	20′
50′	509	358	295	841	066	111	.9001	199	10′
26° 0′	.4538	.4384	2.281	.4877	2.050	1.113	.8988	1.1170	64° 0′
10′	567	410	268	913	035	114	975	141	50′
20′	596	436	254	950	020	116	962	112	40′
30′	.4625	.4462	2.241	.4986	2.006	1.117	.8949	1.1083	30′
40′	654	488	228	.5022	1.991	119	936	054	20′
50′	683	514	215	059	977	121	923	1.1025	10′
27° 0′	.4712	.4540	2.203	.5095	1.963	1.122	.8910	1.0996	63° 0′
10′	741	566	190	132	949	124	897	966	50′
20′	771	592	178	169	935	126	884	937	40′
30′	.4800	.4617	2.166	.5206	1.921	1.127	.8870	1.0908	30′
40′	829	643	154	243	907	129	857	879	20′
50′	858	669	142	280	894	131	843	850	10′
28° 0′	.4887	.4695	2.130	.5317	1.881	1.133	.8829	1.0821	62° 0′
10′	916	720	118	354	868	134	.816	792	50′
20′	945	746	107	392	855	136	802	763	40′
30′	.4974	.4772	2.096	.5430	1.842	1.138	.8788	1.0734	30′
40′	.5003	797	085	467	829	140	774	705	20′
50′	032	823	074	505	816	142	760	676	10′
29° 0′	.5061	.4848	2.063	.5543	1.804	1.143	.8746	1.0647	61° 0′
10′	091	874	052	581	792	145	732	617	50′
20′	120	899	041	619	780	147	718	588	40′
30′	.5149	.4924	2.031	.5658	1.767	1.149	.8704	1.0559	30′
40′	178	950	020	696	756	151	689	530	20′
50′	207	.4975	010	735	744	153	675	501	10′
30° 0′	.5236	.5000	2.000	.5774	1.732	1.155	.8660	1.0472	60° 0′
10′	265	025	1.990	812	720	157	646	443	50′
20′	294	050	980	851	709	159	631	414	40′
30′	.5323	.5075	1.970	.5890	1.698	1.161	.8616	1.0385	30′
40′	352	100	961	930	686	163	601	356	20′
50′	381	125	951	.5969	675	165	587	327	10′
31° 0′	.5411	.5150	1.942	.6009	1.664	1.167	.8572	1.0297	59° 0′
10′	440	175	932	048	653	169	557	268	50′
20′	469	200	923	088	643	171	542	239	40′
30′	.5498	.5225	1.914	.6128	1.632	1.173	.8526	1.0210	30′
40′	527	250	905	168	621	175	511	181	20′
50′	556	275	896	208	611	177	496	152	10′
32° 0′	.5585	.5299	1.887	.6249	1.600	1.179	.8480	1.0123	58° 0′
		Cos	Sec	Cot	Tan	Csc	Sin	Radians	Degrees

TABLE IV   **VALUES OF TRIGONOMETRIC FUNCTIONS** — Cont.

Degrees	Radians	Sin	Csc	Tan	Cot	Sec	Cos		
32° 0′	.5585	.5299	1.887	.6249	1.600	1.179	.8480	1.0123	58° 0′
10′	614	324	878	289	590	181	465	094	50′
20′	643	348	870	330	580	184	450	065	40′
30′	.5672	.5373	1.861	.6371	1.570	1.186	.8434	1.0036	30′
40′	701	398	853	412	560	188	418	1.0007	20′
50′	730	422	844	453	550	190	403	.9977	10′
33° 0′	.5760	.5446	1.836	.6494	1.540	1.192	.8387	.9948	57° 0′
10′	789	471	828	536	530	195	371	919	50′
20′	818	495	820	577	520	197	355	890	40′
30′	.5847	.5519	1.812	.6619	1.511	1.199	.8339	.9861	30′
40′	876	544	804	661	501	202	323	832	20′
50′	905	568	796	703	1.492	204	307	803	10′
34° 0′	.5934	.5592	1.788	.6745	1.483	1.206	.8290	.9774	56° 0′
10′	963	616	781	787	473	209	274	745	50′
20′	992	640	773	830	464	211	258	716	40′
30′	.6021	.5664	1.766	.6873	1.455	1.213	.8241	.9687	30′
40′	050	688	758	916	446	216	225	657	20′
50′	080	712	751	.6959	437	218	208	628	10′
35° 0′	.6109	.5736	1.743	.7002	1.428	1.221	.8192	.9599	55° 0′
10′	138	760	736	046	419	223	175	570	50′
20′	167	783	729	089	411	226	158	541	40′
30′	.6196	.5807	1.722	.7133	1.402	1.228	.8141	.9512	30′
40′	225	831	715	177	393	231	124	483	20′
50′	254	854	708	221	385	233	107	454	10′
36° 0′	.6283	.5878	1.701	.7265	1.376	1.236	.8090	.9425	54° 0′
10′	312	901	695	310	368	239	073	396	50′
20′	341	925	688	355	360	241	056	367	40′
30′	.6370	.5948	1.681	.7400	1.351	1.244	.8039	338	30′
40′	400	972	675	445	343	247	021	308	20′
50′	429	.5995	668	490	335	249	.8004	279	10′
37° 0′	.6458	.6018	1.662	.7536	1.327	1.252	.7986	.9250	53° 0′
10′	487	041	655	581	319	255	969	221	50′
20′	516	065	649	627	311	258	951	192	40′
30′	.6545	.6088	1.643	.7673	1.303	1.260	.7934	.9163	30′
40′	574	111	636	720	295	263	916	134	20′
50′	603	134	630	766	288	266	898	105	10′
38° 0′	.6632	.6157	1.624	.7813	1.280	1.269	.7880	.9076	52° 0′
10′	661	180	618	860	272	272	862	047	50′
20′	690	202	612	907	265	275	844	.9018	40′
30′	.6720	.6225	1.606	.7954	1.257	1.278	.7826	.8988	30′
40′	749	248	601	.8002	250	281	808	959	20′
50′	778	271	595	050	242	284	790	930	10′
39° 0′	.6807	.6293	1.589	.8098	1.235	1.287	.7771	.8901	51° 0′
10′	836	316	583	146	228	290	753	872	50′
20′	865	338	578	195	220	293	735	843	40′
30′	.6894	.6361	1.572	.8243	1.213	1.296	.7716	.8814	30′
40′	923	383	567	292	206	299	698	785	20′
50′	952	406	561	342	199	302	679	756	10′
40° 0′	.6981	.6428	1.556	.8391	1.192	1.305	.7660	.8727	50° 0′
		Cos	Sec	Cot	Tan	Csc	Sin	Radians	Degrees

TABLE IV VALUES OF TRIGONOMETRIC FUNCTIONS — Cont.

Degrees	Radians	Sin	Csc	Tan	Cot	Sec	Cos		Degrees
40° 0′	.6981	.6428	1.556	.8391	1.192	1.305	.7660	.8727	50° 0′
10′	.7010	450	550	441	185	309	642	698	50′
20′	039	472	545	491	178	312	623	668	40′
30′	.7069	.6494	1.540	.8541	1.171	1.315	.7604	.8639	30′
40′	098	517	535	591	164	318	585	610	20′
50′	127	539	529	642	157	322	566	581	10′
41° 0′	.7156	.6561	1.524	.8693	1.150	1.325	.7547	.8552	49° 0′
10′	185	583	519	744	144	328	528	523	50′
20′	214	604	514	796	137	332	509	494	40′
30′	.7243	.6626	1.509	.8847	1.130	1.335	.7490	.8465	30′
40′	272	648	504	899	124	339	470	436	20′
50′	301	670	499	.8952	117	342	451	407	10′
42° 0′	.7330	.6691	1.494	.9004	1.111	1.346	.7431	.8378	48° 0′
10′	359	713	490	057	104	349	412	348	50′
20′	389	734	485	110	098	353	392	319	40′
30′	.7418	.6756	1.480	.9163	1.091	1.356	.7373	.8290	30′
40′	447	777	476	217	085	360	353	261	20′
50′	476	799	471	271	079	364	333	232	10′
43° 0′	.7505	.6820	1.466	.9325	1.072	1.367	.7314	.8203	47° 0′
10′	534	841	462	380	066	371	294	174	50′
20′	563	862	457	435	060	375	274	145	40′
30′	.7592	.6884	1.453	.9490	1.054	1.379	.7254	.8116	30′
40′	621	905	448	545	048	382	234	087	20′
50′	650	926	444	601	042	386	214	058	10′
44° 0′	.7679	.6947	1.440	.9657	1.036	1.390	.7193	.8029	46° 0′
10′	709	967	435	713	030	394	173	.7999	50′
20′	738	.6988	431	770	024	398	153	970	40′
30′	.7767	.7009	1.427	.9827	1.018	1.402	.7133	.7941	30′
40′	796	030	423	884	012	406	112	912	20′
50′	825	050	418	.9942	006	410	092	883	10′
45° 0′	.7854	.7071	1.414	1.000	1.000	1.414	.7071	.7854	45° 0′
	Cos	Sec	Cot	Tan	Csc	Sin	Radians		Degrees

TABLE V **POWERS AND ROOTS**

Number	Square	Square Root	Cube	Cube Root	Number	Square	Square Root	Cube	Cube Root
1	1	1.000	1	1.000	51	2,601	7.141	132,651	3.708
2	4	1.414	8	1.260	52	2,704	7.211	140,608	3.733
3	9	1.732	27	1.442	53	2,809	7.280	148,877	3.756
4	16	2.000	64	1.587	54	2,916	7.348	157,464	3.780
5	25	2.236	125	1.710	55	3,025	7.416	166,375	3.803
6	36	2.449	216	1.817	56	3,136	7.483	175,616	3.826
7	49	2.646	343	1.913	57	3,249	7.550	185,193	3.849
8	64	2.828	512	2.000	58	3,364	7.616	195,112	3.871
9	81	3.000	729	2.080	59	3,481	7.681	205,379	3.893
10	100	3.162	1,000	2.154	60	3,600	7.746	216,000	3.915
11	121	3.317	1,331	2.224	61	3,721	7.810	226,981	3.936
12	144	3.464	1,728	2.289	62	3,844	7.874	238,328	3.958
13	169	3.606	2,197	2.351	63	3,969	7.937	250,047	3.979
14	196	3.742	2,744	2.410	64	4,096	8.000	262,144	4.000
15	225	3.873	3,375	2.466	65	4,225	8.062	274,625	4.021
16	256	4.000	4,096	2.520	66	4,356	8.124	287,496	4.041
17	289	4.123	4,913	2.571	67	4,489	8.185	300,763	4.062
18	324	4.243	5,832	2.621	68	4,624	8.246	314,432	4.082
19	361	4.359	6,859	2.668	69	4,761	8.307	328,509	4.102
20	400	4.472	8,000	2.714	70	4,900	8.367	343,000	4.121
21	441	4.583	9,261	2.759	71	5,041	8.426	357,911	4.141
22	484	4.690	10,648	2.802	72	5,184	8.485	373,248	4.160
23	529	4.796	12,167	2.844	73	5,329	8.544	389,017	4.179
24	576	4.899	13,824	2.884	74	5,476	8.602	405,224	4.198
25	625	5.000	15,625	2.924	75	5,625	8.660	421,875	4.217
26	676	5.099	17,576	2.962	76	5,776	8.718	438,976	4.236
27	729	5.196	19,683	3.000	77	5,929	8.775	456,533	4.254
28	784	5.292	21,952	3.037	78	6,084	8.832	474,552	4.273
29	841	5.385	24,389	3.072	79	6,241	8.888	493,039	4.291
30	900	5.477	27,000	3.107	80	6,400	8.944	512,000	4.309
31	961	5.568	29,791	3.141	81	6,561	9.000	531,441	4.327
32	1,024	5.657	32,768	3.175	82	6,724	9.055	551,368	4.344
33	1,089	5.745	35,937	3.208	83	6,889	9.110	571,787	4.362
34	1,156	5.831	39,304	3.240	84	7,056	9.165	592,704	4.380
35	1,225	5.916	42,875	3.271	85	7,225	9.220	614,125	4.397
36	1,296	6.000	46,656	3.302	86	7,396	9.274	636,056	4.414
37	1,369	6.083	50,653	3.332	87	7,569	9.327	658,503	4.431
38	1,444	6.164	54,872	3.362	88	7,744	9.381	681,472	4.448
39	1,521	6.245	59,319	3.391	89	7,921	9.434	704,969	4.465
40	1,600	6.325	64,000	3.420	90	8,100	9.487	729,000	4.481
41	1,681	6.403	68,921	3.448	91	8,281	9.539	753,571	4.498
42	1,764	6.481	74,088	3.476	92	8,464	9.592	778,688	4.514
43	1,849	6.557	79,507	3.503	93	8,649	9.644	804,357	4.531
44	1,936	6.633	85,184	3.530	94	8,836	9.695	830,584	4.547
45	2,025	6.708	91,125	3.557	95	9,025	9.747	857,375	4.563
46	2,116	6.782	97,336	3.583	96	9,216	9.798	884,736	4.579
47	2,209	6.856	103,823	3.609	97	9,409	9.849	912,673	4.595
48	2,304	6.928	110,592	3.634	98	9,604	9.899	941,192	4.610
49	2,401	7.000	117,649	3.659	99	9,801	9.950	970,299	4.626
50	2,500	7.071	125,000	3.684	100	10,000	10.000	1,000,000	4.642

APPENDIX B

Field Axioms for Real Numbers

The set of real numbers R with the operations of addition and multiplication satisfy the following axioms.

Axiom 1. The Closure Laws

If a and b are real numbers, then
i) $a + b$ is a real number and
ii) $a \cdot b$ is a real number.

Axiom 2. The Commutative Laws

If a and b are real numbers, then
i) $a + b = b + a$ and
ii) $a \cdot b = b \cdot a$.

Axiom 3. The Associative Laws

If a, b, and c are real numbers, then
i) $(a + b) + c = a + (b + c)$ and
ii) $(a \cdot b) \cdot c = a \cdot (b \cdot c)$.

Axiom 4. The Identity Elements

i)   There exists a real number zero, denoted by 0, such that for any real number a
$$a + 0 = 0 + a = a$$

ii) There exists a real number one, denoted by 1, where zero is different from one, such that for any real number a
$$a \cdot 1 = 1 \cdot a = a$$

Axiom 5. The Inverse Elements

i)   For each real number a, there exists a real number, the *additive inverse* of a, denoted by $-a$ such that
$$a + (-a) = (-a) + a = 0$$

ii) For each real number a where $a \neq 0$, there exists a real number, the *multiplicative inverse or reciprocal*, denoted by $1/a$ or a^{-1} such that
$$a \cdot \frac{1}{a} = \frac{1}{a} \cdot a = 1$$

Axiom 6. The Distributive Laws

If a, b and c are real numbers, then
i) $a(b + c) = ab + ac$ and
ii) $(b + c)a = ba + ca$

Any set of elements containing at least two elements with two operations that satisfy these six axioms is called a *field*.

APPENDIX C

Trigonometric and Circular Identities

1. $\sin^2 t + \cos^2 t = 1$

2. $\sin(-t) = -\sin t$

3. $\cos(-t) = \cos t$

4. $\tan t = \dfrac{\sin t}{\cos t}$

5. $\cot t = \dfrac{\cos t}{\sin t}$

6. $\sec t = \dfrac{1}{\cos t}$

7. $\csc t = \dfrac{1}{\sin t}$

8. $\tan t \cot t = 1$

9. $\sec^2 t = 1 + \tan^2 t$

10. $\csc^2 t = 1 + \cot^2 t$

11. $\cos(t + s) = \cos t \cos s - \sin t \sin s$

12. $\cos(t - s) = \cos t \cos s + \sin t \sin s$

13. $\cos\left(\dfrac{\pi}{2} - t\right) = \sin t$

14. $\sin\left(\dfrac{\pi}{2} - t\right) = \cos t$

15. $\sin(t + s) = \sin t \cos s + \cos t \sin s$

16. $\sin(t - s) = \sin t \cos s - \cos t \sin s$

17. $\tan(t + s) = \dfrac{\tan t + \tan s}{1 - \tan t \tan s}$

18. $\tan(t - s) = \dfrac{\tan t - \tan s}{1 + \tan t \tan s}$

19. $\cos 2t = \cos^2 t - \sin^2 t$

20. $\cos 2t = 2\cos^2 t - 1$

21. $\cos 2t = 1 - 2\sin^2 t$

22. $\sin 2t = 2\sin t \cos t$

23. $\tan 2t = \dfrac{2\tan t}{1 - \tan^2 t}$

24. $\cos^2 t = \dfrac{1 + \cos 2t}{2}$

25. $\sin^2 t = \dfrac{1 - \cos 2t}{2}$

Answers to Selected Problems

Chapter 1

PROBLEM SET 1, Page 8

1. a) T   b) F   c) F   d) F   e) F   f) T   g) F   h) T   i) T   j) F
k) F   l) T

3. a) ϕ b) $\{4, 7\}$ c) $\{3, 4, 5, 6, 7\}$ d) $\{3, 4, 5, 6, 7\}$ e) $\{3, 5, 7\}$

5. a) $\{1, 4, 5, 6, 7, 8\}$ b) $\{5, 7\}$ c) $\{7\}$ d) $\{1, 4, 5, 7\}$ e) $\{1, 7\}$

7. a) Disjoint
b) The first set is a proper subset of the second
c) None of the three set relations hold
d) The second set is a proper subset of the first set
e) Equal

PROBLEM SET 2, Page 19

1. a) $-5 = -5.\overline{0} = -\frac{5}{1}$ b) $0 = 0.\overline{0} = \frac{0}{1}$
c) $0.2 = 0.2\overline{0} = \frac{2}{10}$ d) $5\frac{7}{15} = 5.46\overline{6} = \frac{82}{15}$
e) $3.464\overline{646} = 3.464\overline{646} = \frac{343}{99}$ f) $\frac{2}{9} = 0.2\overline{2} = \frac{2}{9}$
g) $\frac{4}{13} = 0.\overline{307692} = \frac{4}{13}$ h) $33\% = 0.33\overline{0} = \frac{33}{100}$
i) $0.499\overline{9} = 0.499\overline{9} = \frac{1}{2}$ j) $7.3621\overline{621} = 7.3621\overline{621} = \frac{73548}{9990}$
k) $0.97\overline{97} = 0.97\overline{97} = \frac{97}{99}$ l) $4.888\overline{8} = 4.888\overline{8} = \frac{44}{9}$

3. a) 8.36 is between 8.3 and 8.4
b) 3.11 is between 3 and π
c) 1.5 is between $\sqrt{2}$ and $\sqrt{3}$

5. a) $1 \leftrightarrow a; 2 \leftrightarrow c; 3 \leftrightarrow d; 4 \leftrightarrow e; 6 \leftrightarrow f$
b) $1 \leftrightarrow -1; 2 \leftrightarrow -2; 3 \leftrightarrow -3; \ldots, n \leftrightarrow -n$
c) Impossible
d) $7 \leftrightarrow 8; 8 \leftrightarrow 9; 9 \leftrightarrow 91$
e) Impossible
f) $1 \leftrightarrow 0; 2 \leftrightarrow -1; 3 \leftrightarrow 1, 4 \leftrightarrow -2, 5 \leftrightarrow 2, 6 \leftrightarrow -3, \ldots$

PROBLEM SET 3, Page 33

1. No. For example, $-3 < 0, -4 < 0$, but $(-3)(-4) = 12 > 0$

5. a) F; $-(-2) = 2 > 0$
b) T; Theorem 3 or Positive Number Axiom
c) F; let $a = 2$ and $b = -5$
d) F; Trichotomy
e) T; Transitive

f) T; Theorem 2
g) T; Theorem 4
h) T; Definition $[-\frac{1}{2} - (-\frac{3}{4}) = \frac{1}{4}$ is positive$]$
i) T; Theorem 4
j) T; Transitivity applied twice
k) T; Theorem 4
l) T; since $a \nless 2$ implies that $a \geq 2$ which in turn implies $a^2 \geq 4$

9. a) [number line, shaded from -1 to 2]   b) [number line, shaded from -4 to 3]

c) [number line, shaded from -1 to 0]   d) [number line, shaded regions near -2, 1, 3, 8]

e) [number line, shaded from 2 onward]   f) [number line, shaded from 3 to 10]

g) [number line, shaded from 0 to 7]   h) [number line, shaded, -4]

11. a) $\frac{1}{7.5}$ hour $< t < \frac{1}{7}$ hour   b) $x < 33\frac{3}{13}$ feet

13. a) $\{x \mid x > \frac{2}{11}\} = (\frac{2}{11}, \infty)$ [number line at $\frac{2}{11}$]

b) $\{x \mid x \geq \frac{9}{4}\} = [\frac{9}{4}, \infty)$ [number line at $\frac{9}{4}$]

c) $\{x \mid x \leq -\frac{1}{2}\} = (-\infty, -\frac{1}{2}]$ [number line at $-\frac{1}{2}$]

d) $\{x \mid \frac{5}{3} \leq x < \frac{17}{3}\} = [\frac{5}{3}, \frac{17}{3})$ [number line from $\frac{5}{3}$ to $\frac{17}{3}$]

e) $\{x \mid \frac{3}{5} \leq x \leq \frac{11}{3}\} = [\frac{3}{5}, \frac{11}{3}]$ [number line from $\frac{3}{5}$ to $\frac{11}{3}$]

f) $\{x \mid x > 3\} = (3, \infty)$ [number line at 3]

g) $\{x \mid -\frac{1}{2} \leq x < \frac{3}{2}\} = [-\frac{1}{2}, \frac{3}{2})$ [number line from $-\frac{1}{2}$ to $\frac{3}{2}$]

h) $\varnothing$

i) $\{x \mid x < 0\} = (-\infty, 0)$ [number line at 0]

j) $\{x \mid x \in R\} = (-\infty, \infty)$ [full number line]

PROBLEM SET 4, Page 46

1. a) 7   b) 1   c) 7   d) -1 e) 12   f) 12   g) 16   h) $\frac{3}{4}$ i) 25
   j) -7
3. a) $\{-4, 4\}$ b) $\{-\frac{7}{3}, 1\}$ c) $\{-4, \frac{3}{2}\}$ d) $\{-2, \frac{12}{7}\}$ e) $\varnothing$ f) $\{-3, 3\}$
   g) $\{1, 3\}$ h) $\{-5, 5\}$ i) $\{-5, 5\}$ j) $\{-5, 5\}$

5. a) $(-\infty, -1) \cup (1, \infty)$

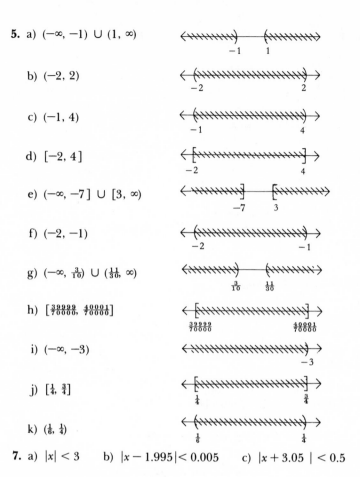

b) $(-2, 2)$

c) $(-1, 4)$

d) $[-2, 4]$

e) $(-\infty, -7] \cup [3, \infty)$

f) $(-2, -1)$

g) $(-\infty, \frac{3}{10}) \cup (\frac{11}{30}, \infty)$

h) $[\frac{39999}{70000}, \frac{40001}{70000}]$

i) $(-\infty, -3)$

j) $[\frac{1}{4}, \frac{3}{4}]$

k) $(\frac{1}{6}, \frac{1}{4})$

7. a) $|x| < 3$ b) $|x - 1.995| < 0.005$ c) $|x + 3.05| < 0.5$

PROBLEM SET 5, Page 58

1. $S \times T = \{(2, a), (2, b), (4, a), (4, b), (6, a), (6, b)\}$
$S \times S = \{(2, 2), (2, 4), (2, 6), (4, 2), (4, 4), (4, 6), (6, 2), (6, 4), (6, 6)\}$
$T \times T = \{(a, a), (a, b), (b, a), (b, b)\}$
$T \times S = \{(a, 2), (a, 4), (a, 6), (b, 2), (b, 4), (b, 6)\}$

a) None have any elements in common; therefore, they are pairwise disjoint; note that $S \times T \neq T \times S$

b) $(S \times T) \cup (S \times S) = \{(2, a), (2, b), (4, a), (4, b), (6, a), (6, b), (2, 2), (2, 4), (2, 6), (4, 2), (4, 4), (4, 6), (6, 2), (6, 4), (6, 6)\}$
$(S \times T) \cup (T \times T) = \{(2, a), (2, b), (4, a), (4, b), (6, a), (6, b), (a, a), (a, b), (b, a), (b, b)\}$
$(S \times T) \cup (T \times S) = \{(2, a), (2, b), (4, a), (4, b), (6, a), (6, b), (a, 2), (a, 4), (a, 6), (b, 2), (b, 4), (b, 6)\}$

c) For the intersections, all equal $\varnothing$ since the sets are disjoint.

3. a) $S \times T = \{(1, 1), (1, 2), (1, 3), (2, 1), (2, 2), (2, 3), (3, 1), (3, 2), (3, 3), (4, 1), (4, 2), (4, 3)\}$

c) $A = \{(1, 1), (2, 2), (3, 3)\}$

d) $B = \{(1, 2), (1, 3), (2, 3)\}$

e) $C = \{(2, 1), (3, 1), (3, 2), (4, 1), (4, 2), (4, 3)\}$

f) $A \cup B \cup C = S \times T$

5. a) I   b) II   c) I   d) None   e) III   f) None

7.

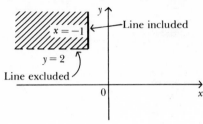

9. a) $\dfrac{\sqrt{41}}{2}$ b) $\sqrt{13}$ c) $3\sqrt{17}$ d) 1

11. b) $\overline{P_1P_2} = 4\sqrt{2}$, $\overline{P_2P_3} = 2\sqrt{2}$; $\overline{P_1P_3} = 6\sqrt{2}$
    c) Since $\overline{P_1P_2} + \overline{P_2P_3} = \overline{P_1P_3}$, P_1, P_2 and P_3 are collinear

13. a) $(3, \frac{5}{2})$ b) $(4, 0)$

REVIEW PROBLEM SET, Page 59

1. a) F   b) T   c) F   d) T   e) F   f) T   g) F   h) T
    i) T   j) T

3. a) T   b) T   c) F   d) T   e) F   f) F

5. a) $\{15, 30, 45, \ldots\}$
    b) $\{6, 12, 18, \ldots\}$
    c) $\{36, 72, 108, \ldots\}$

7. $\frac{1}{9} \cdot \frac{1}{3} = \frac{1}{27} = 0.037\overline{037}$; $\frac{1}{9} + \frac{1}{3} = \frac{4}{9} = 0.444\overline{4}$

9. a) $(4, \infty)$

    b) $(-\infty, -2)$

    c) $[\frac{9}{2}, \infty)$

    d) $(-\infty, -7)$

    e) $[-2, \infty)$

11. a) $(-\infty, \frac{7}{2}] \cup [\frac{13}{2}, \infty)$

    b) $(-\infty, -3) \cup (\frac{7}{3}, \infty)$

    c) $\{3\}$

    d) $(-1, 5)$

e) R

f) $(-2, 7)$

-2 7

g) R

0

h) $(\frac{1}{2}, 1)$

$\frac{1}{2}$ 1

13. a) $\{(a, x), (a, y), (b, x), (b, y), (c, x), (c, y)\}$
b) $\{(x, a), (x, b), (x, c), (y, a), (y, b), (y, c)\}$

15. a) $2\sqrt{10}$ b) $3\sqrt{10}$ c) $4\sqrt{5}$ d) $2\sqrt{29}$ e) 7

Chapter 2

PROBLEM SET 1, Page 69

1. a) and (d) are relations

3. a)

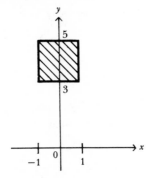

b)

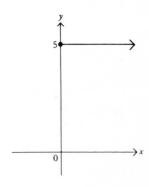

c)

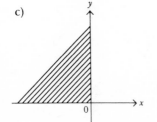

d)

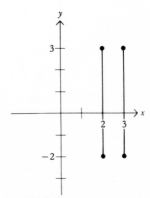

e)

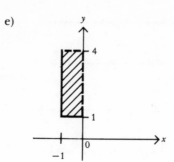

5. a) $\{(x, y)|y = 2x + 1\}$
   Dom $= R$, Range $= R$

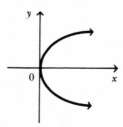

b) $\{(x, y|y = -2\}$
   Dom $= R$, Range $= \{-2\}$

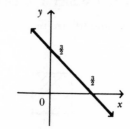

c) $\{(x, y)|y^2 = x\}$
   Dom $= [0, \infty)$, Range $= R$

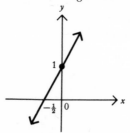

d) $\{(x, y)|2x + 2y = 3\}$
   Dom $= R$, Range $= R$

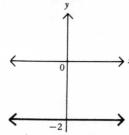

e) $\{(x, y)|y < 2x + 1\}$
   Dom $= R$, Range $= R$

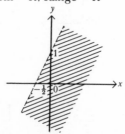

f) $\{(x, y)|x^2 = y^2\}$
   Dom $= R$, Range $= R$

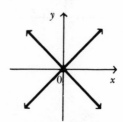

g) $\{(x, y)|x = 7\}$
   Dom $= \{7\}$, Range $= R$

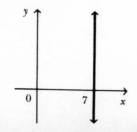

h) $\{(x, y)|x^2 + y^2 = 1\}$
   Dom $=$ Range $= [-1, 1]$

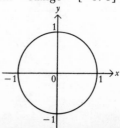

i) $\{(x, y) \mid y = |x+1|\}$
Dom $= R$, Range $= [0, \infty)$

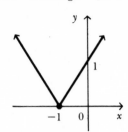

j) $\{(x, y) \mid |x| + |y| = 1\}$
Dom $=$ Range $= [-1, 1]$

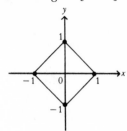

7. All possible sets containing ordered pairs $(1, 1)$, $(2, 1)$, $(1, 2)$, and $(2, 2)$

PROBLEM SET 2, Page 80

1. a) Yes

3. a) Function with Dom $= \{-8, -7, -6,$ $-5, -4, -3, -2\}$ and Range $=$ $\{0, 1, 2, 3\}$.

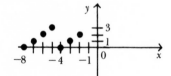

b) Function with Dom $= \{1, 2, 3\}$ and Range $= \{1\}$.

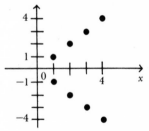

c) Function with Dom $=$ Range $= R$

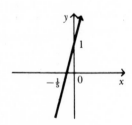

d) *Not* a function; Dom $= \{1, 2, 3, 4\}$ and Range $= \{-1, 1, -2, 2, -3, 3, -4, 4\}$.

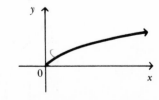

e) *Not* a function:
Dom $=$ Range $= R$

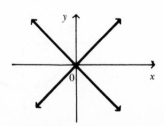

f) Function with
Dom $=$ Range $= [0, \infty)$

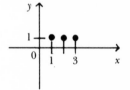

g) Function with Dom = R
   and Range = $(-\infty, 0]$

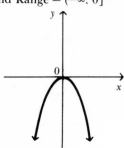

5. Dom = Range = R

a) -3

b) 17

c) -9

d) $-5a + 7$

e) $-15a + 7$

f) $-5b + 7$

g) $-5\sqrt{a} + 7$

h) $-5a^2 + 7$

i) $\dfrac{-25b + 7}{5}$

j) $-5a - 5b + 7$

k) $-5b$

l) $\sqrt{-5a + 8}$

7. a) 5 b) $x^2 + 2x + 2$ c) $x^2 + 2$ d) $x^4 - 2x^2y^2 + y^4 + 1$ e) $x^4 - y^4$
   f) $9x^2 + 6xy + y^2 + 1$

9. a) 0 b) 3 c) $x + \dfrac{h}{2}$ d) $2x + 4 + h$ e) $\dfrac{-1}{x(x + h)}$

PROBLEM SET 3, Page 95

1. a) Even   b) Neither   c) Odd   d) Even   e) Odd   f) Neither

3. a) Origin

b) y-axis

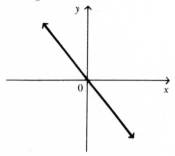

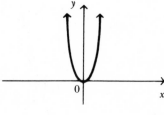

c) Origin

d) y-axis

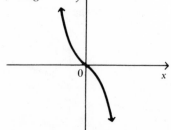

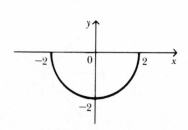

e) y-axis

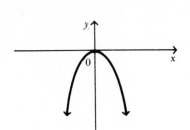

f) y-axis

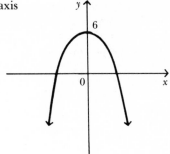

5. Increasing in the interval $[1, 9]$.

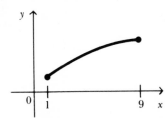

7. a) Decreasing in R b) Increasing in $[0, \infty)$ c) Increasing in $[1, \infty)$
   d) Increasing in $[-2, \infty)$ e) Increasing in R f) Decreasing in R

9. a) Dom = Range = R;
   f is neither even nor odd; no symmetry
   increasing in R;

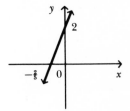

   b) Dom = Range = R;
   odd;
   symmetry w.r.t. origin; increasing in R

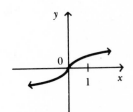

   c) Dom = R, Range = $[0, \infty)$;
   f is neither even nor odd; no symmetry
   increasing in $[1, \infty)$;
   decreasing in $(-\infty, 1]$;

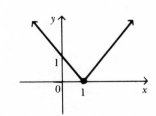

d) Dom = $\{x \mid x \neq \pm 1\}$;
   Range = $(0, \infty) \cup (-\infty, -1]$
   even;
   symmetry w.r.t. y-axis;
   increasing in $(-\infty, -1)$ and $(-1, 0]$;
   decreasing in $[0, 1)$ and $(1, \infty)$

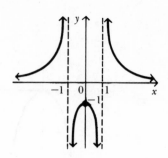

e) Dom = R, Range = I
   no symmetry;
   f is neither increasing nor decreasing

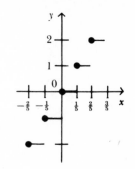

f) Dom = Range = R;
   Decreasing in R;
   f is neither even nor odd;
   no symmetry

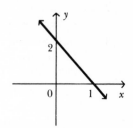

g) Dom = $\{0\}$; Range = $\{0\}$
   symmetry w.r.t. origin
   and y-axis
   even

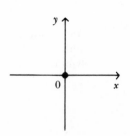

h) Dom = R; Range = $[3, \infty)$
   even;
   symmetry w.r.t. y-axis;
   increasing in $[0, \infty)$;
   decreasing in $(-\infty, 0]$

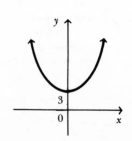

11. a) $\frac{1}{2}, \frac{1}{3}, \frac{1}{4}, \frac{1}{5}, \frac{1}{6}$

Range: subset of rationals

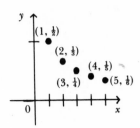

b) $1, \frac{2}{5}, \frac{1}{5}, \frac{2}{17}, \frac{1}{13}$

Range: subset of rationals

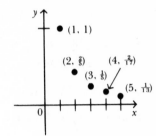

c) $0, 2, 0, 2, 0$

Range $= \{0, 2\}$

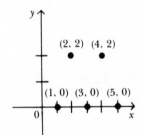

d) $\frac{1}{2}, \frac{1}{6}, \frac{1}{12}, \frac{1}{20}, \frac{1}{30}$

Range: subset of rationals

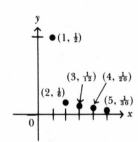

PROBLEM SET 4, Page 105

1. a)

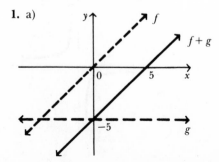

b)

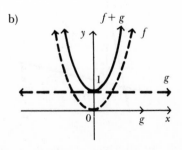

c)

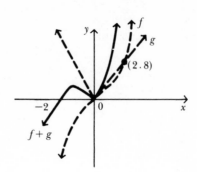

3. a) $9, 8, 13, -\frac{5}{22}$ b) $-x + 12$; $5x - 2$; $-6x^2 - x + 35$; $\dfrac{2x + 5}{7 - 3x}$

c) $\frac{7}{3}$ is excluded from the domain of f/g

5. a) $x^2 + x + 5$ b) $-x^2 + x - 9$ c) $x^3 - 2x^2 + 7x - 14$ d) $\dfrac{x - 2}{x^2 + 7}$

7. a) 2   b) 5   c) 4   d) 2   e) 3   f) 6   g) 5   h) 5   i) 2   j) 3

9. a) $-\frac{7}{2}$ b) Try $g(x) = x$

11. a) $15x + 7$; $-15x - 3$ b) $-15x^2 + x + 2$ c) $\frac{1}{3}$

13. a) $V = 16\pi h$ b) $h = 2t + 4$ c) $V = 32\pi(t + 2)$

PROBLEM SET 5, Page 115

3. a) Yes   b) No   c) No   d) Yes

5. a) $f^{-1}(x) = \dfrac{x + 7}{3}$ b) $f^{-1}(x) = \dfrac{5 - x}{11}$ c) $f^{-1}(x) = \dfrac{4x - 20}{3}$ d) $f^{-1}(x) = \dfrac{\sqrt[3]{x}}{2}$

e) $f^{-1}(x) = -\sqrt{x}$ f) $f^{-1}(x) = \sqrt{-x}$

7. a) $f^{-1}(x) = \dfrac{x - 1}{2}$ b) $f^{-1}(x) = \dfrac{-(x - 2)}{5}$

9. a) No inverse

b) $f^{-1}(x) = \dfrac{x - 5}{7}$ c) $f^{-1}(x) = \dfrac{3}{x}$

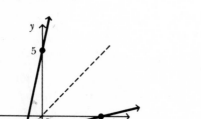

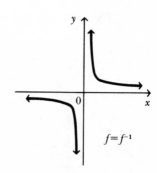

d) No inverse

e) $f^{-1}(x) = \dfrac{x-1}{-3}$ f) $f^{-1}(x) = \sqrt[3]{x-5}$

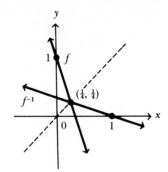 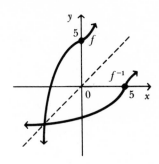

g) $f^{-1}(x) = \sqrt{x-2}$

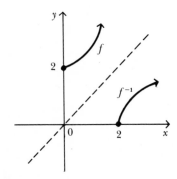

h) No inverse

REVIEW PROBLEM SET, Page 117

1. a) Function; Dom = Range = R
    b) Function; Dom = R; Range = $(-\infty, 25]$
    c) Relation; Dom = $\{1, 2, 6\}$, Range = $\{2, 3, 5, 7\}$
    d) Relation; Dom = Range = R

3. a) 5   b) 2   c) 5   d) $\frac{10}{3}$ e) $3a^2 + 2$ f) $3b^2 + 2$
    g) $3a^2 + 6ab + 3b^2 + 2$ h) $6a + 3b$

5. $(1, 0)$, $(4, 4)$

7. a) No symmetry;
     Dom = Range = R;
     decreasing in R;
     neither even nor odd

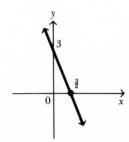

b) No symmetry;
Dom = R; Range = $[0, \infty)$;
decreasing in $(-\infty, -\frac{5}{2}]$,
increasing in $[-\frac{5}{2}, \infty]$;
neither even nor odd

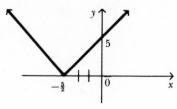

c) Symmetry w.r.t. y-axis
Dom = R, Range = $[0, \infty)$
decreasing in $(-\infty, 0]$,
increasing in $[0, \infty)$;
even

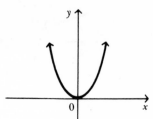

d) No symmetry;
Dom = R, Range = $[0, \infty)$;
decreasing in $(-\infty, 0]$;
neither even nor odd

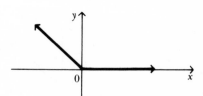

e) Symmetry w.r.t origin;
Dom = Range = R;
decreasing in R; odd

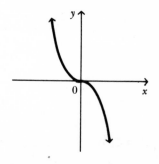

f) Symmetry w.r.t. origin;
Dom = Range = R;
increasing in R; odd

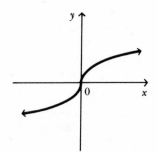

g) Symmetry w.r.t. y-axis;
Dom = R; Range = $(-\infty, 0]$;
increasing in $(-\infty, 0]$,
decreasing in $[0, \infty)$; even

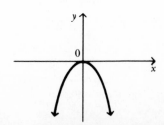

h) Symmetry w.r.t. y-axis;
Dom $= R$, Range $= [1, \infty)$;
decreasing in $(-\infty, 0]$,
increasing in $[0, \infty)$; even

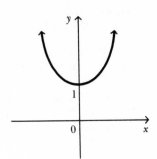

9. a)

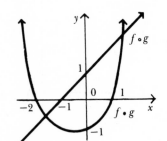

b)

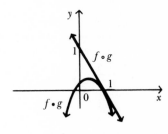

c)

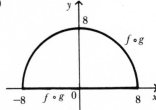

d)

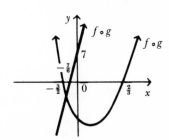

11. a) $-42 + 14x^2 - x^4$; Dom $= R$
    b) $36 - 5x^2$; Dom $= R$
    c) $6 + 25x$; Dom $= R$
    d) $6 - 10x - 25x^2$; Dom $= R$
    e) $\dfrac{6 + 70x + 175x^2}{1 + 10x + 25x^2}$; Dom $= \{-\frac{1}{5}\}^c$
    f) $\dfrac{12 - x^2}{7 - x^2}$; Dom $= \{\sqrt{7}, -\sqrt{7}\}^c$

13. a) $f^{-1}(x) = 2x - 4$
    b) $f^{-1}(x) = x^2, x \geqslant 0$
    c) No inverse
    d) No inverse
    e) $f^{-1}(x) = \dfrac{1 + x}{x}$

Chapter 3

PROBLEM SET 1, Page 139

1. a) Degree 2; Coefficients are $a_2 = -5$, $a_1 = 7$, and $a_0 = \pi$.
   b) Not a polynomial
   c) Degree 0; Coefficient is $a_0 = \frac{1}{5}$
   d) Not a polynomial

e) Degree 3; Coefficients are $a_3 = 5$, $a_2 = -\sqrt{3}$, $a_1 = 2$ and $a_0 = -7$
f) Degree 6; Coefficients are $a_6 = -7$, $a_5 = 1$, $a_4 = 0$, $a_3 = -1$, $a_2 = 0$, $a_1 = 3$ and $a_0 = 0$

3. a) $\overline{P_1P_2} = \sqrt{2}$, $\overline{P_2P_3} = \sqrt{2}$, $\overline{P_1P_3} = 2\sqrt{2}$, therefore, P_1, P_2, and P_3 are collinear
b) $(1, 1)$, $(2, 4)$ and $(3, 2)$ are not collinear

5. a) Slope = 0; not increasing nor decreasing; even function, no inverse; domain = R; Range = R

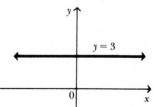

b) Slope = -3; decreasing; neither even nor odd; domain = range = R; $f^{-1}(x) = \dfrac{-x + 5}{3}$

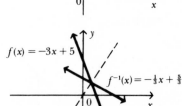

c) Slope = 2; increasing, neither even nor odd; domain = range = R; $f^{-1}(x) = \dfrac{x + 3}{2}$

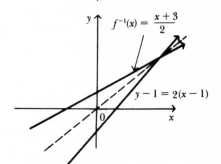

d) Slope = $\frac{6}{5}$; increasing; neither even nor odd; domain = range = R; $f^{-1}(x) = \frac{5}{6}(x - 2)$.

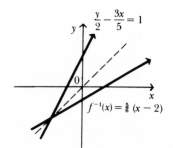

e) Slope = 3; increasing; neither even nor odd; domain = range = R; $f^{-1}(x) = \dfrac{x - 5}{3}$

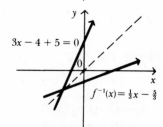

7. a) slope $= \frac{3}{2}$; y intercept $= -6$
   b) slope $= -1$; y intercept $= 3$
   c) slope $= -\frac{2}{5}$; y intercept $= \frac{6}{5}$
   d) slope $= 2$; y intercept $= 5$

9. a) $y = 3x - 2$ b) $y - 5 = -2(x + 2)$
   c) $y = -2$ d) $y - 2 = -3(x - 1)$
   e) $y - 3 = \frac{5}{2}(x - 3)$ f) $y = -\frac{3}{2}x - \frac{11}{2}$
   g) $y = \frac{2}{3}x - 3$

11. a) i) $y = -7x + 17$ ii) $y = \frac{1}{7}x + \frac{19}{7}$ b) i) $y = 7x - 9$ ii) $y = \frac{1}{7}x + \frac{37}{7}$

13. 32

PROBLEM SET 2, Page 153

1. a) $\{(3, 5)\}$ b) $\{(3, 1)\}$
   c) $\{(\frac{23}{4}, \frac{13}{2})\}$ d) $\varnothing$
   e) $\{(x, y) | 3x + y = 4\}$ f) $\{(-\frac{2}{3}, -\frac{7}{3})\}$
   g) $\{(1, 2, 3)\}$ h) $\{(2, 1, -1)\}$
   i) $\{(1, 1, 1)\}$ j) $\{(3, -1, 2)\}$
   k) $\{(4, 5, 6)\}$ l) $\{(2, 0, 0)\}$

3. a), c), d) and e) are in a row reduced echelon form.

PROBLEM SET 3, Page 163

1. a) 17   b) -19 c) 63   d) 135   e) 0   f) $\frac{69}{2}$

3. a) Theorem 2   b) Theorem 3   c) Theorem 1   d) Theorem 3 and 1
   e) Theorem 1 and 3.

5. a) 0   b) -3 c) -20 d) 0   e) 3.

7. a) $\{(\frac{1}{3}, \frac{2}{3})\}$ b) $\{(\frac{2}{5}, \frac{21}{5})\}$
   c) $\{(0, 0)\}$ d) $\varnothing$
   e) $\{(\frac{19}{4}, \frac{-3}{4}, \frac{-29}{4})\}$ f) $\{(2, 0, 1)\}$
   g) $\{(1, 1, 1)]$ h) $\{(-5, \frac{-14}{3}, \frac{-16}{3})\}$
   i) $\{(3, -1, 2)\}$ j) $\{(2, 1, -1)\}$

9. a) Watch $= \$75$, chain $= \$25$, ring $= \$125$
   b) First number $= 6$, second number $= 9$

PROBLEM SET 4, Page 181

1. a) $\{0, \frac{7}{3}\}$ b) $\{-\frac{3}{2}, 11\}$
   c) $\{\frac{-25}{6}, 5\}$ d) $\{-3, 3\}$
   e) $\{2, 4\}$ f) $\{\frac{-3}{4}, -2\}$
   g) $\{-2, 2\}$ h) $\{\frac{-7}{2}, \frac{1}{9}\}$

3. a) $\left\{\dfrac{5 - \sqrt{17}}{4}, \dfrac{5 + \sqrt{17}}{4}\right\}; d = 17$ b) $\{-1, \frac{1}{2}\}; d = 9$
   c) $\{2, 10\}; d = 64$ d) $\{\frac{-11}{4}, \frac{1}{3}\}; d = 1369$
   e) Not real; $d = -3$ f) $\{-\frac{5}{6}, 1\}; d = 121$
   g) $\{4, 14\}; d = 100$ h) Not real; $d = -2672$

5.

	Domain	Range	Extreme point	x intercept	y intercept
a)	R	$[-4, \infty)$	$(0, -4)$	$(\pm\sqrt{2}, 0)$	$(0, -4)$
b)	R	$[-1, \infty)$	$(-1, -1)$	$(0, 0)$ and $(-2, 0)$	$(0, 0)$
c)	R	$[0, \infty)$	$(-2, 0)$	$(-2, 0)$	$(0, 4)$
d)	R	$(-\infty, 0]$	$(-1, 0)$	$(-1, 0)$	$(0, -1)$
e)	R	$(-\infty, -1]$	$(0, -1)$	None	$(0, -1)$
f)	R	$[-\frac{9}{8}, \infty)$	$(\frac{3}{4}, -\frac{9}{8})$	$(0, 0)$ and $(\frac{3}{2}, 0)$	$(0, 0)$
g)	R	$(-\infty, \frac{1}{8}]$	$(\frac{3}{4}, \frac{1}{8})$	$(\frac{1}{2}, 0)$ and $(1, 0)$	$(0, -1)$'
h)	R	$[-\frac{25}{4}, \infty)$	$(-\frac{3}{2}, -\frac{25}{4})$	$(-4, 0)$ and $(1, 0)$	$(0, -4)$
i)	R	$[\frac{35}{12}, \infty)$	$(\frac{1}{6}, \frac{35}{12})$	None	$(0, 3)$
j)	R	$(-\infty, \frac{21}{4}]$	$(\frac{1}{2}, \frac{21}{4})$	$\left(\dfrac{-1 \pm \sqrt{21}}{2}, 0\right)$	$(0, 5)$

7. a) $(-\infty, 1] \cup [\frac{3}{2}, \infty)$ b) $(-\infty, -\frac{1}{2}) \cup (1, \infty)$ c) $\varnothing$ d) $[-4, -2]$ e) $\varnothing$
f) $(-\infty, \frac{3}{2}) \cup (2, \infty)$ g) $\left(\dfrac{5 - \sqrt{13}}{6}, \dfrac{5 + \sqrt{13}}{6}\right)$ h) $(-\frac{3}{2}, 3)$

9. a) $(-3, \frac{1}{2})$ b) $(-\infty, \frac{1}{3}) \cup (2, \infty)$ c) 2 d) $(-3, \frac{3}{2})$ e) $(-\infty, 1] \cup [6, \infty)$
f) R g) $[-3, 1]$ h) $(-4, 4)$

PROBLEM SET 5, Page 193

1. a) $\{\frac{2}{3}\}$ b) $\{-1, \frac{2}{3}\}$ c) $\{1, 2\}$ d) $\{1, 2, 5\}$ e) $\{-2, 2\}$ f) $\{4, -2\}$
g) $\{0, -3, 3\}$ h) $\{0, 1\}$

3. a) $Q(x) = 5x^2 + 13x + 42; R = 122$
b) $Q(x) = 2x^3 + x^2 - 6x + 8; R = -9$
c) $Q(x) = 5x^4 + 7x^3 + 16x^2 + 33x + 59; R = 121$
d) $Q(x) = 2x^3 - 7x^2 + 19x - 32; R = 61$
e) $Q(x) = -4x^5 - 4x^4 - 4x^3 - 9x^2 - 6x - 5; R = 2$
f) $Q(x) = 2x^3 - 5x^2 + 17x - 67; R = 267$

5. $f(-5) = 0, f(-4) = 30, f(-3) = 40, f(-1) = 24, f(0) = 10, f(1) = 0, f(2) = 0,$
$f(3) = 16, f(4) = 54, f(5) = 120$
Factors are $x - 2, x - 1$ and $x + 5$

7. a) $f(1) = 0$ b) $f(-4) = 0$

9. a) x intercepts are $(-1, 0)$ and $(2, 0)$
y intercept is $(0, 4)$

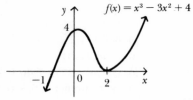

b) x intercepts are $(-2, 0)$, $(1, 0)$ and $(3, 0)$
y intercept is $(0, 6)$

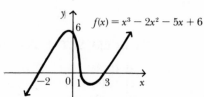

c) x intercepts are $(0, 0)$, $(1, 0)$ and $(-2, 0)$
y intercept is $(0, 0)$

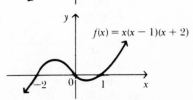

d) x intercepts are $(-1, 0)$, $(\frac{1}{2}, 0)$ and $(2, 0)$
    y intercept is $(0, 2)$

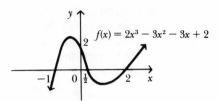

e) x intercept is $(-\frac{1}{2}, 0)$
    y intercept $(0, 1)$

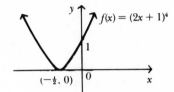

11. a) $\{\frac{1}{3}\}$ b) $\{2\}$ c) $\{1\}$
    d) $\{-4, 2, 3\}$

PROBLEM SET 6, Page 200

1. Domain $= \{x \mid x \neq \frac{2}{3}\}$
    x intercept is $(-\frac{2}{3}, 0)$
    y intercept is $(0, 2)$

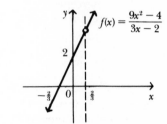

3. Domain $= \{x \mid x \neq -\frac{2}{3}\}$
    Vertical asymptote is $x = -\frac{2}{3}$
    Horizontal asymptote is $y = 0$
    No x intercepts
    y intercept is $(0, \frac{5}{2})$

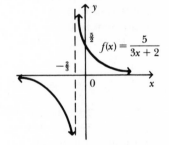

5. Domain $= \{x \mid x \neq 0\}$
    Vertical asymptote is $x = 0$
    Horizontal asymptote is $y = 0$
    No x intercept
    No y intercept

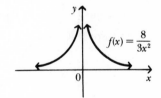

7. Domain $= \{x \mid x \neq -3, x \neq 3\}$
    Vertical asymptotes are $x = -3$ and $x = 3$
    Horizontal asymptote is $y = 0$
    No x intercept
    y intercept is $(0, -\frac{4}{9})$

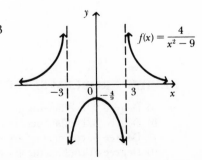

9. Domain $= \{x|x \neq 1\}$
Vertical asymptotes $x = 1$
Other asymptote $y = 2x + 2$
No x intercept
y intercept is $(0, -3)$

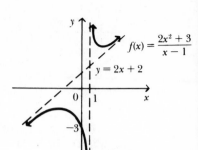

11. Domain $= \{x|x \neq 1, x \neq 2\}$
Vertical asymptote is $x = 1$
Horizontal asymptote is $y = 0$
No x intercept
y intercept is $(0, -3)$

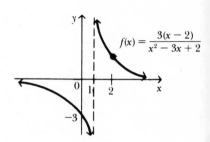

13. Domain $= \{x|x \neq -1\}$
Vertical asymptote is $x = -1$
Horizontal asymptote is $y = 1$
x intercept is $(0, 0)$
y intercept is $(0, 0)$

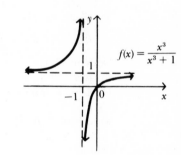

15. Domain $= \{x|x \neq -1, x \neq 1\}$
Vertical asymptote $x = -1$
x intercepts $\left(\dfrac{-1 - \sqrt{2}}{2}, 0\right)$ and $\left(\dfrac{-1 + \sqrt{2}}{2}, 0\right)$
y intercept is $(0, -1)$

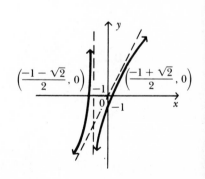

REVIEW PROBLEM SET, Page   201

1. a) Degree 3, coefficients are $a_3 = 4$, $a_2 = 0$, $a_1 = 2$, $a_0 = 6$
    b) Degree 3, coefficients are $a_3 = 2$, $a_2 = 5$, $a_1 = 7$, $a_0 = -3$
    c) Degree 5, coefficients are $a_5 = 7$, $a_4 = -4$, $a_3 = 3$, $a_2 = 2$, $a_1 = -5$, $a_0 = -7$
    d) Degree 7, coefficients are $a_7 = 2$, $a_6 = a_5 = a_4 = a_3 = a_2 = a_1 = 0$, $a_0 = -3$

3.

	Slope	x intercept	y intercept	Increasing	Decreasing	$f^{-1}(x) =$
a)	$\frac{3}{5}$	$(-\frac{35}{9}, 0)$	$(0, \frac{7}{3})$	Yes	No	$\frac{5}{3}x - \frac{35}{9}$
b)	$-\frac{5}{3}$	$(3, 0)$	$(0, 5)$	No	Yes	$-\frac{3}{5}x + 3$
c)	$-\frac{7}{5}$	$(\frac{13}{7}, 0)$	$(0, \frac{13}{5})$	No	Yes	$-\frac{5}{7}x + \frac{13}{7}$
d)	0	None	$(0, 5)$	No	No	None
e)	-1	$(-3, 0)$	$(0, -3)$	No	Yes	$-x - 3$
f)	$-\frac{5}{3}$	$(\frac{11}{5}, 0)$	$(0, \frac{11}{3})$	No	Yes	$-\frac{3}{5}x + \frac{11}{5}$

5. a) $y = 3x - 7$ b) $y = 3$ c) $y = 4x - 14$ d) $y = -\frac{2}{3}x + 2$

7. a) Yes   b) $a^2 + b^2 = 25$ c) $(-3, -5)$
d) $y = \frac{4}{3}x - 6$, $y = \frac{4}{3}x + \frac{32}{3}$, $y = -\frac{3}{4}x + \frac{13}{2}$, $y = \frac{3}{4}x + \frac{1}{4}$, $(3, -2)$, $(-2, 8)$

9. a) $\{(\frac{11}{10}, \frac{42}{5})\}$ b) $\{(\frac{33}{19}, \frac{35}{19})\}$ c) $\left\{(x, y, z) \mid x = \dfrac{1 + z - 2y}{3}\right\}$ d) $\{(-\frac{4}{5}, -\frac{9}{5}, -\frac{13}{5})\}$

11. a) $\{(\frac{11}{10}, \frac{42}{5})\}$ b) $\{(\frac{33}{19}, \frac{35}{19})\}$ c) Not unique   d) $\{(-\frac{4}{5}, -\frac{9}{5}, -\frac{13}{5})\}$

13. a) $\{-1 - \sqrt{7}, -1 + \sqrt{7}\}$ b) $\{-3, 1\}$ c) $\{-\frac{1}{2}, 1\}$ d) $\left\{\dfrac{1 - \sqrt{5}}{2}, \dfrac{1 + \sqrt{5}}{2}\right\}$

15.

	x intercepts	y intercept	Extreme point	Range	Solution of inequality
a)	$(-\frac{1}{2}, 0)$, $(\frac{4}{3}, 0)$	$(0, -4)$	$(\frac{5}{12}, \frac{-121}{24})$	$[\frac{-121}{24}, \infty)$	$[-\frac{1}{2}, \frac{4}{3}]$
b)	$(\frac{1}{2}, 0)$, $(\frac{3}{4}, 0)$	$(0, -3)$	$(\frac{5}{8}, \frac{1}{8})$	$(-\infty, \frac{1}{8}]$	$(-\infty, \frac{1}{2}) \cup (\frac{3}{4}, \infty)$
c)	$(\frac{3}{7}, 0)$, $(\frac{1}{2}, 0)$	$(0, -3)$	$(\frac{13}{28}, \frac{1}{56})$	$(-\infty, \frac{1}{56}]$	$(\frac{3}{7}, \frac{1}{2})$
d)	None	$(0, -2)$	$(1, -1)$	$(-\infty, -1]$	R

17. a) $Q(x) = 3x^2 + 11x + 29$, $R = 55$
b) $Q(x) = 5x^3 + 3x^2 + 14x + 19$, $R = 55$
c) $Q(x) = 2x^2 + 2x + 1$, $R = -19$
d) $Q(x) = x^6 + 2x^5 + 4x^4 + 8x^3 + 16x^2 + 32x + 64$, $R = 123$

19. a) 13   b) 208   c) 4   d) -7

21. a) $\{-1, 2, 3\}$ b) $\{-\frac{1}{2}, \frac{1}{3}, \frac{1}{2}, \frac{3}{4}\}$ c) $\{-\sqrt{3}, \frac{1}{2}, \sqrt{3}, -3\}$

23. a) Domain $= \{x \mid x \neq 5\}$
Vertical asymptote $x = 5$
Horizontal asymptote $y = 0$
y intercept $(0, -\frac{2}{5})$

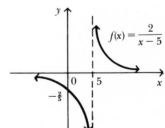

$$f(x) = \frac{2}{x - 5}$$

b) Domain $= \{x \mid x \neq -3\}$
Vertical asymptote $x = -3$
Horizontal asymptote $y = 1$
x intercept $(0, 0)$
y intercept $(0, 0)$

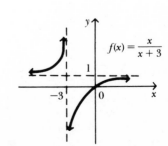

$$f(x) = \frac{x}{x + 3}$$

c) Domain $= \{x | x \neq 3, x \neq -3\}$
Vertical asymptotes $x = -3$ and $x = 3$
Horizontal asymptote $y = 0$
x intercept $(2, 0)$
y intercept $(0, \frac{2}{9})$

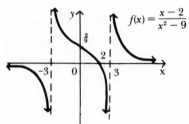

d) Domain $= \{x | x \neq -1\}$
Vertical asymptote $x = -1$
Other asymptote $y = x - 1$
x intercept $(0, 0)$
y intercept $(0, 0)$

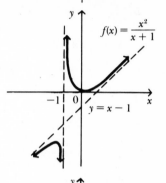

e) Domain $= R$
Horizontal asymptote $y = 1$
y intercept $(0, \frac{2}{3})$

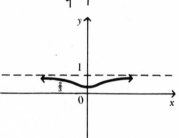

Chapter 4

PROBLEM SET 1, Page 209

1. a) 25   b) 243   c) $\frac{1}{3}$ d) $-\frac{1}{25}$ e) $\frac{1}{3^{11}}$ f) 729   g) 4,096   h) 25

    i) $\frac{9}{64}$ j) $\frac{1}{x^4}$ k) $\frac{1}{x^{13}}$ l) 8   m) 9

3. a) 9   b) $x^{23/20}$ c) $x^{4/3}$ d) $2\sqrt[3]{3}$ e) $3 + 3\sqrt{3}$ f) $\dfrac{4\sqrt[8]{x^7}}{x^2}$

    g) $\frac{1}{4}$ h) 9

5. a) $\sqrt{2}$ b) $5\sqrt{3} + 5\sqrt{2}$ c) $\dfrac{5\sqrt{x} - 5}{x - 1}$ d) $\dfrac{6 + \sqrt{15}}{3}$

    e) $x + \sqrt{x^2 - 1}$ f) $\dfrac{6 + \sqrt{3}\,(3\sqrt{2} + 2) + \sqrt{2}\,(3 - 3\sqrt{5} - \sqrt{5}\sqrt{3})}{12}$

7. a) $\{\frac{5}{3}\}$ b) $\{\frac{3}{2}\}$ c) $\{\frac{-3}{5}\}$ d) $\{1\}$ e) $\{6\}$ f) $\{\frac{5}{2}\}$
    g) $\{5\}$ h) $\{-1\}$

9.

	Domain	Range	Even or Odd	Increasing or Decreasing
a)	$[0, \infty)$	$[0, \infty)$	No	Increasing
b)	$(-\infty, 1]$	$[0, \infty)$	No	Decreasing
c)	R	$[1, \infty)$	Even	Neither
d)	$(-\infty, 0]$	$[0, \infty)$	No	Decreasing
e)	$[0, \infty)$	$[0, \infty)$	No	Increasing

PROBLEM SET 2, Page 213

1.

	Domain	Range	Increasing or Decreasing	Inverse
a)	R	$(0, \infty)$	Increasing	Yes
b)	R	$\{1\}$	Neither	No
c)	R	$(0, \infty)$	Increasing	Yes
d)	R	$(0, \infty)$	Increasing	Yes
e)	R	$(0, \infty)$	Decreasing	Yes
f)	R	$(0, \infty)$	Decreasing	Yes
g)	R	$(0, \infty)$	Increasing	Yes
h)	R	$(0, \infty)$	Decreasing	Yes
i)	R	$(0, \infty)$	Increasing	Yes
j)	R	$(-\infty, 0)$	Decreasing	Yes

3. a) 3   b) 4   c) 3   d) 5   e) any $b > 0$ f) 10

5. a) For $b = 1$, neither; $0 < b < 1$, decreasing; $b > 1$, increasing
b) yes if $b = 1$, even

7. a) 3^{25} b) 5^9 c) 3^{5^x} d) 5^{3^x} e) 5

9. a) 2,997,000   b) 1

11. a) 0.125; 0.1563; $1.953 \cdot 10^{-3}$; $3.8146 \cdot 10^{-6}$; $7.4506 \cdot 10^{-9}$; $8.6736 \cdot 10^{-19}$

PROBLEM SET 3, Page 219

1.

	(a)	(b)
i	$\log_5 25 = 3$	$9^2 = 81$
ii	$\log_4 \frac{1}{16} = -2$	$10^{-4} = 0.0001$
iii	$\log_{32} 2 = \frac{1}{5}$	$(\frac{1}{3})^{-2} = 9$
iv	$\log_{1/3} 9 = -2$	$10^{-1} = \frac{1}{10}$
v	$\log_9 3 = \frac{1}{2}$	$(\sqrt{16})^{1/2} = 2$
vi	$\log_{1/6} \frac{1}{36} = 2$	$36^{3/2} = 216$

3.

	Domain	Range	Increasing or Decreasing	$f^{-1}(x) =$
a)	$(0, \infty)$	R	Increasing	6^x
b)	$(0, \infty)$	R	Increasing	5^x
c)	$(0, \infty)$	R	Decreasing	$(\frac{1}{2})^x$
d)	$(0, \infty)$	R	Decreasing	$(\frac{1}{3})^x$
e)	$(0, \infty)$	R	Increasing	π^x

5. The larger b gets, the "flatter" or "more horizontal" the graph is.

7. For $0 < b < 1$, $y = \log_b x$ is a decreasing function whereas for $b > 1$ it is an increasing function.

9. a) $\{0\}^c$ b) $\{-1\}^c$ c) $(-\infty, 0)$ d) $(\frac{1}{5}, \infty)$ e) $(-2, 2)$

PROBLEM SET 4, Page 223

1. a) 0.6020   b) 1.2552   c) 3   d) 0.0602   e) 3.4771   f) 0.6990
g) 1.7781   h) −0.3010   i) −0.4771   j) 0

3. a) 100,000   b) 4   c) 8   d) 32   e) 1   f) $(0, \infty)$ g) 9   h) 28
i) $\frac{26}{125}$ j) −2

5. a) $\{7\}$ b) $\{\frac{1}{9}\}$ c) $\{\frac{1}{3}\}$ d) $\{7\}$ e) $\{22\}$ f) $\{3, 4\}$

7. a) $\log_5\left(\frac{7}{5}\right)$ b) $\log_2\left(\frac{55}{2}\right)$ c) $\log_c\left(\frac{a}{7}\right)$ d) $\log_7\left(\frac{x+2}{2x}\right)$ e) $\log_x\left(\frac{a}{c}\right)$

f) $\log_a\left(\frac{a^2}{\sqrt[3]{x^2}}\right)$ g) $\log_e\left(\frac{b^7}{4}\right)$

PROBLEM SET 5, Page 233

1. a) $0.1761 + 1$ b) $0.1761 + (-1)$ c) $0.1761 + (-2)$ d) $0.5694 + 0$
e) $0.5694 + (-1)$ f) $0.5740 + 1$ g) $0.5740 + (-3)$ h) $0.1367 + 1$
i) $0.1367 + (-5)$ j) $0.1335 + 8$ k) $0.0645 + 8$ l) $0.0917 + 0$
m) $0.5228 + 1$ n) $0.9007 + 2$ o) $0.4986 + 2$

3. a) 8.20   b) 55.2   c) 0.2551   d) 0.003436   e) 7990   f) 1.22
g) 8.47   h) 1400   i) 0.03126   j) 0.002883   k) 994.7   l) 5.554

5. a) 10.61   b) 9.41   c) 0.033   d) 9.01   e) 222,809.923   f) 1.644

7. a) 2.807   b) 1.6094   c) 0.4017   d) −0.5693   e) 0.609   f) 6.465

9. a) $574.34   b) over 35 years

11. a) 0.1386   b) 138.6 years

PROBLEM SET 6, Page 243

7. a) 3003   b) 1   c) 15   d) 3003   e) 1   f) 6   g) 455   h) n
i) 15

9. a) $x^{20} + 20x^{18}a + 180x^{16}a^2 + 960x^{14}a^3$

b) $64a^6 - 192\,\dfrac{a^5}{b} + 240\,\dfrac{a^4}{b^2} - 160\,\dfrac{a^3}{b^3}$

c) $\left(\dfrac{x}{2}\right)^{7/2} + \dfrac{7}{4}x^3y + 84\left(\dfrac{x}{2}\right)^{5/2}y^2 + 70x^2y^3$

d) $\dfrac{1}{a^{11}} + \dfrac{11x}{2a^{10}} + \dfrac{55x^2}{4a^9} + \dfrac{165x^3}{8a^8}$

e) $a^{12} - 16a^{21/2}x^2 + 112a^9x^4 - 448a^{15/2}x^6$

11. a) $\frac{455}{4096}x^{24}a^3$ b) $-8064y^{10}z^5$ c) $\frac{224}{243}x^6a^{12}$ d) $924x^6a^3$
e) $\frac{28}{243}a^3x^{12}$ f) $2880x^2y^4$

PROBLEM SET 7, Page 249

1. a) 15   b) 7381   c) 15   d) 288   e) $\frac{5}{6}$ f) $\frac{128}{15}$ g) 15   h) 95
i) 500   j) $\frac{25}{4}$

3. a) $\displaystyle\sum_{k=0}^{4}(3k+1)$ b) $\displaystyle\sum_{k=1}^{5}(\tfrac{1}{2})^k$ c) $\displaystyle\sum_{k=1}^{4}(\tfrac{2}{3})^k$ d) $\displaystyle\sum_{k=1}^{4}\left(\dfrac{k}{5k+1}\right)$

5. a) $\frac{32}{99}$ b) $\frac{1}{20}$ c) $\frac{46}{99}$ d) $\frac{8}{111}$ e) $\frac{3558}{999}$ f) $\frac{324,186}{9999}$

REVIEW PROBLEM SET, Page 250

1. a) a^{2n+4} b) 2^{6n+1} c) x^4 d) x^{-3} e) $2x^{1/2}$ f) $\dfrac{x^{17/12}}{2^{1/3} \cdot 3^{1/4}}$

3. a) 11   b) 9   c) 3   d) 4   e) 4

5.

	Domain	Range	Increasing or Decreasing
a)	R	$(0, \infty)$	Decreasing
b)	R	$[2, \infty)$	Decreasing in $(-\infty, 0]$; Increasing in $[0, \infty)$
c)	R	$(0, \infty)$	Increasing
d)	R	R	Increasing
e)	R	$[1, \infty)$	Decreasing in $(-\infty, 0]$; Increasing in $[0, \infty)$

7. a) $\{\frac{3}{2}\}$ b) $\{\frac{-6}{5}\}$ c) $\{0\}$ d) $\left\{\dfrac{\log \frac{3}{2}}{\log 2}, -2\right\}$

9. a) 13.2   b) 2.52   c) 0.92   d) $\sqrt[3]{2.76}$ e) -1.64 f) 0.14
   g) 3.10   h) -0.06

11. a) Domain $= R$, Range $= (0, \infty)$, $f^{-1}(x) = \log_5 x$
   b) Domain $= (0, \infty)$, Range $= R$, $f^{-1}(x) = 5^x$
   c) Domain $= (0, \infty)$, Range $= R$, $f^{-1}(x) = \sqrt{5^x/3}$
   d) Domain $= (-\frac{3}{5}, \infty)$, Range $= R$, $f^{-1}(x) = (5^x - 3)/5$

13. a) $\log 7$ b) $-\log 6$ c) $9 \log x$

15.

a)

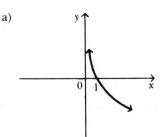

b)

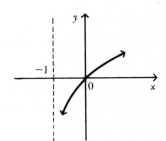

c)

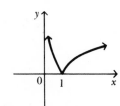

17. $\sqrt{10}$

19. 10

21. $\log_{3/4} (91/100)$

23. a) 5^{x^2+1} b) $\log_3 x$ c) $2^{2x} + 3(2^x) - 2$ d) $\log_2 (x^2 + x + 1)$

25. a) $81x^4 + 108x^3y + 54x^2y^2 + 12xy^3 + y^4$
   b) $243x^5 + 405x^{9/2} + 270x^4 + 90x^{7/2} + 15x^3 + x^{5/2}$
   c) $8x^3 + \dfrac{12x^2}{y} + \dfrac{6x}{y^2} + \dfrac{1}{y^3}$
   d) $x^8 - 8x^6 + 28x^4 - 56x^2 + 70 - 56x^{-2} + 28x^{-4} - 8x^{-6} + x^{-8}$
   e) $729y^6 + 486y^{9/2} + 135y^3 + 20y^{3/2} + \frac{5}{3} + \frac{2}{27}y^{-3/2} + \frac{1}{729}y^{-3}$

f) $x^{27} - 9x^{47/2} + 36x^{20} - 84x^{33/2} + 126x^{13} - 126x^{19/2} + 84x^6 - 36x^{5/2}$
$+ 9x^{-1} - x^{-9/2}$

27. 7.41

29. a) 95　　b) 20　　c) 160　　d) 1246　　e) 3276　　f) $\frac{139}{315}$

Chapter 5

PROBLEM SET 1, Page 265

1. a) $(1, 0)$　　b) $(1, 0)$　　c) $(0, -1)$　　d) $(0, -1)$　　e) $(-1, 0)$　　f) $(0, 1)$

5. a) $f(1)$　　b) $f(4)$　　c) $f(3)$　　d) $f(5.5)$　　e) $f(4.85)$

7. a) $f(0); f(\pi)$　　b) $f(0); f(\pi)$

9. a) I　　b) III　　c) IV　　d) III　　e) III　　f) IV　　g) II　　h) II　　i) IV

11. a) $(-1, 0)$　　b) $(0, 1)$　　c) $(0, 1)$　　d) $\left(\frac{1}{\sqrt{2}}, \frac{-1}{\sqrt{2}}\right)$　　e) $\left(\frac{-1}{\sqrt{2}}, \frac{-1}{\sqrt{2}}\right)$

f) $\left(\frac{-\sqrt{3}}{2}, \frac{1}{2}\right)$　　g) $\left(-\frac{1}{2}, \frac{\sqrt{3}}{2}\right)$　　h) $\left(\frac{-1}{\sqrt{2}}, \frac{-1}{\sqrt{2}}\right)$　　i) $\left(\frac{\sqrt{3}}{2}, -\frac{1}{2}\right)$

j) $\left(\frac{-\sqrt{3}}{2}, -\frac{1}{2}\right)$　　k) $(0, -1)$　　l) $\left(\frac{1}{2}, \frac{\sqrt{3}}{2}\right)$

PROBLEM SET 2, Page 273

3. a) $a = \pm 5$, $\cos t = \frac{3}{5}$, $\sin t = \frac{4}{5}$　　b) $\sin t = -\frac{12}{13}$　　c) $\cos t = \frac{\sqrt{11}}{6}$

d) $\sin t = \frac{-2\sqrt{2}}{3}$　　e) $\sin t > 0$ and $\cos t < 0$ in quadrant II

f) $\frac{-a^4 + 6a^2b^2 - b^4}{(a^2 + b^2)^2}$

5. a) $\frac{\pi}{2}$　　b) $\frac{2\pi}{3}$　　c) 4π　　d) π

7. a) $\cos\left(\frac{13\pi}{6}\right) = \frac{\sqrt{3}}{2}$, $\sin\left(\frac{13\pi}{6}\right) = \frac{1}{2}$

b) $\cos\left(\frac{-5\pi}{4}\right) = \frac{-\sqrt{2}}{2}$, $\sin\left(\frac{-5\pi}{4}\right) = \frac{\sqrt{2}}{2}$

c) $\cos(-5\pi) = -1$, $\sin(-5\pi) = 0$

d) $\cos\left(\frac{-4\pi}{3}\right) = -\frac{1}{2}$, $\sin\left(\frac{-4\pi}{3}\right) = \frac{\sqrt{3}}{2}$

e) $\cos\left(\frac{71\pi}{3}\right) = \frac{1}{2}$, $\sin\left(\frac{71\pi}{3}\right) = -\frac{\sqrt{3}}{2}$

f) $\cos\left(\frac{41\pi}{2}\right) = 0$, $\sin\left(\frac{41\pi}{2}\right) = 1$

g) $\cos\left(\frac{59\pi}{6}\right) = \frac{\sqrt{3}}{2}$, $\sin\left(\frac{59\pi}{6}\right) = -\frac{1}{2}$

h) $\cos\left(\dfrac{-31\pi}{6}\right)=-\dfrac{\sqrt3}{2}$, $\sin\left(\dfrac{-31\pi}{6}\right)=\dfrac12$

i) $\cos\left(\dfrac{67\pi}{4}\right)=-\dfrac{\sqrt2}{2}$, $\sin\left(\dfrac{67\pi}{4}\right)=\dfrac{\sqrt2}{2}$

j) $\cos\left(\dfrac{107\pi}{6}\right)=\dfrac{\sqrt3}{2}$ $\sin\left(\dfrac{107\pi}{6}\right)=-\dfrac12$

9. a) $(g\circ f)(t)=3\cos t;\ (f\circ g)(t)=\cos 3t;\ (f\circ f)(t)=\cos(\cos t)$
   b) Domain of $f\circ g=R$; range of $f\circ g=[-1,1]$
      Domain of $g\circ f=R$; range of $g\circ f=[-3,3]$
   c) $g\circ f, f\circ g$ and $f\circ f$ are even

PROBLEM SET 3, Page 284

1. a) 0.14   b) 1   c) 0.48   d) 1.14   e) 1.40   f) 1.28   g) 0.58
   h) 0.16

3. a) $\sin 0.7=0.6442,\ \cos 0.7=0.7648$
   b) $\sin 1.38=0.9819,\ \cos 1.38=0.1896$
   c) $\sin 0.5=0.4794,\ \cos 0.5=0.8776$
   d) $\sin(-0.35)=-0.3429,\ \cos(-0.35)=0.9394$
   e) $\sin(-0.01)=-0.0100,\ \cos(-0.01)=1.000$

5. a) -0.6288 b) -0.8365
   c) 0.8249 d) -0.1502
   e) -0.2377 f) -0.9950
   g) -0.0849 h) 0.4438
   i) -0.0188 j) 0.8875
   k) 0.9883 l) 0.9450

PROBLEM SET 4, Page 296

1. a)
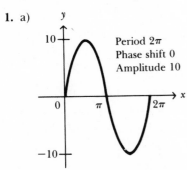
Period 2π
Phase shift 0
Amplitude 10

b)
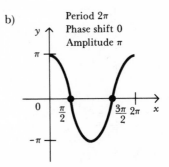
Period 2π
Phase shift 0
Amplitude π

c)
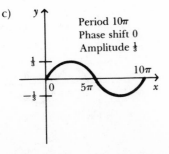
Period 10π
Phase shift 0
Amplitude $\tfrac13$

d)
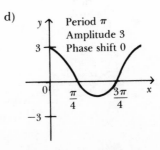
Period π
Amplitude 3
Phase shift 0

e)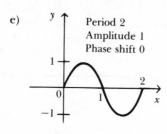

Period 2
Amplitude 1
Phase shift 0

f)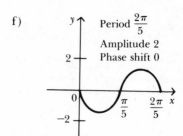

Period $\frac{2\pi}{5}$
Amplitude 2
Phase shift 0

g)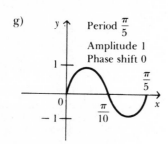

Period $\frac{\pi}{5}$
Amplitude 1
Phase shift 0

h)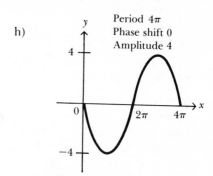

Period 4π
Phase shift 0
Amplitude 4

3. a)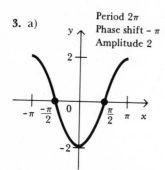

Period 2π
Phase shift $-\pi$
Amplitude 2

b)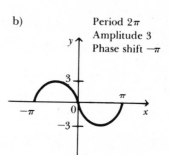

Period 2π
Amplitude 3
Phase shift $-\pi$

c)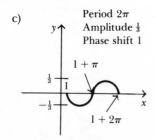

Period 2π
Amplitude $\frac{1}{3}$
Phase shift 1

d)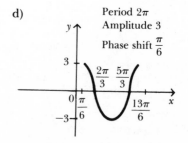

Period 2π
Amplitude 3
Phase shift $\frac{\pi}{6}$

e)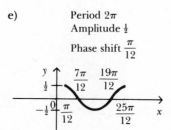

Period 2π
Amplitude $\frac{1}{2}$
Phase shift $\frac{\pi}{12}$

f)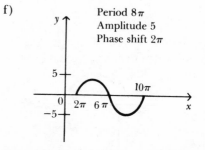

Period 8π
Amplitude 5
Phase shift 2π

g)

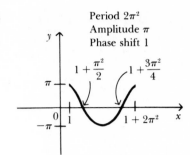

Period 6π
Amplitude 4
Phase shift $-\dfrac{\pi}{2}$

h)

Period $2\pi^2$
Amplitude π
Phase shift 1

i)

Period π
Amplitude 2
Phase shift $\dfrac{3}{2}$

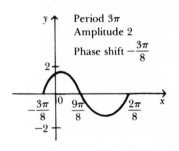

j)

Period 3π
Amplitude 2
Phase shift $-\dfrac{3\pi}{8}$

PROBLEM SET 5, Page 301

1. a) $-\dfrac{\pi}{6}$ b) 0   c) $\dfrac{\pi}{4}$ d) $\dfrac{\pi}{2}$ e) $\dfrac{\pi}{3}$ f) $\dfrac{5\pi}{6}$ g) $-\dfrac{\pi}{4}$ h) π i) $-\dfrac{\pi}{2}$

j) $\dfrac{\pi}{2}$ k) 0.22   l) 0.53   m) -0.89 n) 1.774

3. a) $-\dfrac{\pi}{2}$ b) $\dfrac{\pi}{3}$ c) $\dfrac{\sqrt{2}}{2}$ d) $-\dfrac{\pi}{4}$ e) $\dfrac{\pi}{2}$ f) 0   g) $\dfrac{\sqrt{7}}{4}$ h) $-\frac{7}{25}$

i) $\frac{7}{12}$ j) $\frac{12}{37}$

5. a) $x = \frac{1}{2}\cos y$ b) $x = \frac{1}{3}\sin\dfrac{y}{2}$ c) $x = \frac{1}{4}\sin\left(2 - \dfrac{y}{2}\right)$

d) $x = \frac{1}{2}(\cos y - 1)$ e) $x = 2\cos\left(\dfrac{3y}{4} - \dfrac{1}{4}\right)$

7.   a)

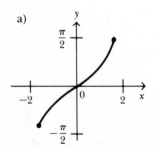

b)

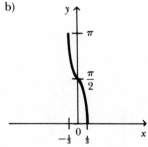

PROBLEM SET 6, Page 317

1. b) i) $\tan\left(\dfrac{11\pi}{3}\right)=-\sqrt{3}$, $\cot\left(\dfrac{11\pi}{3}\right)=-\dfrac{\sqrt{3}}{3}$, $\sec\left(\dfrac{11\pi}{3}\right)=2$, $\csc\left(\dfrac{11\pi}{3}\right)=-\dfrac{2\sqrt{3}}{3}$

ii) $\tan\left(-\dfrac{\pi}{3}\right)=-\sqrt{3}$, $\cot\left(-\dfrac{\pi}{3}\right)=-\dfrac{\sqrt{3}}{3}$, $\sec\left(-\dfrac{\pi}{3}\right)=2$, $\csc\left(-\dfrac{\pi}{3}\right)=-\dfrac{2\sqrt{3}}{3}$

iii) $\tan\left(-\dfrac{3\pi}{4}\right)=1$, $\cot\left(-\dfrac{3\pi}{4}\right)=1$, $\sec\left(-\dfrac{3\pi}{4}\right)=-\sqrt{2}$, $\csc\left(-\dfrac{3\pi}{4}\right)=-\sqrt{2}$

iv) $\tan\left(-\dfrac{2\pi}{3}\right)=\sqrt{3}$, $\cot\left(-\dfrac{2\pi}{3}\right)=\dfrac{\sqrt{3}}{3}$, $\sec\left(-\dfrac{2\pi}{3}\right)=-2$, $\csc\left(-\dfrac{2\pi}{3}\right)=-\dfrac{2\sqrt{3}}{3}$

v) $\tan\left(\dfrac{17\pi}{3}\right)=-\sqrt{3}$, $\cot\left(\dfrac{17\pi}{3}\right)=-\dfrac{\sqrt{3}}{3}$, $\sec\left(\dfrac{17\pi}{3}\right)=2$, $\csc\left(\dfrac{17\pi}{3}\right)=-\dfrac{2\sqrt{3}}{3}$

vi) $\tan\left(\dfrac{85\pi}{6}\right)=\dfrac{\sqrt{3}}{3}$, $\cot\left(\dfrac{85\pi}{6}\right)=\sqrt{3}$, $\sec\left(\dfrac{85\pi}{6}\right)=\dfrac{2\sqrt{3}}{3}$, $\csc\left(\dfrac{85\pi}{6}\right)=2$

vii) $\tan\left(\dfrac{71\pi}{6}\right)=-\dfrac{\sqrt{3}}{3}$, $\cot\left(\dfrac{71\pi}{6}\right)=-\sqrt{3}$, $\sec\left(\dfrac{71\pi}{6}\right)=\dfrac{2\sqrt{3}}{3}$, $\csc\left(\dfrac{71\pi}{6}\right)=-2$

viii) $\tan\left(-\dfrac{11\pi}{4}\right)=1$, $\cot\left(-\dfrac{11\pi}{4}\right)=1$, $\sec\left(-\dfrac{11\pi}{4}\right)=-\sqrt{2}$, $\csc\left(-\dfrac{11\pi}{4}\right)=-\sqrt{2}$

3. a) $\tan(1.14)=2.176$, $\cot(1.14)=0.4596$, $\sec(1.14)=2.395$, $\csc(1.14)=1.101$
   b) $\tan(-0.53)=-0.5859$, $\cot(-0.53)=-1.707$, $\sec(-0.53)=1.159$, $\csc(-0.53)$
   $=-1.978$
   c) $\tan(2.1)=-1.704$, $\cot(2.1)=-0.5870$, $\sec(2.1)=-1.975$, $\csc(2.1)=1.160$
   d) $\tan(9.5)=0.0802$, $\cot(9.5)=12.473$, $\sec(9.5)=-1.003$, $\csc(9.5)=-12.51$
   e) $\tan(-10)=-0.6552$, $\cot(-10)=-1.526$, $\sec(-10)=-1.196$, $\csc(-10)=1.825$
   f) $\tan(-1.6)=32.461$, $\cot(-1.6)=0.0308$, $\sec(-1.6)=-32.476$,
   $\csc(-1.6)=-1.000$
   g) $\tan(3.454)=0.3247$, $\cot(3.454)=3.080$, $\sec(3.454)=-1.051$,
   $\csc(3.454)=-3.238$

5. a) $\sin t=-\dfrac{12}{13}$, $\tan t=-\dfrac{12}{5}$, $\cot t=-\dfrac{5}{12}$, $\sec t=\dfrac{13}{5}$, $\csc t=-\dfrac{13}{12}$
   b) $\cos t=-\dfrac{2\sqrt{2}}{3}$, $\tan t=-\dfrac{\sqrt{2}}{4}$, $\cot t=-2\sqrt{2}$, $\sec t=-\dfrac{3\sqrt{2}}{4}$, $\csc t=3$

   c) $\sin t=\dfrac{1}{5}$, $\cos t=\dfrac{2\sqrt{6}}{5}$, $\tan t=\dfrac{\sqrt{6}}{12}$, $\cot t=2\sqrt{6}$, $\sec t=\dfrac{5\sqrt{6}}{12}$

   d) $\sin t=-\dfrac{\sqrt{3}}{2}$, $\cos t=-\dfrac{1}{2}$, $\tan t=\sqrt{3}$, $\cot t=\dfrac{\sqrt{3}}{3}$, $\csc t=-\dfrac{2\sqrt{3}}{3}$

7. a)

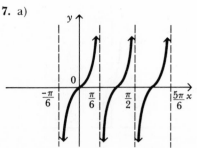

b)

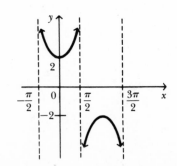

c)

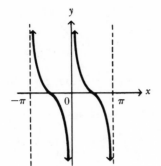

d)

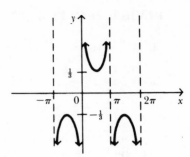

e)

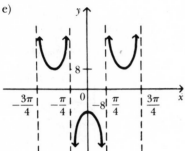

f)

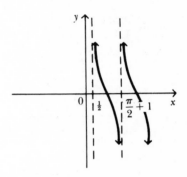

g)

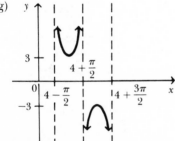

h)

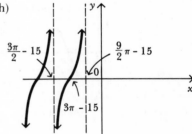

9. a) $\dfrac{\pi}{6}$ b) $-\dfrac{\pi}{4}$ c) 0   d) 0.82

11. a) $\dfrac{\pi}{6}$ b) $\dfrac{5\pi}{6}$ c) $\dfrac{\pi}{4}$ d) 1.27

13. a)

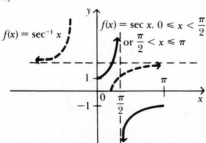

b) $f^{-1}(x) = \sec^{-1} x$

Domain $= (-\infty, -1] \cup [1, \infty)$

Range $= \left[0, \dfrac{\pi}{2}\right) \cup \left(\dfrac{\pi}{2}, \pi\right]$

REVIEW PROBLEM SET, Page 319

1. a) $\dfrac{\pi}{2}$ b) π c) $\dfrac{3\pi}{2}$ d) 0

3. a) $\sin t=-\dfrac{\sqrt3}{2}$, $\cos t=\tfrac12$, $\tan t=-\sqrt3$, $\cot t=-\dfrac{\sqrt3}{3}$, $\sec t=2$, $\csc t=-\dfrac{2\sqrt3}{3}$

   b) $\sin t=\dfrac{\sqrt2}{2}$, $\cos t=-\dfrac{\sqrt2}{2}$, $\tan t=-1$, $\cot t=-1$, $\sec t=-\sqrt2$, $\csc t=\sqrt2$

   c) $\sin t=\tfrac{12}{13}$, $\cos t=\tfrac{5}{13}$, $\tan t=\tfrac{12}{5}$, $\cot t=\tfrac{5}{12}$, $\sec t=\tfrac{13}{5}$, $\csc t=\tfrac{13}{12}$

   d) $\sin t=-\tfrac35$, $\cos t=-\tfrac45$, $\tan t=\tfrac34$, $\cot t=\tfrac43$, $\sec t=-\tfrac54$, $\csc t=-\tfrac53$

5. a) 2   b) $\sqrt3$ c) $\dfrac{\sqrt3}{2}$ d) -1 e) $-\sqrt3$ f) 2   g) $\tfrac12$ h) $\tfrac12$

7. a) 0.2482   b) 0.7446   c) 0.7291   d) 11.081   e) 48.089   f) -0.7895
   g) 1.6970   h) -0.7139 i) 0.3436   j) 0.9284

9. a)

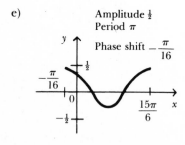

   Amplitude $\tfrac13$
   Period 2π
   Phase shift 0

   b)

   Amplitude 1
   Period 6π
   Phase shift 0

   c)

   Amplitude 2
   Period π
   Phase shift 0

   d)

   Amplitude 4
   Period 2π
   Phase shift $\dfrac{\pi}{6}$

   e)

   Amplitude $\tfrac12$
   Period π
   Phase shift $-\dfrac{\pi}{16}$

   f)

   Amplitude 3
   Period 4π
   Phase shift 2

11. a) f is decreasing in $(0,\ \pi)$

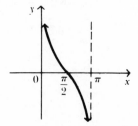

   b) f is increasing in $\left[\dfrac{\pi}{2},\ \pi\right)\cup\left(\pi,\ \dfrac{3\pi}{2}\right]$

   and is decreasing in $\left(0,\ \dfrac{\pi}{2}\right]\cup\left[\dfrac{3\pi}{2},\ 2\pi\right)$

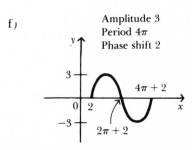

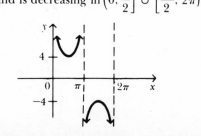

c) f is increasing in $\left(-\dfrac{\pi}{6}, \dfrac{\pi}{6}\right)$

d) f is increasing in $\left[0, \dfrac{\pi}{8}\right) \cup \left(\dfrac{\pi}{8}, \dfrac{\pi}{4}\right]$

and is decreasing in $\left[\dfrac{\pi}{4}, \dfrac{3\pi}{8}\right) \cup \left(\dfrac{3\pi}{8}, \dfrac{\pi}{2}\right]$

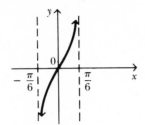

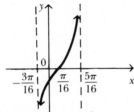

e) f is increasing in $\left(-\dfrac{3\pi}{16}, \dfrac{5\pi}{16}\right)$

13. a) $\dfrac{\pi}{4}$ b) $\dfrac{2\pi}{3}$ c) $\dfrac{\pi}{4}$ d) $\dfrac{5\pi}{6}$ e) $\dfrac{\pi}{6}$ f) $-\dfrac{\pi}{2}$

15. a) $\dfrac{1}{2}$ b) $\sqrt{3}$ c) $\dfrac{\pi}{4}$ d) $-\dfrac{\sqrt{3}}{2}$ e) $\dfrac{\sqrt{2}}{2}$

Chapter 6

PROBLEM SET 1, Page 337

1.

	(a)	(b)
i)	120°	$\dfrac{2\pi}{9}$
ii)	330°	$\dfrac{5\pi}{12}$
iii)	70°	$\dfrac{4\pi}{3}$
iv)	60°30′	$\dfrac{11\pi}{6}$
v)	−105°	$\dfrac{-19\pi}{36}$
vi)	1290°	$\dfrac{-11\pi}{9}$
vii)	−80°	$\dfrac{-111\pi}{45}$
viii)	−630°	$\dfrac{5\pi}{9}$

3.

	S	A
a)	$\dfrac{3\pi}{2}$	$\dfrac{21\pi}{4}$
b)	$\dfrac{8\pi}{5}$	$\dfrac{16\pi}{5}$
c)	$\dfrac{5\pi}{3}$	5π
d)	$\dfrac{55\pi}{4}$	$\dfrac{495\pi}{8}$
e)	$\dfrac{7\pi^2}{180}$	$\dfrac{49\pi^2}{360}$

7.

	$\sin\theta$	$\cos\theta$	$\tan\theta$	$\cot\theta$	$\sec\theta$	$\csc\theta$	θ
a)	0	-1	0	Undefined	-1	Undefined	$180°$
b)	$-\frac{4}{5}$	$-\frac{3}{5}$	$\frac{4}{3}$	$\frac{3}{4}$	$-\frac{5}{3}$	$-\frac{5}{4}$	—
c)	$\dfrac{-10}{\sqrt{149}}$	$\dfrac{7}{\sqrt{149}}$	$\dfrac{-10}{7}$	$\dfrac{-7}{10}$	$\dfrac{\sqrt{149}}{7}$	$\dfrac{-\sqrt{149}}{10}$	—
d)	1	0	Undefined	0	Undefined	1	$90°$
e)	$\dfrac{\sqrt{3}}{2}$	$-\frac{1}{2}$	$-\sqrt{3}$	$\dfrac{-\sqrt{3}}{3}$	-2	$\dfrac{2\sqrt{3}}{3}$	$120°$
f)	$\dfrac{-4}{11}$	$\dfrac{\pm\sqrt{105}}{11}$	$\pm\dfrac{4}{\sqrt{105}}$	$\dfrac{\pm\sqrt{105}}{4}$	$\pm\dfrac{11}{\sqrt{105}}$	$\dfrac{-11}{4}$	—
g)	$\pm\dfrac{1}{\sqrt{10}}$	$\pm\dfrac{3}{\sqrt{10}}$	$\frac{1}{3}$	3	$\dfrac{\pm\sqrt{10}}{3}$	$\pm\sqrt{10}$	—

9. a) $\cos\theta=-\frac{8}{17},\ \tan\theta=-\frac{15}{8},\ \cot\theta=\frac{8}{15},\ \sec\theta=-\frac{17}{8},\ \csc\theta=\frac{17}{15}$

b) $\sin\theta=-\frac{1}{2},\ \tan\theta=-\dfrac{1}{\sqrt{3}},\ \cot\theta=-\sqrt{3},\ \sec\theta=\dfrac{2}{\sqrt{3}},\ \csc\theta=-2$

c) $\sin\theta=-\dfrac{1}{\sqrt{2}},\ \cos\theta=\dfrac{1}{\sqrt{2}},\ \cot\theta=-1,\ \sec\theta=\sqrt{2},\ \csc\theta=-\sqrt{2}$

d) $\sin\theta=-\frac{1}{2},\ \cos\theta=-\dfrac{\sqrt{3}}{2},\ \tan\theta=\dfrac{1}{\sqrt{3}},\ \sec\theta=\dfrac{-2}{\sqrt{3}},\ \csc\theta=-2$

e) $\sin\theta=\frac{12}{13},\ \cos\theta=\frac{-5}{13},\ \tan\theta=\frac{-12}{5},\ \cot\theta=\frac{-5}{12},\ \csc\theta=\frac{13}{12}$

f) $\sin\theta=\dfrac{1}{\sqrt{2}},\ \cos\theta=\dfrac{1}{\sqrt{2}},\ \tan\theta=1,\ \cot\theta=1,\ \sec\theta=\sqrt{2}$

PROBLEM SET 2, Page 346

1.

θ measure		sin θ	cos θ	tan θ	cot θ	sec θ	csc θ
Degree	Radian						
0°	0	0	1	0	undef.	1	undef.
30°	$\dfrac{\pi}{6}$	$\dfrac{1}{2}$	$\dfrac{\sqrt{3}}{2}$	$\dfrac{\sqrt{3}}{3}$	$\sqrt{3}$	$\dfrac{2\sqrt{3}}{3}$	2
45°	$\dfrac{\pi}{4}$	$\dfrac{\sqrt{2}}{2}$	$\dfrac{\sqrt{2}}{2}$	1	1	$\sqrt{2}$	$\sqrt{2}$
60°	$\dfrac{\pi}{3}$	$\dfrac{\sqrt{3}}{2}$	$\dfrac{1}{2}$	$\sqrt{3}$	$\dfrac{\sqrt{3}}{3}$	2	$\dfrac{2\sqrt{3}}{3}$
90°	$\dfrac{\pi}{2}$	1	0	undef.	0	undef.	1
135°	$\dfrac{3\pi}{4}$	$\dfrac{\sqrt{2}}{2}$	$-\dfrac{\sqrt{2}}{2}$	-1	-1	$-\sqrt{2}$	$\sqrt{2}$
150°	$\dfrac{5\pi}{6}$	$\dfrac{1}{2}$	$-\dfrac{\sqrt{3}}{2}$	$-\dfrac{\sqrt{3}}{3}$	$-\sqrt{3}$	$-\dfrac{2\sqrt{3}}{3}$	2
180°	π	0	-1	0	undef.	-1	undef.

3. a) $-\frac{1}{2}$ b) $\dfrac{1}{\sqrt{2}}$ c) $\sqrt{3}$ d) $-\sqrt{3}$ e) 2 f) $\dfrac{2}{\sqrt{3}}$ g) 1

h) $\frac{1}{2}$ i) $\sqrt{2}$

5. a) $\frac{1}{2}$ b) $\frac{1}{2}$ c) $-\dfrac{1}{\sqrt{3}}$ d) -1 e) 2 f) -2

7. a) -0.5736 b) -0.5712

c) -0.8243 d) -0.2247

e) -0.3584 f) -5.759

g) 0.9153 h) -1.063

i) -1.835 j) -0.6099

k) -1.917 l) 4.537

PROBLEM SET 3, Page 354

1. $\sin t = \pm\frac{5}{13}$, $\cos t = \pm\frac{12}{13}$

3. a) $\dfrac{1 + 2\sin t \cos t}{\sin t \cos^3 t}$ b) $\dfrac{\cos t + 1}{1 - \cos t}$ c) $\cos^2 t$

d) $\dfrac{-\cos^4 x}{\sin^2 x}$ e) $\dfrac{\sin^2 x}{\cos^2 x}$ f) $\cos x$

g) $\csc^2 t$ h) $1 + \sin t$

5. a) $\cos \theta$ b) $-\cos \theta$ c) $\cot \theta$ d) $-\cot \theta$

e) $-\cos \theta$ f) $\csc \theta$ g) $\sec \theta$ h) $\tan \theta$

i) $\sin \theta$ j) $-\cos \theta$ k) $-\sin \theta$ l) $-\sin \theta$

7.

	i	ii	iii
a)	$\dfrac{\sqrt{6} + \sqrt{2}}{4}$	$\dfrac{\sqrt{6} - \sqrt{2}}{4}$	$2 + \sqrt{3}$
b)	$\dfrac{-\sqrt{6} - \sqrt{2}}{4}$	$\dfrac{-\sqrt{6} + \sqrt{2}}{4}$	$2 - \sqrt{3}$

9.

	i	ii	iii	iv	v	vi
a)	$-\frac{1}{2}$	$\dfrac{\sqrt{3}}{2}$	$-\dfrac{\sqrt{3}}{2}$	1	$\dfrac{-\sqrt{3}}{2}$	$\dfrac{\sqrt{2 + \sqrt{3}}}{2}$
b)	$\frac{63}{65}$	$\frac{16}{65}$	$\frac{63}{16}$	$\frac{16}{63}$	$\frac{65}{16}$	$\frac{65}{63}$

11. a) $\frac{24}{25}, \frac{7}{25}$ b) $\pm 3\sqrt{7}, \dfrac{\pm\sqrt{7}}{21}$

13. a) $\dfrac{\sqrt{2 - \sqrt{2}}}{2}$ b) $\dfrac{\sqrt{2 + \sqrt{3}}}{2}$ c) $\dfrac{\sqrt{2 + \sqrt{3}}}{2}$ d) $\dfrac{-\sqrt{2 - \sqrt{2}}}{2}$ e) $\dfrac{\sqrt{2 + \sqrt{3}}}{2}$

17. a) $\frac{4}{5}$ b) $\frac{-336}{625}$ c) $\frac{-24}{7}$

PROBLEM SET 4, Page 362

1. a) $\left\{\dfrac{\pi}{4}, \dfrac{3\pi}{4}\right\}$ b) $\left\{\dfrac{\pi}{3}, \dfrac{5\pi}{3}\right\}$ c) $\left\{\dfrac{2\pi}{9}, \dfrac{5\pi}{9}, \dfrac{8\pi}{9}, \dfrac{11\pi}{9}, \dfrac{14\pi}{9}, \dfrac{17\pi}{9}\right\}$ d) $\left\{\dfrac{3\pi}{4}, \dfrac{7\pi}{4}\right\}$

e) $\left\{\dfrac{\pi}{12}, \dfrac{5\pi}{12}, \dfrac{7\pi}{12}, \dfrac{11\pi}{12}, \dfrac{13\pi}{12}, \dfrac{17\pi}{12}, \dfrac{19\pi}{12}, \dfrac{23\pi}{12}\right\}$ f) $\left\{\dfrac{4\pi}{3}, \dfrac{5\pi}{3}\right\}$ g) $\left\{\dfrac{\pi}{2}, \dfrac{3\pi}{2}\right\}$ h) $\varnothing$

i) $\{1.24, 5.04\}$ j) $\{4.23, 5.19\}$ k) $\{0.775, 2.365\}$ l) $\{0.988, 5.292\}$

3. a) $\{240°, 300°\}$ b) $\{152°40', 207°20'\}$ c) $\{30°, 210°\}$ d) $\{60°, 120°\}$
e) $\{30°, 150°, 210°, 330°\}$ f) $\{45°, 225°\}$ g) $\{60°, 120°, 240°, 300°\}$
h) $\{0°, 104°30', 255°30'\}$ i) $\{60°, 180°, 300°\}$ j) $\{45°, 225°\}$

5. a) $\left\{\dfrac{2\pi}{3}, \dfrac{4\pi}{3}\right\}$ b) $\left\{\dfrac{\pi}{6}, \dfrac{5\pi}{6}\right\}$ c) $\left\{\dfrac{\pi}{3}, \dfrac{2\pi}{3}, \dfrac{4\pi}{3}, \dfrac{5\pi}{3}\right\}$ d) $\left\{\dfrac{\pi}{2}, \dfrac{7\pi}{6}, \dfrac{11\pi}{6}\right\}$

e) $\left\{\dfrac{\pi}{2}\right\}$ f) $\{2.419, 3.864\}$ g) $\left\{0, \dfrac{\pi}{6}, \dfrac{5\pi}{6}, \pi\right\}$ h) $\left\{0.644, \dfrac{3\pi}{2}\right\}$

7. a) $\left\{t \mid t = \dfrac{\pi}{4} + 2\pi n,\, n \in I\right\} \cup \left\{t \mid t = \dfrac{5\pi}{4} + 2\pi n,\, n \in I\right\}$

b) $\left\{t \mid t = \dfrac{5\pi}{6} + 2\pi n,\, n \in I\right\} \cup \left\{t \mid t = \dfrac{7\pi}{6} + 2\pi n,\, n \in I\right\}$

c) $\left\{t \mid t = \dfrac{\pi}{3} + \pi n,\, n \in I\right\} \cup \left\{t \mid t = \dfrac{2\pi}{3} + \pi n,\, n \in I\right\}$

d) $\left\{t \mid t = \dfrac{3\pi}{4} + 2\pi n,\, n \in I\right\} \cup \left\{t \mid t = \dfrac{5\pi}{4} + 2\pi n,\, n \in I\right\}$

e) $\left\{t \mid t = \left(\dfrac{2n+1}{2}\right)\pi,\, n \in I\right\} \cup \left\{t \mid t = \dfrac{\pi}{3} + 2\pi n,\, n \in I\right\} \cup \left\{t \mid t = \dfrac{2\pi}{3} + 2\pi n,\, n \in I\right\}$

f) $\{t \mid t = n\pi,\, n \in I\}$

PROBLEM SET 5, Page 375

1. a) $c = \sqrt{34}$, $\beta = 31°$, $\alpha = 59°$
   b) $\beta = 60°$, $c = 20$, $b = \sqrt{300}$
   c) $\alpha = 47°$, $b = 15.85$, $c = 23.24$
   d) $b = \sqrt{95}$, $\beta = 54°20'$, $\alpha = 35°40'$
   e) $\alpha = 75°$, $c = 30.91$, $a = 29.86$
   f) $\beta = 71°30'$, $\alpha = 8°30'$, $c = a\sqrt{10}$
   g) $\alpha = \beta = 45°$, $c = a\sqrt{2}$

3. a) $\beta = 54°30'$, $\alpha = 80°30'$, $a = 21.4$ or $\beta = 125°30'$, $\alpha = 9°30'$, $a = 3.55$
   b) No triangle   c) $\alpha = 30°$, $\beta = 85°$, $b = 20.77$
   d) $\beta = 25°30'$, $\gamma = 82°30'$, $c = 190.23$ e) $\beta = 39°40'$, $\alpha = 95°20'$, $a = 27.20$
   f) $\gamma = 109°50'$, $\alpha = 25°10'$, $c = 6.65$

5. a) $\frac{4}{5}$ b) $\frac{4}{5}$ c) $\frac{5}{4}$ d) $\dfrac{3}{\sqrt{58}}$ e) $\frac{3}{2}$ f) $\dfrac{2}{\sqrt{85}}$

7. a) $2\sqrt{79}$ b) $26°20'$ c) $\alpha = 34°$, $\beta = 101°30'$, $\gamma = 44°30'$ d) $52°20'$
   e) 61.77 f) $90°$

9. a) $\frac{3}{4}$ b) 9.95 c) $114°40'$ d) 19.32 e) $\alpha = 95°30'$, $\beta = 60°$, $\gamma = 24°30'$

11. 3.095 ft

13. 72 ft

15. 97 ft

17. $36°50'$

19. 41.13 and 51.97

REVIEW PROBLEM SET, Page 378

1. a) i) $300°, \dfrac{\pi}{3}$ ii) $-135°, \dfrac{\pi}{4}$ iii) $56°15', \dfrac{5\pi}{16}$ iv) $330°, \dfrac{\pi}{6}$ v) $390°, \dfrac{\pi}{6}$

   vi) $-585°, \dfrac{\pi}{4}$ vii) $480°, \dfrac{\pi}{3}$ viii) $-462°51', \dfrac{3\pi}{7}$ ix) $612°, \dfrac{2\pi}{5}$

   x) $810°, \dfrac{\pi}{2}$ xi) $1485°, \dfrac{\pi}{4}$ xii) $-1845°, \dfrac{\pi}{4}$

   b) i) $\dfrac{29\pi}{18}, 70°$ ii) $-\dfrac{7\pi}{18}, 70°$ iii) $\dfrac{3\pi}{4}, 45°$ iv) $\dfrac{23\pi}{9}, 80°$ v) $\dfrac{19\pi}{18}, 10°$

   vi) $\dfrac{-7\pi}{12}, 75°$ vii) $\dfrac{41\pi}{9}, 80°$ viii) $-\dfrac{\pi}{6}, 30°$ ix) $\dfrac{\pi}{15}, 12°$

3.

	S	A
a)	$\dfrac{35\pi}{12}$	$\dfrac{245\pi}{24}$
b)	$\dfrac{16\pi}{9}$	$\dfrac{16\pi}{9}$
c)	$\dfrac{170\pi}{9}$	$\dfrac{850\pi}{9}$
d)	$\dfrac{2\pi}{3}$	$\dfrac{4\pi}{3}$
e)	$\dfrac{6\pi}{5}$	$\dfrac{18\pi}{5}$

5.

	sin θ	cos θ	tan θ	sec θ	csc θ	cot θ
a)	$\frac{1}{2}$	$\frac{\sqrt{3}}{2}$	$\frac{\sqrt{3}}{3}$	$\frac{2\sqrt{3}}{3}$	2	$\sqrt{3}$
b)	-1	0	Undefined	Undefined	-1	0
c)	0	-1	0	-1	Undefined	Undefined
d)	$\frac{1}{2}$	$\frac{-\sqrt{3}}{2}$	$\frac{-\sqrt{3}}{3}$	$\frac{-2\sqrt{3}}{3}$	2	$-\sqrt{3}$
e)	$-\frac{1}{2}$	$\frac{-\sqrt{3}}{2}$	$\frac{\sqrt{3}}{3}$	$\frac{-2\sqrt{3}}{3}$	-2	$\sqrt{3}$
f)	$\frac{\sqrt{2}}{2}$	$-\frac{\sqrt{2}}{2}$	-1	$-\sqrt{2}$	$\sqrt{2}$	-1
g)	$-\frac{\sqrt{3}}{2}$	$-\frac{1}{2}$	$\sqrt{3}$	-2	$\frac{-2\sqrt{3}}{3}$	$\frac{\sqrt{3}}{3}$
h)	$\frac{\sqrt{2}}{2}$	$-\frac{\sqrt{2}}{2}$	-1	$-\sqrt{2}$	$\sqrt{2}$	-1
i)	$-\frac{\sqrt{2}}{2}$	$\frac{\sqrt{2}}{2}$	-1	$-\sqrt{2}$	$-\sqrt{2}$	-1

7.

	sin θ	cos θ	tan θ	cot θ	sec θ	csc θ
a)	$\frac{4}{5}$	$\frac{3}{5}$	$\frac{4}{3}$	$\frac{3}{4}$	$\frac{5}{3}$	$\frac{5}{4}$
b)	$\frac{12}{13}$	$\frac{5}{13}$	$\frac{12}{5}$	$\frac{5}{12}$	$\frac{13}{5}$	$\frac{13}{12}$
c)	$\frac{25}{\sqrt{674}}$	$\frac{7}{\sqrt{674}}$	$\frac{25}{7}$	$\frac{7}{25}$	$\frac{\sqrt{674}}{7}$	$\frac{\sqrt{674}}{25}$
d)	$\frac{4}{5}$	$\frac{3}{5}$	$\frac{4}{3}$	$\frac{3}{4}$	$\frac{5}{3}$	$\frac{5}{4}$
e)	$\frac{17}{\sqrt{353}}$	$\frac{8}{\sqrt{353}}$	$\frac{17}{8}$	$\frac{8}{17}$	$\frac{\sqrt{353}}{8}$	$\frac{\sqrt{353}}{17}$
f)	$\frac{4}{5}$	$-\frac{3}{5}$	$-\frac{4}{3}$	$-\frac{3}{4}$	$-\frac{5}{3}$	$\frac{5}{4}$
g)	$\frac{3}{5}$	$-\frac{4}{5}$	$\frac{3}{4}$	$\frac{4}{3}$	$-\frac{5}{4}$	$-\frac{5}{3}$
h)	$\frac{6}{\sqrt{61}}$	$\frac{5}{\sqrt{61}}$	$\frac{6}{5}$	$\frac{5}{6}$	$\frac{\sqrt{61}}{5}$	$\frac{\sqrt{61}}{6}$

9. a) $-\frac{4}{7}$ b) $\pm\frac{\sqrt{33}}{7}$ c) $\pm\frac{4}{\sqrt{33}}$ d) $\pm\frac{\sqrt{33}}{4}$ e) $\pm\frac{7}{\sqrt{33}}$

13. a) 1   b) 0   c) $\frac{7}{25}$ d) $\frac{24}{25}$ e) Undefined   f) $\frac{24}{7}$ g) Undefined
h) $\frac{25}{7}$

15. a) $\{270°\}$ b) $\{30°, 330°\}$ c) $\{60°, 240°\}$ d) $\{135°, 225°\}$
e) $\{210°, 330°\}$ f) $\{60°, 240°\}$ g) $\{30°, 150°, 270°\}$

17. a) $\left\{\frac{\pi}{4}, \frac{3\pi}{4}\right\}$ b) $\left\{\frac{\pi}{3}, \frac{2\pi}{3}, \frac{4\pi}{3}, \frac{5\pi}{3}\right\}$ c) $\left\{0, \frac{2\pi}{3}, \frac{4\pi}{3}\right\}$ d) $\left\{0, \pi, \frac{5\pi}{4}, \frac{7\pi}{4}\right\}$

e) $\left\{0, \frac{\pi}{6}, \pi, \frac{7\pi}{6}\right\}$

19. a) $c = \sqrt{74 - 35\sqrt{3}}$ or 3.66, $\alpha = 43°10'$, $\beta = 106°50'$
b) $a = \sqrt{97}$ or 9.85, $\beta = 15°20'$, $\gamma = 44°40'$

c) $\gamma = 60°$, $a = 8.17$, $b = 11.15$
d) $\gamma = 25°$, $\alpha = 45°$, $c = 96.69$
e) $\gamma = 48°40'$, $\alpha = 101°20'$, $a = 7.84$ or $\gamma = 131°20'$, $\alpha = 18°40'$, $a = 2.56$
f) $\beta = 29°50'$, $\gamma = 90°10'$, $c = 15.70$
g) $\gamma = 54°20'$, $\beta = 10°40'$, $b = 0.98$
h) $a = 36.35$, $\beta = 114°30'$, $\gamma = 35°30'$
i) $\alpha = 87°$, $b = 6.70$, $c = 7.78$
j) $b = \sqrt{116 + 40\sqrt{3}}$ or 13.61, $\alpha = 8°30'$, $\beta = 21°30'$

21. 513.48 ft
23. 31.72 ft

Chapter 7

PROBLEM SET 1, Page 396

3. a) i) $\mathbf{w} = -(\mathbf{v}_1 + \mathbf{u}_1)$ ii) $\mathbf{w} = \mathbf{u}_2 - \mathbf{v}_1$ iii) $\mathbf{w} = \mathbf{v}_2 + \mathbf{u}_2$
c) $\mathbf{u}_1 + \mathbf{v}_1 + \mathbf{u}_2 + \mathbf{v}_2 = \mathbf{0}$

5. a) $\langle 0, 20 \rangle$ b) $\langle -9, 8 \rangle$ c) $5 + 2\sqrt{5}$ d) $\sqrt{65}$ e) $\langle 8, 16 \rangle$
f) $\langle 4, 12 \rangle$ g) $\langle 12, -4 \rangle$ h) $\langle -12, -8 \rangle$ i) 15   j) $15 - 6\sqrt{5}$

7. $\dfrac{4}{\sqrt{17}}\mathbf{i} + \dfrac{1}{\sqrt{17}}\mathbf{j}$

9. The following vectors are normalized:

a) $\dfrac{-5}{\sqrt{29}}\mathbf{i} - \dfrac{2}{\sqrt{29}}\mathbf{j}$ b) $\dfrac{-5}{\sqrt{34}}\mathbf{i} - \dfrac{3}{\sqrt{34}}\mathbf{j}$ c) $\dfrac{5}{\sqrt{146}}\mathbf{i} - \dfrac{11}{\sqrt{146}}\mathbf{j}$

d) $\dfrac{4}{\sqrt{41}}\mathbf{i} - \dfrac{5}{\sqrt{41}}\mathbf{j}$ e) $\dfrac{-1}{\sqrt{2}}\mathbf{i} - \dfrac{1}{\sqrt{2}}\mathbf{j}$ f) $\dfrac{2}{\sqrt{5}}\mathbf{i} + \dfrac{1}{\sqrt{5}}\mathbf{j}$

11. a) $(\frac{10}{3}, 2)$ b) $(1, -1)$

PROBLEM SET 2, Page 402

1. a) 10   b) $-10\sqrt{2}$ c) $-10\sqrt{3}$ d) 0
3. a) $0; 90°$ b) $-7; 108°$ c) $0; 90°$ d) $-4; 120°$ e) $2; 78°$
5. a) $-\frac{16}{3}$ b) $\frac{16}{3}$ c) -3
7. a) $\sqrt{|\mathbf{u}|^2 + |\mathbf{v}|^2}$ b) $\sqrt{|\mathbf{u}|^2 + |\mathbf{v}|^2}$ c) $\sqrt{9|\mathbf{u}|^2 + 16|\mathbf{v}|^2}$ d) $\sqrt{9|\mathbf{u}|^2 + 16|\mathbf{v}|^2}$

PROBLEM SET 3, Page 413

1. a) $\dfrac{\sqrt{3}}{2}\mathbf{i} + \frac{1}{2}\mathbf{j}$ b) $-\frac{1}{2}\mathbf{i} + \dfrac{\sqrt{3}}{2}\mathbf{j}$ c) $2\sqrt{3}\mathbf{i} + 2\mathbf{j}$;

$-\left(\dfrac{5 + 7\sqrt{3}}{2}\right)\mathbf{i} + \left(\dfrac{5\sqrt{3} - 7}{2}\right)\mathbf{j}$; $4\mathbf{i} - 4\sqrt{3}\mathbf{j}$; $\left(\dfrac{\sqrt{3} + 1}{2}\right)\mathbf{i} + \left(\dfrac{1 - \sqrt{3}}{2}\right)\mathbf{j}$

d) Yes, $-30°$ direction

3. a) $-\dfrac{5}{\sqrt{2}}\mathbf{i}+\dfrac{1}{\sqrt{2}}\mathbf{j}$ b) $\dfrac{-8}{\sqrt{2}}\mathbf{i}+\dfrac{24}{\sqrt{2}}\mathbf{j}$ c) $\dfrac{23}{\sqrt{2}}\mathbf{i}+\dfrac{29}{\sqrt{2}}\mathbf{j}$

7. One vector form of the line: a) $\langle -3t+4,\, t+3\rangle$ b) $\langle t,3\rangle$ c) $\langle 2,t\rangle$
   d) $\langle t,-3t-1\rangle$ e) $\langle t+1,\, 2-\frac{3}{2}t\rangle$

REVIEW PROBLEM SET, Page 414

1. a) $\left\langle \dfrac{5\sqrt{3}}{2},\, \frac{5}{2}\right\rangle$ b) $\langle 3\sqrt{2},\, 3\sqrt{2}\rangle$ c) $\langle -4\sqrt{3},\, 4\rangle$

3. a) $\langle 9,12\rangle$ b) $\langle 1,-1\rangle$ c) $\langle -2,-5\rangle$ d) $\langle 17,18\rangle$ e) $\langle -5,-2\rangle$

5. a) 0   b) 40   c) 50   d) 50   e) 53   f) 125   g) 60   h) 275

7. a) $\mathbf{u}=10\mathbf{i}-5\mathbf{j};\ |\mathbf{u}|=5\sqrt{5}$ b) $\mathbf{u}=10\mathbf{i}-7\mathbf{j};\ |\mathbf{u}|=\sqrt{149}$
   c) $\mathbf{u}=-\mathbf{i}+4\mathbf{j};\ |\mathbf{u}|=\sqrt{17}$ d) $\mathbf{u}=-10\mathbf{i}+4\mathbf{j};\ |\mathbf{u}|=2\sqrt{29}$
   e) $\mathbf{u}=4\mathbf{i}+2\mathbf{j};\ |\mathbf{u}|=2\sqrt{5}$

9. a) 10   b) 60   c) 132   d) 124   e) -24 f) -97 g) 9

11. a) $90°$ b) $84°$ c) $66°$ d) $102°$

13. a) $\langle \frac{4}{5},\frac{3}{5}\rangle$ b) $\langle \frac{3}{5},\frac{4}{5}\rangle$ c) $\left\langle \dfrac{7}{\sqrt{53}},\, \dfrac{2}{\sqrt{53}}\right\rangle$

15. $90°,\ 45°,\ 45°$

17. a) $-\frac{1}{2}\mathbf{i}+\dfrac{\sqrt{3}}{2}\mathbf{j}$ b) $-\mathbf{i}$ c) $\dfrac{\sqrt{2}}{2}\mathbf{i}-\dfrac{\sqrt{2}}{2}\mathbf{j}$ d) $-0.1736\mathbf{i}+0.9848\mathbf{j}$

19. a) $\frac{1}{2}\mathbf{i}+\dfrac{\sqrt{3}}{2}\mathbf{j}$ b) $60°$ c) $-\dfrac{\sqrt{3}}{2}\mathbf{i}+\frac{1}{2}\mathbf{j}$ d) $\left(\dfrac{2+5\sqrt{3}}{2}\right)\mathbf{i}+\left(\dfrac{-5+2\sqrt{3}}{2}\right)\mathbf{j}$

21. One vector form of the line: a) $\langle 2-5t,\, -6+11t\rangle$ b) $\langle t,\, -1+6t\rangle$

Chapter 8

PROBLEM SET 1, Page 422

1. a) $6+8i$

  b) $2+7i$

  c) $-2-2i$

  d) -7

  e) $33-13i$

  f) $9-37i$

  g) 53

  h) 13

  i) $-i$

  j) $-1+3i$

  k) $-1-3i$

  l) $-1-5i$

  m) $8-6i$

  n) $-8+6i$

  o) $-119+120i$

  p) $-5-\sqrt{2}\,i$

  q) $\dfrac{-6i}{7}$

  r) $\dfrac{-3i}{5}$

  s) $-4i$

  t) $\frac{8}{3}-\frac{5}{3}i$

  u) $-\frac{3}{5}-\frac{7}{5}i$

  v) $\frac{1}{29}-\frac{17}{29}i$

  w) $-\frac{3}{25}+\frac{29}{25}i$

  x) $\frac{41}{34}-\frac{23}{34}i$

  y) $-\dfrac{i}{2}$

  z) $\frac{70}{1369}+\frac{24}{1369}i$

3.

	$\bar{z}$	Re z	Im z	$\dfrac{1}{z}$
a)	$2 - \sqrt{3}\,i$	2	$\sqrt{3}$	$\frac{2}{7} - \dfrac{\sqrt{3}}{7}i$
b)	$1 + \frac{1}{2}i$	1	$-\frac{1}{2}$	$\frac{4}{5} + \frac{2}{5}i$
c)	$5 + 12i$	5	-12	$\frac{5}{169} + \frac{12}{169}i$
d)	$-2i$	0	2	$-\dfrac{i}{2}$

7. a) $13 - 8i$ b) $13 + 10i$ c) $5 - 6i$ d) $5 + 8i$

PROBLEM SET 2, Page 439

	i	ii	iii	iv	v	vi	vii	viii	ix	x
a)	$(2, 120°)$	$(3, 180°)$	$(5, 36°52')$	$(12, 120°)$	$(5\sqrt{2}, 45°)$	$(2, 270°)$	$(6, 120°)$	$(4, 330°)$	$(5\sqrt{2}, 225°)$	$(\sqrt{29}, 111°48')$
b)	$(3\sqrt{3}, 3)$	$(5, 5\sqrt{3})$	$\left(-\frac{7}{2}, \dfrac{7\sqrt{3}}{2}\right)$	$(2\sqrt{3}, -2)$	$(0, 4)$	$(4\sqrt{2}, 4\sqrt{2})$	$(\pi\sqrt{3}, -\pi)$	$(0, 5)$	$(-2\sqrt{3}, -2)$	$(2\sqrt{3}, -2)$

3.

	Modulus	Argument	Polar form
a)	$\sqrt{2}$	$\dfrac{5\pi}{4}$ or $225°$	$\sqrt{2}\left(\cos\dfrac{5\pi}{4} + i\sin\dfrac{5\pi}{4}\right)$
b)	7	0 or $0°$	$7(\cos 0 + i\sin 0)$
c)	2	$\dfrac{3\pi}{2}$ or $270°$	$2\left(\cos\dfrac{3\pi}{2} + i\sin\dfrac{3\pi}{2}\right)$
d)	1	$\dfrac{\pi}{6}$ or $30°$	$1\left(\cos\dfrac{\pi}{6} + i\sin\dfrac{\pi}{6}\right)$
e)	2	$\dfrac{7\pi}{6}$ or $210°$	$2\left(\cos\dfrac{7\pi}{6} + i\sin\dfrac{7\pi}{6}\right)$
f)	5	0.93 or $53°8'$	$5(\cos 0.93 + i\sin 0.93)$
g)	1	$\dfrac{2\pi}{3}$ or $120°$	$1\left(\cos\dfrac{2\pi}{3} + i\sin\dfrac{2\pi}{3}\right)$
h)	2	$\dfrac{3\pi}{2}$ or $270°$	$2\left(\cos\dfrac{3\pi}{2} + i\sin\dfrac{3\pi}{2}\right)$
i)	8	π or $180°$	$8(\cos\pi + i\sin\pi)$

5. a) $1.9696 + 0.3472i$ b) $0.7764 - 2.8977i$ c) $4 + 0 \cdot i$ d) $\sqrt{2} + \sqrt{2}\,i$
   e) $-5\sqrt{2} + 5\sqrt{2}\,i$ f) $0 + 2i$ g) $0 + 7i$

7. $z_1 z_2$:

a) $5(\cos 225° + i\sin 225°) = \dfrac{-5\sqrt{2}}{2} - \dfrac{5\sqrt{2}}{2}i$

b) $6(\cos 90° + i\sin 90°) = 0 + 6i$

c) $8\left(\cos\dfrac{7\pi}{4} + i\sin\dfrac{7\pi}{4}\right) = 4\sqrt{2} - 4\sqrt{2}\,i$

d) $30(\cos 270° + i\sin 270°) = 0 - 30i$

a) $\dfrac{z_1}{z_2}$: $5(\cos 115° + i \sin 115°) = -2.1130 + 4.5315i$

b) $\tfrac{2}{3}(\cos 10° + i \sin 10°) = 0.6566 + 0.1157i$

c) $2\left[\cos\left(-\dfrac{\pi}{4}\right) + i \sin\left(-\dfrac{\pi}{4}\right)\right] = \sqrt{2} - \sqrt{2}\,i$

d) $\tfrac{5}{6}[\cos(-210°) + i \sin(-210°)] = \dfrac{-5\sqrt{3}}{12} + \dfrac{5}{12}\,i$

9. a) $8\sqrt{2}\,(0.9659 - 0.2588i)$ b) $\dfrac{\sqrt{2}}{8}(-0.2588 + 0.9659i)$ c) $64\sqrt{3} - 64i$

PROBLEM SET 3, Page 445

1. a) $\dfrac{-\sqrt{3}}{2} - \tfrac{1}{2}i$ b) $-\tfrac{1}{2} + \dfrac{\sqrt{3}}{2}\,i$ c) $512 - 512\sqrt{3}\,i$ d) $\dfrac{729}{2} + \dfrac{729\sqrt{3}}{2}\,i$

e) 256 f) -1024

3. a) $-125{,}000i$ b) $16 - 16\sqrt{3}\,i$ c) $-8 - 8\sqrt{3}\,i$ d) $\dfrac{-1}{2} + \dfrac{\sqrt{3}}{2}\,i$

e) -2^{30} f) $2^{25}i$ g) -1 h) $-i$

5. a) $\dfrac{\sqrt{2}}{2} + \dfrac{\sqrt{2}}{2}\,i, -\dfrac{\sqrt{2}}{2} - \dfrac{\sqrt{2}}{2}\,i$

b) $\sqrt{3\sqrt{2}}\left(\cos\dfrac{7\pi}{8} + i \sin\dfrac{7\pi}{8}\right),\ \sqrt{3\sqrt{2}}\left(\cos\dfrac{15\pi}{8} + i \sin\dfrac{15\pi}{8}\right)$

c) $2(\cos 0 + i \sin 0),\ 2\left(\cos\dfrac{2\pi}{3} + i \sin\dfrac{2\pi}{3}\right),\ 2\left(\cos\dfrac{4\pi}{3} + i \sin\dfrac{4\pi}{3}\right)$

d) $\cos\dfrac{\pi}{6} + i \sin\dfrac{\pi}{6},\ \cos\dfrac{5\pi}{6} + i \sin\dfrac{5\pi}{6},\ \cos\dfrac{3\pi}{2} + i \sin\dfrac{3\pi}{2}$

e) $2\left(\cos\dfrac{\pi}{4} + i \sin\dfrac{\pi}{4}\right),\ 2\left(\cos\dfrac{3\pi}{4} + i \sin\dfrac{3\pi}{4}\right),\ 2\left(\cos\dfrac{5\pi}{4} + i \sin\dfrac{5\pi}{4}\right),$

$2\left(\cos\dfrac{7\pi}{4} + i \sin\dfrac{7\pi}{4}\right)$

f) $2\left(\cos\dfrac{\pi}{3} + i \sin\dfrac{\pi}{3}\right),\ 2\left(\cos\dfrac{5\pi}{6} + i \sin\dfrac{5\pi}{6}\right),\ 2\left(\cos\dfrac{4\pi}{3} + i \sin\dfrac{4\pi}{3}\right),$

$2\left(\cos\dfrac{11\pi}{6} + i \sin\dfrac{11\pi}{6}\right)$

7. a) $2\left(\cos\dfrac{\pi}{5} + i \sin\dfrac{\pi}{5}\right),\ 2\left(\cos\dfrac{3\pi}{5} + i \sin\dfrac{3\pi}{5}\right),\ 2(\cos\pi + i \sin\pi),$

$2\left(\cos\dfrac{7\pi}{5} + i \sin\dfrac{7\pi}{5}\right),\ 2\left(\cos\dfrac{9\pi}{5} + i \sin\dfrac{9\pi}{5}\right)$

b) All lie on a circle of radius 2 center at the origin, and are equally spaced at angles of $\dfrac{2\pi}{5}$ radians.

PROBLEM SET 4, Page 450

1. a) Yes, 3   b) Yes, 2   c) Yes, 2

3. a) $-1 - i,\ \sqrt{2},\ -\sqrt{2}$ b) $-i,\ \dfrac{-1 \pm \sqrt{33}}{4}$

5. One possible polynomial is: a) $x^4 - 5x^3 + 7x^2 - 5x + 6$
   b) $x^4 - 6x^3 + 14x^2 - 16x + 8$ c) $x^4 - 4x^3 + 24x^2 - 40x + 100$
   d) $x^8 - x^7 + 6x^6 - 6x^5 + 9x^4 - 9x^3 + 4x^2 - 4x$ e) $x^4 - 3x^3 + x^2 + 4$

REVIEW PROBLEM SET, Page 450

1. a) $10 - 5i$ b) 16 c) $1 + 11i$ d) $-1 + 3i$ e) $30 - 16i$ f) $\frac{-15}{13} + \frac{23}{13}i$
   g) $\frac{5}{7} - \dfrac{6\sqrt{3}}{7}i$ h) $-\frac{5}{4} - \frac{3}{4}i$

3.

	i	ii	iii	iv	v	vi
a)	$\left(\dfrac{5\sqrt{2}}{2}, \dfrac{5\sqrt{2}}{2}\right)$	$(0, -2)$	$(-1, -1)$	$(-3, 0)$	$(0, -3)$	$(-1, -\sqrt{3})$
b)	$(3, \pi)$	$\left(4, \dfrac{2\pi}{3}\right)$	$\left(10, \dfrac{7\pi}{6}\right)$	$\left(10\sqrt{2}, \dfrac{\pi}{4}\right)$	$\left(14, \dfrac{-\pi}{2}\right)$	$(15, \pi)$

5.

	Modulus	Argument
a)	$5\sqrt{2}$	$\dfrac{\pi}{4}$
b)	2	$\dfrac{\pi}{6}$
c)	12	$\dfrac{\pi}{6}$
d)	8	$\dfrac{\pi}{2}$
e)	$2\sqrt{2}$	$\dfrac{\pi}{4}$
f)	2	$\dfrac{2\pi}{3}$
g)	5	$53°8'$
h)	4^{10}	$240°$

7. a) Polar form: $z_1 z_2 = 6\left(\cos \dfrac{3\pi}{2} + i \sin \dfrac{3\pi}{2}\right)$, $z_1/z_2 = \frac{2}{3}\left(\cos \dfrac{\pi}{2} + i \sin \dfrac{\pi}{2}\right)$
    Rectangular form: $z_1 z_2 = 0 - 6i$, $z_1/z_2 = 0 + \frac{2}{3}i$
   b) Polar form: $z_1 z_2 = 18(\cos 305° + i \sin 305°)$, $z_1/z_2 = 2(\cos 155° + i \sin 155°)$
    Rectangular form: $z_1 z_2 = 18(0.5736 - 0.8192i)$, $z_1/z_2 = 2(-0.9063 + 0.4226i)$
   c) Polar form: $z_1 z_2 = 12(\cos 322° + i \sin 322°)$, $z_1/z_2 = 3(\cos 102° - i \sin 102°)$
    Rectangular form: $z_1 z_2 = 12(0.7880 - 0.6157i)$, $z_1/z_2 = 3(-0.2079 - 0.9781i)$
   d) Polar form: $z_1 z_2 = 98(\cos 370° + i \sin 370°)$, $z_1/z_2 = 2(\cos 240° + i \sin 240°)$
    Rectangular form: $z_1 z_2 = 98(0.9848 + 0.1736i)$, $z_1/z_2 = 2(-\frac{1}{2} - (\sqrt{3}/2)i)$

9. a) $\frac{1}{2} - (\sqrt{3}/2)i$ b) 2^{20} c) -1 d) 1 e) -2^{30}

11. $5\cos 80°$, $5\sin 80°$, 5

13. a) $4\left(\cos\dfrac{\pi}{3} + i\sin\dfrac{\pi}{3}\right)$, $4(\cos\pi + i\sin\pi)$, $4\left(\cos\dfrac{5\pi}{3} + i\sin\dfrac{5\pi}{3}\right)$

b) $8^{1/4}\left(\cos\dfrac{\pi}{8} - i\sin\dfrac{\pi}{8}\right)$, $8^{1/4}\left(\cos\dfrac{3\pi}{8} + i\sin\dfrac{3\pi}{8}\right)$, $8^{1/4}\left(\cos\dfrac{7\pi}{8} + i\sin\dfrac{7\pi}{8}\right)$,

$8^{1/4}\left(\cos\dfrac{11\pi}{8} + i\sin\dfrac{11\pi}{8}\right)$

c) $2^{1/8}\left(\cos\dfrac{\pi}{16} + i\sin\dfrac{\pi}{16}\right)$, $2^{1/8}\left(\cos\dfrac{9\pi}{16} + i\sin\dfrac{9\pi}{16}\right)$, $2^{1/8}\left(\cos\dfrac{17\pi}{16} + i\sin\dfrac{17\pi}{16}\right)$,

$2^{1/8}\left(\cos\dfrac{25\pi}{16} + i\sin\dfrac{25\pi}{16}\right)$

d) $2^{7/8}\left(\cos\dfrac{\pi}{16} - i\sin\dfrac{\pi}{16}\right)$, $2^{7/8}\left(\cos\dfrac{7\pi}{16} + i\sin\dfrac{7\pi}{16}\right)$,

$2^{7/8}\left(\cos\dfrac{15\pi}{16} + i\sin\dfrac{15\pi}{16}\right)$, $2^{7/8}\left(\cos\dfrac{23\pi}{16} + i\sin\dfrac{23\pi}{16}\right)$

15. One possible polynomial is: a) $f(x) = x^4 + 4x^3 - 5x^2 - 36x - 36$
b) $f(x) = x^3 - 6x^2 + 13x - 10$
c) $f(x) = x^4 + 2x^3 - 17x^2 - 18x + 72$
d) $f(x) = x^5 - 8x^4 + 12x^3 + 26x^2 - 61x + 30$
e) $f(x) = x^4 - 2x^3 + 3x^2 - 2x + 2$

Chapter 9

PROBLEM SET 1, Page 456

1.

	Center	Radius
a)	$(3, 1)$	2
b)	$(-5, 3)$	3
c)	$(1, -2)$	4
d)	$(-2, 3)$	$3\sqrt{2}$
e)	$(\frac{3}{2}, -2)$	$\frac{3}{2}$
f)	$(3, -4)$	$5\sqrt{2}$
g)	$(1, -\frac{3}{2})$	$\frac{1}{2}$

3. a) $(x-2)^2 + (y-1)^2 = 9$ b) $(x+3)^2 + (y+2)^2 = 9$
c) $(x - \frac{32}{13})^2 + (y + \frac{5}{13})^2 = \frac{5365}{169}$ d) $(x+3)^2 + (y+1)^2 = 145$

5. a) $(x-6)^2 + (y+2)^2 = 10$ and $x^2 + (y+4)^2 = 10$
b) $(x - \frac{21}{5})^2 + (y - \frac{28}{5})^2 = 4$
c) $(x+2)^2 + (y+2)^2 = 25$ and $x^2 + (y-2)^2 = 25$

7. $y = \sqrt{9-x^2}$ is a function with domain $[-3, 3]$ and range $[0, 3]$. It is not one to one.
$y = -\sqrt{9-x^2}$ is a function with domain $[-3, 3]$ and range $[-3, 0]$. It is not one to one.
$x^2 + y^2 = 9$ is not a function.

PROBLEM SET 2, Page 462

1. i) a) $(2, -1)$ b) $(1, 5)$ c) $(4, 10)$ d) $(2, -2)$ e) $(-1, 3)$ f) $(0, 0)$
   ii) a) $(9, -1)$ b) $(8, 5)$ c) $(11, 10)$ d) $(9, -2)$ e) $(6, 3)$ f) $(7, 0)$

3. i) a) $(3, -10)$ b) $(-4, 0)$ c) $(-2, 1)$ d) $(\frac{11}{4}, -\frac{25}{4})$ e) $-6, -8)$
   f) $(4, -10)$
   ii) a) $(-1, -11)$ b) $(-8, -1)$ c) $(-6, 0)$ d) $(\frac{-5}{4}, -\frac{29}{4})$ e) $(-10, -9)$
   f) $(0, -11)$

5. a) $\bar{x} = x + 4, \bar{y} = y - \frac{7}{4}$
   b) $\bar{x} = x + \frac{7}{6}, \bar{y} = y - \frac{5}{6}$
   c) $\bar{x} = x - 4, \bar{y} = y - 5$

PROBLEM SET 3, Page 473

1.
	Vertices	Foci
a)	$(4, 0), (-4, 0), (0, 3), (0, -3)$	$(-\sqrt{7}, 0), (\sqrt{7}, 0)$
b)	$(0, -5), (0, 5), (4, 0), (-4, 0)$	$(0, -3), (0, 3)$
c)	$(0, -4), (0, 4), (2, 0), (-2, 0)$	$(0, -2\sqrt{3}), (0, 2\sqrt{3})$
d)	$(-3, 0), (3, 0), (0, -2), (0, 2)$	$(-\sqrt{5}, 0), (\sqrt{5}, 0)$
e)	$(-4, 0), (4, 0), (0, -2), (0, 2)$	$(-2\sqrt{3}, 0), (2\sqrt{3}, 0)$
f)	$(0, -\frac{1}{3}), (0, \frac{1}{3}), (-\frac{1}{5}, 0), (\frac{1}{5}, 0)$	$(0, -\frac{4}{15}), (0, \frac{4}{15})$

3.
	Center	Vertices	Foci
a)	$(1, -2)$	$(9, -2), (-7, -2), (1, 4\sqrt{3} - 2),$ $(1, -4\sqrt{3} - 2)$	$(5, -2), (-3, -2)$
b)	$(3, 1)$	$(3, 6), (3, -4), (5, 1), (1, 1)$	$(3, 1 + \sqrt{21}), (3, 1 - \sqrt{21})$
c)	$(-2, 1)$	$(-7, 1), (3, 1), (-2, 5), (-2, -3)$	$(1, 1), (-5, 1)$
d)	$(1, 2)$	$(3, 2), (-1, 2), (1, 3), (1, 1)$	$(1 + \sqrt{3}, 2), (1 - \sqrt{3}, 2)$
e)	$(-1, 2)$	$(-1, 5), (-1, -1), (1, 2), (-3, 2)$	$(-1, 2 + \sqrt{5}), (-1, 2 - \sqrt{5})$
f)	$(3, -2)$	$(0, -2), (6, -2), (3, 0), (3, -4)$	$(3 - \sqrt{5}, -2), (3 + \sqrt{5}, -2)$

5. a) $\dfrac{(x-3)^2}{4} + \dfrac{(y+2)^2}{25} = 1$ b) $\dfrac{(x-6)^2}{36} + \dfrac{(y+1)^2}{9} = 1$ c) $\dfrac{(x-3)^2}{4} + \dfrac{(y-1)^2}{25} = 1$

7. a) $\dfrac{x^2}{25} + \dfrac{y^2}{16} = 1$ b) $\dfrac{(x-2)^2}{4} + \dfrac{(y-4)^2}{3} = 1$

   c) $\dfrac{x^2}{9} + \dfrac{y^2}{4} = 1$ d) $\dfrac{(x+3)^2}{1} + \dfrac{(y-1)^2}{25} = 1$

9. $y = \sqrt{1 - \dfrac{x^2}{4}}$ and $y = -\sqrt{1 - \dfrac{x^2}{4}}$ are functions, whereas $y^2 = 1 - \dfrac{x^2}{4}$ is not a function

PROBLEM SET 4, Page 480

1.
	Vertices	Foci	Asymptotes
a)	$(7, 0), (-7, 0)$	$(-\sqrt{85}, 0), (\sqrt{85}, 0)$	$y = \pm\frac{6}{7}x$
b)	$(0, \sqrt{3}), (0, -\sqrt{3})$	$(0, \sqrt{15}), (0, -\sqrt{15})$	$y = \pm\frac{1}{2}x$
c)	$(0, 3), (0, -3)$	$(0, 3\sqrt{2}), (0, -3\sqrt{2})$	$y = \pm x$
d)	$(\frac{1}{6}, 0), (-\frac{1}{6}, 0)$	$(\sqrt{5}/6, 0), (-\sqrt{5}/6, 0)$	$y = \pm 2x$
e)	$(3, 0), (-3, 0)$	$(\sqrt{10}, 0), (-\sqrt{10}, 0)$	$y = \pm\frac{1}{3}x$
f)	$(1, 0), (-1, 0)$	$(\sqrt{10}, 0), (-\sqrt{10}, 0)$	$y = \pm 3x$

3.

	Vertices	Foci	Asymptotes	Center
a)	$(10, 4)$, $(0, 4)$	$(5 + \sqrt{29}, 4)$, $(5 - \sqrt{29}, 4)$	$y - 4 = \pm\frac{2}{5}(x - 5)$	$(5, 4)$
b)	$(2, 3)$, $(-6, 3)$	$(3, 3)$, $(-7, 3)$	$y - 3 = \pm\frac{3}{4}(x + 2)$	$(-2, 3)$
c)	$(-3, 6)$, $(-3, -2)$	$(-3, \sqrt{41} + 2)$, $(-3, -\sqrt{41} + 2)$	$y - 2 = \pm\frac{4}{5}(x + 3)$	$(-3, 2)$
d)	$(-1, -1 + \sqrt{3})$, $(-1, -1-\sqrt{3})$	$\left(-1, -1 + \dfrac{\sqrt{15}}{2}\right)$, $\left(-1, -1 - \dfrac{\sqrt{15}}{2}\right)$	$y + 1 = \pm 2(x + 1)$	$(-1, -1)$
e)	$(4, 3)$, $(4, -1)$	$(4, 1 + \sqrt{7})$, $(4, 1 - \sqrt{7})$	$y - 1 = \pm\dfrac{2}{\sqrt{3}}(x - 4)$	$(4, 1)$
f)	$(\frac{9}{2}, 2)$, $(\frac{7}{2}, 2)$	$\left(4 - \dfrac{\sqrt{13}}{6}, 2\right)$, $\left(4 + \dfrac{\sqrt{13}}{6}, 2\right)$	$y - 2 = \pm\frac{2}{3}(x - 4)$	$(4, 2)$

5. $\dfrac{(y - 5)^2}{1} - \dfrac{(x + 1)^2}{3} = 1$

7. a) $\dfrac{(y - 3)^2}{3} - \dfrac{(x - 2)^2}{1} = 1$ b) $\dfrac{(y - 1)^2}{9} - \dfrac{(x + 2)^2}{16} = 1$

9. $H = \begin{bmatrix} \dfrac{x - h}{a} & \dfrac{y - k}{b} \\ \dfrac{y - k}{b} & \dfrac{x - h}{a} \end{bmatrix}$ and $\det H = 1$

PROBLEM SET 5, Page 488

1.

	Vertex	Focus	Directrix	Focal chord
a)	$(0, 0)$	$(2, 0)$	$x = -2$	8
b)	$(0, 0)$	$(0, -\frac{1}{2})$	$y = \frac{1}{2}$	2
c)	$(0, 0)$	$(0, 1)$	$y = -1$	4
d)	$(0, 0)$	$(-1, 0)$	$x = 1$	4
e)	$(0, 0)$	$(-\frac{5}{4}, 0)$	$x = \frac{5}{4}$	5
f)	$(0, 0)$	$(0, \frac{3}{8})$	$y = \frac{-3}{8}$	$\frac{3}{2}$

3.

	Vertex	Focus	Directrix	Focal chord
a)	$(2, 1)$	$(2, -\frac{1}{2})$	$y = \frac{5}{2}$	6
b)	$(-1, -3)$	$(0, -3)$	$x = -2$	4
c)	$(-3, -1)$	$(-\frac{7}{2}, -1)$	$x = \frac{-5}{2}$	2
d)	$(-3, -2)$	$(-3, -\frac{7}{4})$	$y = \frac{-9}{4}$	1
e)	$(1, 1)$	$(1, 2)$	$y = 0$	4
f)	$(-2, 3)$	$(-2, \frac{7}{4})$	$y = \frac{17}{4}$	5

5. a) $(y - 2)^2 = 8(x - 1)$ b) $x^2 = 4(y - 3)$ c) $(y - 4)^2 = 8(x - 3)$
d) $(x - 2)^2 = -32(y + 5)$

7. Symmetric w.r.t. the line determined by the vertex and focus

9. $y = \frac{3}{5}x + \frac{6}{5}$

11.

Case	Matrix P	Equation
$y^2 = 4cx$	$\begin{bmatrix} y & x \\ 4c & y \end{bmatrix}$	$\det P = 0$
$y^2 = -4cx$	$\begin{bmatrix} y & x \\ -4c & y \end{bmatrix}$	$\det P = 0$
$x^2 = 4cy$	$\begin{bmatrix} x & y \\ 4c & x \end{bmatrix}$	$\det P = 0$
$x^2 = -4cy$	$\begin{bmatrix} x & y \\ -4c & x \end{bmatrix}$	$\det P = 0$

PROBLEM SET 6, Page 494

1. a) $16(y - \frac{15}{4})^2 - 20(x-3)^2 = 125$ b) $\dfrac{x^2}{16} - \dfrac{y^2}{9} = 1$ c) $\dfrac{x^2}{25} + \dfrac{y^2}{16} = 1$

  d) $12(x-1)^2 + 9(y - \frac{10}{3})^2 = 64$ e) $y^2 = -4(x-3)$

3.

	Eccentricity	Directrix
a)	$e = \frac{3}{5}$	$y = \pm \frac{16}{3}$
b)	$e = \sqrt{2}/2$	$x = \pm \sqrt{2}$
c)	$e = \sqrt{\frac{2}{3}}$	$x = \pm \sqrt{6}$
d)	$e = 1$	$x = \frac{53}{12}$
e)	$e = \frac{5}{3}$	$y = \pm \frac{9}{5}$
f)	$e = \sqrt{2}$	$y = \pm \sqrt{2}/4$
g)	$e = \sqrt{2/3}$	$x = \frac{5}{6} \pm \sqrt{1489}/2$
h)	$e = 1$	$y = \frac{-229}{4}$

5. Center $= (-1, -\frac{3}{2})$

  Foci $= (-1, -\frac{3}{2} - \sqrt{29}), (-1, -\frac{3}{2} + \sqrt{29})$

  Vertices $= (-1, -\frac{13}{2}), (-1, \frac{7}{2})$

  Asymptotes $y + \frac{3}{2} = \pm \frac{5}{2}(x+1)$

  Eccentricity $e = \sqrt{29}/5$

REVIEW PROBLEM SET, Page 494

1. a) $(x-4)^2 + (y+3)^2 = 25$ b) $(x + \frac{7}{2})^2 + (y - \frac{5}{2})^2 = \frac{125}{2}$

  c) $(x-1)^2 + (y-2)^2 = 5$

3. a) $(x+4)^2 = -6(y - \frac{3}{2})$ b) $(y-3)^2 = -5(x - \frac{5}{4})$

  c) $(x-1)^2 = 3(y + \frac{13}{4})$ d) $y^2 = 8(x-1)$

5. a) $\dfrac{x^2}{25} + \dfrac{y^2}{16} = 1$ b) $\dfrac{x^2}{9} + \dfrac{y^2}{8} = 1$ c) $\dfrac{x^2}{144} + \dfrac{y^2}{108} = 1$ d) $\dfrac{x^2}{25} + \dfrac{y^2}{16} = 1$

7. a) $\dfrac{x^2}{9} + \dfrac{y^2}{5} = 1$ b) $\dfrac{x^2}{20} + \dfrac{y^2}{36} = 1$ c) $\dfrac{x^2}{100} + \dfrac{y^2}{64} = 1$ d) $25x^2 + 21y^2 = 189$

9. a) $\dfrac{x^2}{4} - \dfrac{y^2}{5} = 1$ b) $\dfrac{x^2}{64} - \dfrac{y^2}{36} = 1$ c) $\dfrac{x^2}{225} - \dfrac{y^2}{144} = 1$ d) $\dfrac{y^2}{56} - \dfrac{x^2}{56} = 1$

11. Asymptotes: $y = \pm \frac{3}{2}x$

13. Yes   **15.** $3(x-5)^2 - y^2 = 12$ **17.** 91.32 million miles

Index